Lecture Notes in Mathematics

Edited by A. Dold and B. Eckmann

798

Analytic Functions Kozubnik 1979

Proceedings of a Conference Held in Kozubnik, Poland, April 19–25, 1979

Edited by J. Ławrynowicz

Springer-Verlag
Berlin Heidelberg New York 1980

Editor
Julian Ławrynowicz
Institute of Mathematics of the
Polish Academy of Sciences
Łódź Branch
Kilińskiego 86
90-012 Łódź
Poland

AMS Subject Classifications (1980): 20Hxx, 30-XX, 31-XX, 32-XX, 33-XX, 35-XX, 46-XX, 49-XX, 58-XX

ISBN 3-540-09985-9 Springer-Verlag Berlin Heidelberg New York
ISBN 0-387-09985-9 Springer-Verlag New York Heidelberg Berlin

Library of Congress Cataloging in Publication Data. Conference on Analytic Functions, 7th, Kozubnik, Poland, 1979. Analytic functions, Kozubnik 1979. (Lecture notes in mathematics; 798) "Sponsored and organized by the Institute of Mathematics of the Polish Academy of Sciences in collaboration with the Institutes of Mathematics of the Łódź University and the Silesian University, Katowice." Bibliography: p. Includes index. 1. Analytic functions--Congresses. I. Ławrynowicz, J., 1939- II. Polska Akademia Nauk. Instytut Matematyczny. III. Łódź, Poland. Uniwersytet. Instytut Matematyki. IV. Uniwersytet Śląski w Katowicach. Instytut Matematyki. V. Title. VI. Series: Lecture notes in mathematics (Berlin); 798. QA3.L28 no. 798. [QA331]. 510s. [515.9] 80-14622

Printing and binding: Beltz Offsetdruck, Hemsbach/Bergstr.
2141/3140-543210

FOREWORD

These Proceedings contain selected papers from those submitted by a part of mathematicians lecturing at the 7th Conference on Analytic Functions held in Poland at Kozubnik (Carpathian Mountains, Province Bielsko-Biała) during the seven days from 19 to 25 April 1979. These papers form extended versions of their lectures. The Conference had been originally planned at Wisła (the same region) from May 21 to June 6, 1979, but the organizers had to change the place and time because of accomodation difficulties.

According to the tradition of the preceding six conferences (held in Łódź 1954, Lublin 1958, Kraków 1962, Łódź 1966, Lublin 1970, and Kraków 1974) the topics chosen are rather homogeneous. A considerable part of the papers is concerned with extremal methods and their applications to various branches of complex analysis: one and several complex variables, quasiconformal mappings and complex manifolds. This is however not a rule and the organizers decided to accept also papers on other subjects in complex analysis if they were of good quality.

The Organizing Committee consisted of: C. Andreian-Cazacu (Bucharest), A. Andreotti (Pisa), Z. Charzyński (Łódź), A. A. Gončar (Moscow), J. Górski (Katowice) – Vice-Chairman, H. Grauert (Göttingen), J. Krzyż (Lublin), J. Ławrynowicz (Łódź) – Chairman, O. Lehto (Helsinki), F. Leja (Kraków) – Honorary Chairman († 11th October 1979), P. Lelong (Paris), S. N. Mergeljan (Erevan), J. Siciak (Kraków), W. Tutschke (Halle an der Saale), V. S. Vladimirov (Moscow), and W. Kucharz (Katowice) – Secretary. The Conference was attended by 171 participants (97 from Poland) representing 20 countries.

The Conference was sponsored and organized by the Institute of Mathematics of the Polish Academy of Sciences in collaboration with the Institutes of Mathematics of the Łódź University and the Silesian University, Katowice.

The Organizing Committee of the Conference expresses its gratitude to the Springer-Verlag for kind consent of publishing the Proceedings in the series "Lecture Notes in Mathematics".

Łódź, October 1979 Julian Ławrynowicz

CONTENTS

LIST OF SEMINARS HELD DURING THE CONFERENCE

H. GRUNSKY (Würzburg) [Chairman]: Seminar on extremal problems for analytic functions of one complex variable

L. V. AHLFORS (Cambridge, MA) [Chairman]: Seminar on quasiconformal mappings

G. M. HENKIN (Moskva) [Chairman]: Seminar on functions of several complex variables (including the theory of analytic functions in topological vector spaces)

J. EELLS (Coventry) [Chairman]: Seminar on analysis on complex manifolds

During the seminars new problems were posed and discussed.

LECTURES NOT INCLUDED IN THIS VOLUME

(* = one hour lecture)

A. K. BAHTIN (Kiev): О коэффициентах однолистных функций

Galina P. BAHTINA (Kiev): Экстремальные свойства неналегающих областей

Türkan BAŞGÖZE (Ankara): On α-spiral functions

A. BAYOUMI (Uppsala) * The Levi problem and the radius of convergence of holomorphic functions on some metric vector spaces

Z. BOGUCKI and J. ZDERKIEWICZ (Lublin): О корнях уравнений $f(z) = p\,f(a)$, $|f(z)| = |p\,f(a)|$ в некоторых классах аналитических функций

B. BOJARSKI (Warszawa) * : Remarks on the index of elliptic operators on compact surfaces

B. BOJARSKI, T. IWANIEC, and R. KOPIECKI (Warszawa) * : Some stability theorems for quasiconformal mappings

M. BRANDT (Berlin): Mapping theorems for finitely connected domains

P. CARAMAN (Iaşi): Conformal capacity in $\mathbb{R}^\infty$

Z. CHARZYŃSKI (Łódź): Special interpolation of the Čebyšev type

E. M. ČIRKA (Moskva) * : Boundary properties of analytic sets

Anne CUMENGE (Toulouse): Extension de fonctions analytiques avec estimation

K. CZAJA (Częstochowa): Two results from analytic geometry

J. DAVIDOV (Sofia): Bounded families of holomorphic mappings between complex manifolds

S. DIMIEV (Sofia): Applications presque symplectiques sur les variétés presque hermitiennes

G. DINI and Carla PARRINI (Firenze): Removable singularity sets for Cauchy-Riemann distributions on flat domains

B. DITTMAR (Halle an der Saale): Übertragung eines Extremalproblems von M. Schiffer und N. S. Hawley für quasikonforma Abbildungen

Krystyna DOBROWOLSKA and I. DZIUBIŃSKI (Łódź): On starlike and convex functions of many variables

P. DOLBEAULT (Paris) * : Sur les bords d'ensembles analytiques complexes

Ju. N. DROŽŽINOV and V. S. VLADIMIROV (Moskva) * : Multidimensional tauberian theorem in complex domain

I. DZIUBIŃSKI and R. SITARSKI (Łódź): On classes of holomorphic functions of many variables starlike and convex on some hypersurfaces

J. FUKA (Praha): On an extremal problem for matrices

B. GAVEAU (Paris): Intégrales de volume et valeurs frontières des applications holomorphes

Tatiana GENTCHEVA (Sofia): Entire functions of exponential type bounded on the real axis

F. HASLINGER (Wien): A dual relationship between generalized Abel-Gončarov bases and certain Pincherle bases

G. M. HENKIN (Moskva) * : Интегральные представления форм в псевдовогнутых областях и некоторые приложения

G. M. HENKIN (Moskva) and J. LEITERER (Berlin) * : Global integral formulas for solving the $\bar{\partial}$-problem on Stein manifolds

V. HRISTOV (Sofia): Some results about the Carathéodory and Kobayashi pseudometrics

E. JANIEC (Łódź): Typically real bounded functions

M. JARNICKI (Kraków): On some Fréchet spaces of analytic functions

J. KAJA (Rzeszów): Differentiable characterization of the boundary of the boundary of a polynomially convex hull of the circular compact in $\mathbb{C}^2$

S. L. KALLA (Maracaibo): On the H-functions

J. KAMIŃSKI (Gdańsk): Some growth problems for certain α-convex functions

Virjinia S. KIRIAKOVA (Sofia): An explanation of the Stokes phenomenon in complex domain

S. KIRSCH (Halle an der Saale): Verallgemeinerter transfiniter Durchmesser im Zusammenhang mit einer quasikonformen Normalabbildung

C. O. KISELMAN (Uppsala) *: Growth of plurisubharmonic functions

L. KOCZAN and J. SZYNAL (Lublin): The region of variability of some functional in the class of bounded analytic functions

V. KRIVOV (Moskva) *: Некоторые свойства экстремальных дифференциальных форм в дефиниции обобщенного модуля

R. KÜHNAU (Halle an der Saale) *: Charakterisierung ebener Unterschallströmungen durch ein Extremalproblem in einer Klasse quasikonformer Abbildungen

L. LEMPERT (Budapest) *: Fatou-type theorems for analytic sets

P. LICZBERSKI (Łódź): Ein Extremalproblem für die schlichten und beschränkten Abbildungen von $\mathbb{C}^n$ in $\mathbb{C}^n$

W. MAJCHRZAK (Łódź): An extremal arclength problem in some classes of univalent and p-symmetric functions

L. MIKOŁAJCZYK and S. WALCZAK (Łódź): Application of the extremum principle to investigating certain extremal problems

A. NOWAKOWSKI (Łódź): Sufficient conditions in classes of holomorphic and univalent functions

Elena I. OBOLAŠVILI (Tbilisi) *: Комплексное представление общего решения уравнений сферической теории упругости

A. PIERZCHALSKI (Łódź): The capacity of small spherical rings on Riemannian manifolds

S. I. PINČUK (Čelabinsk) *: Analytic continuation of holomorphic mappings and holomorphic equivalence problem in $\mathbb{C}^n$

U. PIRL and C. MICHEL (Berlin): Standard domains of finite connectivity with respect to conformal mapping with four fixpoints

I. RAMADANOV (Sofia): On the connection between the solution of a Monge-Ampère equation and the Bergman kernel for simpler cases of domains

M. O. READE (Ann Arbor, MI): A uniqueness theorem concerning close-to-convex functions

L. REICH (Graz): Continuous iteration of locally and formally biholomorphic mappings

H. RENELT (Halle an der Saale): Über Integraltransformationen, die analytische Funktionen in Lösungen elliptischer Differentialgleichungssysteme überführen

P. RUSEV (Sofia): On the representation of analytic functions by means of series in Laguerre polynomials

A. SADULLAEV (Taškent): Defect divisors of holomorphic mappings

M. SAKAI (Hiroshima): Estimates for the Gaussian curvature of the span metric

I. A. ŠEVČUK (Kiev): Smoothness on compacts

Henryka ŚMIAŁEK (Łódź): Schroeder's functions

J. STANKIEWICZ (Lublin): Quasisubordination and quasimajorization

D. SUNDARARAMAN (Madras): Holomorphic maps of compact complex manifolds

D. SUNDARARAMAN (Madras): Versal deformations of principal bundles over a compact complex manifold

O. SUZUKI (Tokyo): The Riemann-Hilbert problem and Fuchs relations in several complex variables

O. SUZUKI (Tokyo) *: Variational methods on kähler manifolds

Maria SZAPIEL (Lublin): Subordination in the class of typically real functions

W. SZAPIEL (Lublin): Integral representations for some classes of holomorphic functions

A. SZWANKOWSKI (Łódź): Estimation of the functional $|a_3 - \alpha a_2^2|$ in the family S of holomorphic and univalent functions for α complex

P. M. TAMRAZOV (Kiev) *: Capacities and potentials in complex analysis

P. G. TODOROV (Plovdiv): New explicit formulas for the coefficients of Grunsky of univalent functions

Ju. Ju. TROHIMČUK (Kiev) *: Дифференциальные свойства функций

Z. D. USMANOV (Dušanbe): Регулярность решений обобщенной системы Коши-Римана с особой точкой

W. TUTSCHKE (Halle an der Saale) *: Boundary value problems for non-linear partial differential equations in both one and several complex variables

F.-H. VASILESCU (Bucureşti): The stability of the Euler characteristic for Hilbert complexes

Donka VASSILEVA (Sofia): Функции с положительной вещественной частью

M. VUORINEN (Helsinki) *: On cluster sets and boundary behavior of quasiregular mappings in $\mathbb{R}^n$

B. WAJNRYB (Jerusalem): On nonramified endomorphisms of an affine plane

P. WALCZAK (Łódź): A remark on groups of isometries of nearly Kähler manifolds

J. WANIURSKI (Lublin): Univalent polynomials with quasiconformal extension

T. WINIARSKI (Kraków): Inverse of polynomial mappings in $\mathbb{C}^n$

K. WŁODARCZYK (Łódź): Power inequalities for pairs of vector functions and their applications

Z. WRONICZ (Kraków): On approximation by analytic splines

Ju. B. ZELINSKIĬ (Kiev) *: Некоторые применения многозначных отображений в комплексном анализе

ERGODIC PROPERTIES OF GROUPS OF MÖBIUS TRANSFORMATIONS

Lars V. Ahlfors* (Cambridge, MA)

This is an expository paper whose main purpose is to draw attention to an important paper by Dennis Sullivan [S] . His paper deals with many aspects of ergodic theory, but I shall speak mainly about a particular result that is fairly close to my own work. My presentation will not differ essentially from that of Sullivan, except for my use of language which I hope is accessible to analysts without Sullivan's broad background. There is a very minor improvement in that my version of the proof applies to any dimension.

1. We shall be interested in the group G of Möbius transformations in $\underline{R}^n$, $n \geq 2$, that preserve the unit ball $B = \{|x| < 1\}$ and hence also the sphere $S = \{|x| = 1\}$ and the Poincaré metric $ds = |dx|/(1-|x|^2)$. Throughout the paper we shall denote the Jacobian matrix of $\gamma \in G$ at x by $\gamma'(x)$. It is a positive multiple of an orthogonal matrix. We shall denote this multiple, which measures the change of linear scale, by $|\gamma'(x)|$, so that $\gamma'(x)/|\gamma'(x)| \in O(n)$.

Every $\gamma \in G$ can be written in a simple canonical form. For this purpose let $a = \gamma^{-1}0$ and let σ_a denote reflection in the non-euclidean plane equidistant from 0 and a . Then $\sigma_a \in G$ and $\sigma_a(a) = 0$, from which it follows that $\gamma = \beta\sigma_a$ with $\beta \in O(n)$. Sullivan's proof consists essentially in a very thorough investigation of the mappings σ_a.

2. Let Γ be a discrete subgroup of G . We are interested mainly in its action on S . It is well known that S splits into a set of discontinuity Ω and its complement, the limit set Λ . In classical terminology Γ is of the first kind if Ω is empty, of the second kind if Ω is non-empty.

It is a less familiar fact that S can also be written uniquely, up to null-sets, as a disjoint union of a dissipative set $\mathcal{D}$ and a recurrent set $\mathcal{R}$. A set is dissipative if it has a measurable funda-

Research supported by NSF Grant MCS 77 07782.

mental set, and the recurrent set has the property that every measurable $E \in \mathcal{R}$ with $m(E) > 0$ satisfies $m(E \cap \gamma E) > 0$ for infinitely many $\gamma \in \Gamma$. Although this is a fairly elementary theorem it seems to have been overlooked by the specialists in Kleinian group. It is evident that $\mathcal{D}$ and $\mathcal{R}$ are invariant under Γ, and that $\mathcal{R} \subset \Lambda$. For $n = 2$ and 3 it is known that $\mathcal{R} = \Lambda$ if Γ is finitely generated. It is not known whether the corresponding fact is true for $n > 3$.

A vector field on $\mathcal{R}$ is a measurable vector-valued function $f: \mathcal{R} \to R^n$ such that $f(x)$ is orthogonal to x for all $x \in \mathcal{R}$. It is invariant under Γ if

$$(1) \qquad f(\gamma x) = (\gamma'(x)/|\gamma'(x)|)f(x)$$

for a.e. $x \in \mathcal{R}$ and all $\gamma \in \Gamma$. Note that the length $|f(x)|$ is then an invariant scalar. The vector field is trivial if $f(x) = 0$ a.e.

THEOREM (Sullivan). *There is no non-trivial invariant vector field on* $\mathcal{R}$.

Clearly, if f is invariant and non-trivial, then $|f|$ is constant and $\neq 0$ on an invariant subset $\mathcal{R}_0 \subset \mathcal{R}$ of positive measure. Therefore, to prove Sullivan's theorem it is sufficient to assume that f is invariant and $|f(x)| = 1$ on an invariant subset $\mathcal{R}_0 \subset \mathcal{R}$ of positive measure, and derive a contradiction.

3. The behaviour of Γ, especially its action on $S = \partial B$, is quite different depending on whether the series

$$(2) \qquad \sum_{\gamma \in \Gamma} (1 - |\gamma 0|^2)^{n-1}$$

converges or diverges, although Sullivan's theorem, as quoted above, happens to be true in both cases. However, Sullivan gives entirely different proofs in the two cases.

In the divergence case he proves that the diagonal action of Γ on $S \times S$ is ergodic in the sense that a measurable invariant subset is either a null-set or the complement of a null-set. He then shows that the non-existence of a measurable invariant vector-field on S is a rather direct consequence of this result. This part of Sullivan's paper is a remarkable complement to earlier work of H. Hopf (1939) who had proved that the action on $S \times S$ is ergodic provided that almost all geodesics on the quotient manifold B/Γ do not tend to the ideal boundary. However, it was not known, except for $n = 2$, whether this hypothesis is a consequence of the divergence of the series. Sullivan turns the argument around and shows that the ergodicity implies Hopf's hypothesis, and that the divergence follows from that hypothesis. In other words, he has settled the problem of ergodicity by showing that the following properties are equivalent: (i) divergence of the series; (ii) ergodicity on $S \times S$; (iii) almost no geodesics tend to the boundary.

In this paper we shall analyze only that part of Sullivan's proof which deals with the convergence case. To carry out this program we shall assume from now on that the series (2) converges.

4. As a standard notation we shall write $x^* = x/|x|^2$ for the mirror image of x with respect to S. The Jacobian $(x^*)'$ is the matrix with elements $|x|^{-2}(\delta_{ij} - 2x_i x_j/|x|^2)$. It will be convenient to introduce the notation $Q(x)$ for the matrix with elements $x_i x_j/|x|^2$ so that $(x^*)' = |x|^{-2}(1 - 2Q(x))$ where for simplicity 1 stands for the unit matrix 1_n. We note that $Q(x)^2 = Q(x)$ and $(1 - 2Q(x))^2 = 1$ so that $1 - 2Q(x) \in O(n)$.

The non-euclidean plane midway between 0 and a lies on the <u>isometric sphere</u> characterized by $|\gamma'(x)| = 1$. It has the center a^* and radius $(|a^*|^2 - 1)^{1/2}$. The reflection in the isometric sphere is therefore given by

$$\sigma_a(x) = a^* + (|a^*|^2 - 1)(x - a^*)^* . \tag{3}$$

For $x \in S$ the correspondence is even more obvious, for it is given by the fact that a^*, x and $\sigma_a x$ are in a straight line. Derivation of (3) leads to

$$\sigma_a'(x) = \frac{1 - |a|^2}{|a|^2|x-a^*|^2}\,(1 - 2Q(x-a^*)) \tag{4}$$

and consequently

$$|\gamma'(x)| = \frac{1 - |a|^2}{|a|^2|x-a^*|^2} \tag{5}$$

and

$$\gamma'(x)/|\gamma'(x)| = \beta(1 - 2Q(x-a^*)) . \tag{6}$$

We observe that $|a||x-a^*| = |x-a|$ when $x \in S$.

5. Given γ and a number $k > 0$ we divide S into three zones, the near zone $N_k\gamma$, the middle zone $M_k\gamma$, and the far zone $F_k\gamma$, defined as follows:

$$\begin{aligned} &x \in N_k\gamma \quad \text{if} \quad |x-a^*| \le k(|a^*|^2 - 1) , \\ &x \in M_k\gamma \quad \text{if} \quad k(|a^*|^2 - 1) < |x-a^*| < k^{-1}, \\ &x \in F_k\gamma \quad \text{if} \quad |x-a^*| \ge k^{-1} . \end{aligned} \tag{7}$$

As before, $a = \gamma^{-1}0$. All three zones are non-empty provided that $1 + |a^*| > k^{-1} > (|a^*|^2 - 1)^{1/2}$, as we shall assume. By elementary geometry $|x-a^*| \cdot |\gamma_a x-a^*| = |a^*|^2-1$. Therefore γ maps $M_k\gamma$ on itself and interchanges $N_k\gamma$ and $F_k\gamma \cdot N_k\gamma$ and $F_k\gamma$ are spherical caps. By the center of a cap we shall mean its midpoint on the sphere, and its radius is the distance from the center to the rim of the cap.

It is easy to see that for fixed k the radius of $N_k\gamma$ is of the order $O(1-|a|)$. The $(n-1)$-dimensional measure of $N_k\gamma$ is of the order $(1-|a|)^{n-1}$ and the convergence of (2) implies that the sum of the measures of all the $N_k\gamma$, $\gamma \in \Gamma$, is finite.

6. An important ingredient of Sullivan's proof is a lemma which we shall formulate slightly differently.

LEMMA 1. *For any measurable set* $Y \subset \mathcal{R}_o$, $m(Y) > 0$, *and any* $k > 0$ *there exist* $\gamma \in \Gamma$ *and* y *such that*

(i) y *and* $\gamma y \in Y$,

(ii) $y \in M_k\gamma$.

P r o o f. Because the sum of the $m(N_k\gamma)$ is finite one can omit finitely many γ so that the union of the remaining $N_k\gamma$ has measure $< \frac{1}{2} m(Y)$. It follows that $X = Y - \bigcup N_k\gamma - \bigcup \beta N_k\gamma$ has positive measure (the unions are over the remaining γ) . By recurrence there is a remaining $\gamma \in \Gamma$ such that $X \cap \gamma^{-1}X$ is not empty. If $y \in X \cap \gamma^{-1}X$, then y and γy are in X , hence in Y . Moreover $y \notin N_k\gamma$ and $\gamma y = \beta\sigma_a y \notin \beta N_k\gamma$. The last condition is equivalent to $\sigma_a y \notin N_k\gamma$, $y = \sigma_a^2 y \notin \sigma_a N_k\gamma = F_k\gamma$. Hence $y \in M_k\gamma$, and the lemma is proved.

7. Suppose that f is an invariant vector field on $\mathcal{R}_o$ with $|f| = 1$. Cover S with a finite number of spherical caps $B(y,\eta) \cap S$, $y \in S$, with a fixed radius η ; henceforth, when $y \in S$ we shall let $B(y,\rho)$ denote the spherical cap with center y and radius ρ , that is to say the ball relatively to the metric of S . There is one $B = B(y,\eta)$ whose inverse image $f^{-1}(Y)$ has positive measure. If $x,y,$ γx and γy are all in Y it follows by (1) that

$$(8) \qquad \left| \frac{\gamma'(x)}{|\gamma'(x)|} f(x) - \frac{\gamma'(y)}{|\gamma'(y)|} f(y) \right| = |f(\gamma x) - f(\gamma y)| < 2\eta$$

and with the help of (6)

$$|(1-2Q(x-a^*)(f(x) - f(y)) + 2(Q(x-a^*) - Q(y-a^*))f(y)| < 2\eta$$

from which we derive, since $1-2Q(x-a^*) \in O(n)$,

(9) $$|(Q(x-a^*) - Q(y-a^*))f(y)| < 2\eta \ .$$

We shall show that η, a, x, y can be chosen so that (9) is a contradiction.

8. For this purpose we need to study the first and second derivatives of $Q(x)$, or rather of the vector-valued function $Q(x)f(y)$ for a fixed y . We shall use the notations $DF(x)[u] = \Sigma_k u_k D_k F(x)$ and $D^2F(x)[u] = \Sigma_{hk} u_h u_k D_h D_k F(x)$ and (fu) for the inner product. Routine computation gives

$$D(Qf)(x)[u] = |x|^{-2}((fu) - 2(fx)(ux)/|x|^2)x + |x|^{-2}(fx)u,$$

$$D^2(Qf)(x)[u] = |x|^{-4}(-2(fx)|u|^2 - 4(fu)(ux) + 8(fx)(ux)^2/|x|^2)x + |x|^{-2}(2(fu) - 4(fx)(ux)/|x|^2)u$$

and, surprisingly,

(10) $$|D(Qf)(x)[u]| = |x|^{-2}|(fx)u - (fu)x|$$

(11) $$|D^2(Qf)(x)[u]| = 2|x|^{-3}|u||(fx)u - (fu)x| \ .$$

In these formulas Q is short for $Q(x)$, f is $f(y)$, and $u \in \underline{R}^n$.

The Taylor formula with integral remainder (see [C] p. 70) yields

$$(Q(x+u) - Q(x))f(y) = D(Qf)(x)[u] + \int_0^1 (1-t)D^2(Qf)(x+tu)[u]dt$$

and subject to the condition $|u| < \alpha|x|$, $0 < \alpha < 1$, it follows by use of (10) and (11) that

(12) $$|(Q(x+u) - Q(x))f(y)| > |x|^{-2}|(fx)u - (fu)x| - 2(\alpha/1-\alpha)^2 \ .$$

For any two $x,y \in S$ and a given $a = \gamma^{-1}0$ let us introduce the notations $|y-a^*| = r$, $y-a^* = rv$, $\rho = \alpha r$, $x-y = \rho u$. Thus $|v| = 1$, and $|u| < 1$ if and only if $x \in B(y,\rho)$. On replacing x by $y-a^*$ and u by ρu in (12) we obtain, for $|u| < 1$,

(13) $$|(Q(x-a^*) - Q(y-a^*))f(y)| > \alpha|(fv)u - (fu)v| - 2(\alpha/1-\alpha)^2 \ .$$

This inequality will be compared with (9) .

9. In order to derive a contradiction we shall first find a lower bound for $|(fv)u - (fu)v|$ which applies to the majority of the points $x \in B(y,\rho)$. We shall make use of the identity

$$(14) \qquad |(fv)u - (fu)v|^2 = (u, f - (fv)v)^2 + (fv)^2(|u|^2 - (uv)^2)$$

where for greater clarity one of the inner products has been written in the (,) notation.

We write $h = f - (fv)v$ and denote the projections of u and h on the tangent plane at y by u' and h'. In other words, $u' = u - (uy)y$ and $h' = h - (hy)y$. Simple computations show that $(uy) = -\frac{1}{2}\rho|u|^2$ and $|h|^2 = 1 - (fv)^2$, $(hy) = -(fv)(vy)$ (recall that $(fy) = 0$). It follows that

$$(15) \qquad |h'|^2 = 1 - (fv)^2(1 + (vy)^2)$$

and

$$(16) \qquad (u'h') = (uh) - \frac{1}{2}\rho|u|^2(fv)(yv) .$$

From (16) one derives

$$(17) \qquad (uh)^2 \geq \frac{1}{2}(u'h')^2 - \rho^2/4 .$$

Similarly, if the projection of v on the tangent plane is denoted by v' one obtains

$$(18) \qquad |u|^2 - (vu)^2 \geq |u'|^2 - (v'u')^2 - \rho .$$

u' varies over a ball $B^{n-1}(\rho_o)$ with $\rho_o^2 = 1 - \rho^2/4$. There is a number λ_o, depending only on n, such that $(u'h')^2 > \lambda_o^2\rho_o^2|h'|^2$ on precisely three quarters of the points $u' \in B^{n-1}(\rho_o)$, measured by volume, and a number λ_1 such that $|u'|^2 - (v'u')^2 > \lambda_1^2\rho_o^2$ on another three quarters. If λ_o and λ_1 are replaced by $\lambda < \min(\lambda_o,\lambda_1)$ both conditions are fulfilled on more than half the ball, and for all points u' in that set (14), together with (15), (17) and (18), implies

$$(19) \qquad |(fv)u - (fu)v|^2 > \frac{1}{2}\rho_o^2\lambda^2(1 - (vy)^2) - \rho .$$

We assume that $y \in M_k\gamma$. With the same notations as before $(vy) = r^{-1}(1 - (ya^*))$ and $r^2 = |a^*|^2 - 1 + 2(1 - (ya^*))$. This implies $|(vy)| \leq \frac{1}{2}(r + (|a^*|^2 - 1)/r) < k^{-1}$ by (7). At the same time $\rho < r < k^{-1}$, $\rho_o^2 > 1 - 1/4k^2$. Therefore, by choosing k large enough we can make sure, as a consequence of (19), that

$$(20) \qquad |(fv)u - (fu)v| > \frac{1}{2}\lambda$$

for more than half the u'. From u' we pass to u by the inverse of the projection map. The point u lies on the sphere $S(\rho^{-1}y,\rho^{-1})$. The cosine of the angle between y and the normal to the sphere is $(xy) = 1 - r^2/2 > 1 - 1/2k^2$. Therefore, if the u' fill out a ratio $> 1/2$, then the u fill a ratio greater than $\frac{1}{2}(1 - 2/2k^2)$, and the same is true of the corresponding $x \in B(y,\rho)$. If k is sufficiently large this ratio is still $> 1/2$.

LEMMA 2. *There exist positive constants λ and k_o such that $y \in M_k\gamma$ with $k > k_o$ and if $\alpha = \rho/r < \min(1/2,\lambda/32)$, then*

(21) $\quad |(Q(x-a^*) - Q(y-a^*))\, f(y)| > \frac{1}{4}\alpha\lambda$

for more than half the points $x \in B(y,\rho)$.

10. Lemma 1 applied to $Y = f^{-1}(B)$ produces γ and y such that y and $\gamma y \in Y$ and $y \in M_k\gamma$. We wish to show that for suitable choice of ρ, r and k more than half the points $x \in B(y,\rho)$ also satisfy x, $\gamma x \in Y$.

There is no reason why an arbitrary $y \in Y$ should have this property. Let us recall, however, that a.e. $y \in Y$ is a point of density in the sense that

(22) $\quad \lim_{\rho\to 0} m(B(y,\rho) \cap Y)/m(B(y,\rho)) = 1$.

By Egoroff's theorem the limit in (22) is uniform on a subset $Y' \subset Y$ with $m(Y') > 0$. In other words, for any $0 < \varepsilon < 1$ there exists $\delta = \delta(\varepsilon,\eta) > 0$ such that

(23) $\quad m(B(y,\rho) \cap Y) > (1-\varepsilon)m(B(y,\rho))$

as soon as $y \in Y'$ and $\rho < \delta$; note that δ depends on η because Y was defined in terms of η (Sec. 7). By applying Lemma 1 to Y' rather than Y we can prove the following stronger result:

LEMMA 3. *There is a constant α_o, $0 < \alpha_o < 1$, and $k_1 = k_1(\eta)>0$, such that with the same notations as before, if $\alpha < \alpha_o$ and $k > k_1$, then there exist $\gamma \in \Gamma$ and $y \in Y \cap M_k\gamma$ such that $\gamma y \in Y$ and $x,\gamma x \in Y$ for more than half the points $x \in B(y,\rho)$.*

Proof. Lemma 1 applied to Y' yields $y \in Y' \cap M_k\gamma$ and $\gamma y \in Y'$. If k is large enough, then $\rho < k^{-1} < \delta$, so that (23) will be valid. In other words, a proportion $1-\varepsilon$ of the points $x \in B(y,\rho)$ are in Y.

It remains to estimate the proportion of points $x \in B(y,\rho)$ with $\gamma x \in Y$. We note that the ratio

(24) $|\gamma'(y)|/|\gamma'(x)| = |x-a^*|^2/|y-a^*|^2$

varies between $(1-\alpha)^2$ and $(1+\alpha)^2$.

If $x \in B(y,\rho)$ so that $|x-y| < \alpha|y-a^*|$, then $(1-\alpha)|y-a^*| < |x-a^*| < (1+\alpha)|y-a^*|$ and thus, because $y \in M_k\gamma$, $(1-\alpha)k(|a^*|^2 - 1) < |x-a^*| < [(1-\alpha)k]^{-1}$. Hence $x \in M_{(1-\alpha)k}\gamma$ and therefore also $\sigma_a x \in M_{(1-\alpha)k}\gamma$. It follows that $|\gamma x-\gamma y| = |\sigma_a x-\sigma_a y| < k^{-1}+(1-\alpha)^{-1}k^{-1} < \delta$ for sufficiently large k. We conclude that (23) is applicable to the largest ball $B(\gamma y,\rho') \subset \gamma B(y,\rho)$, and we obtain

(25) $m(B(\gamma y,\rho') \cap Y) > (1-\varepsilon)m(B(\gamma y,\rho'))$.

Let ρ'' denote the radius of $\gamma B(y,\rho)$. It follows easily from (23) that $\rho'/\rho'' > [(1-\alpha)/(1+\alpha)]^2$. Because $\rho' < \rho''$ it is true that $m(B(\gamma y,\rho')) > (\rho'/\rho'')^{n-1}m(B(y,\rho'')) > [(1-\alpha)/(1+\alpha)]^{2n-2}\, m(\gamma B(y,\rho))$. In view of (25) we conclude that at least the proportion $(1-\eta)[(1-\alpha/(1+\alpha)]^{2n-2}$ of the points in $\gamma B(y,\rho)$ are also in Y. The mapping by γ^{-1} changes this ratio by at most another factor $[(1-\alpha)/(1+\alpha)]^{2n-2}$. In other words, at least the ratio $(1-\varepsilon)[(1-\alpha)/(1+\alpha)]^{4n-4}$ of the points $x \in B(y,\rho)$ are such that $\gamma x \in Y$.

If two sets cover the fractions $1-\varepsilon$ and $1-\varepsilon'$ of a third, then their intersection covers at least the fraction $1-\varepsilon-\varepsilon'$. We choose $\varepsilon = 1/8$ and subsequently α_o so that $[(1-\alpha_o)/(1+\alpha_o)]^{4n-4} = 5/7$. If $\alpha < \alpha_o$ this implies $1-\varepsilon' = (1-\varepsilon)[(1-\alpha)/(1+\alpha)]^{4n-4} > 5/8$, $1-\varepsilon-\varepsilon' > 1/2$ and Lemma 3 is proved, for all our conclusions are valid as soon as k is greater than a certain k_1.

<u>11</u>. We are ready for the contradiction. Choose $\alpha < \min(\alpha_o,\lambda/32)$, $0 < \eta < \alpha\lambda/8$, and $k > \max(k_o,k_1)$. With this choice there exist $y,\gamma \in \Gamma$ and $\rho > 0$ such that the conclusions of Lemma 2 and Lemma 3 are valid, each for more than half the points $x \in B(y,\rho)$. As a consequence there is an x to which both conclusions apply. In these circumstances the inequalities (9) and (21) are both fulfilled, and we are led to the contradiction

$$2\eta < \alpha\lambda/4 < |(Q(x-a^*) - Q(y-a^*))f(y)| < 2\eta .$$

The contradiction shows that no invariant vector field can exist.

References

[S] DENNIS SULLIVAN: On the ergodic theory at infinity of an arbitrary discrete group of hyperbolic motions, Proceedings of the Stony Brook Conference on Riemann Surfaces and Kleinian Groups, June 1978.

[H] EBERHARD HOPF: Statistik der geodetischen Linien in Mannigfaltigkeiten negativer Krümmung, Berichte der Akademie der Wissenschaften Leipzig, Math.-Phys.-Klasse, 91, 1939, pp. 261-304.

[C] HENRI CARTAN: Differential Calculus, Hermann and Houghton-Mifflin Company, Boston 1971.

Special notice. Sullivan informs me that Klaus Schmidt has recently proved a stronger property of the recurrent set which makes it possible to prove Lemma 1 without assuming that Γ is of convergence type. Thus Sullivan's theorem can be proved in full generality without separating cases.

Mathematics Department, Harvard University
Cambridge, MA 01638, USA

TRACES OF PLURIHARMONIC FUNCTIONS

Paolo de Bartolomeis and Giuseppe Tomassini (Firenze)

Contents

0. Introduction

Let S be a real oriented hypersurface in a complex manifold X which divides X into two open sets X^+ and X^-.
In this paper we characterize in terms of tangential differential operators on S the distributions T on S which are "jumps" or traces (in the sense of currents) of pluriharmonic functions in X^+ and X^-.

The starting point of our investigation is the <u>non tangential</u> characterizing equation $\bar{\partial}_b \partial T = 0$, which can be deduced from the theory of boundary values of holomorphic forms [5]. If S is not Levi-flat (i.e. not characteristic for $\bar{\partial}\partial$ operator) we construct a second order tangential differential operator ω on S such that if T is the trace on S of a pluriharmonic function h, then $\omega(T) = \partial h$. This enables us to prove that, under suitable topological assumptions, the tangential equation $\bar{\partial}_b \omega(T) = 0$ characterizes the "jumps" on S of pluriharmonic functions on $X \setminus S$ (Riemann-Hilbert problem).

From this result, we deduce the solvability of Cauchy-Dirichlet problem (traces problem) when S is either compact or its Levi-form has at least one positive eigenvalue.

1. Preliminaries and notations

In the present paper X will be a complex manifold of dimension

$n \geq 2$ and $S \subset X$ a real oriented connected C^∞ hypersurface. We suppose that S is defined by $\rho = 0$ where $\rho: X \to R$ is a C^∞ function such that $d\rho \neq 0$ on S. We say that such a ρ is a defining function for S.

S divides X into two open sets X^+ and X^- defined respectively by $\rho > 0$ and $\rho < 0$; we can also assume that there exists an $\varepsilon_o > 0$ such that, if $|\varepsilon| < \varepsilon_o$ and S_ε is the level hypersurface defined by $\rho = \varepsilon$, there is a diffeomorphism $\pi_\varepsilon: S_\varepsilon \to S$; we denote by X^+_ε and X^-_ε the open sets defined by $\rho > \varepsilon$ and $\rho < -\varepsilon$ respectively.

We will use the standard notations for currents and distributions spaces; in particular we fix the orientation on S in such a way that $d[X^+] = [S]$. Furthermore, we list the following definitions:

i) Let $L: \mathcal{E}^{(r)}(X) \to \mathcal{E}^{(r)}(S)$ be the restriction operator; we set: $\mathcal{E}^{(p,q)}(S) = L(\mathcal{E}^{(p,q)}(X))$ and $\mathcal{D}^{(p,q)}(S) = L(\mathcal{D}^{(p,q)}(X))$.

ii) Let $K \in \mathcal{D}'^{(r)}(S)$: $K \wedge [S]$ will be the $(r+1)$-current on X defined by: $\langle K \wedge [S], \varphi \rangle = \langle K, L(\varphi) \rangle$, $\varphi \in \mathcal{D}^{(2n-r-1)}(X)$.

iii) We say that $K \in \mathcal{D}'^{(r)}(S)$ is an (r,o)-current (resp. (o,r)) if $(K \wedge [S])^{p,o} = 0$ for $p \leq r$ (resp. $(K \wedge [S])^{o,p} = 0$ for $p \leq r$); we denote by $\mathcal{D}'^{(r,o)}(S)$ (resp. $\mathcal{D}'^{(o,r)}(S)$) the relative space.

iv) If $K \in \mathcal{D}'^{(r,o)}(S)$, $K \wedge [S]^{1,o}$ is the $(r+1,o)$-current on X defined by: $\langle K \wedge [S]^{1,o}, \varphi \rangle = \langle K, L(\varphi) \rangle$, $\varphi \in \mathcal{D}^{(n-r-1,n)}(X)$ and $K \wedge [S]^{o,1}$ is the $(r,1)$-current on X defined by: $\langle K \wedge [S]^{o,1}, \varphi \rangle = \langle K, L(\varphi) \rangle$, $\varphi \in \mathcal{D}^{(n-r,n-1)}(X)$

and also

v) Let $\alpha \in \mathcal{E}^{(r)}(X^+)$; we say that α admits trace $K \in \mathcal{D}'^{(r)}(S)$ on S in the sense of currents if for every $\varphi \in \mathcal{D}^{(2n-r-1)}(S)$ we have: $\lim_{\varepsilon \to o^+} \int_{S_\varepsilon} \alpha \wedge \pi^*_\varepsilon(\varphi) = \lim_{\varepsilon \to o^+} \int_S \pi_{\varepsilon *}(\alpha) \wedge \varphi$

$$= \lim_{\varepsilon \to o^+} \langle \pi_{\varepsilon *}(\alpha), \varphi \rangle = \langle K, \varphi \rangle$$

(cf. [5]); we set $K = \gamma_+(\alpha)$; in the same manner we define $\gamma_-(\alpha)$ if $\alpha \in \mathcal{E}^{(r)}(X^-)$.

2. Tangential operators on S

Now let $X = D$ be a domain in $\mathbb{C}^n$, $n \geq 2$.
Let $(\ , \ )$ be the usual Hermitian product in $\mathbb{C}^n$ extended punctually to $\mathcal{E}^{(p,q)}(D)$. Without loss of generality we can assume that $(\partial\rho, \partial\rho) \equiv 1/2$ on S. Put

$$\mathcal{N}^{(p,q)}(S) = \{\alpha \in \mathcal{E}^{(p,q)}(S) \mid \alpha = \varphi \wedge \partial\rho \text{ on } S\}, \quad p \geq 1,$$

$$\mathcal{T}^{(p,q)}(S) = \{\alpha \in \mathcal{E}^{(p,q)}(S) \mid (\alpha,\beta) = 0 \text{ on } S \ \forall \beta \in \mathcal{N}^{(p,q)}(S)\},$$

and

$$\bar{\mathcal{N}}^{(p,q)}(S) = \{\alpha \in \mathcal{E}^{(p,q)}(S) \mid \alpha = \varphi \wedge \bar{\partial}\rho \text{ on } S\}, \quad q \geq 1,$$

$$\bar{\mathcal{T}}^{(p,q)}(S) = \{\alpha \in \mathcal{E}^{(p,q)}(S) \mid (\alpha,\beta) = 0 \text{ on } S \ \forall \beta \in \bar{\mathcal{N}}^{(p,q)}(S)\}.$$

We have the decompositions

$$\mathcal{E}^{(p,q)}(S) = \mathcal{N}^{(p,q)}(S) \oplus \mathcal{T}^{(p,q)}(S),$$

$$\mathcal{E}^{(p,q)}(S) = \bar{\mathcal{N}}^{(p,q)}(S) \oplus \bar{\mathcal{T}}^{(p,q)}(S)$$

and we denote by

$$\tau : \mathcal{E}^{(p,q)}(S) \to \mathcal{T}^{(p,q)}(S), \quad \bar{\tau} : \mathcal{E}^{(p,q)}(S) \to \bar{\mathcal{T}}^{(p,q)}(S)$$

the natural projections;
we set: $\partial_b\alpha = \tau(\partial\alpha)$ and $\bar{\partial}_b\alpha = \bar{\tau}(\bar{\partial}\alpha)$ (cf. [4]).

In particular, on S we have:

a) if $f \in \mathcal{E}^{(0,0)}(D)$, the following formulas hold:

$$\partial_b f = \partial f - 2N(f)\partial\rho,$$

$$\bar{\partial}_b f = \bar{\partial} f - 2\bar{N}(f)\bar{\partial}\rho,$$

where $N(f) = (\partial f, \partial\rho) = \sum_{j=1}^{n} \frac{\partial f}{\partial z_j} \frac{\partial\rho}{\partial \bar{z}_j}$ and $\bar{N}(f) = (\bar{\partial} f, \bar{\partial}\rho)$

$$= \sum_{j=1}^{n} \frac{\partial f}{\partial \bar{z}_j} \frac{\partial\rho}{\partial z_j} ;$$

b) if $\beta \in \mathcal{E}^{(1,0)}(D)$, $\beta = \sum_{j=1}^{n} \beta_j \, dz_j$, the following formulas hold:

$$\partial_b\beta = \partial\beta - 2N(\beta) \wedge \partial\rho,$$

$$\bar{\partial}_b\beta = \sum_{j=1}^{n} \bar{\partial}_b\beta_j \wedge dz_j,$$

where $N(\beta) = \sum_{j=1}^{n} (\frac{\partial\rho}{\partial\bar{z}_j} \partial\beta_j - N(\beta_j)dz_j)$;

by conjugation, we obtain the analougous formulas for (o,1)-forms.

A local linear operator $\Phi : \mathcal{E}^{(p,q)}(D) \to \mathcal{E}^{(r,s)}(D)$ is said to be <u>tangential on</u> S if from $L(f) = 0$ it follows $L(\Phi(f)) = 0$, a tangential operator Φ induces a new operator: $\mathcal{E}^{(p,q)}(S) \to \mathcal{E}^{(r,s)}(S)$ (which will be denoted again by Φ). By definition ∂_b and $\bar{\partial}_b$ are tangential operators on S .

Let again $f \in \mathcal{E}^{(o,o)}(D)$: it is easy to check that:

i) in the point of S , $N(f)$ is the complex normal derivative of f and $N(f) + \bar{N}(f)$ is the real normal derivative of f,

ii) if $f \in C^\infty(D,\mathbb{R})$, then $\overline{N(f)} = \bar{N}(f)$

iii) $f \to N(f) - \bar{N}(f)$ is a tangential operator on S .

We deduce also the following

<u>LEMMA 2.1</u>. <u>Let</u> β <u>be a holomorphic</u> (1,o)-<u>form on</u> D <u>such that</u> $L(\beta) = \partial_b f$, <u>where</u> $f \in C^\infty(S,\mathbb{R})$; <u>then</u> $\partial\beta = 0$.

∂_b and $\bar{\partial}_b$ can be extended in a natural way to currents on S ; it is easy to recognize that:

$K \in \mathcal{D}'^{(r,o)}(S)$ satisfies $\partial_b K = 0$

if and only if $\partial[K \wedge [S]^{1,o}] = 0$

and

$K \in \mathcal{D}'^{(r,o)}(S)$ satisfies $\bar{\partial}_b K = 0$

if and only if $\bar{\partial}[K \wedge [S]^{o,1}] = 0$.

Let now $U \subset D$ be an open set and $\mathcal{P}(U)$ be the space of pluriharmonic functions in U (i.e. $f \in \mathcal{P}(U)$ if $f\ \bar{\partial}\partial f = 0$) .

We have the following:

<u>Proposition 2.2</u>. <u>Suppose</u> $\bar{\partial}\partial\rho \neq 0$ <u>on</u> S ; <u>then there exists a differential operator</u> $R : \mathcal{E}^{(o,o)}(S) \to \mathcal{E}^{(o,o)}(S)$ <u>such that if</u> $h \in \mathcal{P}(D)$ <u>then</u> $R(L(h)) = L(N(h) + \bar{N}(h))$.

<u>P r o o f</u>. We set $\delta_b = (\partial_b + \bar{\partial}_b)$, $*\delta_b = (\bar{\tau}\partial_b + \tau\bar{\partial}_b)$, $\delta_b^c = i(\bar{\partial}_b - \partial_b)$; using e.g. special moving frames in the sense of [3] and [4] , if $h \in \mathcal{P}(D)$ we obtain the formula

(##) $\quad *\delta_b\delta_b^c h = 2i[N(h) + \bar{N}(h)]\tau(\bar{\partial}_b\partial\rho)$.

Since $i\tau(\bar{\partial}_b\partial\rho)$, which is a real operator, represents the restriction to S of the Levi-form of ρ , in our hypothesis, we can "divide" (##) by $i\tau(\bar{\partial}_b\partial\rho)$ in order to obtain R .
Thus, if h is pluriharmonic, $N(h) + \bar{N}(h)$, ∂h and $\bar{\partial}h$ can be expressed in terms of tangential operators.

R e m a r k s. a) In [6] a similar formula is proved in a more laborious way.

b) The above constructions can be carried on a complex manifold equipped with a hermitian structure on $T_{\mathbb{C}}(X) \oplus \overline{T_{\mathbb{C}}(X)}$.

3. Traces of pluriharmonic functions

We deduce from Proposition 2.2 that if h is a pluriharmonic function on X and S is not Levi-flat, then the following formula holds:

(*) $\quad \partial h = \partial_b h + [N(h) - \bar{N}(h)]\partial\rho + R(h)\partial\rho$.

Let $\omega : \mathcal{E}^{(0,0)}(S) \to \mathcal{E}^{(1,0)}(S)$ defined by the right member of (*) and let $f \in \mathcal{E}^{(0,0)}(S)$; then:

a) if $\tilde{f} \in \mathcal{E}^{(0,0)}(X)$ and $L(\tilde{f}) = f$ one has:

$$\omega(f) \wedge [S]^{1,0} = \partial[f[S]^{1,0}] = \partial\tilde{f} \wedge [S]^{1,0} = \partial_b\tilde{f} \wedge [S]^{1,0} ,$$

in particular $\partial_b\omega(f) = 0$;

b) if f is the trace of a function F which is pluriharmonic in X^+ and C^∞ up to S , then $\partial F \wedge [S] = \omega(f) \wedge [S]$ and thus $\bar{\partial}_b\omega(f) = 0$;

c) if $f \in C^\infty(S,\mathbb{R})$ then there exists $\tilde{f} \in C^\infty(X,\mathbb{R})$ such that $L(\tilde{f}) = f$ and $\omega(f) \wedge [S] = \partial\tilde{f} \wedge [S]$;

d) ω can be extended to $\mathcal{D}'^{(0,0)}(S)$, taking values in $\mathcal{D}'^{(1,0)}(S)$ with the same properties.

We note that similar operators for balls in $\mathbb{C}^n$ can be found in [1] and [2] .

Let $\alpha \in \mathcal{E}^{(p,q)}(X \setminus S)$ such that $\gamma_+(\alpha)$ and $\gamma_-(\alpha)$ exist: we say that $\gamma_+(\alpha) - \gamma_-(\alpha)$ is the <u>jump</u> of α on S . We are able now to give the following solution to the Riemann-Hilbert problem for the $\bar{\partial}\partial$ operator:

THEOREM 3.3. Suppose X is a Stein manifold such that $H^2(X,\mathbb{C}) = 0$ and $\bar{\partial}\partial\rho \neq 0$ on S. Let T be a real distribution on S; then the following statements are equivalent:

i) $\bar{\partial}_b\omega(T) = 0$,

ii) there exists $F \in \mathcal{P}(X-S)$ such that $\gamma_+(F) - \gamma_-(F) = T$.

P r o o f. ii) implies i) as follows from the previous remarks. Conversely, assume $\bar{\partial}[\omega(T) \wedge [S]^{0,1}] = 0$.
Since X is Stein, there exists $\tilde{K} \in \mathcal{D}'^{(1,0)}(X)$ such that $\bar{\partial}\tilde{K} = \omega(T) \wedge [S]^{0,1}$; let $K_\pm = \tilde{K}_{|X^\pm}$.
K_+ and K_- are holomorphic $(1,0)$-forms in X^+ and X^- respectively, $\gamma_+(K_+)$, $\gamma_-(K_-)$ exist and $\gamma_+(K_+) - \gamma_-(K_-) = \omega(T)$; furthermore $\gamma_+(\partial K_+)$ and $\gamma_-(\partial K_-)$ exist too (cf. [5]). Using Lemma 2.1 we observe that the $(2,0)$-current $\hat{K} = \partial K_+ \wedge [X^+] + \partial K_- \wedge [X^-]$ is holomorphic and ∂-closed and so, as $H^2(X,\mathbb{C}) = 0$, it is possible to find $\check{K} \in \mathcal{E}^{(1,0)}(X)$, holomorphic, such that $\partial\check{K} = \hat{K}$ and also $\tilde{G} \in \mathcal{D}'^{(0,0)}(X)$ such that $\partial\tilde{G} = K_+ \wedge [X^+] + K_- \wedge [X^-] - \check{K} + T \wedge [S]^{1,0}$.

If we set $G = \tilde{G}_{|X \setminus S}$, we have $\bar{\partial}\partial G = 0$ on $X \setminus S$; thus we obtain the following:

a) G is a pluriharmonic function on $X \setminus S$,

b) since G can be extended as distribution across S and satisfies $\bar{\partial}\partial G = 0$, then $\gamma_+(G)$ and $\gamma_-(G)$ exist (cf. [5]),

c) one has $\partial G = \tilde{K} - \check{K}$ on $X \setminus S$.

It is easy to show now that $\gamma_+(G) - \gamma_-(G) + T$ is ∂_b-closed and thus there exists an antiholomorphic function H on $X \setminus S$ such that $\gamma_+(H) - \gamma_-(H) = \gamma_+(G) - \gamma_-(G) + T$. It follows that $F = H - G$ is a pluriharmonic function on $X \setminus S$ such that $\gamma_+(F) - \gamma_-(F) = T$; furthermore, F can be choosen to be real: this follows directly from

LEMMA 3.2. Suppose S is not Levi-flat and let $f_\pm \in \mathcal{P}(X^\pm)$ admitting traces $\gamma_+(f_+)$ and $\gamma_-(f_-)$ on S; if $\gamma_+(f_+) = \gamma_-(f_-)$, then there exists $f \in \mathcal{P}(X)$ such that $F_{|X^\pm} = f_\pm$.

P r o o f. Since the problem is local, we can suppose X is a domain of $\mathbb{C}^n$. Then $(\partial/\partial z_j)f_\pm$, $1 \leq j \leq n$, are holomorphic in $X^\pm$,

$\gamma_+(\partial/\partial z_j)f_+$ and $\gamma_-(\partial/\partial z_j)f_-$ exist and Proposition 2.2 assures that $\gamma_+(\partial/\partial z_j)f_+ = \gamma_-(\partial/\partial z_j)f_-$, $1 \le j \le n$. Hence we have essentially reduced our proof to the case where f is holomorphic, and this follows from [5] .

Remark. Classical results of potential theory assure that if T in Theorem 3.1 is a continuous function, then $F_\pm = F_{|X^\pm}$ are continuous up to S and if T is a C^k function, $1 \le k \le \infty$, then $F_\pm$ are $C^{k-\varepsilon}$, $\varepsilon > 0$, up to S .

The previous results and cohomological arguments enable us to prove the following

THEOREM 3.3. *Suppose X is a Stein manifold, $\bar\partial\partial\rho \neq 0$ on S and $H^1(S,\mathbb{R}) = 0$; then the equation $\bar\partial_b\omega(T) = 0$ characterizes the jumps of pluriharmonic functions in $X \setminus S$.*

We can use the previous results to study the Cauchy-Dirichlet problem, i.e. the trace problem for $\bar\partial\partial$ operator. Suppose we have fixed $p \in S$ so that there exists a neighbourhood W of p in S where the Levi form of ρ has an eigenvalue of the same sign, e.g. > 0. The (1,o)-form K_- , constructed as in Theorem 3.1 , extends across S as β to a Stein neighbourhood U of p in X ; it is clear that, near p , one has $\gamma_+(K_+ - \beta) = \omega(T)$ and $\gamma_+(\partial K_+ - \partial\beta) = 0$. Using similar arguments as in Theorem 3.1 we can prove

THEOREM 3.4. *Let T be a real distribution on S . Then if T satisfies $\bar\partial_b\omega(T) = 0$, for every $p \in S$ such that p is pseudoconvex there exists a neighbourhood U of p in X such that if $U_+ = U \cap X^+$ it is possible to find $F \in \mathcal{P}(U_+)$ such that $\gamma_+(F) = T$ (an analogous result holds for pseudoconcavity points).*

In the assumption of Theorems 3.1 and 3.3 the jump is locally a trace (possibly from different sides).

In the same manner we can deduce

PROPOSITION 3.5. *Suppose that the Levi form of ρ has at least one positive eigenvalue at every point $p \in S$. Then there exists a neighbourhood U of S such that if $U_+ = U \cap X^+$, the equation $\bar\partial_b\omega(T) = 0$ characterizes the distributions on S which are traces of pluriharmonic functions in U_+ .*

We also have:

PROPOSITION 3.6. Suppose X is a Stein manifold, X^+ is relatively compact and $\bar\partial\partial\rho \neq 0$ on S . If T is a real distribution on S , then the following statements are equivalent:

i) $\bar\partial_b \omega(T) = 0$,

ii) there exists $F \in \mathcal{P}(X^+)$ such that $\gamma_+(F) = T$.

In a similar manner, we can deal with the Levi-flat case and find in a direct way characterizing conditions for jumps of pluriharmonic functions and non linear conditions for boundary values (cf. [5]) .

References

[1] AUDIBERT T. : Opérateurs differentiels sur la sphère de $\mathbb{C}^n$ caractérisant les restrictions des fonctions pluriharmoniques, Th. 3.ème c. Univ. de Provence, U.E.R. Math.

[2] BEDFORD E., FEDERBUSH P. : Pluriharmonic boundary values, Tohoku Math.J. 26 (1974), 505-511.

[3] KOHN J.J. : Harmonic integrals on strongly pseudo-convex manifolds I - II, Ann. of Math. 78 (1963), 112-148 and 79 (1964), 450-472.

[4] KOHN J.J., ROSSI U. : On the extension of holomorphic functions from the boundary of a complex manifold Ann. of. Math. 81 (1965), 451-473.

[5] ŁOJASIEWICZ S., TOMASSINI G. : Valeurs au bord des formes holomorphes, in: Several Complex Variables, Proceedings of International Conferences, Cortona, Italy 1976-1977, Scuola Normale Superiore, Pisa (1978), 222-245.

[6] RIZZA G.B. : Dirichlet problem for n-harmonic functions and related geometrical properties Math. Ann. 130 (1955), 202-218.

Istituto Matematico "Ulisse Dini"
Università di Firenze, Viale Morgagni 67/A
I-50134 Firenze, Italia

SOME BANACH ALGEBRAS OF ANALYTIC FEYNMAN INTEGRABLE FUNCTIONALS

Robert Horton Cameron and David Arne Storvick* (Minneapolis, MN)

Contents

1. Introduction

This paper presents three Banach algebras of analytic Feynman integrable functionals and gives formulae for their Feynman integrals. Because these formulae do not involve analytic extension, they could provide consistent definitions for the Feynman integral without reference to the term "analytic extension". Our Banach algebras are similar to the spaces of Fresnel integrals of Albeverio and Höegh-Krohn [1]. However, in our "Fourier transforms" of measures we use complex exponentials of bilinear functionals which are not inner products. As an example, consider the following functionals which are of interest in quantum mechanics. Let $F(\vec{x}(\cdot)) \equiv F(x_1(\cdot), \ldots, x_\nu(\cdot))$ be a functional defined on $C^\nu \equiv C^\nu[a,b]$, (ν - dimensional Wiener space), so that $x_j(t)$ is continuous on $[a,b]$, and $x_j(a) = 0$. Let

$$(1.1) \quad F(\vec{x}) \equiv \exp\left\{ \int_a^b \theta(t\,;\,\vec{x}(t))dt \right\}$$

*Research sponsored by the National Science Foundation Grant MCS 77-02116

where $\theta(t,\vec{u})$ is a function which for each t is a Fourier transform of a bounded complex measure:

$$(1.2) \quad \theta(t,\vec{u}) = \int_{\mathbb{R}^\nu} \exp\{i \sum_{k=1}^{\nu} u_k v_k\} \, d\sigma_t(\vec{v}) .$$

Here $\sigma_t(\cdot)$ is a uniformly bounded family of complex measures on $\mathbb{R}^\nu$, $\|\sigma_t\| \leq M$, such that for each Borel set E in $[a,b] \times \mathbb{R}^\nu$, $\sigma_t(E^{(t)})$ is a measurable function of t on $[a,b]$. The symbol $E^{(t)}$ denotes the t-section of E:

$$(1.3) \quad E^{(t)} = \{\vec{v} \mid \vec{v} \in \mathbb{R}^\nu , \; t \times \vec{v} \in E\} .$$

It will be shown below that F is an element of the Banach algebra S'', (defined below) and hence that the Feynman integral of this functional F can be expressed as a sum of finite dimensional Lebesgue-Stieltjes integrals. Thus for real $q \neq 0$, the analytic Feynman integral with parameter q of the functional F is

$$(1.4) \quad \int_{C^\nu}^{\mathrm{anf}_q} F(\vec{x}) d\vec{x} = \int_{C^\nu}^{\mathrm{anf}_q} \exp\{ \int_a^b \theta(t ; \vec{x}(t)) \, dt\} \, d\vec{x}$$

$$= 1 + \sum_{n=1}^{\infty} \int_{\Delta_n \times \mathbb{R}^{n\nu}} \exp\{ \frac{1}{2qi} \sum_{k=1}^{\nu} \sum_{\ell=1}^{n} \sum_{j=1}^{\ell} (2 - \delta_{j,\ell}) v_{k,j} v_{k,\ell} (t_j - a)\}$$

$$d\sigma_{t_1}(\vec{v}_1) \cdots d\sigma_{t_n}(\vec{v}_n) \, dt_1 \cdots dt_n .$$

Here $\Delta_n \equiv \{\vec{t} \mid a < t_1 < t_2 < \cdots < t_n \leq b\}$. We note that if ψ is the Fourier transform of a bounded measure on $\mathbb{R}^\nu$,

$$(1.5) \quad \exp\{ \int_a^b \theta(t,\vec{x}(t) dt\} \, \psi(\vec{x}(b))$$

is also an element of S'' and its Feynman integral can be expressed as a sum of integrals similar to that for F.

In this paper we shall use a definition of analytic Feynman integral which is similar to that used in [3].

<u>Definition</u>. Let F be a functional such that the ν-dimensional Wiener integral

$$(1.6) \qquad J(\lambda) \equiv \int_{C^{\nu}[a,b]} F(\lambda^{-1/2}\vec{x})d\vec{x}$$

exists for all real $\lambda > 0$. If there exists a function $J^*(\lambda)$ analytic in the half-plane $\operatorname{Re}\lambda > 0$ such that $J^*(\lambda) = J(\lambda)$ for all real $\lambda > 0$, then we define J^* to be the <u>analytic Wiener integral of</u> F <u>over</u> $C^{\nu}[a,b]$ <u>with parameter</u> λ, and for $\operatorname{Re}\lambda > 0$ we write

$$(1.7) \qquad \int_{C^{\nu}[a,b]}^{\text{anw}_\lambda} F(\vec{x})d\vec{x} \equiv J^*(\lambda) \quad .$$

<u>Definition</u>. Let q be a real parameter $(q \neq 0)$ and let F be a functional whose analytic Wiener integral exists for $\operatorname{Re}\lambda > 0$. Then if the following limit exists, we call it the <u>analytic Feynman integral of</u> F <u>over</u> $C^{\nu}[a,b]$ with parameter q, and we write

$$(1.8) \qquad \int_{C^{\nu}[a,b]}^{\text{anf}_q} F(\vec{x})d\vec{x} \equiv \lim_{\substack{\lambda \to -iq \\ \operatorname{Re}\lambda > 0}} \int_{C^{\nu}[a,b]}^{\text{anw}_\lambda} F(\vec{x})d\vec{x} \quad .$$

In the example given in equation (1.1), the exponent of the exponential is bounded because in this case θ is given by (1.2).

It will be shown in Example 3 of Section 7 that the functional F given in (1.5) is an element of the space S when the functional θ is given by

$$(1.9) \qquad \theta(t,\vec{u}) \equiv -\sum_{j=1}^{\nu} \rho_j u_j^2 \ , \ \rho_j > 0 \ ,$$

even though this new θ is unbounded. The fact that this exponent θ is unbounded makes it appear doubtful that the functional (1.5) is a Fresnel integrable functional in the sense of Albeverio and Höegh-Krohn [1].

The formula for the analytic Feynman integral turns out to be even simpler in this case than in the case when θ satisfies (1.2).

Another Banach algebra of Feynman integrable functionals was given by Johnson and Skoug in [6].

The authors acknowledge with appreciation improvements in the manuscript suggested by Johnson and Skoug.

2. THE SPACE S

We now define the largest of our three spaces of analytic Feynman integrable functionals S, S', S". We shall show that they satisfy $S'' \subset S' \subset S$ and with appropriate norms form Banach algebras. We begin with the largest space S, and first define a σ-algebra $\mathfrak{a}$ in L_2^ν and a class of complex measures $\mathcal{M}$ on L_2^ν. (In this paper L_2 always means real L_2).

Notation. Let $\mathfrak{a}$ be the σ-algebra of subsets of $L_2^\nu[a,b]$ generated by the class of sets of the form

$$\{\vec{v} \mid \vec{v} \in L_2^\nu[a,b], \int_a^b v_j(t)\varphi_j(t)dt < \lambda_j, j = 1,2,\ldots,\nu\},$$

where $\vec{\varphi} \equiv (\varphi_1,\ldots,\varphi_\nu)$ ranges over all elements of $L_2^\nu[a,b]$, and $\vec{\lambda} \equiv (\lambda_1,\ldots,\lambda_\nu)$ ranges over $\mathbb{R}^\nu$.

The σ-algebra $\mathfrak{a}$ of subsets of $L_2^\nu[a,b]$ is translation invariant.

Definition. Let $\mathcal{M} \equiv \mathcal{M}(L_2^\nu[a,b])$ be the class of complex measures of finite variation defined on $L_2^\nu[a,b]$ with $\mathfrak{a}$ as its σ-algebra of measurable sets If $\mu \in \mathcal{M}$, we set $\|\mu\| = \operatorname{var} \mu$ over L_2^ν.

Remark. It is clear that $\mathcal{M}$ is a linear class of measures and is invariant under change of scale.

Terminology. We shall say that two functionals F(x) and G(x) are *equal s - almost everywhere* if for each ρ > 0 the equation F(ρx) = G(ρx) holds for almost all x ε C[a, b], in other words, if F(x) = G(x) except for a scale-invariant null set. We denote this equivalence relation between functionals by

$$F \approx G.$$

The definition of S also involves the P.W.Z. (Paley - Wiener - Zygmund) integral [4] which is defined as follows.

Definition. Let $\varphi_1, \varphi_2, \ldots$ be a C.O.N. (complete orthonormal) set of real functions of bounded variation on [a,b]. Let $v \in L_2[a,b]$ and

$$v_n(t) = \sum_{j=1}^{n} \varphi_j(t) \int_a^b v(s)\varphi_j(s)ds .$$

Then the P.W.Z. integral is defined by

$$\int_a^b v(s)\tilde{d}x(s) \equiv (\varphi)\int_a^b v(s)\tilde{d}x(s) \equiv \lim_{n\to\infty} \int_a^b v_n(s)\, dx(s)$$

for all $x \in C[a,b]$ for which the above limit exists.

Note. It was shown in [5] that this integral exists for almost all $x \in C[a,b]$ and is essentially independent of the choice of $\varphi_1, \varphi_2, \ldots$. Moreover if v is bounded variation, it is essentially equivalent to the Riemann-Stieltjes integral. Clearly "almost all" may be replaced by "s-almost all" in this statement.

Definition. Let $S \equiv S(L_2^\nu)$ be the space of functionals expressable in the form

$$(2.1)\qquad F(\vec{x}) \equiv \int_{L_2^\nu} \exp\{i \sum_{j=1}^{\nu} \int_a^b v_j(t)\, \tilde{d}x_j(t)\}\, d\mu(\vec{v})$$

for s-almost all $\vec{x} \in C^\nu[a,b]$, where $\mu \in \mathfrak{M} \equiv \mathfrak{M}(L_2^\nu)$. (Note. It is assumed that the P.W.Z. integral $\int_a^b v_j(t)\tilde{d}x_j(t)$ is based on the same C.O.N. sequence $\{\varphi_n\}$ for all choices of v).

Lemma 2.1 If $\mu \in \mathfrak{M}$ and $\{\varphi_n\}$ is a complete orthonormal sequence of functions of bounded variation on $[a,b]$, then $(\varphi)\int_a^b v_j(t)\tilde{d}x_j(t)$ is measurable in $\vec{x}\times\vec{v}$ with respect to the measure $m_w^\nu \times \mu$ on $C^\nu[a,b] \times L_2^\nu$. Here m_w^ν denotes Wiener measure in $C^\nu[a,b]$. (This implies, of course, that the set where the P.W.Z. integral exists is measurable). Also, for s-almost every $\vec{x} \in C^\nu[a,b]$, the P.W.Z. integral $(\varphi)\int_a^b v_j(t)\tilde{d}x_j(t)$ exists for μ-almost all $\vec{v}$ in L_2^ν.

Proof. By definition

$$(2.2)\qquad (\varphi)\int_a^b v_j(t)\tilde{d}x_j(t) = \sum_{n=1}^{\infty} \int_a^b v_j(t)\,\varphi_n(t)dt \int_a^b \varphi_n(t)dx_j(t) \quad ,$$

and it exists whenever the series on the right converges. Morever, we know from [4] that for each $v \in L_2^\nu$, the series on the right converges for almost all $x \in C^\nu[a,b]$.

By the definition of $\mathcal{M}$, $\int_a^b v_j(t)\varphi_n(t)dt$ is μ-measurable as a function of $\vec{v}$ on L_2^ν, and since it is independent of $\vec{x}$, it is $m_w^\nu \times \mu$ measurable as a function of $\vec{x}$ and $\vec{v}$ on $C^\nu \times L_2^\nu$. Similarly $\int_a^b \varphi_n(t)dx_j(t)$ is Wiener measurable as a function of $\vec{x}$ on $C^\nu[a,b]$ and hence it is $m_w^\nu \times \mu$ measurable as a function of $\vec{x}$ and $\vec{v}$ on $C^\nu \times L_2^\nu$. Thus each term of the right hand member of (2.2) is $m_w^\nu \times \mu$ measurable on $C^\nu \times L_2^\nu$. It follows that the series converges on a measurable set and converges to a measurable functional, so the left member of (2.2) is measurable with respect to $m_w^\nu \times \mu$.

Since the set where the left member of (2.2) exists is $m_w^\nu \times \mu$ measurable, its complement is measurable, and since every section of this set obtained by holding $\vec{v}$ fixed is a null set, the entire complement is a null set. Thus for almost every $\vec{x} \in C^\nu[a,b]$, the integral $(\varphi)\int_a^b v_j(t)\tilde{d}x_j(t)$ exists for almost all $\vec{v}$ in L_2^ν. Clearly this holds for s-almost every $\vec{x} \in C^\nu[a,b]$, and the lemma is proved.

<u>Corollary 1.</u> <u>Let</u> $\mu \in \mathcal{M}$ <u>and let</u> $\{\varphi_n\}_{\mu=1,2,\ldots}$ <u>be a</u> C.O.N. <u>set of functions of</u> B.V. <u>on</u> $[a,b]$, <u>then the function</u> F <u>defined below exists for s-almost every</u> $\vec{x} \in C^\nu[a,b]$:

$$(2.3) \qquad F(\vec{x}) \equiv \int_{L_2^\nu} \exp\{i \sum_{i=1}^{\nu} (\varphi)\int_a^b v_j(t)\tilde{d}x_j(t)\} d\mu(\vec{v}) .$$

<u>Moreover</u> $F(\vec{x})$ <u>is integrable in</u> $\vec{x}$ <u>on</u> $C^\nu[a,b]$.

<u>Corollary 2.</u> <u>If the</u> C.O N. $\{\varphi_n\}_{n=1,2\ldots}$ <u>in the hypothesis of Lemma 2.1 is changed to a new</u> C.O.N. <u>of</u> B.V. <u>on</u> $[a,b]$, <u>the new</u> F <u>is s-almost everywhere equal to the old one.</u>

In view of Corollary (2), we will sometimes omit the symbol (φ) before the P.W.Z. integral in the right member of (2.3). The integral will then be defined up to an equivalence class.

Note. If $F \in S$, $\lambda > 0$, and $G(\vec{x}) \equiv F(\lambda \vec{x})$, we have $G \in S$.

Lemma 2.2 *Let $\mu_1, \mu_2, \ldots$ be a sequence of real non-negative elements of $\mathcal{M}$ such that $\sum_{n=1}^{\infty} \|\mu_n\| < \infty$. For all $E \in \mathcal{G}$, let $\mu(E) \equiv \sum_{p=1}^{\infty} \mu_p(E)$.*

Then $\mu \in \mathcal{M}$ and $\|\mu\| \leqq \sum_{p=1}^{\infty} \|\mu_p\|$.

Proof. Let $\{E_q\}$ be a disjoint sequence of elements of $\mathcal{G}$ and let $E = \bigcup_{q=1}^{\infty} E_q$. Then $E \in \mathcal{G}$. Moreover

$$\mu_p(E) = \mu_p\left(\bigcup_{q=1}^{\infty} E_q\right) = \sum_{q=1}^{\infty} \mu_p(E_q) \quad \text{for } p = 1, 2, \ldots,$$

Thus

$$\mu(E) = \sum_{p=1}^{\infty} \mu_p(E) = \sum_{p=1}^{\infty} \sum_{q=1}^{\infty} \mu_p(E_q) =$$

$$\sum_{q=1}^{\infty} \sum_{p=1}^{\infty} \mu_p(E_q) = \sum_{q=1}^{\infty} \mu(E_q) .$$

Thus μ is a measure on the σ-algebra $\mathcal{G}$. Moreover $\|\mu\| = \mu(L_2^{\nu}) = \sum_{p=1}^{\infty} \mu_p(L_2^{\nu}) = \sum_{p=1}^{\infty} \|\mu_p\| < \infty$. Hence $\mu \in \mathcal{M}$ and the lemma is proved.

Because a bounded complex measure may be decomposed into its real and imaginary parts and they in turn decomposed into positive and negative parts, Lemma 1.2 can be extended by omitting the restriction that $\mu_1, \mu_2, \ldots$ be real and non-negative.

Lemma 2.3 *Let $\mu_1, \mu_2, \ldots$ be a sequence of elements of $\mathcal{M}$ such that $\sum_{n=1}^{\infty} \|\mu_n\| < \infty$. If $E \in \mathcal{G}$, let $\mu(E) = \sum_{p=1}^{\infty} \mu_p(E)$. Then $\mu \in \mathcal{M}$ and $\|\mu\| \leqq 4 \sum_{p=1}^{\infty} \|\mu_p\|$.*

Lemma 2.4 *The space S is a linear space.*

Proof. Let $F_1, F_2 \in S$ and let

$$(2.5) \quad F_p(\vec{x}) \approx \int_{L_2^{\nu}} \exp\left\{i \sum_{j=1}^{\nu} \int_a^b v_j(t)\, \tilde{d} x_j(t)\right\} d\mu_p(\vec{v}) \quad \text{for } p = 1, 2$$

where $\mu_1, \mu_2 \in \mathcal{M}$. Let μ be defined by

$\mu(E) = c_1\mu_1(E) + c_2\mu_2(E)$ for each $E \in \mathcal{G}$ where c_1 and c_2 and complex numbers. Since $\mathcal{M}$ is linear, $\mu \in \mathcal{M}$. Then if we set

$$F(\vec{x}) \equiv c_1 F_1(\vec{x}) + c_2 F_2(\vec{x})$$

we have

$$F(\vec{x}) \approx \int_{L_2^\nu} \exp\{i \sum_{j=1}^{\nu} \int_a^b v_j(t)\, \tilde{d} x_j(t)\} d\mu(\vec{v}) .$$

Clearly $F \in S$ and thus S is a linear space.

We next prove a uniqueness theorem.

<u>Theorem 2.1</u> <u>If</u> $F \in S$, <u>there is a unique measure</u> $\mu \in \mathcal{M}$ <u>such that</u>

$$(2.6) \qquad F(\vec{x}) = \int_{L_2^\nu} \exp\{i \sum_{j=1}^{\nu} \int_a^b v_j(t)\, \tilde{d} x_j(t)\} d\mu(\vec{v})$$

<u>for almost all</u> $\vec{x} \in C^\nu[a,b]$. <u>Moreover</u> (2.6) <u>provides a 1-1 correspondence between</u> $\mathcal{M}$ <u>and</u> S. <u>Finally, if</u> F, $G \in S$ <u>and</u> $F(\vec{x}) = G(\vec{x})$ <u>for almost all</u> x, <u>then</u> $F \approx G$.

<u>Proof</u>. Suppose that there are two measures μ_1 and μ_2 such that (2.6) holds for F with μ_1 and (2.6) holds for F with μ_2. Then if $\mu_3 \equiv \mu_1 - \mu_2$, $\mu_3 \in \mathcal{M}$, and for almost all $\vec{x} \in C^\nu$, we have

$$(2.7) \qquad 0 \equiv \int_{L_2^\nu} \exp\{i \sum_{j=1}^{\nu} \int_a^b v_j(t)\, \tilde{d} x_j(t)\} d\mu_3(\vec{v}) .$$

If $-\infty \leq \alpha < \beta < \infty$, let

$$(2.8) \qquad g_{\alpha,\beta}(s) \equiv \chi_{(\alpha,\beta)}(s) ,$$

and if $0 < \delta < (\beta - \alpha)/2$, let $g_{\alpha,\beta,\delta}$ be the trapezoidal approximation

$$(2.9) \qquad g_{\alpha,\beta,\delta}(s) \equiv \begin{cases} (s-\alpha)/\delta & \text{for } \alpha \leqq s \leqq \alpha + \delta \\ 1 & \text{for } \alpha + \delta \leqq s \leqq \beta - \delta \\ (\beta - s)/\delta & \text{for } \beta - \delta \leqq s \leqq \beta \\ 0 & \text{elsewhere,} \end{cases}$$

and let

$$(2.10) \qquad g^*_{\alpha,\beta,\delta}(u) \equiv g_{\alpha,\beta,\delta}(u) \quad e^{-u^2/2} .$$

Then $g^*_{\alpha,\beta,\delta}$ is a continuous bounded function of class L_1 on $\mathbb{R}^1$ having a piecewise continuous derivative of bounded variation and vanishing outside a finite interval. Consequently its Fourier transform $G^*_{\alpha,\beta,\delta}$ is a continuous bounded function of class L_1 on $\mathbb{R}^1$ and we have

$$(2.11)\quad G^*_{\alpha,\beta,\delta}(u) \equiv (2\pi)^{-1/2} \int_{-\infty}^{\infty} e^{isu}\, g^*_{\alpha,\beta,\delta}(s)ds \qquad \text{for } u\in\mathbb{R}^1 ,$$

and

$$(2.12)\quad g^*_{\alpha,\beta,\delta}(s) = (2\pi)^{-1/2} \int_{-\infty}^{\infty} e^{-isu}\, G^*_{\alpha,\beta,\delta}(u)du \qquad \text{for } s\in\mathbb{R}^1 .$$

Now if $f_j(t)\in L_2$ and $\int_a^b f_j^2(t)dt=1$, we have by the P.W.Z. theorem that

$$\prod_{j=1}^{\nu} \{G^*_{\alpha,\beta_j,\delta}[-\int_a^b f_j(t)\tilde{d}x_j(t)]\exp\{\tfrac{1}{2}[\int_a^b f_j(t)\tilde{d}x_j(t)]^2\}\}$$

is integrable over $C^\nu[a,b]$. Thus from (2.7),

$$0 = \int_{C^\nu[a,b]} \prod_{j=1}^{\nu} \{G^*_{\alpha,\beta_j,\delta}[-\int_a^b f_j(t)\tilde{d}x_j(t)]\exp\{1/2[\int_a^b f_j(t)\tilde{d}x_j(t)]^2\}\}$$

$$\int_{L_2^\nu} \exp\{i\sum_{j=1}^{\nu}\int_a^b v_j(t)\tilde{d}x_j(t)\}d\mu_3(\vec{v})d\vec{x}$$

$$= \int_{L_2^\nu}\int_{C^\nu[a,b]} \prod_{j=1}^{\nu} \{G^*_{\alpha,\beta_j,\delta}[-\int_a^b f_j(t)\tilde{d}x_j(t)]\exp\{\tfrac{1}{2}[\int_a^b f_j(t)\tilde{d}x_j(t)]^2\}\}$$

$$\exp\{i\sum_{j=1}^{\nu}\int_a^b v_j(t)\tilde{d}x_j(t)\}d\vec{x}\,d\mu_3(\vec{v}) .$$

In order to evaluate the Wiener integrals by the P.W.Z. theorem, we orthonormalize $f_j(t)$ and $v_j(t)$ as follows: let

$$(2.13)\quad \theta_j(t) = v_j(t) - f_j(t)\int_a^b v_j(s)f_j(s)ds \text{ , so that } \int_a^b \theta_j(t)f_j(t)dt = 0$$

and

$$(2.14)\quad p_j^2 \equiv \int_a^b \theta_j^2(t)dt = \int_a^b v_j^2(t)dt - [\int_a^b v_j(t)f_j(t)dt]^2 \text{ , } p_j \geqq 0 .$$

Let $\varphi_j(t)=\theta_j(t)/p_j$ for $p_j>0$, let $\varphi_j=0$ if $p_j=0$, then for each j , f_j and φ_j are orthogonal and for almost all $t\in[a,b]$,

(2.15) $$v_j(t) = p_j\varphi_j(t) + f_j(t)\int_a^b v_j(s)\, f_j(s)\,ds$$

and

$$\int_a^b v_j(t)\,\tilde{d}x_j(t) = p_j\int_a^b \varphi_j(t)\tilde{d}x_j(t) + \int_a^b v_j(s)\,f_j(s)\,ds\int_a^b f_j(t)\tilde{d}x_j(t) \quad .$$

Thus we have

(2.16) $$0 = \int_{L_2^\nu}\int_C \overset{(\nu)}{\cdots}\int_C \prod_{j=1}^{\nu}\{G^*_{\alpha,\beta_j,\delta}[-\int_a^b f_j(t)\tilde{d}x_j(t)] \cdot$$

$$\exp\{\tfrac{1}{2}[\int_a^b f_j(t)\tilde{d}x_j(t)]^2 + i\int_a^b v_j(s)f_j(s)ds\int_a^b f_j(t)\tilde{d}x_j(t)\} \cdot$$

$$\exp\{i\sum_{j=1}^{\nu} p_j\int_a^b \varphi_j(t)\tilde{d}x_j(t)\}\}\,dx_1\ldots dx_\nu\,d\mu_3(\vec{v})$$

$$= (2\pi)^{-\nu}\int_{L_2^\nu}\int_{\mathbb{R}^{2\nu}} \prod_{j=1}^{\nu}\{G^*_{\alpha,\beta_j,\delta}[-u_j]\exp\{iu_j\int_a^b v_j(s)f_j(s)\,ds\}$$

$$\exp\{i\sum_{j=1}^{\nu} w_j p_j\}\ \exp\{-\sum_{j=1}^{\nu}\frac{w_j^2}{2}\}du_1\ldots du_\nu\,dw_1\ldots dw_\nu\,d\mu_3(\vec{v})$$

$$= \int_{L_2^\nu}\prod_{i=1}^{\nu}\{g^*_{\alpha,\beta_j,\delta}(\int_a^b v_j(s)f_j(s)ds)\}\exp\{-\sum_{j=1}^{\nu}\tfrac{1}{2}p_j^2\}\,d\mu_3(\vec{v})$$

$$= \int_{L_2^\nu}\prod_{i=1}^{\nu}\{g^*_{\alpha,\beta_j,\delta}(\int_a^b v_j(s)f_j(s)ds)\}\exp\{\tfrac{1}{2}\sum_{j=1}^{\nu}[\int_a^b v_j(t)f_j(t)dt]^2$$

$$-\tfrac{1}{2}\sum_{j=1}^{\nu}\int_a^b v_j^2(t)dt\}\,d\mu_3(\vec{v})$$

$$= \int_{L_2^\nu}\prod_{j=1}^{\nu}\{g_{\alpha,\beta_j,\delta}(\int_a^b v_j(s)f_j(s)\,ds)\}\quad\exp\{-\tfrac{1}{2}\sum_{j=1}^{\nu}\int_a^b v_j^2(t)\,dt\}\,d\mu_3(\vec{v}).$$

Thus

(2.17) $$\int_{L_2^\nu}\prod_{j=1}^{\nu}\{g_{\alpha,\beta_j,\delta}(\int_a^b v_j(s)f_j(s)\,ds)\}\,d\mu_4(\vec{v}) = 0 \quad ,$$

where μ_4 is the measure defined for each set $E \in \mathcal{Q}$ by

(2.18) $$\mu_4(E) = \int_E \exp\{-\tfrac{1}{2}\sum_{j=1}^{\nu}\int_a^b v_j^2(t)\,dt\}\,d\mu_3(\vec{v}) \quad .$$

It is clear that $\mu_4 \in \mathfrak{M}$.

Since $|g_{\alpha,\beta,\delta}(s)| \leqq 1$ for all s and δ we may take the limit of both sides of (2.17) as $\delta \to 0^+$ and obtain by dominated convergence

$$\int_{L_2^\nu} \prod_{j=1}^{\nu} \{g_{\alpha,\beta_j}(\int_a^b v_j(s)f_j(s)ds)\} d\mu_4(\vec{v}) = 0$$

Again by dominated convergence, taking the limit as $\alpha \to -\infty$, we obtain

$$\int_{L_2^\nu} \prod_{j=1}^{\nu} \{g_{-\infty,\beta_j}(\int_a^b v_j(s)f_j(s)ds\} d\mu_4(\vec{v}) = 0 \quad ,$$

so if

$$(2.19) \quad E = \{\vec{v} | \int_a^b v_j(s)f_j(s)ds\} < \beta_j \quad \text{for} \quad j = 1,2, \ldots ,\nu\} \quad ,$$

we have

$$(2.20) \quad \mu_4(E) = 0 \quad .$$

Since $\mathfrak{C}$ is generated by sets of the form (2.19), it follows that (2.20) holds for all $E \in \mathfrak{C}$.

Now solving (2.18) for μ_3 , we have for each $E \in \mathfrak{C}$

$$\mu_3(E) = \int_E \exp\{\frac{1}{2} \sum_{j=1}^{\nu} \int_a^b v_j^2(t)dt\} d\mu_4(\vec{v}) \quad ,$$

so we obtain from (2.20) that for $E \in \mathfrak{C}$

$$\mu_3(E) \equiv 0 \ ,$$

and hence for $E \in \mathfrak{C}$

$$\mu_1(E) = \mu_2(E) \ ,$$

and the theorem is proved.

<u>Corollary</u>. Let $F(\cdot,y) \in S$ for each $y \in \mathcal{Y}$, where $\mathcal{Y}$ is a measure space with a measure σ , and let $F(\vec{x},y)$ be measurable in $(\vec{x},y)$ on $C^\nu[a,b] \times \mathcal{Y}$. Let μ_y be the family of measures corresponding to $F(\cdot,y)$, so that for each $y \in \mathcal{Y}$, $\mu_y \in \mathfrak{M}$ and

$$F(\vec{x},y) = \int_{L_2^\nu} \exp\{i \sum_{j=1}^{\nu} \int_a^b v_j(t)\tilde{d}x_j(t)\} d\mu_y(\vec{v})$$

for almost all $\vec{x} \in C^{\nu}[a,b]$. Then it follows that for each set $E \in G$, $\mu_y(E)$ is measurable as a function of y on $\mathcal{Y}$.

Proof. This corollary follows from the fact that the method of proof of Theorem 2.1 has provided a method for explicitly constructing μ_y in terms of $F(\cdot,y)$.

Definition. If $F \in S$, we define the norm of F by

$$(2.22) \quad \|F\| = \|\mu\|$$

where μ is associated with F by the relation (2.6). It follows from the uniqueness theorem that $\|F\|$ is uniquely determined by F .

Lemma 2.5 . Let $G_n \in S$ for $n = 1,2, \ldots,$ and let

$$(2.24) \quad \sum_{n=1}^{\infty} \|G_n\| < \infty .$$

Then it follows that the sum on the right below converges absolutely and uniformly, and the functional G defined below is also an element of S :

$$(2.25) \quad G(\vec{x}) \equiv \sum_{n=1}^{\infty} G_n(\vec{x})$$

for s-almost all $\vec{x} \in C^{\nu}[a,b]$. Moreover

$$(2.26) \quad \|G\| \leqq 4 \sum_{n=1}^{\infty} \|G_n\|$$

and for each $N > 0$,

$$(2.27) \quad \left\|G - \sum_{k=1}^{N} G_k\right\| \leqq 4 \sum_{k=N+1}^{\infty} \|G_n\| .$$

Proof. Since $|G_n(\vec{x})| \leqq \|G_n\|$, the absolute and uniform convergence of second member of (2.25) follows from (2.24) for all $\vec{x}$ for which all the $G_n(\vec{x})$ are defined, and this includes s-almost all $\vec{x} \in C^{\nu}$. [The identity of the exceptional set of course depends on the choice of the C.O.N. sequence used in defining the P.W.Z. integral in (2.1)]. Thus G is uniquely defined by (2.25) up to a scale-invariant Wiener null set.

By definition of S there exist elements $\mu_n \in \mathfrak{M}$ such that for $n = 1,2, \ldots,$

(2.28) $G_n(\vec{x}) \approx \int_{L_2^\nu} J(\vec{v})\, d\mu_n(\vec{v})$,

where for each $\vec{v}$ in L_2^ν and s-almost every $\vec{x} \in C^\nu[a,b]$,

(2.29) $J(\vec{v}) \equiv \exp\{i \sum_{j=1}^{\nu} \int_a^b v_j(t)\tilde{d}x_j(t)\} \equiv J(\vec{x},\vec{v})$;

so that

(2.30) $\|\mu_n\| = \|G_n\|$

Thus we have

(2.31) $\Sigma \|\mu_n\| = \sum_{n=1}^{\infty} \|G_n\| < \infty$

and $\{\mu_n\}$ satisfy the hypothesis of Lemma 2.3. Hence there exists $\mu \in \mathfrak{M}$ such that

(2.32 a) $\mu(E) = \sum_{n=1}^{\infty} \mu_n(E)$ for $E \in \mathfrak{a}$

and

(2.32 b) $\|\mu\| \leqq 4 \sum_{n=1}^{\infty} \|\mu_n\|$.

Let a C.O.N. sequence $\{\varphi_n\}$ in L_2 with each μ_n of B.V. be given and let all P.W.Z. integrals be based on it. Let $\vec{x}$ be an element of $C^\nu[a,b]$ for which (2.25) holds and for which $J(\vec{x},\vec{v})$ exists for ν-almost all $\vec{v}$ in L^ν_2 for each $j=1, \ldots, \nu$.

(2.33) $J_k(\vec{v}) \equiv e^{i m 2^{-k}\pi} \equiv J(\vec{x},\vec{v})$

when

(2.34) $(m-1)2^{-k}\pi < \operatorname{Arg} J(\vec{v}) \leqq m\, 2^{-k}\pi$, $m = -2^k+1, \ldots, 2^k$; $k = 1,2, \ldots$.

Then clearly $J_k(\vec{v})$ is a simple functional with respect to each of the measures $\mu_1, \mu_2, \ldots$ and μ , and

(2.35) $\lim_{k\to\infty} J_k(\vec{v}) = J(\vec{v})$

uniformly for almost all $\vec{v} \in L_2^\nu$. Hence

(2.36) $\lim_{k\to\infty} \int_{L_2^\nu} J_k(\vec{v})\, d\mu_n(\vec{v}) = G_n(\vec{x})$ $n = 1,2, \ldots$

and

$$(2.37)\quad \lim_{k\to\infty} \int_{L_2^\nu} J_k(\vec{v})\,d\mu(\vec{v}) = \int_{L_2^\nu} J(\vec{v})\,d\mu(\vec{v}) \quad .$$

Now since J_k is a simple functional, it follows from (2.32a) that

$$(2.38)\quad \int_{L_2^\nu} J_k(\vec{v})\,d\mu(\vec{v}) = \sum_{n=1}^{\infty} \int_{L_2^\nu} J_k(\vec{v})\,d\mu_n(\vec{v}) \quad .$$

But the n-th term on the right is dominated by $\|\mu_n\|$ and $\Sigma\|\mu_n\|$ converges. Hence we may take limits on both sides of (2.38) using (2.36) and (2.37) and obtain

$$(2.39)\quad \int_{L_2^\nu} J(\vec{v})\,d\mu(\vec{v}) = \sum_{n=1}^{\infty} G_n(\vec{x}) = G(\vec{x})$$

for all $\vec{x}$ for which (2.25) holds, and for which $\int_a^b v_j(t)\tilde{d}x$ exists for ν-almost all $\vec{v}$ in L_2^ν for each $j=1;\ldots,\nu$. But by Lemma 2.1 this includes s-almost all $\vec{x}$. Thus $G \in S$.

Finally, note that from (2.39) and (2.32b) and (2.28) we have

$$(2.40)\quad \|G\| = \|\mu\| \leqq 4 \sum_{n=1}^{\infty} \|\mu_n\| = 4 \sum_{n=1}^{\infty} \|G_n\| \quad .$$

Thus (2.26) holds. Inequality (2.27) is established in the same way.

<u>Theorem 2.2</u>. <u>The space</u> S <u>is a linear, normed, complete space,</u> (Banach space).

<u>Proof</u>. The linearity of S was proved in Lemma (2.4), it only remains to be shown that S is complete under the norm given above.

Let $\{F_n\}$ be a Cauchy sequence of elements of S, and let $\{\mu_n\}$ be a sequence of measures from $\mathfrak{M}$ so that F_n and μ_n are related by (2.1).

Let us define a sequence of subscripts $\{n_k\}$ such that $1 = n_1 < n_2 < \ldots < n_k < \ldots$ and

$$\|F_{n_k} - F_{n_{k+1}}\| < \frac{1}{2^k} \quad \text{for } k = 2,3,\ \ldots \ .$$

Let $G_1 \equiv F_1$, and $G_k \equiv F_{n_k} - F_{n_{k-1}}$ for $k = 2,3,\ \ldots.$ Then $\sum_{k=1}^{\infty} \|G_k\| < \infty$ and each $G_k \in S$. By the previous lemma there exist a $G \in S$ such that $G(\vec{x}) = \sum_{k=1}^{\infty} G_k(\vec{x})$ for s-almost all $\vec{x} \in C^\nu[a,b]$. Thus for s-almost all $\vec{x}$

$$G(\vec{x}) = \lim_{N\to\infty} \sum_{k=1}^{N} G_k(\vec{x})$$

$$= \lim_{N\to\infty} \sum_{k=2}^{N} [F_{n_k}(\vec{x}) - F_{n_{k-1}}(\vec{x})] + F_1(\vec{x})$$

$$= \lim_{N\to\infty} F_{n_N}(\vec{x}) - F_{n_1}(\vec{x}) + F_1(\vec{x})$$

$$= \lim_{N\to\infty} F_{n_N}(\vec{x}) \quad .$$

Moreover F_{n_k} converges to G in the norm topology since by the lemma,

$$\|G - F_{n_N}\| = \|G - \sum_{k=1}^{N} G_k\| \leq 4 \sum_{N+1}^{\infty} \|G_k\| ,$$

and since $\{F_n\}$ is a Cauchy sequence it follows that

$$\lim_{n\to\infty} \|G - F_n\| = 0 \quad .$$

Since $G \in S$, the theorem is proved.

<u>Theorem 2.3.</u> <u>$S \equiv S(L_2^\nu)$ is a Banach Algebra.</u>

<u>Proof</u> Let $F, G \in S$. In particular, let

$$F(\vec{x}) \equiv \int_{L_2^\nu[a,b]} \exp\{i \sum_{j=1}^{\nu} \int_a^b v_j(t)\tilde{d}x_j(t)\} d\mu_1(\vec{v})$$

$$G(\vec{x}) \equiv \int_{L_2^\nu} \exp\{i \sum_{j=1}^{\nu} \int_a^b u_j(t)\tilde{d}x_j(t)\} d\mu_2(\vec{u})$$

for s-almost all $\vec{x} \in C^\nu[a,b]$. Let $H(\vec{x}) = F(\vec{x})\, G(\vec{x})$. Then

$$H(\vec{x}) \approx \int_{L_2^\nu} \int_{L_2^\nu} \exp\{i \sum_{j=1}^{\nu} \int_a^b [v_j(t) + u_j(t)]\tilde{d}x_j(t)\} d\mu_1(\vec{v})\, d\mu_2(\vec{u})$$

Let $\vec{w} = \vec{u} + \vec{v}$. Then

$$H(\vec{x}) \approx \int_{L_2^\nu} \int_{L_2^\nu} \exp\{i \sum_{j=1}^{\nu} \int_a^b w_j(t)\tilde{d}x_j(t)\} d_w\, \mu_1(\vec{w} - \vec{u}) d\mu_2(\vec{u})$$

where $d_w\mu_1(\vec{w} - \vec{u})$ means $d(T_{\vec{u}}\mu_1)(\vec{w})$, where $T_{\vec{u}}E = \{\vec{v}(\cdot) | \vec{u}(\cdot) + \vec{v}(\cdot) \in E\}$,

and we have

$$H(\vec{x}) \approx \int_{L_2^\nu} \exp\{i \sum_{j=1}^{\nu} \int_a^b w_j(t)\tilde{d}x_j(t)\} d\mu_3(\vec{w})$$

where

$$\mu_3(E) = \int_{L_2^\nu} \mu_1(T_{\vec{u}(\cdot)}E)\, d\mu_2(\vec{u}(\cdot)) \quad .$$

Now $\quad |\mu_3(E)| \leqq \int_{L_2^\nu} |\mu_1(T_{\vec{u}(\cdot)}E)|\,|d\mu_2(\vec{u}(\cdot))|$.

Let $L_2^\nu = \bigcup_{k=1}^{q} E_k$, where $\{E_k\}$ is a disjoint sequence of elements of $\mathcal{C}$.
Then we have

$$\sum_{k=1}^{q} |\mu_3(E_k)| \leqq \sum_{k=1}^{q} \int_{L_2^\nu} |\mu_1(T_{\vec{u}(\cdot)}E_k|\,|d\mu_2(\vec{u}(\cdot)|$$

$$= \int_{L_2^\nu} [\sum_{k=1}^{q} |\mu_1(T_{\vec{u}(\cdot)}E_k)|]\,|d\mu_2(\vec{u}(\cdot))|$$

$$\leqq \int_{L_2^\nu} \|\mu_1\|\,|d\mu_2(\vec{u}(\cdot))|$$

$$= \|\mu_1\|\,\|\mu_2\| \quad .$$

By taking the supremum on the left, we have

$$\|\mu_3\| \leqq \|\mu_1\|\,\|\mu_2\| \quad .$$

Thus

$$\|H\| \leqq \|F\|\,\|G\|$$

and the theorem is proved.

3. THE SPACE S'

Definition. Let $\mathfrak{B} \equiv \mathfrak{B}[a,b]$ the space of real right continuous functions of bounded variation on $[a,b]$ that vanish at b . Let $\mathfrak{B}^\nu = \overset{\nu}{X}\mathfrak{B}$.

Notation. Let $\mathcal{C}'$ be the σ-algebra of subsets of $\mathcal{B}^\nu \equiv \mathcal{B}^\nu[a,b]$ generated by the class of sets of the form

$$\{\vec{v} \mid \vec{v} \in \mathcal{B}^\nu \ , \ \int_a^b v_j(t)\varphi_j(t)dt < \lambda_j \ , \quad j = 1, \ldots, \nu\} \ ,$$

where $\vec{\varphi} \equiv (\varphi_1, \ldots, \varphi_\nu)$ ranges over all elements of L_2^ν and $\vec{\lambda} \equiv (\lambda_1, \ldots, \lambda_\nu)$ ranges over Q^ν .

We note as in the case of $\mathcal{C}$, that $\mathcal{C}'$ is translation invariant.

Definition. Let $\mathcal{M}' \equiv \mathcal{M}'(\mathcal{B}^\nu[a,b])$ be the class of complex measures of finite variation defined on $\mathcal{B}^\nu[a,b]$ with $\mathcal{C}'$ as its class of measurable sets. If $\mu \in \mathcal{M}'$, we set $\|\mu\| = \text{var}\,\mu$ over $\mathcal{B}^\nu$

Remark. It is clear that $\mathcal{M}'$ is a linear class of measures.

Definition. Let $S' \equiv S'(\mathcal{B}^\nu)$ be the space of functionals of the form

$$(3.1) \qquad F(\vec{x}) \equiv \int_{\mathcal{B}^\nu} \exp\{i \sum_{j=1}^{\nu} \int_a^b v_j(t)dx_j(t)\} \, d\mu(\vec{v})$$

for $\vec{x} \in C^\nu[a,b]$ where $\mu \in \mathcal{M}'$.

Theorem 3.0. $S' \subset S$.

Proof. Let $F \in S'$, then there exists $\mu' \in \mathcal{M}'$ such that (3.1) holds. Since if $\vec{v} \in \mathcal{B}^\nu$, $v_j(t)$ is of bounded variation on $[a,b]$, $v_j \in L_2[a,b]$ and $\vec{v} \in L_2^\nu$, so $\mathcal{B}^\nu \subset L_2^\nu$. Let $E \in \mathcal{C}$. Let $E' = E \cap \mathcal{B}^\nu$ so $E' \in \mathcal{C}'$. Let us define a measure on $L_2^\nu[a,b]$ by

$$(3.2) \qquad \mu(E) \equiv \mu'(E \cap \mathcal{B}^\nu) \quad \text{for all} \quad E \in \mathcal{C} \ .$$

Now

$$F(\vec{x}) = \int_{\mathcal{B}^\nu} \exp\{i \sum_{j=1}^{\nu} \int_a^b v_j(t)dx_j(t)\} \, d\mu'(\vec{v})$$

$$= \int_{\mathcal{B}^\nu} \exp\{i \sum_{j=1}^{\nu} \int_a^b v_j(t)\tilde{d}x_j(t)\} \, d\mu'(\vec{v})$$

$$= \int_{L_2^\nu} \exp\{i \sum_{j=1}^{\nu} \int_a^b v_j(t)\tilde{d}x_j(t)\} d\mu(\vec{v})$$

for s-almost all $\vec{x} \in C^{\nu}[a,b]$, and the lemma is proved.

Note. We do not assume that $\mathcal{B}^{\nu}$ is a measurable subset of L_2^{ν}.

The question whether $\mathcal{S}'$ is a proper subset of $\mathcal{S}$ remains open. The question whether the example given in equation (2.41) at the end of the previous section is an element of $\mathcal{S}'$ also remains open.

Lemma 3.1. If $F \in S'$, there is a unique element $\mu' \in \mathcal{M}'$ such that (3.1) holds for F and μ'. This follows from the uniqueness theorem for S.

Lemma 3.2. If $F \in S'$ and μ' is the measure in $\mathcal{M}'$ related to F by (3.1), then $\|F\| = \|\mu'\|$.

Proof. By definition, $\|F\| = \|\mu\|$ where μ is the unique element of $\mathcal{M}$ related to F by (2.1). Then, it follows that

$$\mu(E) = \mu'(\mathcal{B}^{\nu} \cap E) \qquad \text{for } E \in G .$$

Now

$$\operatorname*{var}_{L_2^{\nu}} \mu = \operatorname*{var}_{\mathcal{B}^{\nu}} \mu + \operatorname*{var}_{L_2^{\nu} - \mathcal{B}^{\nu}} \mu = \operatorname*{var}_{\mathcal{B}^{\nu}} \mu' + 0$$

so

$$\|\mu\| = \|\mu'\|$$

and the lemma is proved

Theorem 3.1. The space S' is a linear normed complete space, (Banach space).

Here we use the same norm for elements of S' as for elements of S, since $S' \subset S$. The proof of this theorem is identical to the proof of the Theorem 2.2 and its supporting lemma.

Theorem 3.2. The space S' is a Banach algebra.

Again the proof is identical with the proof of the theorem that S is a Banach algebra.

4. THE SPACE S"

Notation. Let

(4.1) $\Delta_n \equiv \{(t_1,\ldots,t_n) | a < t_1 < t_2 < \ldots < t_n \leq b\}$, and if $(t_1,\ldots,t_n) \in \Delta_n$, let $t_0 \equiv a$.

Definition. Let $\mathcal{M}_n'' \equiv \mathcal{M}_n''(\Delta_n \times \mathbb{R}^{n\nu})$ be the class of bounded complex Borel measures on $\Delta_n \times \mathbb{R}^{n\nu}$. Here we take $\|\mu\| \equiv \operatorname{var}\mu$, the total variation of μ over $\Delta_n \times \mathbb{R}^{n\nu}$. Let $\hat{\mathcal{M}}_n \equiv \hat{\mathcal{M}}_n((a,b]^n \times \mathbb{R}^{n\nu})$ be the corresponding class of measures on $(a,b]^n \times \mathbb{R}^{n\nu}$.

Definition. Let ν and n be positive integers and let $S_n'' = S_n''(\Delta_n \times \mathbb{R}^{n\nu})$ be the space of functionals of the form

$$(4.2) \quad F(\vec{x}) \equiv \int_{\Delta_n \times \mathbb{R}^{n\nu}} \exp\{i \sum_{k=1}^{\nu} \sum_{j=1}^{n} v_{k,j}\, x_k(t_j)\}\, d\mu(\vec{t} \times \vec{v})$$

for $\vec{x} \in C^\nu[a,b]$, where $\mu \in \mathcal{M}_n''$ and $\vec{v}$ represents the matrix $\{v_{k,j}\}_{k=1,\ldots,\nu;\, j=1,\ldots,n}$. Here we take

$$\|F\|_n'' \equiv \inf(\|\mu\|) ,$$

where the infimum is taken over all μ's so that F and μ are related by (4.2).

Lemma 4.0. For each $n = 1,2,\ldots$, we have $S_n \subset S_{n+1}''$, and if $F \in S_n''$,

$$\|F\|_n'' \geq \|F\|_{n+1}''$$

Proof. Let $F \in S_n''$ and let $\varepsilon > 0$ be given. Choose $\mu \in \mathcal{M}_n''$ such that (4.2) holds and such that

$$\|\mu\| < \|F\|_n'' + \varepsilon .$$

Let T be a transformation of $\Delta_n \times \mathbb{R}^{n\nu}$ into $\Delta_{n+1} \times \mathbb{R}^{(n+1)\nu}$ such that if $\vec{t} \times \vec{v} \in \Delta_n \times \mathbb{R}^{n\nu}$, then

$$T(\vec{t} \times \vec{v}) = \vec{t}' \times \vec{v}' \in \Delta_{n+1} \times \mathbb{R}^{(n+1)\nu} , \text{ where}$$

$$t_j' = \begin{cases} \frac{1}{2}(t_1 + a) & \text{if } j = 1 \\ t_{j-1} & \text{if } j = 2,3,\ldots,n+1 \end{cases}$$

and

$$v'_{k,j} = \begin{cases} 0 & \text{if } k = 1,2,\ldots,\nu \text{ and } j = 1 \\ v_{k,j-1} & \text{if } k = 1,\ldots,\nu \text{ and } j = 2,3,\ldots n+1 \end{cases} .$$

Let μ^* be a measure on $\Delta_{n+1}\times R^{(n+1)\nu}$ such that if E is a Borel subset of $\Delta_{n+1}\times R^{(n+1)\nu}$, then

$$\mu^*(E) = \mu(T^{-1}E) .$$

Then transforming $\vec{t}\times\vec{v}$ into $\vec{t}'\times\vec{v}'$ by the transformation T, in the second member of (4.2), we have

$$F(x) \equiv \int_{\Delta_{n+1}\times R^{(n+1)\nu}} \exp\{i \sum_{k=1}^{\nu}\sum_{\ell=1}^{n+1} v'_{k,\ell}\, x_k(t'_\ell)\} d\mu^*(\vec{t}'\times\vec{v}')$$

$$= \int_{\Delta_{n+1}\times R^{(n+1)\nu}} \exp\{i \sum_{k=1}^{\nu}\sum_{\ell=2}^{n+1} v'_{k,\ell}\, x_k(t'_\ell)\} d\mu^*(\vec{t}\times\vec{v}') \in S''_{n+1} .$$

Moreover, since T is a one-to-one transformation of $\Delta_n\times R^{n\nu}$ onto a subset ρ of $\Delta_{n+1}\times R^{(n+1)\nu}$,

$$\|\mu^*\| = \operatorname{var}_{\Delta_{n+1}\times R^{(n+1)\nu}} \mu^* = \operatorname{var}_{\rho} \mu^* = \operatorname{var}_{\Delta_n\times R^{n\nu}} \mu = \|\mu\| < \|F\|''_n + \epsilon$$

Hence $\|F\|''_{n+1} < \|F\|''_n + \epsilon$, and the lemma is proved.

It follows immediately from the lemma that if n and k are positive integers that $S''_n \subset S''_{n+k}$ and if $F \in S''_n$ then $\|F\|''_n \geq \|F\|''_{n+k}$.

<u>Definition</u>. Let $S^\wedge_n \equiv S^\wedge_n((a,b]^n \times R^{n\nu})$ be the space of functionals F such that

$$(4.3) \quad F(\vec{x}) \equiv \int_{(a,b]^n\times R^{n\nu}} \exp\{i \sum_{k=1}^{\nu}\sum_{j=1}^{n} v_{k,j}\, x_k(t_j)\} d\mu(\vec{t}\times\vec{v})$$

where $\mu \in \mathcal{M}^\wedge_n$ and μ is related to F by (4.3). Clearly

$$S''_n \subset S^\wedge_n$$

and for $F \in S''_n$

$$\|F\|^\wedge_n \leqq \|F\|''_n$$

since every measure in $\mathcal{M}''_n$ can be extended to be in $\mathcal{M}^\wedge_n$ without increasing its variation.

We note that if $F \in S''_n$, $|F(\vec{x})| \leq \|F\|''_n$ for all $\vec{x} \in C^{\nu}[a,b]$. For completeness we define S''_o to be constant functionals and define their norms to be their absolute values.

Definition. Let $\mathcal{M}'' \equiv \mathcal{M}''(\Sigma \Delta_n \times R^{n\nu})$ be the class of sequences of measures $\{\mu_n\}_{n=1,2,\ldots}$ such that $\mu_n \in \mathcal{M}''_n$ and $\sum_{n=1}^{\infty} \|\mu_n\| < \infty$.

Definition. Let $S'' = S''(\Sigma \Delta_n \times R^{n\nu})$ be the space of functionals F defined on $C^{\nu}[a,b]$ where F is of the form

$$(4.4) \quad F(\vec{x}) \equiv \sum_{n=1}^{\infty} F_n(\vec{x}) , \quad F_n \in S''_n$$

and where $\sum_1^{\infty} \|F_n\|''_n < \infty$. The norm of an element $F \in S''$ is defined by

$$\|F\|'' = \inf \left(\sum_1^{\infty} \|F_n\|_n \right)$$

where the infimum is taken over all representation of F given by (4.4). For $F \in S''$, the series in (4.4) converges absolutely and uniformly over $C^{\nu}[a,b]$. Moreover if $F \in S''$,

$$|F(x)| \leq \|F\|'' \quad \text{for all } \vec{x} \in C^{\nu}[a,b] .$$

Assertion. In the space S''_n, the functional $F \in S''_n$ does not determine the measure μ uniquely We present an example when $\nu = 1$, $n = 1$. Let τ be given, $a < \tau \leq b$. Let a measure μ_τ be defined over the space $[a,b] \times R^1$, thus for $E \subset [a,b] \times R'$,

$$\mu_\tau(E) = \begin{cases} 1 & \text{if the point } (\tau, 0) \in E \\ 0 & \text{if the point } (\tau, 0) \notin E \end{cases}$$

Let us define a functional $F_\tau(x)$ by (4.2) using the measure μ_τ , thus

$$F_\tau(x) = \int_{[a,b] \times R^1} \exp\{i\,v\,x(t)\} d\mu_\tau (t \times v)$$

$$= \exp\{i \cdot 0 \cdot x(\tau)\} = 1$$

Thus $F_\tau(x)$ is independent of τ , (independent of x also). Hence each member of the family of measures $\{\mu_\tau\}$ for $a < \tau < b$ produces the same functional.

Note. We observe that a vector function is discontinuous at a point if at least one of its components is discontinuous at the point.

Lemma 4.1 For each $n \geq 0$,

(4.4) $S_n'' \subset S'$

and if $F \in S_n''$,

(4.5) $\|F\| \leq \|F\|_n''$.

(Here we emphasize S_n'' is based on Δ_n, not $(a,b]^n$.)

Proof. The case $n = 0$ is trivial so let $n \geq 1$ be given and let F given by (4.2) be an element of S_n'', where $\mu \in \mathcal{M}_n''$.

Let $\mathcal{B}_n^\nu$ be the subset of $\mathcal{B}^\nu$ consisting of ν-tuples of right continuous step functions $\vec{u}(\cdot) = [u_1(\cdot), u_2(\cdot), \ldots, u_\nu(\cdot)]$ where $\vec{u}(\cdot)$ has no more than n points of discontinuity and where $\vec{u}(b) = \vec{0}$.

We now establish a relationship between $\Delta_n \times R^{n\nu}$ and $\mathcal{B}_n^\nu$ as follows. Corresponding to each point $\vec{t}, \vec{v}$ of $\Delta_n \times R^{n\nu}$ we define a vector function $\vec{u}(\cdot)$ with components $u_1(\cdot), u_2(\cdot), \ldots, u_\nu(\cdot)$ such that each $u_k(\cdot)$ is the right continuous

(4.6) $$u_k(s) = \begin{cases} \sum_{j=p}^{n} v_{k,j} & \text{for } t_{p-1} \leq s < t_p,\ p = 1, \ldots, n \\ 0 & \text{for } t_n \leq s \leq b \end{cases}$$

step function. This correspondence defines a function Φ thus:

(4.7) $\Phi(\vec{t}, \vec{v}, \cdot) = \vec{u}(\cdot)$.

Let $\tilde{\mathcal{A}}_n$ be the class of subsets $E \subset \mathcal{B}_n^\nu$ such that $\Phi^{-1}(E)$ is a Borel subset of $\Delta_n \times R^{n\nu}$.

Remark. It is clear that $\tilde{\mathcal{A}}_n$ is a σ-algebra of sets.

Let us define a measure σ on the subsets $\tilde{\mathcal{A}}_n$ of $\mathcal{B}_n^\nu$ as follows: If E is an element of $\tilde{\mathcal{A}}_n$, let

(4.8) $\sigma(E) \equiv \mu(\Phi^{-1}(E))$.

It is clear that σ is a bounded complex measure on $\tilde{G}_n$. Since the correspondence Φ is many-to-one, and the norms of the measures $\|\mu\|$ and $\|\sigma\|$ are defined as variations of the measures, it is clear that

$$\|\mu\| \geq \|\sigma\| \quad . \tag{4.9}$$

We now define a complex measure $\hat{\sigma}$ on the elements of $\mathcal{B}^\nu$ as follows:

$$\hat{\sigma}(E) \equiv \sigma(\mathcal{B}_n^\nu \cap E)$$

whenever the latter exists. The fact that $\hat{\sigma}(E)$ is defined on G' follows from this assertion:

<u>Assertion:</u> <u>If</u> $E \in G'$ <u>and</u> $E^* = E \cap \mathcal{B}_n^\nu$ <u>then</u> $E^* \in \tilde{G}_n$.

We shall sketch the proof of the assertion, and begin by considering sets E which belong to the class of sets used to generate G'. Thus let

$$E \equiv \{\vec{u}(\cdot) \mid \int_a^b u_j(t)\varphi_j(t)dt < \lambda_j \quad \text{for} \quad j = 1,\ldots,\nu\}$$

where $\vec{\varphi} \in L_2^\nu$ and $\vec{\lambda} \in R^\nu$. Let $E^* = E \cap \mathcal{B}_n^\nu$ and let $Q = \Phi^{-1}(E^*)$. Then it follows from the definitions of E and Φ that

$$Q = \{(\vec{t} \times \vec{v}) \mid \vec{t} \times \vec{v} \in \Delta_n \times R^{n\nu} \ , \ \sum_{j=1}^{n} v_{\ell,j}\Theta_\ell(t_j) < \lambda_\ell \quad \text{for} \quad \ell = 1,2,\ldots,\nu\} \ ,$$

where

$$\Theta_\ell(t) = \int_a^t \varphi_\ell(s)ds \quad .$$

Since $\Theta_\ell(t)$ is continuous on $[a,b]$, Q is a Borel set, and $E^* \in \tilde{G}_n$. Thus the assertion holds for those elements E of G' of the form that we are at present considering. But since G' is generated by sets E of this form, it follows that the assertion holds for all elements of G'.

It follows from the assertion that $\hat{\sigma}$ is a complex measure defined on G'.

Clearly $\hat{\sigma} \in \mathcal{M}'$ and $\|\hat{\sigma}\| \leq \|\sigma\|$. We define for natural numbers m,

$$J_m(w) \equiv \exp\{\tfrac{q}{m}2\pi i\} \quad \text{where} \quad \frac{2\pi(q-1)}{m} < w \leq \frac{2\pi q}{m} \quad \text{where} \quad q = 1,2,\ldots,m \ ,$$

and note that $\lim_{m\to\infty} J_m(w) = \exp(iw)$ boundedly. Then for each $\vec{x} \in C^\nu[a,b]$ by bounded convergence,

$$\lim_{m\to\infty} F_m(\vec{x}) = F(\vec{x}) \quad ,$$

where

$$F_m(\vec{x}) \equiv \int_{\Delta_n \times R^{n\nu}} J_m\{\sum_{k=1}^{\nu} \sum_{j=1}^{n} v_{k,j}\, x_k(t_j)\} d\mu(\vec{t}\times\vec{v}) \quad .$$

Then if

$$\mathcal{R}_{m,q} \equiv \{(\vec{t}\times\vec{v}) \in \Delta_n \times R^{n\nu} \mid (q-1)\frac{2\pi}{m} < \sum_{k=1}^{\nu}\sum_{j=1}^{n} v_{k,j}\, x_k(t_j) \leq q\frac{2\pi}{m}\}$$

we have

$$F_m(\vec{x}) = \sum_{q=1}^{m} \exp(\frac{2\pi q i}{m})\mu(\mathcal{R}_{m,q}) .$$

$$= \sum_{q=1}^{m} \exp(\frac{2\pi i q}{m})\hat{\sigma}(E_{m,q})$$

$$= \int_{\mathcal{B}^{\nu}} J_m\{(\sum_{k=1}^{\nu} \int_a^b u_k(t) d\, x_k(t)\} d\hat{\sigma}(\vec{u}(\cdot))$$

where

$$E_{m,q} = \{\vec{u}(\cdot) \in \mathcal{B}^{\nu} \mid \frac{(q-1)\,2\pi}{m} < \sum_{k=1}^{\nu} \int_a^b u_k(t) d\, x_k(t) \leq \frac{q}{m} 2\pi\} \quad .$$

Thus we have by bounded convergence that

$$F(\vec{x}) = \lim_{m\to\infty} F_m(\vec{x}) = \int_{B^{\nu}} \exp\{i \sum_{k=1}^{\nu} \int_a^b u_k(t) d\, x_k(t)\} d\hat{\sigma}(\vec{u}(\cdot)) \quad .$$

Hence $S_n'' \subset S'$. By (4.9) and Lemma 3.1, $\|\hat{\sigma}\| \leq \inf\|\mu\|$ where the inf. is taken over all μ such that F and μ are related by (4.2). Hence $\|F\| \leq \|F\|_n''$ and the lemma is proved.

Lemma 4.2. The space S'' is contained in the space S'; $S'' \subset S'$; and if $F \in S''$, $\|F\| \leq \|F\|''$.

Proof. By definition of S'', there exist functionals $F_n \in S_n''$ such that $F(\vec{x}) = \sum_1^{\infty} F_n(\vec{x})$ and $\sum_1^{\infty} \|F_n\|_n'' < \infty$. It follows from inequality (4.5) that $\sum_1^{\infty} \|F_n\| < \infty$. Since S' is a Banach space, $F \in S'$. Because

$$\|F\| \leq \|\sum_{n=1}^{\infty} F_n\| \leq \sum_1^{\infty} \|F_n\| \leq \sum_1^{\infty} \|F_n\|_n'' \quad ,$$

we have

$$\|F\| \leqq \inf \sum_1^\infty \|F_n\|''_n = \|F\|'' \qquad \text{and}$$

the lemma is proved.

<u>Lemma 4.3. The space</u> $\mathcal{M}''_n(\Delta_n \times R^{n\nu})$ <u>is complete under the norm</u>

$$\|\mu\| = \int_{\Delta_n \times R^{n\nu}} |d\mu(\vec{t} \times \vec{v})| \quad .$$

<u>Proof</u>. Let $\{\mu_p\} \subset \mathcal{M}''_n(\Delta_n \times R^{n\nu})$ be a Cauchy sequence, $\lim_{p,q \to \infty} \|\mu_p - \mu_q\| = 0$.

<u>Case</u> I. Let μ_p be real and suppose $\sum_{p=2}^\infty \|\mu_p - \mu_{p-1}\| < \infty$. We set $\mu_n = \sum_{p=1}^n \sigma_p$; i.e. $\sigma_1 = \mu_1$, $\sigma_n = \mu_n - \mu_{n-1}$ for $n = 2, 3, \ldots$, so that $\sum_1^\infty \|\sigma_n\| < \infty$. Then if we decompose the measure σ_n by considering the positive and negative variations we have

$$\sigma_n = \sigma_n^+ - \sigma_n^- \quad \text{and} \quad \sum_1^\infty \|\sigma_n^+\| < \infty \quad \text{and} \quad \sum_1^\infty \|\sigma_n^-\| < \infty \quad .$$

Suppose $E = \bigcup_{j=1}^\infty E_j$ and $E_j \cap E_k = \emptyset$; $E_j \subset \Delta_n \times R^{n\nu}$, E_j a Borel set.

Then $\sigma_n^+(E) = \sigma_n^+(\bigcup_1^\infty E_j) = \sum_{j=1}^\infty \sigma_n^+(E_j)$; and if $F(E) \equiv \sum_{n=1}^\infty \sigma_n^+(E)$, we have

$$F(E) = \sum_{n=1}^\infty \sum_{j=1}^\infty \sigma_n^+(E_j)$$

$$= \sum_{j=1}^\infty \sum_{n=1}^\infty \sigma_n^+(E_j) = \sum_{j=1}^\infty F(E_j) \quad ,$$

and so F is a measure. Similarly G is a measure, where

$$G(E) \equiv \sum_{n=1}^\infty \sigma_n^-(E) \quad .$$

Then if we set

$$H \equiv F - G \quad ,$$

H is a signed measure and

$$H^+(E) \leqq F(E) \quad \text{and} \quad H^-(E) \leqq G(E) \quad .$$

We next show that $\lim_{N \to \infty} \|H - \sum_{n=1}^N \sigma_n\| = 0$. For any E ,

$$H(E) - \sum_{n=1}^{N} \sigma_n(E) = F(E) - G(E) - \sum_{n=1}^{N} \sigma_n^+(E) + \sum_{n=1}^{N} \sigma_n^-(E)$$

$$= F(E) - \sum_{n=1}^{N} \sigma_n^+(E) - (G(E) - \sum_{n=1}^{N} \sigma_n^-(E))$$

$$= \sum_{n=N+1}^{\infty} \sigma_n^+(E) - \sum_{n=N+1}^{\infty} \sigma_n^-(E) .$$

So we have

$$|H(E) - \mu_N(E)| \equiv | H(E) - \sum_{n=1}^{N} \sigma_n(E)| \leqq \sum_{N+1}^{\infty} \sigma_n^+(E) + \sum_{N+1}^{\infty} \sigma_n^-(E) .$$

If $S \equiv \Delta_n \times R^{n\nu} = \bigcup_{k=1}^{\infty} E_k$, where the E_k are disjoint Borel sets, we have

$$\sum_{k=1}^{\infty} |H(E_k) - \mu_N(E_k)| \leqq \sum_{k=1}^{\infty} \sum_{n=N+1}^{\infty} (\sigma_n^+(E_k) + \sigma_n^-(E_k))$$

$$= \sum_{n=N+1}^{\infty} (\sigma_n^+(S) + \sigma_n^-(S)) = \sum_{n=N+1}^{\infty} \|\sigma_n\| .$$

Thus taking the supremum over all partitions $\{E_k\}$ we have

$$\|H - \mu_N\| \leqq \sum_{n=N+1}^{\infty} \|\sigma_n\|$$

and hence

$$\lim_{N\to\infty} \|H - \mu_N\| = 0 .$$

Thus completeness has been proved in Case I.

Case II. Let μ_p be real and $\lim_{p,q\to\infty} \|\mu_p - \mu_q\| = 0$. We choose p_1 such that if $p \geqq p_1$ and $q \geqq p_1$

$$\|\mu_p - \mu_q\| \leq \frac{1}{2} ,$$

and inductively choose $p_k > p_{k-1}$ such that if $p \geq p_k$, $q \geq p_k$

$$\|\mu_p - \mu_q\| \leq \frac{1}{2^k} .$$

Then we have for $k = 1,2, \ldots$..

$$\|\mu_{p_{k-1}} - \mu_{p_k}\| < \frac{1}{2^{k-1}}$$

so that

$$\sum_{k=1}^{\infty} \|\mu_{p_k} - \mu_{p_{k-1}}\| < 1 \ .$$

Thus by Case I the subsequence $\{\mu_{p_k}\}$ has a limit μ which is a real bounded measure on S. Because the original sequence $\{\mu_p\}$ is a Cauchy sequence Case II is proved.

Case III. Let μ_p be complex. The conclusion follows by decomposing each measure into its real and imaginary parts and applying Case II.

Lemma 4.4. The space $S_n'' = S_n''(\Delta_n \times R^{n\nu})$ is complete under the norm $\|F\|_n''$.

Proof. Because elements of S_n'' are expressible by formula (4.2) and $\|F\|_n'' \equiv \inf(\|\mu\|)$, Lemma 4.4 follows immediately from Lemma 4.3.

Lemma 4.5. The space $S'' = S''(\Sigma \Delta_n \times R^{n\nu})$ is complete.

Proof. Let $\{F^\ell\}$ be a Cauchy sequence of elements of S''. Let $G^\ell \equiv F^\ell - F^{\ell-1}$ for $\ell > 0$, $G^1 = F^1$.

Case I. Suppose $\sum_{\ell=1}^{\infty} \|G^\ell\|'' < \infty$. For each $n > 0$, there exist $\{G_n^\ell\} \subset S_n''$ such that $G^\ell(x) = \sum_{n=0}^{\infty} G_n^\ell(x)$ for $\ell = 1, 2, \ldots$ and

$$\|G^\ell\|'' + \frac{1}{2^\ell} > \sum_{n=0}^{\infty} \|G_n^\ell\|_n'' \ .$$

Thus

$$(4.10) \quad \sum_{n=0}^{\infty} \sum_{\ell=1}^{\infty} \|G_n^\ell\|_n'' = \sum_{\ell=1}^{\infty} \sum_{n=0}^{\infty} \|G_n^\ell\|_n'' < 1 + \sum_{\ell=1}^{\infty} \|G^\ell\|'' < \infty \ ,$$

and so for each n,

$$\sum_{\ell=1}^{\infty} \|G_n^\ell\|_n'' < \infty \ .$$

Let

$$G_n(\vec{x}) \equiv \sum_{\ell=1}^{\infty} G_n^\ell(\vec{x})$$

which converges absolutely and uniformly by the previous convergence. By the completeness of S_n'', $G_n \in S_n''$ and $\|G_n\|_n'' \leq \sum_{\ell=1}^{\infty} \|G_n^\ell\|_n'' < \infty$. By inequality

(4.10),

$$\sum_{n=0}^{\infty} \|G_n\|_n'' \leq \sum_{n=0}^{\infty} \sum_{\ell=1}^{\infty} \|G_n^{\ell}\|_n'' < \infty .$$

Thus by the definition of S'',

$$F \equiv \sum_{n=0}^{\infty} G_n \in S'' .$$

Now for each $\vec{x} \in C^{\nu}[a,b]$,

$$F(\vec{x}) = \sum_{n=0}^{\infty} G_n(\vec{x}) = \sum_{n=0}^{\infty} \sum_{\ell=1}^{\infty} G_n^{\ell}(\vec{x}) = \sum_{\ell=1}^{\infty} \sum_{n=0}^{\infty} G_n^{\ell}(\vec{x})$$

$$= \sum_{\ell=1}^{\infty} G^{\ell}(\vec{x}) = \sum_{\ell=2}^{\infty} (F^{\ell}(\vec{x}) - F^{\ell-1}(\vec{x})) + F^1(\vec{x}) = \lim_{\ell \to \infty} F^{\ell}(\vec{x}) .$$

Thus we have

$$\|F - F^{\ell}\|'' = \|F - \sum_{j=1}^{\ell} G^j\|'' = \|\sum_{j=\ell+1}^{\infty} G^j\|'' = \|\sum_{j=\ell+1}^{\infty} \sum_{n=0}^{\infty} G_n^j\|''$$

$$= \|\sum_{n=0}^{\infty} \sum_{j=\ell+1}^{\infty} G_n^j\|'' = \|\sum_{n=0}^{\infty} H_n^{\ell}\|''$$

where

$$H_n^{\ell}(\vec{x}) = \sum_{j=\ell+1}^{\infty} G_n^j(\vec{x})$$

and $H_n^{\ell} \in S_n''$ because the space is complete. From the definition of the norm in S'' we have

$$\|\sum_{n=0}^{\infty} H_n^{\ell}\|'' \leq \sum_{n=0}^{\infty} \|H_n^{\ell}\|_n'' \leq \sum_{n=0}^{\infty} \|\sum_{j=\ell+1}^{\infty} G_n^j\|_n'' \leq \sum_{n=0}^{\infty} \sum_{j=\ell+1}^{\infty} \|G_n^j\|_n''$$

where the last inequality holds because of the completeness of S_n''. Thus we have

$$\|F - F^{\ell}\|'' \leq \sum_{n=0}^{\infty} \sum_{j=\ell+1}^{\infty} \|G_n^j\|_n'' = \sum_{j=\ell+1}^{\infty} [\sum_{n=0}^{\infty} \|G_n^j\|_n'']$$

and since the last member has a limit of zero as $\ell \to \infty$, we have $\|F - F^{\ell}\|'' \underset{\ell \to \infty}{\to} 0$.

Case II. (General case) Because $\{F^{\ell}\}$ is a Cauchy sequence, there exists an index ℓ_1 such that for $j,k \geq \ell_1$, $\|F^j - F^k\|'' < 1/2$. Continuing inductively there exists for each n an index $\ell_n > \ell_{n-1}$ such that for $j,k \geq \ell_n$, $\|F^j - F^k\|'' < 1/2^n$. Set $\hat{F}^n \equiv F^{\ell_n}$ for $n = 1,2,\ldots.$ Then $\{\hat{F}^n\}$ is a subsequence of $\{F^{\ell}\}$, and if we set $G^n \equiv \hat{F}^{n+1} - \hat{F}^n$, for $n \geq 1$,

we have

$$\sum_{n=1}^{\infty} \|G^n\|' = \sum_{n=1}^{\infty} \|\hat{F}^{n+1} - \hat{F}^n\|'' < \sum_{1}^{\infty} \frac{1}{2^n} < \infty \quad .$$

Thus $\{\hat{F}^n\}$ satisfies the extra hypotheses of Case I, and hence there exists $F \in S(\Sigma\Delta_n \times R^{n\nu})$ such that

$$\lim_{n\to\infty} \|F^{\ell_n} - F\|'' = \lim_{n\to\infty} \|\hat{F}^n - F\|'' = 0$$

Since $\{F^{\ell}\}$ is a Cauchy sequence, it follows that

$$\lim_{\ell\to\infty} \|F^{\ell} - F\|'' = 0$$

and the lemma is proved.

Notation. Let

$$(4.11) \quad J_{n,\nu} \equiv J_{n,\nu}(\vec{x},\vec{t},\vec{v}) = \exp\{i \sum_{k=1}^{\nu} \sum_{j=1}^{n} v_{k,j}\, x_k(t_j)\}$$

Lemma 4.6 (First Decomposition Lemma).

If $G \in \hat{S}_n$, we can express G in the form

$$(4.12) \quad G = F + H$$

where $F \in S''_n$ and $H \in \hat{S}_{n-1}$ if $n > 1$ and $H \equiv 0$ if $n = 1$, and where

$$(4.13) \quad \|G\|^{\wedge}_n \geqq \|F\|''_n + \|H\|^{\wedge}_{n-1} \quad .$$

Proof. The conclusion follows easily if $n = 1$, so we assume $n > 1$. Let $\epsilon > 0$ be given and choose $\mu \in \hat{\mathcal{M}}_n$ such that (4.3) holds for G and μ and such that

$$(4.14) \quad \|\mu\| < \|G\|^{\wedge}_n + \epsilon \quad .$$

Let

$$(4.15) \quad \mathcal{J} = \{\vec{t}\times\vec{v} \,|\, (\vec{t}\times\vec{v}) \in (a,b)^n \times R^{n\nu}, \; t_j = t_k \text{ for some } j,k \text{ such that } 1 \leqq j < k \leqq n\} \quad .$$

Let

$$(4.16) \quad F(\vec{x}) \equiv \int_{(a,b)^n \times R^{n\nu} - \mathcal{J}} J_{n,\nu}\, d\mu(\vec{t}\times\vec{v}) \quad .$$

and let

$$(4.17) \quad H(\vec{x}) \equiv \int_{\mathcal{J}} J_{n,\nu}\, d\mu(\vec{t}\times\vec{v}) \,,$$

so that (4.12) holds.

In order to show that $F \in S''_n$, let $\rho_1, \rho_2, \ldots, \rho_{n!}$ denote the $n!$ permutations of $1,2,\ldots,n$, and let ρ_1 be the identity permutation. Let ρ_q take $1,2,\ldots,n$ into $p_{q,1}, p_{q,2}, \ldots, p_{q,n}$. Let $\rho_q\Delta_n$ denote the new n-dimensional tetrahedron obtained by replacing each point $\vec{t}$ by the point obtained by replacing each t_j by $t_{p_{q,j}}$. Let ρ_q^{-1} denote the inverse permutation, so that $\rho_q^{-1}\rho_q\Delta_n = \Delta_n$.

Now it is clear that

$$(a,b]^n \times R^{n\nu} = \mathcal{J} \cup \bigcup_{q=1}^{n!} (\rho_q\Delta_n) \times R^{n\nu} ,$$

where the union on the right is disjoint, and hence

$$(4.18) \quad F(\vec{x}) = \sum_{q=1}^{n!} \int_{\rho_q\Delta_n \times R^{n\nu}} J_{n,\nu} \, d\mu(\vec{t}\times\vec{v}) = \sum_{q=1}^{n!} F_q(\vec{x})$$

where

$$(4.19) \quad F_q(\vec{x}) \equiv \int_{\Delta_n \times R^{n\nu}} J_{n,\nu} \, d\rho_q^{-1}\mu(\vec{t}\times\vec{v})$$

and where $\rho_q^{-1}\mu$ is the measure such that for Borel sets E

$$(4.20) \quad \rho_q^{-1}\mu(E) \equiv \mu(\rho_q^{-1}E), \ \rho_q(t_j \times v_{k,j}) = t_{p_{q,j}} \times v_{k,j} .$$

Clearly each term of the sum in the last member of (4.18) is an element of S''_n, and hence $F \in S''_n$.

Next, in order to show that $H \in \hat{S}_{n-1}$, we define triangular arrays of subsets of $\mathcal{J}$ by a two dimensional inductive process. Thus

Let

$$\mathcal{J}_{1,2} \equiv \{(\vec{t}\times\vec{v}) \mid (\vec{t}\times\vec{v}) \in \mathcal{J}, \ t_1 = t_2\} ; \ \mathcal{I}_{1,2} \equiv \mathcal{J} - \mathcal{J}_{1,2} ,$$

$$\mathcal{J}_{1,3} \equiv \{(\vec{t}\times\vec{v}) \mid (\vec{t}\times\vec{v}) \in \mathcal{I}_{1,2}, \ t_1 = t_3\} ; \ \mathcal{I}_{1,3} \equiv \mathcal{I}_{1,2} - \mathcal{J}_{1,3}$$

$$\ldots$$

$$(4.21) \quad \mathcal{J}_{1,n} \equiv \{(\vec{t}\times\vec{v}) \mid (\vec{t}\times\vec{v}) \in \mathcal{I}_{1,n-1}, \ t_1 = t_n\} ; \ \mathcal{I}_{1,n} \equiv \mathcal{I}_{1,n-1} - \mathcal{J}_{1,n}$$

let

$$\mathcal{J}_{2,3} \equiv \{(\vec{t}\times\vec{v}) \mid (\vec{t}\times\vec{v}) \in \mathcal{I}_{1,n}, \ t_2 = t_3\} ; \ \mathcal{I}_{2,3} \equiv \mathcal{I}_{1,n} - \mathcal{J}_{2,3}$$

$$\mathcal{J}_{2,4} \equiv \{(\vec{t}\times\vec{v}) \mid (\vec{t}\times\vec{v}) \in \mathcal{I}_{2,3}, \ t_2 = t_4\} ; \ \mathcal{I}_{2,4} \equiv \mathcal{I}_{2,3} - \mathcal{J}_{2,4}$$

$$\ldots$$

$$\mathcal{J}_{2,n} \equiv \{(\vec{t}\times\vec{v}) \mid (\vec{t}\times\vec{v}) \in \mathcal{I}_{2,n-1}, \ t_2 = t_n\} ; \ \mathcal{I}_{2,n} \equiv \mathcal{I}_{2,n-1} - \mathcal{J}_{2,n}$$

$$\ldots$$

and let

$$\mathcal{I}_{n-1,n} \equiv \{(\vec{t}\times\vec{v}) \mid (\vec{t}\times\vec{v}) \in \mathcal{J}_{n-2,n},\ t_{n-1} = t_n\}\ ;\ \mathcal{J}_{n-1,n} = \mathcal{J}_{n-2,n} - \mathcal{I}_{n-1,n} \quad .$$

Now for each $\mathcal{I}_{p,q}$, we have $t_j \neq t_k$ for $j<p$ and all $k>j$ and also for $j=p$ and all k satisfying $p<k<q$, but $t_p = t_q$, and for each $\mathcal{J}_{p,q}$, we have $t_j \neq t_k$ for $j<p$ and all $k>j$ and also for $j=p$ and all k satisfying $p<k\leq q$. Thus $\mathcal{J}_{n-1,n}$ is empty, since in this set all the t_j must be disjoint, and this is impossible since $\mathcal{J}_{n-1,n} \subset \mathcal{I}$. Hence we have

$$\mathcal{I}_{n-1,n} = \mathcal{J}_{n-2,n} \quad ,$$

and

$$(4.22)\quad \mathcal{I} = \bigcup_{p=1}^{n-1} \bigcup_{q=p+1}^{n} \mathcal{I}_{p,q} \quad ,$$

and since the $\mathcal{I}_{p,q}$ are disjoint, we have

$$(4.23)\quad H(\vec{x}) = \sum_{p=1}^{n-1} \sum_{q=p+1}^{n} \int_{\mathcal{I}_{p,q}} J_{n,\nu}\, d\mu(\vec{t}\times\vec{v}) \quad .$$

We next show that each term of the right member of (4.23) is an element of $\hat{S}_{n-1}$, and will thus prove that $H \in \hat{S}_{n-1}$.

<u>Definition.</u> If $E \subset (a,b]^{n-1} \times R^{(n-1)\nu}$, E a Borel set, let

$$\mathfrak{J}_{p,q} E \equiv \{(\vec{t},\vec{v}) \mid (\vec{t},\vec{v}) \in \mathcal{I}_{p,q}\ ,\ (\vec{t}',\vec{v}') \in E \text{ where}$$

$$(4.24)\quad t_1' = t_1, t_2' = t_2, \ldots t_{q-1}' = t_{q-1}\ ,\ t_q' = t_{q+1}, \ldots\ t_{n-1}' = t_n \text{ and}$$

$$v_{k,1}' = v_{k,1},\ v_{k,2}' = v_{k,2}, \ldots,$$

$$v_{k,p-1}' = v_{k,p-1},\ v_{k,p}' = v_{k,p} + v_{k,q},\ v_{k,p+1}' = v_{k,p+1}, \ldots, v_{k,q-1}' = v_{k,q-1},$$

$$v_{k,q}' = v_{k,q+1}, \ldots, v_{k,n-1}' = v_{k,n} \quad \text{for } k = 1, \ldots, \nu\} \quad .$$

We next define some new measures, $\mu_{p,q}$ for $1 \leq p < q \leq n$ thus: If E is a Borel set in $(a,b]^{n-1} \times R^{(n-1)\nu}$, let

$$\mu_{p,q}(E) = \mu(\mathfrak{J}_{p,q} E) \quad .$$

Then we have

$$(4.25)\quad H_{p,q} \equiv \int_{\mathcal{I}_{p,q}} J_{n,\nu}\, d\mu(\vec{t}\times\vec{v}) = \int_{(a,b]^{n-1}\times R^{(n-1)\nu}} J_{n-1,\nu}\, d\mu_{p,q}(\vec{t}\times\vec{v}) \quad .$$

But the $H_{p,q} \in \hat{S}_{n-1}$ and hence by (4.23), we have $H \in \hat{S}_{n-1}$.

It remains to establish (4.13). If E is a Borel subset of $(a,b]^n \times R^{n\nu}$, let $\mu^*(E) \equiv \mu[E \cap \mathcal{J}]$ and $\eta(E) \equiv \mu(E) - \mu^*(E) = \mu[E \cap [(a,b] \times R^{n\nu} - \mathcal{J}]$ then $F(\vec{x}) = \int_{(a,b]^n \times R^{n\nu}} J_{n\nu}\, d\eta(\vec{t} \times \vec{v})$ and $H(\vec{x}) = \int_{(a,b]^n \times R^{n\nu}} J_{n\nu}\, d\mu^*(\vec{t} \times \vec{v})$ and $\mu(E) = \eta(E) + \mu^*(E)$. Since $\mathcal{J}$ and $(a,b]^n \times R^{n\nu} - \mathcal{J}$ are disjoint,

$$(4.26) \quad \|\mu\| = \|\eta\| + \|\mu^*\| .$$

Similarly let η_q be the measure defined by

$$\eta_q(E) = \eta(E \cap [\rho_q^{-1} \Delta_n) \times R^{n\nu}]) \qquad \text{for } q=1,2,\dots,n!,$$

and since

$$(a,b]^n \times R^{n\nu} - \mathcal{J} = \bigcup_{q=1}^{n!} (\rho_q \Delta_n) \times R^{n\nu}$$

and the sets $(\rho_q \Delta_n) \times R^{n\nu}$ for $q=1,\dots,n!$ are disjoint, we have

$$(4.27) \quad \|\eta\| = \sum_{q=1}^{n!} \|\eta_q\| .$$

Again, since the $\mathcal{J}_{p,q}$ are disjoint, for E a fixed Borel set in $(a,b]^{n-1} \times R^{(n-1)\nu}$, the sets $J_{p,q}E$ for $1 \leq p < q \leq 1$ are disjoint. Hence, using (4.22), we have

$$(4.28) \quad \sum_{p=1}^{n-1} \sum_{q=p+1}^{n} \|\mu_{p,q}\| = \| \sum_{p=1}^{n-1} \sum_{q=p+1}^{n} \mu_{p,q} \| = \operatorname{var}_{\mathcal{J}} \mu = \|\mu^*\| .$$

Thus using (4.17) we have from (4.14) and (4.26), (4.27), (4.28) that

$$\|G_n\|^\wedge + \epsilon > \|\mu\| = \sum_{q=1}^{n!} \|\eta_q\| + \sum_{p=1}^{n-1} \sum_{q=p+1}^{n} \|\mu_{p,q}\| \geqq \sum_{q=1}^{n!} \|F_q\|_n'' + \sum_{p=1}^{n-1} \sum_{q=p+1}^{n} \|H_{p,q}\|_{n-}^\wedge$$

$$\geqq \|F\|_n'' + \|H\|_{n-1}^\wedge ,$$ so the lemma is proved.

<u>Lemma 4.7</u> (Second Decomposition Lemma). <u>If</u> $G \in \hat{S}_n$, <u>we can express</u> G <u>in the form</u>

$$G = \sum_{q=1}^{n} F_q ,$$

<u>where</u> $F_q \in S_q''$, $q=1,2,\dots,n$. <u>Moreover</u> $\|G\|_n^\wedge \geq \sum_{q=1}^{n} \|F_q\|_q''$.

Proof. This lemma follows from repeated application of the preceding lemma.

Lemma 4.8 For each $n = 1,2,3 \ldots$ we have $S_n^\wedge = S_n''$, and if $F \in S_n''$,

$$\|F\|_n^\wedge = \|F\|_n'' .$$

Proof. It was pointed out just after the definition of $S_n^\wedge$ that $S_n'' \subset S_n^\wedge$ and $\|F\|_n^\wedge \leqq \|F\|_n''$. It remains therefore to show that $S_n^\wedge \subset S_n''$ and that if $F \in S_n''$,

$$(4.29) \quad \|F_n\|_n^\wedge \geqq \|F\|_n'' .$$

To show that $S_n^\wedge \subset S_n''$, let $G \in S_n^\wedge$. Then by the second decomposition lemma there exists $F_q \in S_q''$ for each $q=1,2, \ldots n$ such that

$$(4.30) \quad G = \sum_{q=1}^{n} F_q .$$

But by the remark following Lemma 4.0, $F_q \in S_n''$ for $q = 1, \ldots, n$, and since S_n is linear, the right member of (4.30) is in S_n'' and hence $G \in S_n''$. Since G was any member of $S_n^\wedge$, it follows that $S_n^\wedge \subset S_n''$, and hence that $S_n^\wedge = S_n''$.

Again if $F \in S_n''$, we have $F \in S_n^\wedge$, and by the second decomposition lemma there exist $F_q \in S_q''$ for each $q = 1,2, \ldots, n$ such that

$$F = \sum_{q=1}^{n} F_q$$

and such that

$$\|F\|_n^\wedge \geqq \sum_{q=1}^{n} \|F_q\|_q'' .$$

Again by the remark following Lemma 4.0 we have

$$\|F_q\|_q'' \geqq \|F_q\|_n'' ,$$

and hence

$$\|F\|_n^\wedge \geqq \sum_{q=1}^{n} \|F_q\|_n'' \geqq \left\| \sum_{q=1}^{n} F_q \right\|_n'' = \|F\|_n'' .$$

Thus the lemma is proved.

<u>Lemma 4.9</u>. If $F \in S''_m$ and $G \in S''_n$ and $H(\vec{x}) = F(\vec{x})\,G(\vec{x})$ for $x \in C^\nu$, then $H \in S''_{m+n}$, and $\|H\|''_{m+n} \leq \|F\|''_m \|G\|''_n$.

<u>Proof</u>. Let $\epsilon > 0$ be given. By the Lemma 4.8

$$F(\vec{x}) = \int_{(a,b]^m \times R^{m\nu}} \exp\{i \sum_{k=1}^{\nu} \sum_{j=1}^{m} v_{k,j}\, x_k(t_j)\} d\mu_1(\vec{t} \times \vec{v})$$

$$G(\vec{x}) = \int_{(a,b]^n \times R^{n\nu}} \exp\{i \sum_{k=1}^{\nu} \sum_{\ell=1}^{n} v'_{k,j}\, x_k(t'_\ell)\} d\mu_2(\vec{t'} \times \vec{v'})$$

where $\mu_1 \in \mathcal{M}^\wedge_m$, and $\mu_2 \in \mathcal{M}^\wedge_n$ and μ_1 and μ_2 are chosen so that

$$\|\mu_1\| \leq \|F\|''_m + \epsilon \quad \text{and} \quad \|\mu_2\| \leq \|G\|''_n + \epsilon .$$

Let $\tau_p = \begin{cases} t_p & \text{if } p = 1, \dots, m \\ t'_{p-m} & \text{if } p = m+1, \dots, m+n \end{cases}$,

and let

$$w_{k,p} = \begin{cases} v_{k,p} & \text{if } k = 1, \dots, \nu \ ; \ p = 1, \dots, m \\ v'_{k,p-m} & \text{if } k = 1, \dots, \nu \ ; \ p = m+1, \dots, m+n \end{cases}$$

and let $\mu_3 \equiv \mu_1 \times \mu_2$. Here we permute the variables so that μ_3 is defined on $(a,b]^{m+n} \times R^{(m+n)\nu}$. Then

$$H(\vec{x}) = \int_{(a,b]^{m+n} \times R^{(m+n)\nu}} \exp\{i \sum_{k=1}^{\nu} \sum_{p=1}^{m+n} w_{k,p}\, x_k(\tau_p)\} d\mu_3(\vec{\tau} \times \vec{w}) .$$

Thus $H \in S^\wedge_{m+n} \equiv S''_{m+n}$; since $\mu_3 \equiv \mu_1 \times \mu_2$,

$$\|\mu_3\| \equiv \operatorname{var} \mu_3 = (\operatorname{var} \mu_1)(\operatorname{var} \mu_2) = \|\mu_1\|\,\|\mu_2\| .$$

Thus

$$\|H\|''_{m+n} \leq \|\mu_3\| = \|\mu_1\|\,\|\mu_2\| \leq (\|F\|''_m + \epsilon)(\|G\|''_n + \epsilon) ,$$

and since $\epsilon > 0$ is arbitrary,

$$\|H\|''_{m+n} \leq \|F\|''_m \|G\|''_n$$

and the lemma is proved.

<u>Theorem 4.1</u> <u>The space S'' is a Banach algebra with norm $\|\cdot\|''$</u>.

Proof. Let $F, G \in S''$ and $\epsilon > 0$ be given; then by definition of S'',

$$F(\vec{x}) = \sum_1^\infty F_n(\vec{x}) \quad \text{and} \quad G(\vec{x}) = \sum_1^\infty G_n(\vec{x}) \quad \text{where } F_n, G_n \in S''_n, \text{ and}$$

$$\sum_1^\infty \|F_n\|''_n < \|F\|'' + \epsilon \quad \text{and} \quad \sum_1^\infty \|G_n\|'' < \|G\|'' + \epsilon$$

and the series representing F and G converge absolutely and uniformly over $C^\nu[a,b]$. Let $H(\vec{x}) = F(\vec{x})\, G(\vec{x})$, so $H(\vec{x}) = [\sum_{n=0}^\infty F_n(\vec{x})][\sum_{m=0}^\infty G_m(\vec{x})]$

$$= \sum_{n=0}^\infty \sum_{m=0}^\infty F_n(\vec{x})\, G_m(\vec{x}) .$$

Let $H_{n,m}(\vec{x}) = F_n(\vec{x})\, G_m(\vec{x})$. By our Lemma 4.9, $H_{n,m} \in S''_{n+m}$. Thus

$$H(\vec{x}) = \sum_{n=0}^\infty \sum_{m=0}^\infty H_{n,m}(\vec{x}) = \sum_{n=0}^\infty \sum_{p=n}^\infty H_{n,p-n}(\vec{x}) = \sum_{p=0}^\infty \sum_{n=0}^p H_{n,p-n}(\vec{x}) .$$

But for each n satisfying $0 \leq n \leq p$, by the preceding lemma we have

$$H_{n,p-n} \in S''_p .$$

Let $K_p(\vec{x}) = \sum_{n=0}^p H_{n,p-n}(\vec{x})$, so $K_p(\vec{x}) \in S''_p$. Now $H(\vec{x}) = \sum_{p=0}^\infty K_p(\vec{x})$,

and the series converges absolutely and uniformly. Now

$$\|H\|'' \leqq \sum_{p=0}^\infty \|K_p\|''_p \leqq \sum_{p=0}^\infty \sum_{n=0}^p \|H_{n,p-n}\|''_p$$

$$\leqq \sum_{p=0}^\infty \sum_{n=0}^p \|F_n\|''_n \|G_{p-n}\|''_{p-n}$$

$$= \sum_{n=0}^\infty \sum_{p=n}^\infty \|F_n\|''_n \|G_{p-n}\|''_{p-n}$$

$$= \sum_{n=0}^\infty \sum_{m=0}^\infty \|F_n\|''_n \|G_m\|''_m$$

$$= [\sum_{n=0}^\infty \|F_n\|''_n]\,[\sum_{m=0}^\infty \|G_m\|''_m] < (\|F\|'' + \epsilon)(\|G\|'' + \epsilon)$$

Therefore since $\epsilon > 0$ is arbitrary, $H \in S''$ and S'' is a Banach algebra and the theorem is proved.

5. FORMULAE FOR FEYNMAN INTEGRALS

In this section we prove the existence of the analytic Feynman integral for every element of each of our Banach algebras. We evaluate these Feynman integrals in terms of formulae that do not involve analytic continuation. We also prove a theorem for interchanging summation and analytic Feynman integration.

Theorem 5.1. Let $\mu \in \mathcal{M}(L_2^\nu)$ and let $F \in S(L_2^\nu)$ be the Fourier transform of μ, thus

$$(5.1)\qquad F(\vec{x}) = \int_{L_2^\nu} \exp\{i \sum_{j=1}^{\nu} \int_a^b v_j(t)\, \tilde{d} x_j(t)\} d\mu(\vec{v}) .$$

Then F is analytic Feynman integrable on $C^\nu[a,b]$, and if q is real, $q \neq 0$,

$$(5.2)\qquad \int_{C^\nu}^{\text{anf}_q} F(\vec{x})\, d\vec{x} = \int_{L_2^\nu} \exp\{ \frac{1}{2qi} \sum_{j=1}^{\nu} \int_a^b [v_j(t)]^2\, dt\} d\mu(\vec{v}) .$$

Proof. Let us first compute the Wiener integral of $F(\lambda^{-\frac{1}{2}}\vec{x})$ for positive λ. By Lemma 2.1 and the Fubini theorem we have the existence of the integrals below and the equality,

$$J \equiv \int_{C^\nu} F(\lambda^{-1/2}\vec{x})\, d\vec{x} = \int_{C^\nu}\int_{L_2^\nu} \exp\{i\lambda^{-\frac{1}{2}} \sum_{j=1}^{\nu} \int_a^b v_j(t)\,\tilde{d} x_j(t)\} d\mu(\vec{v})\, d\vec{x}$$

$$= \int_{L_2^\nu}\int_{C^\nu} \exp\{i\lambda^{-\frac{1}{2}} \sum_{j=1}^{\nu} \int_a^b v_j(t)\,\tilde{d} x_j(t)\} d\vec{x}\, d\mu(\vec{v}) .$$

Rewriting the above integral and evaluating the interior integral we have

$$J = \int_{L_2^\nu} \prod_{j=1}^{\nu} [\int_C \exp\{i\lambda^{-1/2} \int_a^b v_j(t)\,\tilde{d} x_j(t)\} d x_j] d\mu(\tilde{v})$$

$$= \int_{L_2^\nu} [(2\pi)^{-\nu/2} \prod_{j=1}^{\nu} \int_{-\infty}^{\infty} \exp\{i\lambda^{-1/2}\|v_j\|u_j\}\exp\{-u_j^2/2\}du_j] d\mu(\vec{v})$$

$$= \int_{L_2^\nu} \exp\{-1/2 \sum_{j=1}^{\nu} \frac{\|v_j\|^2}{\lambda}\} d\mu(\vec{v}) .$$

Next we extend this result to the right half plane $\Re e\, \lambda > o$. The integrand of the last member of the above equation is an analytic function of λ on $\Re e\, \lambda > o$ and is bounded by one. Thus the right member is analytic for $\Re e\, \lambda > o$ and hence the first member has an analytic extension to $\Re e\, \lambda > o$. Thus we have for $\Re e\, \lambda > o$

$$\int_{C^\nu}^{anw_\lambda} F(\vec{x})d\vec{x} = \int_{L_2^\nu} \exp\{-1/2 \sum_{j=1}^{\nu} \frac{\|v_j\|^2}{\lambda}\} d\mu(\vec{v}) .$$

An application of the dominated convergence theorem enables us to pass to the limit as $\lambda \to -iq$, $q \neq o$ and obtain equation (2) and the theorem is proved.

Theorem 1 has some special cases which are worth stating explicitly.

<u>Theorem 5.2.</u> <u>Let</u> $\mu \in \mathcal{M}'(\mathcal{B}^\nu)$ <u>and let</u> $F \in S'(\mathcal{B}^\nu)$ <u>be the Fourier transform of</u> μ , <u>thus</u>

$$(5.3) \qquad F(\vec{x}) = \int_{\mathcal{B}^\nu} \exp\{i \sum_{j=1}^{\nu} \int_a^b v_j(t)\, dx_j(t)\} d\mu(\vec{v}) .$$

Then F is analytic Feynman integrable on $C^\nu[a,b]$, and if q is real, $q \neq o$,

$$(5.4) \qquad \int_{C^\nu}^{anf_q} F(\vec{x})d\vec{x} = \int_{\mathcal{B}^\nu} \exp\{\frac{1}{2qi} \sum_{j=1}^{\nu} \int_a^b [v_j(t)]^2\, dt\} d\mu(\vec{v}) .$$

<u>Proof</u>. The measure $\mu \in \mathcal{M}(\mathcal{B}^\nu)$ can be extended to give a measure $\mu^* \subset \mathcal{M}(L_2^\nu)$ by defining for each $E \in \mathcal{G}$

$$\mu^*(E) \equiv \mu(E \cap \mathcal{B}^\nu) .$$

This transforms equation (5.3) into equation (5.1) and hence the conclusion (5.2) of Theorem 5.2 holds, (5.4) follows immediately and the theorem is proved.

Theorem 5.3. *Let* $\mu \in \mathcal{M}_n''(\Delta_n \times R^{n\nu})$ *and let* $F \in S_n''(\Delta_n \times R^{n\nu})$ *be the Fourier transform of* μ, *thus*

$$(5.5)\qquad F(\vec{x}) = \int_{\Delta_n \times R^{n\nu}} \exp\{i \sum_{k=1}^{\nu} \sum_{j=1}^{n} x_k(t_j) u_{k,j}\} d\mu(\vec{t} \times \vec{\mu}) \quad .$$

Then F *is analytic Feynman integrable on* $C^\nu[a,b]$ *and if* q *is real,* $q \neq 0$,

$$(5.6)\qquad \int_{C^\nu}^{\text{anf}_q} F(\vec{x}) d\vec{x} = \int_{\Delta_n \times R^{n\nu}} \exp\{\frac{1}{2qi} \sum_{k=1}^{\nu} \sum_{j=1}^{n} (t_j - t_{j-1})[\sum_{\ell=j}^{n} u_{k,\ell}]^2\} d\mu(\vec{t} \times \vec{u})$$

$$= \int_{\Delta_n \times R^{n\nu}} \exp\{\frac{1}{2qi} \sum_{k=1}^{\nu} \sum_{\ell=1}^{n} \sum_{j=1}^{\ell} (2 - \delta_{j\ell}) u_{k,j} u_{k,\ell} (t_j - a)\} d\mu(\vec{t} \times \vec{u})$$

where $t_0 \equiv a$.

Proof. Here we are considering $\Delta_n \times R^{n\nu}$ as related to $\mathcal{B}^\nu$ and our measure μ on $\Delta_n \times R^{n\nu}$ as related to a measure σ on $\mathcal{B}^\nu$ as in equation (4.8). In particular equation (4.7) defines a transformation from the space $\Delta_n \times R^{n\nu}$ to the subset of $\mathcal{B}^\nu$ on which the measure is concentrated By Theorem (5.2) and equation (4.6) we have

$$\int_{C^\nu[a,b]}^{\text{anf}_q} F(\vec{x}) d\vec{x} = \int_{\mathcal{B}^\nu} \exp\{\frac{1}{2qi} \sum_{k=1}^{\nu} \int_a^b [v_k(t)]^2 dt\} d\sigma(\vec{v})$$

$$= \int_{\Delta_n \times R^{n\nu}} \exp\{ \frac{1}{2qi} \sum_{k=1}^{\nu} \sum_{j=1}^{n} \int_{t_{j-1}}^{t_j} [\sum_{\ell=j}^{n} u_{k,\ell}]^2 dt\} d\mu(\vec{t} \times \vec{u})$$

and the theorem is proved.

Theorem 5.4 *Let* $F_n \in S$ *for* $n = 1,2,\ldots$, *and let*

$$(5.7)\qquad \sum_{n=1}^{\infty} \|F_n\| < \infty \quad .$$

Then $F \in S$ where for almost all $\vec{x} \in C^{\nu}[a,b]$,

$$(5.8) \quad F(\vec{x}) \equiv \sum_{n=1}^{\infty} F_n(\vec{x}) ,$$

and

$$(5.9) \quad \int_{C^{\nu}[a,b]}^{\mathrm{anf}_q} F(\vec{x})d\vec{x} = \sum_{n=1}^{\infty} \int_{C^{\nu}[a,b]}^{\mathrm{anf}_q} F_n(\vec{x})d\vec{x} .$$

Proof. For each $n = 1,2, \ldots$ let $\mu_n \in \mathcal{M}$ be chosen so that F_n is related to μ_n by equation (2.1) for almost all $\vec{x} \in C^{\nu}[a,b]$. We define a measure μ on L_2^{ν} as follows: if $E \in \mathcal{G} \subset L_2^{\nu}$

$$(5.10) \quad \mu(E) \equiv \sum_{n=1}^{\infty} \mu_n(E) .$$

The above series converges absolutely since it follows from (5.7) that

$$(5.11) \quad \sum_{n=1}^{\infty} \|\mu_n\| = \sum_{n=1}^{\infty} \|F_n\| < \infty .$$

Moreover μ is a measure, since if $E_n \in \mathcal{G}$ for $k = 1,2,\ldots$ and the E_k are disjoint, then

$$\mu(\bigcup_{k=1}^{\infty} E_k) = \sum_{n=1}^{\infty} \mu_n(\bigcup_{k=1}^{\infty} E_k) = \sum_{n=1}^{\infty} \sum_{k=1}^{\infty} \mu_n(E_k)$$

$$= \sum_{k=1}^{\infty} \sum_{n=1}^{\infty} \mu_n(E_k) = \sum_{k=1}^{\infty} \mu(E_k) .$$

Here the first double sum converges absolutely because of (5.11). Moreover μ is a bounded measure because of (5.10) and (5.11) and we have $\mu \in \mathcal{M}$ and

$$(5.12) \quad \|\mu\| \leqq \sum_{n=1}^{\infty} \|F_n\| < \infty .$$

Now by (5.8) and (2.1) and (5.10) and (5.11) we have

$$F(\vec{x}) = \sum_{n=1}^{\infty} \int_{L_2^{\nu}} \exp\{i \sum_{k=1}^{\nu} \int_a^b v_k(t)\tilde{d}x_k(t)\}d\mu_n(\vec{v}(\cdot))$$

$$= \int_{L_2^{\nu}} \exp\{i \sum_{k=1}^{\nu} \int_a^b v_k(t)\tilde{d}x_k(t)\}d\mu(\vec{v}(\cdot))$$

for almost all $x \in C^{\nu}$; and $F \in S$. Moreover from Theorem 5.1 ,

$$\int_{C^{\nu}}^{\mathrm{anf}_q} F(\vec{x})d\vec{x} = \int_{L_2^{\nu}} \exp\{\frac{1}{2qi} \sum_{j=1}^{\nu} \int_a^b [v_j(t)]^2 dt\} d\mu(\vec{v})$$

$$= \sum_{n=1}^{\infty} \int_{L_2^{\nu}} \exp\{\frac{1}{2qi} \sum_{j=1}^{\nu} \int_a^b [v_j(t)]^2 dt\} d\mu_n(\vec{v})$$

$$= \sum_{n=1}^{\infty} \int_{C^{\nu}}^{\mathrm{anf}_q} F_n(\vec{x}) d\vec{x} ,$$

and the theorem is proved.

Theorem 5.5. Let $\{\mu_n\} \in \mathcal{M}''(\Sigma\Delta_n \times R^{n\nu})$ and let $F \in S''(\Sigma\Delta_n \times R^{n\nu})$ be given by

$$(5.13) \quad F(\vec{x}) = \sum_{n=1}^{\infty} \int_{\Delta_n \times R^{n\nu}} \exp\{i \sum_{k=1}^{\nu} \sum_{j=1}^{n} u_{k,j} x_k(t)\} d\mu_n(\vec{t} \times \vec{u}) .$$

Then F is analytic Feynman integrable on $C^{\nu}[a,b]$ and if q is real, $q \neq 0$,

$$(5.14) \quad \int_{C^{\nu}}^{\mathrm{anf}_q} F(\vec{x})d\vec{x} = \sum_{n=1}^{\infty} \int_{\Delta_n \times R^{n\nu}} \exp\{\frac{1}{2qi} \sum_{k=1}^{\nu} \sum_{\ell=1}^{n} \sum_{j=1}^{\ell} (2-\delta_{j\ell}) u_{k,j} u_{k,\ell} (t_j - a)\} d\mu_n(\vec{t} \times \vec{u}) .$$

Proof. This theorem follows immediately from Theorems 5.3 and 5.4.

6. A FUBINI THEOREM OF ANALYTIC FEYNMAN INTEGRALS

We first prove an unsymmetric Fubini theorem. (See [2]).

Theorem 6.1. Let $\mathcal{C}^*$ be a σ-algebra of subsets of a space V and let μ_y be a family of complex measures on $\mathcal{C}^*$. If $E \in \mathcal{C}^*$, let $\mu_y(E)$ be measurable with respect to σ on the measure space $\mathcal{Y}$, where σ is a complex measure on $\mathcal{Y}$. Let $\mathrm{var}_V \mu_y$ be dominated by a σ-integrable function of y on $\mathcal{Y}$. Then for each $E \in \mathcal{C}^*$,

(6.1) $m(E) \equiv \int_{\mathcal{Y}} \mu_y(E)\,d\sigma(y)$

is a bounded complex measure on $\mathcal{A}^*$. _Moreover, if_ f _is a bounded complex function on_ V _measurable on_ $\mathcal{A}^*$, _then_ f _is integrable with respect to_ m _and each member of the equation below exists and they are equal:_

(6.2) $$\int_{\mathcal{Y}} [\int_V f(v)\,d\mu_y(v)]\,d\sigma(y) = \int_V f(v)\,dm(v)$$

$$\equiv \int_V f(v)\,d[\int_{\mathcal{Y}} \mu_y(v)\,d\sigma(y)] \quad .$$

Proof. Let $g(\cdot)$ be a real non-negative function integrable with respect to σ on $\mathcal{Y}$ so that

(6.3) $$\int_{\mathcal{Y}} g(y)\,|d\sigma(y)| < \infty$$

and let

(6.4) $$\mathrm{var}_V \mu_y \leqq g(y) \quad .$$

Let $E \in \mathcal{A}^*$. Then by (6.4) ,

(6.5) $$|\mu_y(E)| \leqq g(y)$$

and by (6.3) and (6.5), $m(E)$ exists and

(6.6) $$|m(E)| \leqq \int_{\mathcal{Y}} g(y)\,|d\sigma(y)| \quad .$$

Moreover if $E_1, E_2, \ldots$ are disjoint elements of $\mathcal{A}^*$ such that

$$E = \bigcup_{n=1}^{\infty} E_n \quad ,$$

we have

(6.7) $$\sum_{n=1}^{\infty} m(E_n) = \sum_{n=1}^{\infty} \int_{\mathcal{Y}} \mu_y(E_n)\,d\sigma(y)$$
$$= \int_{\mathcal{Y}} [\sum_{n=1}^{\infty} \mu_y(E_n)]\,d\sigma(y) = \int_{\mathcal{Y}} \mu_y(E)\,d\sigma(y) = m(E) \quad .$$

Here the interchange of summation and integration is justified by (6.3) and the fact that

$$\sum_{n=1}^{\infty} |\mu_y(E_n)| \leqq \mathrm{var}_V \mu_y \leqq g(y) \quad .$$

Hence by (6.7), m is a complex measure on $\mathcal{G}^*$, and by (6.6), m is bounded on $\mathcal{G}^*$.

<u>Case I.</u> Let $f(v) = \chi_{E_0}(v)$, where $E_0 \in \mathcal{G}^*$. Now clearly (6.2) holds, since each of its members equals $m(E_0)$, and thus Case I is established.

<u>Case II.</u> By taking finite linear combinations of characteristic functions and passing to the limit by dominated convergence we obtain the general case and the theorem is proved.

<u>Theorem</u> 6.2. <u>Let</u> σ <u>be a finite complex measure on a measure space</u> $\mathcal{Y}$. <u>Let</u> $F(\vec{x},y)$ <u>be defined and measurable on</u> $C^\nu \times \mathcal{Y}$. <u>For each</u> $y \in \mathcal{Y}$, <u>let</u> $F(\cdot,y) \in S$, <u>and let</u> $\|F(\cdot,y)\|$ <u>be dominated by a</u> σ-<u>integrable function of</u> y <u>on</u> $\mathcal{Y}$. <u>Then for real</u> $q \neq 0$, <u>the following integrals exist and are equal:</u>

$$(6.8) \qquad \int_{\mathcal{Y}} [\int_{C^\nu}^{\mathrm{anf}_q} F(\vec{x},y)\,d\vec{x}\,]\,d\sigma(y) = \int_{C^\nu}^{\mathrm{anf}_q} [\int_{\mathcal{Y}} F(\vec{x},y)\,d\sigma(y)\,]\,d\vec{x} \quad .$$

<u>Proof</u>. For each $y \in \mathcal{Y}$, let μ_y be the element of such that for $y \in \mathcal{Y}$ and almost all $\vec{x} \in C^\nu$,

$$(6.9) \qquad F(\vec{x},y) = \int_{L_2^\nu} \exp\{i \sum_{j=1}^{\nu} \int_a^b v_j(t)\,\tilde{d}x_j(t)\}\,d\mu_y(\vec{v}) \quad .$$

From Theorem 2.1 it follows that μ_y is uniquely defined for each $y \in \mathcal{Y}$, and from the corollary to Theorem 2.1, it follows that for each $E \in \mathcal{G}$, $\mu_y(E)$ is a σ-measurable function on $\mathcal{Y}$. Since by (2.22), we have $\mathrm{var}_V \mu_y = \|\mu_y\| = \|F(\cdot,y)\|$, so that by hypothesis $\mathrm{var}_V \mu_y$ is dominated by a σ-integrable function of y on $\mathcal{Y}$.

By Theorem 5.1 we have

$$\int_{C^\nu}^{\mathrm{anf}_q} F(\vec{x},y)\,d\vec{x}$$

$$= \int_{C^\nu}^{\text{anf}_q} \int_{L_2^\nu} \exp\{i \sum_{j=1}^{\nu} v_j(t)\tilde{d}x_j(t)\}\, d\mu_y(\vec{v})\, d\vec{x}$$

$$= \int_{L_2^\nu} \exp\{\frac{1}{2qi} \sum_{j=1}^{\nu} \int_a^b [v_j(t)]^2\, dt\}\, d\mu_y(\vec{v}) .$$

We integrate the first and last members with respect to $\sigma(y)$ and apply the unsymmetric Fubini theorem to obtain

$$I \equiv \int_{\mathcal{Y}} \int_{C^\nu}^{\text{anf}_q} F(\vec{x},y)\, d\vec{x}\, d\sigma(y)$$

$$= \int_{\mathcal{Y}} \int_{L_2^\nu} \exp\{\frac{1}{2qi} \sum_{j=1}^{\nu} \int_a^b [v_j(t)]^2\, dt\}\, d\mu_y(\vec{v})\, d\sigma(y)$$

$$= \int_{L_2^\nu} \exp\{\frac{1}{2qi} \sum_{j=1}^{\nu} \int_a^b [v_j(t)]^2\, dt\}\, dm(\vec{v})$$

where the measure m is defined by

$$m(E) \equiv \int_{\mathcal{Y}} \mu_y(E)\, d\sigma(y)$$

for each $E \in \mathcal{G}$. By a second application of Theorem 5.1 we obtain

$$I = \int_{C^\nu}^{\text{anf}_q} \int_{L_2^\nu} \exp\{i \sum_{j=1}^{\nu} v_j(t)\tilde{d}x_j(t)\}\, dm(\vec{v})\, d\vec{x} .$$

An application of the unsymmetric Fubini theorem to the interior integral above gives us

$$\int_{L_2^\nu} \exp\{i \sum_{j=1}^{\nu} v_j(t)\tilde{d}x_j(t)\}\, dm(\vec{v})$$

$$= \int_{\mathcal{Y}} \int_{L_2^\nu} \exp\{i \sum_{j=1}^{\nu} v_j(t)\tilde{d}x_j(t)\}\, d\mu_y(\vec{v})\, d\sigma(y) .$$

Thus using (6.9) we have

$$I = \int_{C^\nu}^{\mathrm{anf}_q} \int_{\mathcal{Y}} \int_{L_2^\nu} \exp\{i \sum_{j=1}^{\nu} \int v_j(t)\,dx_j(t)\}\, d\mu_y(\vec{v})\, d\sigma(y)\, d\vec{x}$$

$$= \int_{C^\nu}^{\mathrm{anf}_q} \int_{\mathcal{Y}} F(\vec{x},y)\, d\sigma(y)\, d\vec{x}$$

and the theorem is proved.

7. EXAMPLES

Example 1:

We now show that the example (1.1) given in the introduction is an element of S". Since S" is a Banach algebra, we need only show that the exponent

$$(7.1) \qquad f(\vec{x}) \equiv \int_a^b \theta(t,\vec{x}(t))\,dt \in S'' .$$

To see this, note that

$$(7.2) \qquad f(\vec{x}) = \int_a^b \int_{R^\nu} \exp\{i \sum_{k=1}^{\nu} v_k x_k(t)\}\, d\sigma_t(\vec{v})\,dt$$

$$= \int_{[a,b]\times R^\nu} \exp\{i \sum_{k=1}^{\nu} v_k x_k(t)\}\, d\mu(t\times\vec{v})$$

where the measure μ is defined by

$$\mu(E) \equiv \int_a^b \sigma_t(E^{(t)})\,dt$$

for each Borel set E in $(a,b]\times R^\nu$, using (1.3).

To see that μ is a measure, let $\{E_n\}$ be an infinite disjoint sequence of Borel sets in $[a,b]\times R^\nu$, and note that for each $t \in (a,b]$, $\{E_n^{(t)}\}$ is an infinite disjoint set of Borel sets in R^ν. Thus

$$\sigma_t\Big(\bigcup_{n=1}^{\infty} E_n^{(t)}\Big) = \sum_{n=1}^{\infty} \sigma_t(E_n^{(t)}) \qquad \text{for } a<t\leqq b$$

and

$$\sum_{n=1}^{\infty} |\sigma_t(E_n^{(t)})| \leqq M \qquad \text{for } a<t\leqq b .$$

By dominated convergence,

$$\mu\left(\bigcup_{n=1}^{\infty} E_n\right) = \int_a^b \sigma_t\left(\bigcup_{n=1}^{\infty} E_n^{(t)}\right)dt = \int_a^b \sum_{n=1}^{\infty} \sigma_t(E_n^{(t)})dt$$

$$= \sum_{n=1}^{\infty} \int_a^b \sigma_t(E_n^{(t)})dt = \sum_{n=1}^{\infty} \mu(E_n) \quad ,$$

and the fact that μ is a measure is proved. From this (7.2) readily follows, and (7.1) follows and so $F \in S''$.

We next show that the formula (1.4) given in the introduction is valid. We begin with the functional

$$(7.3) \quad F(\vec{x}) = \exp\{\int_a^b \theta(t; \vec{x}(t))dt\}$$

$$= \sum_{n=0}^{\infty} \frac{1}{n!} \{\int_a^b \theta(t; \vec{x}(t))dt\}^n$$

$$= 1 + \sum_{n=1}^{\infty} \frac{1}{n!} \int_{[a,b]^n} \prod_{j=1}^{n} \theta(t_j; x(t_j))d\vec{t}$$

$$= 1 + \sum_{1}^{\infty} \int_{\Delta_n} \prod_{j=1}^{n} \theta(t_j; \vec{x}(t_j))d\vec{t}$$

$$= 1 + \sum_{1}^{\infty} \int_{\Delta_n} \prod_{j=1}^{n} [\int_{R^\nu} \exp\{i \sum_{k=1}^{\nu} v_{k,j} x_k(t_j)\} d\sigma_{t_j}(\vec{v}_j)] d\vec{t}$$

$$= 1 + \sum_{n=1}^{\infty} \int_{\Delta_n \times R^{\nu n}} \exp\{i \sum_{j=1}^{n} \sum_{k=1}^{\nu} v_{k,j} x_k(t_j)\} d\mu_n(\vec{t} \times \vec{\vec{v}})$$

where $d\mu_n(\vec{t} \times \vec{\vec{v}}) = \prod_{j=1}^{n} (d\sigma_{t_j}(\vec{v}_j)dt_j)$.

Thus applying Theorem 5.5 to $F(\vec{x})$ we obtain equation (1.4).

Example 2: We shall show that the following functional

$$(7.4) \quad G(x) \equiv \exp\{-\frac{1}{2}\int_a^b [\int_t^b \rho(s)dx(s)]^2 dt\}$$

is an element of S , though the exponent is not in S . Here ρ denotes a positive increasing continuous function on $[a,b]$.

Proof. Define a measure μ over $L_2[a,b]$ by using a modified Wiener measure as follows: for each set $E \in \mathcal{B} \subset L_2[a,b]$ let

$$\mu(E) = m_w(E^*) \quad \text{where}$$

$$E^* \equiv \{v| \quad \rho(\cdot)\,v(\cdot) \in E \ , \ v \in C[a,b]\}$$

and m_w denotes Wiener measure.

Consider the functional

$$(7.5) \quad F(x) \equiv \int_{L_2[a,b]} \exp\{i \int_a^b v(t)\tilde{d}\,x(t)\}\,d\mu(v) \quad .$$

It is easy to see that $\mu \in \mathcal{M}$ and hence $F \in S \equiv S(L_2)$. Now let

$$\theta(t) \equiv \int_t^b \rho(s)\,dx(s)$$

and note that

$$F(x) = \int_{C[a,b]} \exp\{i \int_a^b \omega(t)\rho(t)\tilde{d}x(t)\}\,dm_w(\omega)$$

$$= \int_{C[a,b]} \exp\{i \int_a^b \theta(t)\tilde{d}\omega(t)\}\,dm_w(\omega)$$

$$= \int_{C[a,b]} \exp\{i\|\theta\| \int_a^b \frac{\theta(t)}{\|\theta\|}\,\tilde{d}\,\omega(t)\}\,d\,m_w(\omega)$$

$$= \frac{1}{\sqrt{2\pi}} \int_{-\infty}^{\infty} \exp\{i\,\|\theta\|\,u - \frac{u^2}{2}\}\,du$$

$$= \exp\{-\frac{1}{2}\,\|\theta\|^2\}$$

$$= \exp\{-\frac{1}{2}\int_a^b [\theta(t)]^2 dt\}$$

$$= \exp\{-\frac{1}{2}\int_a^b [\int_t^b \rho(s)\,d\,x(s)]^2\,d\,t\} = G(x) \quad .$$

Thus $G = F \in S$. However the exponent in the definition of G is not in S since it is unbounded.

Example 3:

Consider the functional

$$(7.6) \ F(\vec{x}) \equiv \exp\{-\sum_{j=1}^{\nu} \rho_j \int_a^b [x_j(t)]^2 d\,t\}\,\psi(\vec{x}(b))$$

where ρ_j is real and positive and where

$$(7.7)\quad \psi(u_1, \ldots, u_\nu) = \int_{R^\nu} \exp\{ \sum_{j=1}^{\nu} i\, u_j v_j \} d\mu(\vec{v})$$

and μ is a complex measure of bounded variation on R^ν. We shall show that $F \in S$, and in the case $\nu = 1$ obtain a formula for its analytic Feynman integral. The general case does not appear to present additional complications. To show that $F \in S$, write

$$F(\vec{x}) \equiv \psi(\vec{x}(b)) \prod_{j=1}^{\nu} F_j(\vec{x})$$

where

$$F_j(\vec{x}) \equiv F_j(x_j) \equiv \exp\{-\rho_j \int_a^b [x_j(t)]^2 \, dt\} \ .$$

We shall show that $\psi(\vec{x}(b)) \in S_1'' \subset S$ and that $F_j(\vec{x}) \in S$ for $j = 1, 2, \ldots, \nu$, and it will then follow that $F \in S$, since S is a Banach algebra.

To show that $\psi(\vec{x}(b)) \in S_1''$, let us define a measure μ^* on $(a,b] \times R^\nu$ as follows:

For each Borel set $E \subset (a,b] \times R^\nu$, set

$$\mu^*(E) \equiv \mu(E^{(b)})$$

where μ is the Borel measure on R^ν in terms of which ψ is defined in (7.7) and where $E^{(b)}$ is the b-section of E:

$$E^{(b)} \equiv \{\vec{v} \mid (b \times \vec{v}) \in E\} \ .$$

Then we obtain from (7.7) and (4.2) that

$$\begin{aligned} \psi(\vec{x}(b)) &= \int_{R^\nu} \exp\{ i \sum_{j=1}^{\nu} v_j x_j(b) \} \, d\mu(\vec{v}) \\ &= \int_{(a,b] \times R^\nu} \exp\{ i \sum_{j=1}^{\nu} v_j x_j(t) \} \, d\mu^*(t \times \vec{v}) \in S_1'' \ . \end{aligned}$$

To show that $F_j \in S$, let τ_j be the transformation

$$\tau_j x_j(\cdot) \equiv x_j[a + b - (\cdot)]$$

which maps $L_2[a,b]$ onto itself. Let μ_j be the measure on $L_2[a,b]$ defined as follows.

If $G \subset L_2[a,b]$, we shall say that G is μ_j-measurable if $(\tau_j G) \cap C[a,b]$ is Wiener measurable, and we define

$$\mu_j(G) = m_w[(\tau_j G) \cap C[a,b]] \quad .$$

It can be seen that $\mu_j \in \mathfrak{M}$. Now let us calculate the functional F_j^* (which is seen to be an element of S by its definition):

$$F_j^*(x) \equiv \int_{L_2[a,b]} \exp\{i\sqrt{2\rho_j} \int_a^b v(t)\tilde{d}x_j(t)\}d\mu_j(v)$$

$$= \int_{C[a,b]} \exp\{i\sqrt{2\rho_j} \int_a^b v(a+b-t)\tilde{d}x_j(t)\}dm_w(v)$$

$$= \int_{C[a,b]} \exp\{-i\sqrt{2\rho_j} \int_a^b x_j(t)\tilde{d}v(a+b-t)\}dm_w(v)$$

$$= \int_{C[a,b]} \exp\{i\sqrt{2\rho_j} \int_a^b x_j(a+b-s)\tilde{d}v(s)\}dm_w(v) \quad .$$

Let

$$z(s) = x(a+b-s)[\int_a^b [x(a+b-s']^2 ds']^{-1/2} = x(a+b-s)[\int_a^b [x(t)]^2 dt]^{-1/2}$$

so that z is normalized. Then by the Paley-Wiener-Zygmund theorem,

$$F_j^*(x) = \int_{C[a,b]} \exp\{i\,[2\rho_j \int_a^b (x_j(t))^2 dt]^{\frac{1}{2}} \int_a^b z(s)\tilde{d}v(s)\}dm_w(v)$$

$$= \frac{1}{\sqrt{2\pi}} \int_{-\infty}^{\infty} \exp\{i\,[2\rho_j \int_a^b (x_j(t))^2 dt]^{\frac{1}{2}} u - \frac{u^2}{2}\}du$$

$$= \exp\{-\rho_j \int_a^b [x_j(t)]^2 dt\} = F_j(x_j) = F_j(\vec{x}) \quad .$$

Thus $F_j \in S$ and so $F \in S$.

According to Theorem 5.1, the analytic Feynman integral of F is given by (5.2) after μ is determined by (5.1). However, in the present case, it is easier to calculate the analytic Feynman integral directly from its definition. We shall do this for the case $\nu = 1$. We therefore assume that

$$F(x) \equiv \exp\{-\rho \int_a^b [x(t)]^2 dt\}\,\psi(x(b)) \; ,$$

where $\Psi(u)$ is bounded and measurable for $u \in R^1$. Then for real $\lambda > 0$, we have

$$\int_{C[a,b]} F(\lambda^{-1/2}x)\,dx = \int_{C[a,b]} \exp\{-\frac{\rho}{\lambda}\int_a^b [x(t)]^2 dt\}\,\psi(\lambda^{-1/2}x(b))\,dx \quad .$$

We integrate the right member above by applying Theorem 1a of [4]. (In applying this theorem, we modify it to bring it into line with modern notation. These modifications replace $C[0,1]$ by $C[a,b]$ and bring the exponentials in the definition of the Wiener integral into line with the standard definition of the Brownian notion process). Setting $\sigma = (\frac{2\rho}{\lambda})^{1/2}$, we have

$$\int_{C[a,b]} F(\lambda^{-1/2}x)\,dx =$$

$$\{\cosh[\sigma(b-a)]\}^{-1/2} \int_{C[a,b]} \psi\{\lambda^{-1/2}x(b) + \sigma\lambda^{-1/2}\int_a^b \operatorname{sech}[\sigma(s-b)]\tanh[\sigma(s-b)]x(s)ds\}$$

$$= \{\cosh[\sigma(b-a)]\}^{-1/2} \int_{C[a,b]} \psi\{\lambda^{-1/2}\int_a^b \operatorname{sech}[\sigma(s-b)]\,dx(s)\}\,dx \quad .$$

To evaluate the last integral, we normalize $\operatorname{sech}[\sigma(s-b)]$ and apply the Paley-Wiener-Zygmund formula and obtain

$$\int_{C[a,b]} F(\lambda^{-1/2}x)\,dx = \{2\pi\cosh[\sigma(b-a)]\}^{-1/2} \int_{-\infty}^{\infty} \psi\{\lambda^{-1/2}[\sigma^{-1}\tanh[\sigma(b-a)]]^{1/2}u\}\exp\{-u^2/2\}\,du$$

$$= \{2\pi\cosh[(b-a)(2\rho/\lambda)^{1/2}]\}^{-1/2} \int_{-\infty}^{\infty} \psi\{[(2\rho\lambda)^{-1/2}\tanh[(b-a)(2\rho/\lambda)^{1/2}]]^{1/2}u\}\exp\{-u^2/2\}d$$

$$= \{2\pi(2\rho\lambda)^{-1/2}\cosh[(b-a)(2\rho/\lambda)^{1/2}]\tanh[(b-a)(2\rho/\lambda)^{1/2}]\}^{-1/2}$$

$$\cdot\int_{-\infty}^{\infty} \psi(v)\exp\{-\frac{1}{2}[(2\rho\lambda)^{-1/2}\tanh[(b-a)(2\rho/\lambda)^{1/2}]]^{-1}v^2\}\,dv \quad .$$

Now $|\arg\coth z| < \pi/4$ when $|\arg z| \leqq \pi/4$, $z \neq 0$, so when $\operatorname{Re}\lambda \geqq 0$ $\lambda \neq 0$, we have $\operatorname{Re} Q(\lambda) > 0$ where $Q(\lambda) \equiv (2\rho\lambda)^{1/2}\coth[(b-a)(2\rho/\lambda)^{1/2}]$. Further we observe that $\operatorname{Re} Q(\lambda)$ has a positive lower bound in the neighborhood of $\lambda = -iq$, q real, $q \neq 0$. Thus the last integral is analytic in λ for $\operatorname{Re}\lambda > 0$ and is continuous for $\operatorname{Re}\lambda \geqq 0$, $\lambda \neq 0$. Hence the analytic Feynman integral of F exists for real $q \neq 0$ and we obtain it by setting $\lambda = -iq$ thus

$$\int_{C[a,b]}^{\text{anf}_q} F(x)\,dx = \{\pi(\tfrac{2i}{\rho q})^{1/2}\cosh[(b-a)(\tfrac{2\rho i}{q})^{1/2}]\tanh[(b-a)(\tfrac{2\rho i}{q})^{1/2}]\}^{-1/2}$$

$$\cdot\int_{-\infty}^{\infty} \psi(v)\exp\{-\tfrac{1}{2}(-2i\rho q)^{1/2}\coth[(b-a)(\tfrac{2\rho i}{q})^{1/2}]v^2\}dv \quad .$$

(It is understood that all radicals are to be given their principal values).

Bibliography

[1] S. Albeverio and R. Hoegh-Krohn, Mathematical theory of Feynman path integrals, Lecture Notes in Mathematics 523, Springer-Verlag Heidelberg-Berlin-New York 1976.

[2] R. H. Cameron and W. T. Martin, An unsymmetric Fubini theorem, Bull. Amer. Math. Soc. 47 (1941), 121-125.

[3] R. H. Cameron and D. A. Storvick, An L_2 analytic Fourier-Feynman transform, Michigan Math. J. 23 (1976), 1-30.

[4] R. H. Cameron and W. T. Martin, Evaluation of various Wiener integrals by use of certain Sturn-Liouville differential equations, Bull. Amer. Math. Soc. 51 (1945), 73-89.

[5] R.E.A.C. Paley, N. Wiener and A. Zygmund, Notes or random functions, Math. Z. 37 (1933), 647-688.

[6] G. Johnson and D. Skoug, A Banach algebra of Feynman integrable functionals with application to an integral equation formally equivalent to Schroedinger's equation, J. Functional anal., 12, No. 2 (1973), 129-152.

School of Mathematics
University of Minnesota
Minneapolis, MN 55455, USA

p-CAPACITY AND CONFORMAL CAPACITY IN INFINITE DIMENSIONAL SPACES

Petru Caraman (Iaşi)

Contents

Introduction

In this paper, we try to show that, in the case of a product measure space or an abstract Wiener space (if the B-norm involved in its definition is obtained by means of a self-adjoint, strictly positive Hilbert-Schmidt operator), the concepts of p-capacity and conformal capacity proposed by us may be considered as a natural extension by a limiting proces with respect to the dimension of the space.

And now, it is a pleasure to me to express my gratitude to my colleague Viorel Barbu for his suggestions concerning Lemma 11, as well as the use of the minimax theorem (quoted as Proposition3) for the proof of Lemma 2.

We remind that the p-capacity of a Borel set $E \subset R^n$ (the Euclidean n-space) is

$$\operatorname{cap}_p E = \inf \int_{R^n} |\nabla u(x)|^p dx,$$

where $\nabla u = (\frac{\partial u}{\partial x^1}, \ldots, \frac{\partial u}{\partial x^n})$ is the gradient of u and the infimum is taken over all real functions $u \in C_o^1$ (i.e. continuously differentiable and with compact support S_u) and such that the restriction $u_{|E} = 1$. In the case of the conformal capacity (i.e. when $p=n$), S_u is supposed to be contained in a fixed ball.

As it is well known, it is possible to obtain an equivalent definition if the infimum in the expression of the p-capacity $(p \leq n)$

is taken with respect to another class of admissible functions, i.e. $u: R^n \longrightarrow [0,1]$, S_u is compact, $u_{|E} = 1$ and u is ACL (absolutely continuous on lines), which means that, for every interval I, u is AC (absolutely continuous) on almost every line segment parallel to the coordinate axes and contained in I, i.e. the orthogonal projection of the segments on which u is not AC is a set of $(n-1)$-dimensional Lebesgue measure zero.

Remark. Since the property of a function u of being Lipschitzian is weaker than $u \in C_0^1$, but stronger than being ACL, we are allowed to conclude that we obtain an equivalent definition of the p-capacity $(p \leq n)$ if u is supposed to be Lipschitzian with a constant $K > 0$, instead of being for instance ACL.

In general, a capacity C is a non-negative, extended real-valued set function, whose domain is a class $\mathcal{N}$ of subsets of a topological space X (usually the algebra of Borel sets of X), which contains the compact sets and is closed under countable unions. Furthermore, C is supposed to satisfy also the following conditions:

(i) $C(\emptyset) = 0$, $\emptyset$ the empty set;

(ii) $E_1, E_2 \in \mathcal{N}$ and $E_1 \subset E_2 \Longrightarrow C(E_1) \leq C(E_2)$ (monotonicity);

(iii) $E_m \in \mathcal{N}$ $(m=1,2,\ldots) \Longrightarrow C(\bigcup_{m=1}^{\infty} E_m) \leq \sum_{m=1}^{\infty} C(E_m)$ (subadditivity).

The p-capacity (and in particular the conformal capacity) are capacities in the above sense. It is easy to see that conditions (i) and (ii) are verified. For (iii), see W.Ziemer [10]. He establishes that the p-capacity of a Borel set E is equal to the p-module $M_p(\Gamma_E)$ of all arcs joining E to the point at infinity of the compactification $\overline{R^n}$ of R^n. Then, if $E = \bigcup_{m=1}^{\infty} E_m$ and Γ_{E_m} is the arc family corresponding to E_m $(m = 1,2\ldots)$, since, according to B.Fuglede [4],

$$M_p(\Gamma_E) \leq \sum_{m=1}^{\infty} M_p(\Gamma_{E_m}),$$

W.Ziemer deduced that

$$\operatorname{cap}_p E = M_p(\Gamma_E) \leq \sum_{m=1}^{\infty} M_p(\Gamma_{E_m}) = \sum_{m=1}^{\infty} \operatorname{cap}_p E_m.$$

We proved property (iii) in the infinite dimensional case in a more direct way.

1. p-capacity in a product space

We begin by reminding a few concepts and results needed in the sequel.

A measure space is a triple $(X, \mathcal{N}, \mu)$, where X is a set, $\mathcal{N}$ is an algebra of subsets of X and $\mu \geq 0$ is a (σ-additive) σ-finite measure on $\mathcal{N}$. A function $f : X \longrightarrow R$ is said to be $\mathcal{N}$-measurable on X if for every open set $D \subset \bar{R}, f^{-1}(D) \in \mathcal{N}$.

Next let us introduce (according to E.Hewitt and K.Stromberg [7], Chap. III, § 12, p.164) the notion of abstract Lebesgue integral.

We recall that a mesurable dissection of X is any finite, pairwise disjoint family $\{E_1, \ldots, E_m\} \subset \mathcal{N}$ such that $\bigcup_{m=1}^{\infty} E_m = X$.

Let $(X, \mathcal{N}, \mu)$ be a mesure space and f any function from X into $[0,\infty]$. Define

$$L(f) = \sup_{\{E_k\}}\{ \sum_{k=1}^{n} \inf_{E_k} f(x)\mu(E_k); \{E_1, \ldots, E_n\} \text{ is a measurable dissection of } X\}.$$

Here $\inf \emptyset = 0$. For an extended real-valued function f, we define $f^+ = \max[f,0], f^- = -\min[f,0], f = f^+ - f^-$. The abstract Lebesgue integral (or simply the integral) of f is $L(f) = L(f^+) - L(f^-)$ provided that at least one of the numbers $L(f^+)$ or $L(f^-)$ is finite. If $L(f^+) = L(f^-) = \infty$, then, we do not define $L(f)$. The functional L is ordinarily written in integral notation

$$L(f) = \int_X f(x)\,d\mu(x) = \int_X f d\mu = \int f d\mu.$$

And then, let $L_1^r(X, \mathcal{N}, \mu)$ be the set of all $\mathcal{N}$-mesurable real-valued functions f defined μ-a.e. (μ-almost everywhere) on X (i.e. everywhere exept in a set of measure μ zero) such that $\int f d\mu$ exists and is finite.

Now, let $X_n, \mathcal{M}_n, \mu_n$ $(n = 1,2,\ldots)$ be a sequence of measurable spaces, where X_n are topological spaces, $\mathcal{M}_n$ is the σ-algebra of the Borel sets of X_n and $\mu_n(X_n) = 1$ $(n = 1,2,\ldots)$. Then, let us consider the measure space $(X, \mathcal{M}, \mu)$, where $X = \prod_{n=1}^{\infty} X_n$ (the Cartesian product), $\mathcal{M}$ is the σ-algebra generated by the sets of the form $E = E_{J_n} \times X_{J'_n}$ with $E_k \in \mathcal{M}_k$, $E_{J_n} = \prod_{k=1}^{n} E_k$, $X_{J'_n} = \prod_{k=n+1}^{\infty} X_k$, $J_n = \{1, \ldots, n\}$, $J'_n = \{n+1, n+2, \ldots\}$, and μ is defined by

PROPOSITION 1. *There is a unique (σ-additive) measure μ on $\mathcal{M}$ such that if $E_{J_n} \in \mathcal{M}_{J_n}$, then*

$$\mu(E_{J_n} \times X_{J'_n}) = \mu_{J_n}(E_{J_n})$$

and if $E_{J_n} = \prod_{k=1}^{n} E_k$, with $E_k \in \mathcal{M}_k$, then

$$\mu_{J_n}(E_{J_n}) = \prod_{k=1}^{n} \mu_k(E_k)$$

{E.Hewitt and K.Stromberg [7], Chap.VI, Lemma (22.5), p.431, Theorem (22.7), p.432 and Theorem (22.8), p.433}.

The connection between the n-dimensional integral and the infinite dimensional one is given by

JESSEN'S THEOREM. *Let* $(X, \mathcal{M}, \mu)$ *be the measure space from above and* $f \in L^r(X, \mathcal{M}, \mu)$; *then*

$$\int_X f(x)\,d\mu(x) = \lim_{n\to\infty} \int_{X_{J_n}} f(x_{J_n}, \tilde{x}_{J_n}{}')\,d\mu_{J_n}(x_{J_n})$$

holds μ-*a.e. in* X, *where* $x_{J_n} = (x_1, \ldots, x_n), \tilde{x}_{J_n}{}' = (\tilde{x}_{n+1}, \tilde{x}_{n+2}, \ldots)$ *and* $d\mu_{J_n}(x_{J_n}) = d\mu_1(x_1)\ldots d\mu_n(x_n)$ {E.Hewitt and K.Stromberg [7], Chap.VI, Theorem (22.22), p.443}.

And now, in order to establish the preceding theorem in more general hypotheses, let us remind some other preliminary results.

LEBESGUE THEOREM. *Let* $(X, \mathcal{N}, \mu)$ *be a measure space and* $\{f_m\}$ *a sequence of nonnegative extended real-valued* $\mathcal{N}$-*measurable functions on* X. *Then*

$$\int \Big(\sum_{m=1}^{\infty} f_m\Big) d\mu = \sum_{m=1}^{\infty} \int f_m d\mu$$

{For the proof, see E.Hewitt an K.Stromberg [7], Chap.III, Theorem (12.21), p.171}.

Let us denote by χ_E the characteristic function of a set $E \subset X$, defined as

$$\chi_E(x) = \begin{cases} 1 \text{ if } x \in E \\ 0 \text{ if } x \in CE \text{ (the complement of } E). \end{cases}$$

COROLLARY. *Let* f *be a nonnegative, extended real-valued* $\mathcal{N}$-*measurable function on* X *and*

$$E = \bigcup_{m=1}^{\infty} E_m, E_p \cap E_q = \emptyset \quad \text{for} \quad p \neq q\ ,\ E_m \in \mathcal{M}_m \ (m = 1, 2, \ldots).$$

Then,

$$\int_E f d\mu = \sum_{m=1}^{\infty} \int_{E_m} f d\mu .$$

Indeed, denoting $f\chi_{E_m} = f_m$, we have

$$\int_E f d\mu = \int_X f\chi_E d\mu = \int_X \sum_{m=1}^{\infty} f\chi_{E_m} d\mu = \int_X \sum_{m=1}^{\infty} f_m d\mu = \sum_{m=1}^{\infty} \int_X f_m d\mu = \sum_{m=1}^{\infty} \int_X f\chi_{E_m} d\mu = \sum_{m=1}^{\infty} \int_{E_m} f d\mu,$$

as desired.

Let $(X,\mathcal{N})$ be a measurable space and let μ,ν be measures on $(X,\mathcal{N})$. We say that ν is AC (absolutely continuous) with respect to μ, if $\mu(E)=0 \Longrightarrow \nu(E)=0 \ \forall E \in \mathcal{N}$ ($\forall$ means "for every") and we write $\nu \ll \mu$.

LEBESGUE-RADON-NIKODYM THEOREM. *Let $(X,\mathcal{N},\mu)$ be a σ-finite measure space and let η be any measure on $(X,\mathcal{N})$ such that $\eta \ll \mu$. Then there is a nonnegative, finite-valued, $\mathcal{N}$-measurable function f_o on X such that*

(1) $\quad \eta(E) = \int_E f_o d\mu$

$\forall E \in \mathcal{N}$. *Moreover, f_o is unique in the sense that if g_o is any nonnegative, extended real-valued, $\mathcal{N}$-measurable function for which (1) holds, then $g_o(x) = f_o(x)$ μ-a.e.* {E.Hewitt and K.Stromberg [7], Chap.VI, Corollary (19.28) of Theorem (19.27), p.320}.

The essentially unique function f_o appearing in the preceding theorem is often called the *Lebesgue-Radon-Nikodym derivative of* η *with respect to* μ, and the notation $d\eta/d\mu$ is used to denote f_o, as well as also the formula $d\eta = f_o d\mu$ or $\eta = f_o\mu$.

FUBINI THEOREM. *Let $(X,\mathcal{M}_\mu,\mu)$ and $(Y,\mathcal{M}_\eta,\eta)$ be σ-finite measures spaces and let $(X\times Y, \mathcal{M}_\mu \times \mathcal{M}_\eta, \mu\otimes\eta)$ be the product measure space, where $\mathcal{M}_\mu \times \mathcal{M}_\eta$ is the σ-algebra of the sets of $X\times Y$ generated by the algebra of the measurable rectangles $E_1 \times E_2 \subset X\times Y$ with $E_1 \in \mathcal{M}_\mu, E_2 \in \mathcal{M}_\eta$, and where the product measure $\mu\otimes\eta$ is defined on $\mathcal{M}_\mu \times \mathcal{M}_\eta$ by $\mu\otimes\eta(E_1\times E_2) = \mu(E_1)\eta(E_2)$. If f is a nonnegative, extended real-valued $\mathcal{M}_\mu\times\mathcal{M}_\eta$-measurable function on $X\times Y$, then the equality*

$$\int_{X\times Y} f(x,y)\,d(\mu\otimes\eta)(x,y) = \int_Y\int_X f(x,y)\,d\mu(x)\,d\eta(y) = \int_X\int_Y f(x,y)\,d\eta(y)\,d\mu(x)$$

holds {E.Hewitt and K.Stromberg [7], Chap.VI, Theorem(21.12).pp.384,385}.

Let $\mathcal{M}_{J_n}$ be the σ-algebra of sets of $X_{J_n} = \prod_{k=1}^{n} X_k$ generated by the measurable rectangles $E_{J_n} = \prod_{k=1}^{n} E_k$, with $E_k \in \mathcal{M}_k (k=1,\dots,n)$ and let $\mathcal{M}_{J'_n}$ be the σ-algebra of sets of $X_{J'_n}$ generated by the sets of the form $E_{J'_n} = E_{J_p \cap J'_n} \times X_{J'_p}$, where $p \geq n$ and $E_{J_p \cap J'_n} = \prod_{k=n+1}^{p} E_k$, with

$E_k \in \mathcal{M}_k (k=n+1,\dots,p)$. With these notations, we have

PROPOSITION 2. Let μ_{J_n} and $\mu_{J'_n}$ be the measures on the σ-algebras $\mathcal{M}_{J_n}$ and $\mathcal{M}_{J'_n}$ from above. Identifying $\mathcal{M}$ and $\mathcal{M}_{J_n} \times \mathcal{M}_{J'_n}$, we have $\mu=\mu_{J_n} \otimes \mu_{J'_n}$ {E.Hewitt and K.Stromberg [7], Chap.VI, Lemma (22.12), p.438}.

$x,y \in X = \prod_{n=1}^{\infty} X_n$ are said to be ultimately equal if there is some $n_o \in N$ such that $x_n = y_n$ $\forall n \geq n_o$, where $x=(x_1,x_2,\dots), y=(y_1y_2,\dots)$.

THE ZERO-ONE LAW. Let $(X,\mathcal{M},\mu)$ be a measure space and $E_o \in \mathcal{M}$ be such that $x \in E_o$ iff all y ultimately equal to x are also in E_o. Then $\mu(E_o)$ is zero or one {E.Hewitt and K.Stromberg [7], Chap.VI Theorem (22.21), p.443}.

LEMMA 1. Let $\{X_n, \mathcal{M}_n, \mu_n\}$ be a sequence of measure spaces and $(X,\mathcal{M},\mu)$ the corresponding measure space from above; then for every extended, real valued, $\mathcal{M}$-measurable function f on X, the relation

$$\int_X |f| d\mu = \lim_{n \to \infty} \int_{X_{J_n}} |f(x_{J_n}, \tilde{x}_{J'_n})| d\mu_{J_n}(x_{J_n}) \tag{2}$$

holds μ-a.e.

In order to establish this generalization of Jessen's theorem, we shall follow E.Hewitt and K.Stromberg's proof (mentioned above) in the more restrictive case $f \in L_1^r(X,\mathcal{M},\mu)$ considered by Jessen, and we shall point out differences.

Let $\mathcal{M}^{(n)}$ be the family of all subsets of X of the form $X_{J_n} \times E_{J'_n}$ with $E_{J'_n} \in \mathcal{M}_{J'_n}$. It is evident that $\mathcal{M} \supset \mathcal{M}^{(1)} \supset \mathcal{M}^{(2)}, \dots$ and each $\mathcal{M}^{(n)}$ is a σ-algebra of sets of X. Any set

$$E_o \in \mathcal{M}_o = \bigcap_{n=1}^{\infty} \mathcal{M}^{(n)}$$

cleary satisfies the hypotheses of the zero-one law and so has measure 0 or 1. Define η on $\mathcal{M}$ by

$$\eta(E) = \int_E |f| d\mu. \tag{3}$$

Clearly, $0 \leq \eta(E) \leq \infty$ and $\eta(\emptyset) = 0$. η is also σ-additive since, if

$$E = \bigcup_{n=1}^{\infty} E_n,$$

with $E_m \cap E_n = 0$ $(m \neq n)$, then, according to the preceding corollary,

$$\eta(E) = \int_E |f|\,d\mu = \sum_{n=1}^{\infty} \int_{E_n} |f|\,d\mu = \sum_{n=1}^{\infty} \eta(E_n),$$

and thus, η is a measure.

Arguing as in Jessen's theorem, we conclude that

$$f_n(\tilde{x}) = f_n(\tilde{x}_{J'_n}) = \int |f(x_{J_n}, \tilde{x}_{J'_n})|\,d\mu_{J_n}(x_{J_n})$$

is $\mathcal{M}^{(n)}$-measurable [of course $f_n(\tilde{x}_{J'_n})$ may be formally written as $f_n(\tilde{x})$]. Now, we intend to show that $f_n(x)$ is a Lebesgue-Radon-Nikodym derivative of $\eta^{(n)}$ with respect to $\mu^{(n)}$ [where $\eta^{(n)}$ and $\mu^{(n)}$ are the restrictions to $\mathcal{M}^{(n)}$ of η and μ, respectively], i.e. to establish that

$$\eta(E) = \int_E f_n(x)\,d\mu(x) \tag{4}$$

for all $E \in \mathcal{M}^{(n)}$. Write

$$E = X_{J_n} \times E_{J'_n}, \quad \text{with } E_{J'_n} \in \mathcal{M}_{J'_n}.$$

Applying the preceding proposition and Fubini theorem, and taking into account that $\chi_{X_{J_n} \times E_{J'_n}}(x_{J_n}, x_{J'_n}) = \chi_{E_{J'_n}}(x_{J'_n})$, we obtain

$$\eta(E) = \int_X \chi_E |f|\,d\mu = \int_{X_{J_n} \times X_{J'_n}} \chi_{X_{J_n} \times E_{J'_n}}(x_{J_n}, x_{J'_n}) |f(x_{J_n}, x_{J'_n})|\,d(\mu_{J_n} \otimes \mu_{J'_n})(x_{J_n}, x_{J'_n}) = \int_{X_{J_n}} \int_{X_{J'_n}} \chi_{X_{J_n} \times E_{J'_n}}(x_{J_n}, x_{J'_n}) |f(x_{J_n}, x_{J'_n})|\,d\mu_{J_n}(x_{J_n})\,d\mu_{J'_n}(x_{J'_n})$$
$$= \int_{X_{J'_n}} \chi_{E_{J'_n}}(x_{J'_n}) \int_{X_{J_n}} |f(x_{J_n}, x_{J'_n})|\,d\mu_{J_n}(x_{J_n})\,d\mu_{J'_n}(x_{J'_n})$$
$$= \int_{X_{J'_n}} \chi_{E_{J'_n}}(x_{J'_n}) f_n(x_{J'_n})\,d\mu_{J'_n}(x_{J'_n}), \tag{5}$$

but on the other side

$$\int_E f_n(x)\,d\mu(x) = \int_X \chi_E(x) f_n(x)\,d\mu(x) \tag{6}$$

$$=\int_{X_{J_n}}\int_{X_{J'_n}} \chi_{X_{J_n}\times E_{J'_n}}(x_{J_n},x_{J'_n})f_n(x_{J'_n})d\mu_{J_n}(x_{J_n})d\mu_{J'_n}(x_{J'_n})$$

$$=\int_{X_{J'_n}} \chi_{E_{J'_n}}(x_{J'_n})f_n(x_{J'_n})d\mu_{J'_n}(x_{J'_n}),$$

hence, and since the last integrals in (5) and (6) coincide, we obtain (4), allowing us to conclude that f_n is a Lebesgue-Radon-Nikodym derivative of η with respect to μ for $\mathcal{M}^{(n)}$ and, arguing as in Jessen's theorem, it follows that the limit

$$\lim_{n\to\infty} f_n(x) = f_o(x)$$

exists μ-a.e. and is a Lebesgue-Radon-Nikodym derivative of η with respect to μ on $\mathcal{M}_o$. Hence, since f_o is $\mathcal{M}$-measurable and μ assumes only values 0 and 1 on $\mathcal{M}_o$, it is easy to see that there is a number $\alpha, 0 \le \alpha \le \infty$, such that $f_o(x) = \alpha$ $\forall x$ in a set of μ-measure 1. Thus, on account of (1) and since f_o is a Lebesgue-Radon-Nikodym derivative of η with respect to μ on $\mathcal{M}_o$, we have $\eta(X)=\int|f|d\mu$ $=\int f_o d\mu=\alpha$, implying (2), as desired.

COROLLARY. *In the hypotheses of the preceding lemma, we have* μ-*a.e.*

$$(7)\qquad [\int|f|^p d\mu]^{\frac{1}{p}} = \lim_n[\int|f(x_{J_n},\tilde{x}_{J'_n})|^p d\mu_{J_n}(x_{J_n})]^{\frac{1}{p}}.$$

PROPOSITION 3. *Let* F *be a compact set in a topological space, let* E *be an arbitrary set and let* Φ *be a real-valued function defined on* $F\times E$ *and continuous on* F $\forall y\in E$. *Then the following statements are equivalent*:

(a) $\forall\beta\in R$ *and* $y_1,\dots,y_n\in E$ *such that* $\beta>\max_{x\in F}\min_{1\le i\le m}\Phi(x,y_i)$, *there is an* $y_o\in E$ *such that* $\beta>\max_{x\in F}\Phi(x,y_o)$;

(b) Φ *satisfies the relation*

$$\max_{x\in F}\inf_{y\in E}\Phi(x,y)=\inf_{y\in E}\max_{x\in F}\Phi(x,y).$$

(See for instance the monograph of V.Barbu and T.Precupanu [3], Chap.2, Theorem 3.4,p.141.)

LEMMA 2. *Let* $\{X_n,\mathcal{M}_n,\mu_n\}$ *be a sequence of measure spaces and*

$(X, \mathcal{M}, \mu)$ *the corresponding measure space from above. If* $\mathcal{B}$ *is a class of* $\mathcal{M}$*-measurable functions, then*

$$(8)\qquad \lim_{n\to\infty}\inf_{\mathcal{B}}\ \operatorname{ess\,inf}_X \int |f(x_{J_n}, x_{J'_n})|\, d\mu_{J_n}(x_{J_n}) = \inf_{\mathcal{B}} \int |f|\, d\mu,$$

where

$$\operatorname{ess\,inf}_X |f(x)| = \sup_E [\inf_{X-E} |f(x)|],$$

with the supremum taken over all sets $E \subset X$ *such that* $\mu(E) = 0$.

Let us show first that

$$(9)\qquad \sup_n \operatorname{ess\,inf}_X \int |f(x_{J_n}, \tilde{x}_{J'_n})|\, d\mu_{J_n}(x_{J_n})$$

$$= \lim_{n\to\infty} \operatorname{ess\,inf}_X \int |f(x_{J_n}, \tilde{x}_{J'_n})|\, d\mu_{J_n}(x_{J_n}) = \int |f|\, d\mu.$$

Indeed, clearly, $\forall n \in N$,

$$\int |f(x_{J_n}, \tilde{x}_{J'_n})|\, d\mu_{J_n}(x_{J_n}) \le \sup_n \int |f(x_{J_n}, \tilde{x}_{J'_n})|\, d\mu_{J_n}(x_{J_n});$$

hence

$$(10)\qquad \lim_{n\to\infty} \int |f(x_{J_n}, \tilde{x}_{J'_n})|\, d\mu_{J_n}(x_{J_n}) \le \sup_n \int |f(x_{J_n}, \tilde{x}_{J'_n})|\, d\mu_{J_n}(x_{J_n})$$

and, on account of the preceding lemma,

$$\int |f|\, d\mu = \lim_{n\to\infty} \int |f(x_{J_n}, \tilde{x}_{J'_n})|\, d\mu_{J_n}(x_{J_n}) \le \sup_n \int |f(x_{J_n}, \tilde{x}_{J'_n})|\, d\mu_{J_n}(x_{J_n})$$

μ-a.e., i.e. in a set CE_f, where $\mu(E_f) = 0$. But then

$$(11)\qquad \int |f|\, d\mu = \lim_{n\to\infty} \int |f(x_{J_n}, \tilde{x}_{J'_n})|\, d\mu_{J_n}(x_{J_n}) \le \inf_{X-E_f} \sup_n \int |f(x_{J_n}, \tilde{x}_{J'_n})|\, d\mu_{J_n}(x_{J_n})$$

$$\le \operatorname{ess\,inf}_X \sup_n \int |f(x_{J_n}, \tilde{x}_{J'_n})|\, d\mu_{J_n}(x_{J_n})$$

in CE_f, whence, since $\int |f|\, d\mu$, $\inf_{X-E} \sup_n \int |f(x_{J_n}, \tilde{x}_{J'_n})|\, d\mu_{J_n}(x_{J_n})$ and

$\operatorname{ess\,inf}_X \sup_n \int |f(x_{J_n}, \tilde{x}_{J'_n})|\, d\mu_{J_n}(x_{J_n})$ are nonnegative numbers,

$$(12)\qquad \alpha_o = \int |f|\, d\mu \le \inf_{X-E_f} \sup_n \int |f(x_{J_n}, \tilde{x}_{J'_n})|\, d\mu_{J_n}(x_{J_n})$$

$$\leqq \operatorname{ess\,inf}_X \sup_n \int |f(x_{J_n}, \tilde{x}_{J'_n})| d\mu_{J_n}(x_{J_n}).$$

Now, by means of the preceding proposition, we shall establish the relation

$$(13)\qquad \sup_n \inf_{X-E_f} \int |f(x_{J_n}, \tilde{x}_{J'_n})| d\mu_{J_n}(x_{J_n}) = \inf_{X-E_f} \sup_n \int |f(x_{J_n}, \tilde{x}_{J'_n})| d\mu_{J_n}(x_{J_n}).$$

Let us show that the hypotheses of the preceding proposition are satisfied. For this, we denote $\alpha_n = 1/n$, $F = \{0\}\{\alpha_n; n=1,2,\ldots\}$ and

$$\Psi(\alpha,\tilde{x}) = \Psi(\alpha,f,\tilde{x}_{J'_n}) = \begin{cases} \int |f(x_{J_n}, \tilde{x}_{J'_n})| d\mu_{J_n}(x_{J_n}) & \text{for } \alpha = \alpha_n, \\ \lim\limits_{n\to\infty} \int |f(x_{J_n}, \tilde{x}_{J'_n})| d\mu_{J_n}(x_{J_n}) & \text{for } \alpha = 0. \end{cases}$$

Clearly, F is compact and $\sup_n = \max_\alpha$. Now, suppose that we have

$$\beta > \max_\alpha \min_{1\leq i\leq m} \int |f(x_{J_n}, x^i_{J'_n})| d\mu_{J_n}(x_{J_n}).$$

Then, there are two possibilities: I. The maximum is obtained for a certain $\alpha = \alpha_{n_o}$ $(n_o < \infty)$, or II. It is obtained for $\alpha = 0$.

I. In the first case, it is easy to see that

$$\beta > \max_\alpha \min_i \int |f(x_{J_{\frac{1}{\alpha}}}, x^i_{J'_{\frac{1}{\alpha}}})| d\mu_{J_{\frac{1}{\alpha}}}(x_{J_{\frac{1}{\alpha}}}) = \min_i \int |f(x_{J_{n_o}}, x^i_{J'_{n_o}})| d\mu_{J_{n_o}}(x_{J_{n_o}})$$

$$= \int |f(x_{J_{n_o}}, x^{i_o}_{J'_{n_o}})| d\mu_{J_{n_o}}(X_{J_{n_o}}) = \max_\alpha \int |f(x_{J_{\frac{1}{\alpha}}}, x^{i_o}_{J'_{\frac{1}{\alpha}}})| d\mu_{J_{\frac{1}{\alpha}}}(x_{J_{\frac{1}{\alpha}}}),$$

where $x^{i_o}_{J'_{n_o}}$ is the point of $X^{i_o}_{J'_{n_o}}$ at which the minimum with respect to i is attained, and thus, the hypotheses of the preceding proposition are satisfied in this case.

II. In the second case, there is a sequence $\{n_{k'}\}, n_k \in N$ (N being the set of positive integers) such that

$$\beta > \max_\alpha \min_{1\leq i\leq m} \int |f(x_{J_{\frac{1}{\alpha}}}, x^i_{J_{\frac{1}{\alpha}}})| d\mu_{J_{\frac{1}{\alpha}}}(x_{J_{\frac{1}{\alpha}}}) = \varlimsup_{n\to\infty} \min_i \int |f(x_{J_n}, x^i_{J'_n})| d\mu_{J_n}(x_{J_n})$$

$$= \lim_k \min_i \int |f(x_{J_{n_k}}, x^i_{J'_{n_k}})| d\mu_{J_{n_k}}(x_{J_{n_k}}).$$

Let us denote by $x^{n_k}_{J'_{n_k}}$ the point of $X^{n_k}_{J'_{n_k}}$ at which

$$\min_i \int |f(x_{J_{n_k}}, x^i_{J'_{n_k}})| d\mu_{J_{n_k}}(x_{J_{n_k}}) = \int |f(x_{J_{n_k}}, x^{n_k}_{J'_{n_k}})| d\mu_{J_{n_k}}(x_{J_{n_k}}) \quad (k=1,2,\ldots).$$

The sequence of these points is of the form $\{x^{n_k}_{J'_{n_k}}\} = \{x^1_{J'_{1_k}}\} \cup \ldots \cup \{x^m_{J'_{n_k}}\}$,

where at least one of the m subsequences, let us say $\{x^i_{J'_{i_k}}\}$, is infinite. But, since the limit

$$\lim_{k\to\infty} \int |f(x_{J_{n_k}}, x^{n_k}_{J'_{n_k}})| d\mu_{J_{n_k}}(x_{J_{n_k}})$$

$$= \lim_{k\to\infty} \min_i \int |f(x_{J_{n_k}}, x^i_{J'_{n_k}})| \, d\mu_{J_{n_k}}(x_{J_{n_k}})$$

exists, then, any of its subsequences has the same limit, so that we are allowed to conclude that, for instance,

$$\beta > \max_\alpha \min_i \int |f(x_{J_{\frac{1}{\alpha}}}, x^i_{J'_{\frac{1}{\alpha}}})| d\mu_{J_{\frac{1}{\alpha}}}(x_{J_{\frac{1}{\alpha}}}) = \lim_{k\to\infty} \int |f(x_{J_{n_k}}, x^{n_k}_{J'_{n_k}})| d\mu_{J_{n_k}}(x_{J_{n_k}})$$

$$= \lim_{k\to\infty} \int |f(x_{J_{i_k}}, x^i_{J'_{i_k}})| d\mu_{J_{i_k}}(x_{J_{i_k}}) = \lim_{n\to\infty} \int | f(x_{J_n}, x^i_{J'_n})| d\mu_{J_n}(x_{J_n})$$

$$= \max_\alpha \int |f(x_{J_{\frac{1}{\alpha}}}, x^i_{J'_{\frac{1}{\alpha}}})| d\mu_{J_{\frac{1}{\alpha}}}(x_{J_{\frac{1}{\alpha}}}).$$

Thus, the conditions of the preceding lemma are verified also in this case, so that in each of them, we deduce relation (13), which, on account of (12), yields

$$(14) \qquad \alpha_0 \leq \sup_n \operatorname{ess} \inf_X \int |f(x_{J_n}, \tilde{x}_{J'_n})| d\mu_{J'_n}) \, d\mu_{J_n}(x_{J_n}).$$

Now, in order to prove the opposite inequality, we observe that

$$\inf_{X-E_0} \int |f(x_{J_n}, \tilde{x}_{J'_n})| d\mu_{J_n}(x_{J_n}) \leq \int | f(x_{J_n}, \tilde{x}_{J'_n})| d\mu_{J_n}(x_{J_n})$$

$$= \int |f(x_{J_n}, \tilde{x}_{n+1}, \tilde{x}_{J'n+1})| d\mu_{J_n}(x_{J_n})$$

in CE_0; hence, integrating with respect to $\tilde{x}_{n+1}$ and taking into account

$\mu_n(X_n) = 1$, we obtain

$$\inf_{X-E_o} \int |f(x_{J_n}, \tilde{x}_{J'_n})| d\mu_{J_n}(x_{J_n}) \le \int |f(x_{J_{n+1}}, \tilde{x}_{J'_{n+1}})| d\mu_{J_{n+1}}(x_{J_{n+1}}),$$

in CE_o, whence

$$\inf_{X-E_o} \int |f(x_{J_n}, \tilde{x}_{J'_n})| d\mu_{J_n}(x_{J_n}) \le \inf_{X-E_o} \int |f(x_{J_{n+1}}, \tilde{x}_{J'_{n+1}})| d\mu_{J_{n+1}}(x_{J_{n+1}}),$$

for instance, $\forall E_o \subset X$ with $\mu(E_o) = 0$, implying

$$\operatorname{ess\,inf}_X \int |f(x_{J_n}, \tilde{x}_{J'_n})| d\mu_{J_n}(x_{J_n}) \le \operatorname{ess\,inf}_X \int |f(x_{J_{n+1}}, \tilde{x}_{J'_{n+1}})| d\mu_{J_{n+1}}(x_{J_{n+1}})$$

$\forall n \in N$, i.e. $\operatorname{ess\,inf}_X \int |f(x_{J_n}, \tilde{x}_{J'_n})| d\mu_{J_n}(x_{J_n})$ is a non-decreasing sequence, so that

$$(15) \quad \sup_n \operatorname{ess\,inf}_X \int |f(x_{J_n}, \tilde{x}_{J'_n})| d\mu_{J_n}(x_{J_n})$$

$$= \lim_{n\to\infty} \operatorname{ess\,inf}_X \int |f(x_{J_n}, \tilde{x}_{J'_n})| d\mu_{J_n}(x_{J_n})$$

and, on account of (11) and (14), we obtain

$$\lim_{n\to\infty} \operatorname{ess\,inf}_X \int |f(x_{J_n}, \tilde{x}_{J'_n})| d\mu_{J_n}(x_{J_n}) \le \lim_{n\to\infty} \int |f(x_{J_n}, \tilde{x}_{J'_n})| d\mu_{J_n}(x_{J_n})$$

$$= \int |f| d\mu \le \sup_n \operatorname{ess\,inf}_X \int |f(x_{J_n}, \tilde{x}_{J'_n})| d\mu_{J_n}(x_{J_n})$$

$$= \lim_{n\to\infty} \operatorname{ess\,inf}_X \int |f(x_{J_n}, \tilde{x}_{J'_n})| d\mu_{J_n}(x_{J_n})$$

in CE_f, from which it follows that

$$\lim_{n\to\infty} \operatorname{ess\,inf}_X \int |f(x_{J_n}, \tilde{x}_{J'_n})| d\mu_{J_n}(x_{J_n}) = \lim_{n\to\infty} \int |f(x_{J_n}, \tilde{x}_{J'_n})| d\mu_{J_n}(x_{J_n})$$

$$= \sup_n \operatorname{ess\,inf}_X \int |f(x_{J_n}, \tilde{x}_{J'_n})| d\mu_{J_n}(x_{J_n})$$

in CE_f and then (9), as desired.

But, we are able to get even more. Indeed, arguing as above for (14) and (15) and taking into account that, if a sequence has a limit, any of its infinite subsequence has the same limit, we derive that

$$\alpha_o = \lim_{n\to\infty} \operatorname{ess\,inf}_X \int |f(x_{J_n}, \tilde{x}_{J'_n})| d\mu_{J_n}(x_{J_n}) = \lim_{k\to\infty} \operatorname{ess\,inf}_X \int |f(x_{J_{n_k}}, \tilde{x}_{J'_{n_k}})|$$

$$\times d\mu_{J_{n_k}}(x_{J_{n_k}}) \le \sup_k \operatorname{ess\,inf}_X \int |f(x_{J_{n_k}},\tilde{x}_{J'_{n_k}})|\, d\mu_{J_{n_k}}(x_{J_{n_k}})$$

$$\le \sup_n \operatorname{ess\,inf}_X \int |f(x_{J_n},\tilde{x}_{J'_n})|\, d\mu_{J_n}(x_{J_n})$$

$$= \lim_{n\to\infty} \operatorname{ess\,inf}_X \int |f(x_{J_n},\tilde{x}_{J'_n})|\, d\mu_{J_n}(x_{J_n}),$$

hence

$$\sup_k \operatorname{ess\,inf}_X \int |f(x_{J_{n_k}},\tilde{x}_{J'_{n_k}})|\, d\mu_{J_{n_k}}(x_{J_{n_k}})$$

$$= \lim_{k\to\infty} \operatorname{ess\,inf}_X \int |f(x_{J_{n_k}},\tilde{x}_{J'_{n_k}})|\, d\mu_{J_{n_k}}(x_{J_{n_k}}).$$

Next, from (9), we get also that

(16) $$\inf_{\mathcal{E}} \lim_{n\to\infty} \operatorname{ess\,inf}_X \int |f(x_{J_n},\tilde{x}_{J'_n})|\, d\mu_{J_n}(x_{J_n}) = \inf_{\mathcal{E}} \int |f|\, d\mu.$$

Now, we shall prove, by means of the preceding proposition, that

(17) $$\inf_{\mathcal{E}} \sup_k \operatorname{ess\,inf}_X \int |f(x_{J_{n_k}},\tilde{x}_{J'_{n_k}})|\, d\mu_{J_{n_k}}(x_{J_{n_k}})$$

$$= \sup_k \inf_{\mathcal{E}} \operatorname{ess\,inf}_X \int |f(x_{J_{n_k}},\tilde{x}_{J'_{n_k}})|\, d\mu_{J_{n_k}}(x_{J_{n_k}}).$$

For simplicity sake, we shall esatblish this relation in the more particular case in which, instead of the subsequence $\{n_k\}$, we consider the sequence of positive integers $n=1,2,\ldots$. Let us show first that we are in the hypotheses of the preceding proposition. Let us denote $\alpha_n = \frac{1}{n}$ and let us consider the compact set $F=\{0\}\cup\{\alpha_n, n=1,2,\ldots\}$. And now, let us introduce the function

$$\Phi(\alpha,f) = \begin{cases} \operatorname{ess\,inf}_X \int |f(x_{J_n},\tilde{x}_{J'_n})|\, d\mu_{J_n}(x_{J_n}) & \text{for } \alpha_n=\alpha_n, \\ \int |f|\, d\mu & \text{for } \alpha=0. \end{cases}$$

And then, let us indentify $\mathcal{E}$ with E and let us consider a number $\beta>0$ and $f_1,\ldots,f_m \in \mathcal{E}$ so that

$$\beta > \max_{\alpha\in F} \min_{1\le i\le m} \Phi(\alpha,f_i).$$

Let us prove now that there exists a function $f_o \in \mathcal{E}$ with the property that

$$\beta > \max_{\alpha\in F} \Phi(\alpha,f_o).$$

From (9), we deduce that

$$\Phi(\alpha,f_i) \le \Phi(0,f_i) = \max_{\alpha\in F} \Phi(\alpha,f_i) = \lim_{n\to\infty} \Phi(\alpha_n,f_i);$$

hence

$$\max_{\alpha\in F} \min_{1\le i\le m} \Phi(\alpha,f_i) = \lim_{n\to\infty} \min_{1\le i\le m} \Phi(\alpha_n,f_i) = \min_{1\le i\le m} \Phi(0,f_i).$$

Indeed, let us suppose that the maximum is attained for $\alpha = \alpha_p$, i.e.

$$\max_{\alpha\in F} \min_{1\le i\le m} \Phi(\alpha,f_i) = \min_{1\le i\le m} \Phi(\alpha_p,f_i) = \Phi(\alpha_p,f_{i_o}),$$

but

$$\Phi(\alpha_p,f_i) \le \Phi(0,f_i) \quad (i=1,\ldots,m);$$

hence

$$\Phi(\alpha_p,f_{i_o}) \le \min_{1\le i\le m} \Phi(0,f_i),$$

so that the maximum is attained for $\alpha = 0$.

And now, let us define f_n by means of the relation

$$\min_{1\le i\le m} \Phi(\alpha_n,f_i) = \Phi(\alpha_n,f_n).$$

Clearly,

$$\{\Phi(\alpha_n,f_n)\} = \{\Phi(\alpha_{1_k},f_1)\}\cup\ldots\cup\{\Phi(\alpha_{m_k},f_m),$$

where $\{\alpha_n\} = \{\alpha_{1_k}\}\cup\ldots\cup\{\alpha_{m_k}\}$, in other words, $\{\Phi(\alpha_{i_k},f_i)\}$ for instance contains those elements of $\{\Phi(\alpha_n,f_n)\}$ for which $n=i_k \Longrightarrow f_n = f_i$. Since $\{\Phi(\alpha_n,f_n)\}$ is an infinite sequence, it follows that at least one of the m subsequences, let us say $\{\Phi(\alpha_{i_k},f_i)\}$ will be infinite and the existence of the limit for $\{\Phi(\alpha_n,f_n)\}$ will imply the existence of the limit for any of its infinite subsequences and the equality of all these limits, so that

$$\beta > \max_{\alpha\in F} \min_{1\le i\le m} \Phi(\alpha,f_i) = \lim_{n\to\infty} \Phi(\alpha_n,f_n) = \lim_{k\to\infty} \Phi(\alpha_{i_k},f_i) = \max_{\alpha\in F} \Phi(\alpha,f_i),$$

where the last equality is a consequence of the existence of the limit of the sequence $\{\Phi(\alpha_n,f_n)\}$; thus, codition (a) of the preceding proposition is verified for $f_o = f_i$, and then relation (17) is true. But, this implies

(18) $$\lim_{n\to\infty} \inf_{\mathcal{E}} \operatorname{ess\,inf}_{X} \int |f(x_{J_n}, \tilde{x}_{J'_n})| d\mu_{J_n}(x_{J_n}) = \inf_{\mathcal{E}} \int |f| d\mu = C_0,$$

since if not, it would be posible to find a sequence $\{n_k\}$ such that

$$|C_0 - \inf_{\mathcal{E}} \operatorname{ess\,inf}_{X} \int |f(x_{J_{n_k}}, \tilde{x}_{J'_{n_k}})| d\mu_{J_{n_k}}(x_{J_{n_k}})| \geq \varepsilon,$$

contradicting (16), which, on account of (17), implies for every subsequence $\{n_k\}$,

$$\sup_{k} \inf_{\mathcal{E}} \operatorname{ess\,inf}_{X} \int |f(x_{J_{n_k}}, \tilde{x}_{J'_{n_k}})| d\mu_{J_{n_k}}(x_{J_{n_k}})$$

$$= \inf_{\mathcal{E}} \sup_{k} \operatorname{ess\,inf}_{X} \int |f(x_{J_{n_k}}, \tilde{x}_{J'_{n_k}})| d\mu_{J_{n_k}}(x_{J_{n_k}}) = \inf_{\mathcal{E}} \int |f| d\mu = C_0.$$

This contradiction establishes (18), and thus our lemma is completely proved.

COROLLARY 1. *In the hypotheses of the preceding lemma,*

$$\lim_{n\to\infty} \inf_{\mathcal{E}} \operatorname{ess\,inf}_{X} \int |f(x_{J_n}, \tilde{x}_{J'_n})|^p d\mu_{J_n}(x_{J_n}) = \inf_{\mathcal{E}} \int |f|^p d\mu \quad (p > 0).$$

COROLLARY 2. *In the hypotheses of the preceding lemma,*

$$\lim_{n\to\infty} \inf_{\mathcal{E}} \operatorname{ess\,inf}_{X} \left[\int |f(x_{J_n}, \tilde{x}_{J'_n})|^p d\mu_{J_n}(x_{J_n})\right]^{\frac{1}{p}} = \inf_{\mathcal{E}} \left(\int |f|^p d\mu\right)^{\frac{1}{p}} \quad (p > 0).$$

Now, let us consider the class of exeptional sets introduced by N.Aronszajn [1].

1° Let B be a real seperable Banach space, $\mathcal{M}_B$ the σ-algebra of Borel sets of B (with respect to the B-norm) and, for $0 \neq a \in B$, let $\mathcal{A}(a) = \{E \in \mathcal{M}_B; \forall x \in B, m_1[E \cap (x+Ra)] = 0\}$, where m_1 is Lebesgue measure on the line $x + Ra$.

2° For every sequence $\{a_m\} \subset B$ with $a_m \neq 0$,
$\mathcal{A}\{a_m\} = \{E \in \mathcal{M}_B; E = \bigcup_m E_m,\ E_m \in \mathcal{A}(a_m)\}$.

3° $\mathcal{A} \equiv \mathcal{A}_B = \cap \mathcal{A}\{a_m\}$, the intersection being taken over all sequences $\{a_m\}$ complete in B.

We remind that a sequence $\{a_m\}$ B is *complete* iff $\overline{[\{a_m\}]} = B$, where $[E]$ denotes the linear span of E, i.e. the intersection of all the subpaces of B containing E.

PROPOSITION 4. *The class $\mathcal{A}$ of exeptional sets satisfies the following conditions:*

a) $\mathcal{A}$ <u>is σ-additive</u>;

b) $\mathcal{A}$ <u>is hereditary</u>, i.e. $E_1 \subset E, E \in \mathcal{A} \Longrightarrow E_1 \in \mathcal{A}$;

c) $\mathcal{A}$ <u>does not contain any open subset of</u> B (N.Aronszajn [1], pp.151,154).

If S is a finite dimensional subspace of B, let us consider the class of exeptional sets : $\mathcal{A}(S) = \{E \in \mathcal{M}_B ; \forall x \in B, m_{J_n}[E \cap (x+S)] = 0\}$, where m_{J_n} is Lebesgue n-dimensional measure.

<u>PROPOSITION 5</u>. $S = [\{a_n\}] \Longrightarrow \mathcal{A}(S) = \mathcal{A}\{a_n\}$ (N.Aronszajn [1], Proposition 1, p.151).

Next, let us consider the Gâteaux differential of a function $u : B \longrightarrow R$ at a point $x \in B$ and in the direction e_s

$$(19) \qquad Du(x;e_s) = \lim_{\rho \to 0} \frac{u(x+\rho e_s) - u(x)}{\rho} = \frac{\partial u(x)}{\partial s}.$$

In particular, if $\{e_n\}$ is an orthonormal basis and $e_s = e_n$, then

$$Du(x;e_n) = \frac{\partial u(x)}{\partial x^n}.$$

This remark justifies the notation $Du(x;e_s) = \nabla u(x)(e_s)$; hence

$$\nabla u(X)(e_n) = \frac{\partial u(x)}{\partial x^n} \quad (n=1,2,\ldots).$$

We have also $Du(x;e_s) = \langle e_s, \nabla u(x) \rangle$, where this time $\langle x,y \rangle$ means $y(x)$ with $x \in B$ and $y \in B^*$ (the dual of B), and ∇u is the gradient of u (see for instance V.Barbu and T.Precupanu [3], Chap.2, p.93).

We remind that Du(x) is called by N.Aronszajn <u>a differential</u> if it is linear and the convergence in (19) is uniform in e_s on each compact. If the convergence is uniform on each bounded set of B, then Du(x) is the <u>Fréchet differential</u> of u. It is easy to see that the condition of Fréchet differentability is more restrictive than the differentability in Aronszajn's sense.

As a particular case of 2 results of N.Aronszajn, we have:

<u>PROPOSITION 6</u>. <u>If</u> $u:G \longrightarrow R$ (G <u>open in</u> B) <u>is Lipschitzian (with respect to the</u> B-<u>norm</u>), <u>then the set where</u> Du(x) <u>does not exists is in</u> $\mathcal{A}$ (N.Aronszajn [1], Corollary 1 of Lemma 1,p.165).

<u>PROPOSITION 7</u>. <u>If</u> B <u>is a real separable Banach space</u>, G <u>is open in</u> B <u>and</u> $u:B \longrightarrow R$ <u>is Lipschitzian</u>, <u>then</u> Du(x) <u>is a differential except in sets of</u> $\mathcal{A}$ (N.Aronszajn [1], Theorem 1,p.166).

We recall that if $\mathcal{M}_B$ is the σ-algebra of Borel sets of B, then a Borel measure is any measure defined in $(B, \mathcal{M}_B)$.

A σ-finite measure μ on B is said to be AC (absolutely continuous) relatively to $\mathcal{A}$ if every set in $\mathcal{A}$ is of μ-measure zero.

Gauss measure μ_1 in the real axis R is the set function from the class of Borel sets of R into $[0,\infty)$ defined as

$$\mu_1(E) = \frac{1}{\sqrt{2\pi}} \int_E e^{-\frac{x^2}{2}} dm_1(x),$$

where $E \subset R$.

Let us introduce now the following new class of exeptional sets:

1'° For $0 \neq a \in B$, let $\mathcal{G}(a) = \{E \in \mathcal{M}_B; \forall x \in B, \mu_1[E \cap (x+Ra)] = 0\}$.

2'° $\forall \{a_m\} \subset B$ with $a_m \neq 0$, $\mathcal{G}\{a_m\} = \{E \in \mathcal{M}_B; E = \bigcup_m E_m, E_m \in \mathcal{G}(a_m)\}$.

3'° $\mathcal{G} \equiv \mathcal{G}_B = \cap \mathcal{G}\{a_m\}$, the intersection being taken over all sequences $\{a_m\}$ complete in B.

LEMMA 3. $\mathcal{G}(a) = \mathcal{A}(a)$, $\mathcal{G}\{a_m\} = \mathcal{A}\{a_m\}$ and $\mathcal{G} = \mathcal{A}$.

Clearly, it is enough to show that $\mathcal{G}(a) = \mathcal{A}(a)$ and for this, it is sufficient to prove that Gauss and Lebesgue measures are equivalent on the real axis (i.e. each of them is AC with respect to the other). But, the expression of Gauss measure from above yields that $m_1(E) = 0 \Longrightarrow \mu_1(E) = 0$ (i.e. $\mu_1 << m_1$). In order to see the converse, we observe (again from the above expression of Gauss measure) that

$$d\mu_1(x) = \frac{1}{\sqrt{2\pi}} e^{-\frac{|x|^2}{2}} dm_1(x),$$

hence

$$dm_1(x) = \sqrt{2\pi}\, e^{\frac{|x|^2}{2}} d\mu_1(x)$$

and then

$$m_1(E) = \sqrt{2\pi} \int_E e^{\frac{|x|^2}{2}} d\mu_1(x),$$

allowing us to conclude that also $\mu_1(E) = 0 \Longrightarrow m_1(E) = 0$ (i.e. $m_1 << \mu_1$), that is $\mu_1 \sim m_1$ (m_1 is equivalent to μ_1). Thus Aronszajn classes of exeptional sets coincide with the classes defined above by the conditions 1'°, 2'°, 3'°. We shall use further only Aronszajn's notation.

Let us consider now the sequence of measure spaces $\{X_n, \mathcal{M}_n, \mu_n\}$, where $X_n = R$ and μ_n is Gauss measure on R, and then, let $(X, \mathcal{M}, \mu)$ be the measure space from above , where

$$X = \prod_{n=1}^{\infty} X_n$$

and μ is the corresponding measure given by Proposition 1. Next, let $\{e_n\}$ be an orthonormal basis on X such that

$$\forall x \in X, x = (x_1, x_2, \ldots) = \sum_{n=1}^{\infty} x_n e_n.$$

Now, let us denote by $H \subset X$ the set characterized by

$$H = \{x \in X;\ \sum_{n=1}^{\infty} x_n^2 < \infty\}.$$

For the elements $x, y \in H$, we define the scalar product $<x,y> = x_1y_1 + x_2y_2 + \ldots$ and the norm $|x| = \sqrt{<x,x>}$. It is easy to see that H provided with the topology of the norm is a Hilbert space.

LEMMA 4. <u>Let</u> $(X, \mathcal{M}, \mu)$ <u>be the measure space from above</u>, $H \subset X$ <u>the corresponding Hilbert space and</u> $\mathcal{M}_H$ <u>the</u> σ-<u>algebra of Borel sets of</u> H <u>(with respect to the</u> H-<u>norm); then</u> $\mathcal{M}_H \subset \mathcal{M}$.

Let E_{J_n} be a Borel set of $X_{J_n} = R^n$ and let us consider the cylinders

$$E^H = \{x \in H; (<x,e_1>, \ldots, <x,e_n>) \in E_{J_n}\},\ E^X = \{x \in X; (x_1, \ldots, x_n) \in E_{J_n}\}$$

$$= E_{J_n} \times X_{J'_n}$$

Let us denote by R_H and R_X the rings of the cylinders E^H and E^X, respectively. Then $\mathcal{M}_H$ and $\mathcal{M}$ will be the σ-algebras generated by R_H and R_X, respectively (see A.Balakrishnan [2]). But then, in order to prove the theorem, it is enough to show that the closed unit ball $\bar{B} = \{x \in H; |x| \le 1\} \in \mathcal{M}$. Let us prove that if $Z_n = \{x \in X;\ \sum_{k=1}^{n} x_k^2 \le 1\} \in \mathcal{M}$, then $\bar{B} = \cap_n Z_n \in \mathcal{M}$. Clearly, $\bar{B} \subset \cap_n Z_n$. In order to obtain the converse, suppose, to prove it is false, that there exists a point $x \in \cap_n Z_n$ with $|x| = \sum_{k=1}^{\infty} x_k^2 > 1$. But then, there is an $n \in N$ sufficiently large such that $\sum_{k=1}^{n} x_k^2 > 1$, implying that $x \notin Z_n$, this contradiction allowing us to conclude that $\bar{B} = \cap_n Z_n \in \mathcal{M}$, and arguing as in Balakrishnan ([2],p.122) also that $\mathcal{M}_H \subset \mathcal{M}$.

LEMMA 5. <u>Let</u> $(X, \mathcal{M}, \mu)$ <u>be the measure space from above and</u> H <u>the corresponding Hilbert space, then</u> μ <u>is</u> AC <u>relatively to</u> $\mathcal{A}\{e_n\}$, <u>where</u> $\mathcal{A}\{e_n\} = \{E \in \mathcal{M}_H;\ E = \bigcup_{n=1}^{\infty} E_n, E_n \in \mathcal{A}(e_n)\}$.

Suppose $E \in \mathcal{A}\{e_n\}$. Then, clearly, $\chi_E(x) \le \sum_{n=1}^{\infty} \chi_{E_n}(x)$. Hence, on account of Lebesgue and Fubini theorems, as well as of the preceding lemma, which implies that $E \in \mathcal{M}$, we deduce that

$$(20)\quad \begin{aligned} \mu(E) &= \int_X \chi_E d\mu = \int_H \chi_E\, d\mu \le \int_H \sum_{n=1}^{\infty} \chi_{E_n} d\mu = \sum_{n=1}^{\infty} \int_H \chi_{E_n} d\mu \\ &= \sum_{n=1}^{\infty} \int_{H_{n'}} \int_{X_n} \chi_E(x_n, x_{n'}) d\mu_n(x_n) d\mu_{n'}(x_{n'}), \end{aligned}$$

where

$$x_{n'} = (x_1, \dots, x_{n-1}, x_{n+1}, x_{n+2}, \dots) \quad \text{and} \quad H_{n'} = \{x_{n'} \prod_{k \ne n} X_k;\ \sum_{k \ne n} x_k^2 < \infty\}.$$

But, since $E_n \in \mathcal{A}(e_n)$, then $\forall x_{n'} \in H_{n'}$,

$$\int_{X_n} \chi_{E_n}(x_n, x_{n'}) d\mu_n(x_n) = \mu_n[(x_{n'} + Re_n)] \cap E_n = 0,$$

so that (20) implies

$$\mu(E) = \sum_{n=1}^{\infty} \int_{H_{n'}} \int_{X_n} \chi_{E_n}(x_n, x_{n'}) d\mu_n(x_n) d\mu_{n'}(x_{n'}) = 0,$$

as desired.

The Gauss measure in H is the set function μ_H from R_H into $[0,\infty)$ defined as follows: If $E = E^H \in R_H$, then

$$(21)\quad \mu_H(E) = \frac{1}{(\sqrt{2\pi})^n} \int_{E_{J_n}} e^{-\frac{|x_{J_n}|^2}{2}} dm_{J_n}(x_{J_n}),$$

where $E_{J_n} \subset R^n$ and m_J is n-dimensional Lebesgue measure.

COROLLARY 1. Let $(X, \mathcal{M}, \mu)$ be the measure space from above (with $X_n = R$ and μ_n the Gauss measure on X_n), $H \subset X$ the corresponding Hilbert space, $u : X \longrightarrow R$ a Lipschitzian function with respect to the H-norm and the support $S_u \subset H$. Then $Du(x)$ exists μ-a.e.

COROLLARY 2. Let $(X, \mathcal{M}, \mu)$ be the measure space from above, H the corresponding Hilbert space and $u : X \longrightarrow R$ Lipschitzian (with respect to the Hilbert norm), and the support $S_u \subset H$. Then u is differentiable (in the sense of Aronszajn) μ-a.e.

These two corollaries are direct consequences of the preceding two propositions, on account of the preceding lemma, since $\mathcal{A} \subset \mathcal{A}\{e_n\}$ and then, if μ is AC with respect to $\mathcal{A}\{e_n\}$, it is a fortiori AC with respect to $\mathcal{A}$ allowing us to deduce that the conclusions of the preceding two propositions hold μ-a.e.

LEMMA 6. Let B be real separable Banach space, e_s a unit vector of direction $s, u : B \longrightarrow R_+$ (R_+ the nonnegative semi-axis) a function

Gâteaux differentiable at x and ∇u the gradient of u, then

$$||\nabla u(x)|| = \sup_s \left|\frac{\partial u(x)}{\partial s}\right|,$$

where $||\cdot||$ is the B-norm.

Indeed, from

$$Du(x,e_s) = \frac{\partial u(x)}{\partial s} = \langle e_s, \nabla u(x)\rangle$$

(see, for instance V. Barbu and T. Precupanu [3], p.93) and since $|\langle x,x\rangle| \le ||x|| \cdot ||y||$, we deduce that

$$\left|\frac{\partial u(x)}{\partial s}\right| = |\langle e_s, \nabla u(x)\rangle| \le ||e_s|| \cdot ||\nabla u(x)|| = ||\nabla u(x)||$$

and the equality is reached for $e_s = \nabla u(x)/||\nabla u(x)||$.

Let $\{X_n, \mathcal{M}_n, \mu_n\}$ be a sequence of measure spaces, $(X, \mathcal{M}, \mu)$ the corresponding measure space from above and $E \subset B(R)$ an $\mathcal{M}$-measurable set. Let us denote by $\mathcal{U}$ the class of admissible functions $u : X \longrightarrow [0,1]$ for E characterized by the following properties: u is Lipschitzian with Lipschitz constant $K > 0$, the support $S_u \subset B(R)$ and $u_{|E} = 1$.

LEMMA 7. Let $(X, \mathcal{M}, \mu)$ be the measure space from above, $E \subset B(R)$ an $\mathcal{M}$-measurable set and $\mathcal{U}$ the class of admissible functions for E'; then

$$\inf_{\mathcal{U}} [\int |\nabla u(x)|^p d\mu(x)]^{\frac{1}{p}} = \lim_{n\to\infty} \inf_{\mathcal{U}} \operatorname{ess}_X \inf [\int |\nabla u(x_{J_n}, \tilde{x}_{J'_n})|^p d\mu_{J_n}(x_{J_n})]^{\frac{1}{p}}.$$

This is a consequence of Lemma 2. We have only to prove that $|\nabla u|$ exists μ-a.e. and is $\mathcal{M}$-measurable. But, since S_u is bounded, it follows that $S_u \subset H$ and we are in the hypotheses of the preceding corollary, so that u is differentiable (in the sense of Aronszajn) μ-a.e. and, on account of the preceding lemma, from the existence μ-a.e. of $Du(x)$, we conclude the existence μ-a.e. of $|\nabla u(x)| = \sup_s \left|\frac{\partial u(x)}{\partial s}\right| = \sup_s |Du(x;e_s)|$.

In order to prove that $|\nabla u|$ is $\mathcal{M}$-measurable, we observe that since u is $\mathcal{M}_H$-measurable, then, on account of Lemma 4, u is $\mathcal{M}$-measurable; hence, also the differential quotient $(1/\rho)[u(x+\rho e_s)-u(x)]$ will be, but if x is a point of differentiability (in the sense of Aronszajn) of u, the existence of the limit of the differential quotient for $\rho \longrightarrow 0$ yields a fortiori the existence of the limit for a sequence $\{\rho_m\}$ with $\rho_m \longrightarrow 0$ for $m \longrightarrow \infty$, implying the $\mathcal{M}$-measurability of

$Du(x;e_s)$ and then also of $|\nabla u(x)| = \sup_s |Du(x;e_s)|$, as desired.

Now, taking into account the preceding lemma, let us give the following characterization of the p-capacity in a measure space $(X,\mathcal{M},\mu)$.

The p-capacity of an $\mathcal{M}$-measurable set $E \subset B(R)$ of a measure space $(X,\mathcal{M},\mu)$, corresponding to a sequence of measure spaces $\{X_n,\mathcal{M}_n,\mu_n\}$, with $X_n = R$ and μ_n the Gauss measure, is given by

$$(22)\qquad \operatorname{cap}_p E = \inf_{\mathcal{U}} (\int |\nabla u|^p d\mu)^{\frac{1}{p}}.$$

In order to justify the preceding characterization of the p-capacity, let us consider a Borel set $E_{J_n} \subset B(R)$, let $\mathcal{U}_n$ be the corresponding family of admissible functions for E_{J_n} in R^n and $\operatorname{cap}_p^{(n)}$ the p-capacity in R^n. Then, it is easy to see that

$$\operatorname{ess}_X\inf\, [|\nabla u(x_{J_n},\tilde{x}_{J_n'})|^p d\mu_{J_n}(x_{J_n})]^{\frac{1}{p}} = \operatorname{ess}_X\inf[\int |\nabla u(x_{J_n})|^p d\mu_{J_n}(x_{J_n})]^{\frac{1}{p}}$$
$$= [\int |\nabla u(x_{J_n})|^p d\mu_{J_n}(x_{J_n})]^{\frac{1}{p}}.$$

Thus,

$$\inf_{\mathcal{U}_n} \operatorname{ess}_X\inf[\int |\nabla u(x_{J_n},\tilde{x}_{J_n'})|^p d\mu_{J_n}(x_{J_n})]^{\frac{1}{p}} = \inf_{\mathcal{U}_n}[\int |\nabla u(x_{J_n})|^p d\mu_{J_n}(x_{J_n})]^{\frac{1}{p}}$$
$$= [\operatorname{cap}_p^{(n)} E_{J_n}]^{\frac{1}{p}},$$

so that the concept of p-capacity proposed by us is obtained by a limiting process from the p-capacity in R^n. But also the converse is true, i.e. given a Borel set $E_{J_n} \subset B(R)$ and considering the class $\mathcal{U}_n$ of admissible functions for E_{J_n} (defined in R^n), the p-capacity proposed by us reduces to the p-capacity of $R^n = X_{J_n}$, since, on account Fubini's theorem,

$$\inf_{\mathcal{U}_n}(\int |\nabla u|^p d\mu)^{\frac{1}{p}} = \inf_{\mathcal{U}_n}[\int_{X_{J_n'}} \int_{R^n} |\nabla u(x_{J_n})|^p d\mu_{J_n}(x_{J_n}) d\mu_{J_n'}(x_{J_n'})]^{\frac{1}{p}}$$
$$= \inf_{\mathcal{U}_n}[\int |\nabla u(x_{J_n})|^p d\mu_{J_n}(x_{J_n})]^{\frac{1}{p}} = (\operatorname{cap}_p^{(n)} E_n)^{\frac{1}{p}}$$

Now, in order to establish that the p-capacity defined above is a capacity, i.e. satisfies conditions (i), (ii), (iii) given in the introduction, let us prove some preliminary results.

MINKOWSKI INEQUALITY. For $1 \leq p < \infty$ and $f,g \in L_p$, we have

$$\|f+g\|_p \leq \|f\|_p + \|g\|_p. \tag{23}$$

[For the proof, see E.Hewitt and K.Stromberg [7], Chap.IV, Theorem (13.7), p.191].

COROLLARY 1. For $1 \leq p < \infty$ and f,g μ-measurable, the inequality (23) still holds.

If $\|f\|_p = \infty$ or $\|g\|_p = \infty$, then (23) is trivial, while if $f,g \in L_p$, we are in the hypotheses of the preceding theorem and again (23) is true.

COROLLARY 2. For $1 \leq p < \infty$ and f_n μ-mesurable $(n = 1,\dots,k)$,

$$\|\sum_{n=1}^{k} f_n\|_p \leq \sum_{n=1}^{k} \|f_n\|_p.$$

COROLLARY 3. For $1 \leq p < \infty$ and f_n μ-mesurable $(n = 1,2,\dots)$,

$$\|\sum_{n=1}^{\infty} f_n\|_p \leq \sum_{n=1}^{\infty} \|f_n\|_p$$

Indeed, on account of the preceding corollary,

$$\|\sum_{n=1}^{k} f_n\|_p \leq \sum_{n=1}^{k} \|f_n\|_p \leq \sum_{n=1}^{\infty} \|f_n\|_p$$

and since the inequality is satisfied for any $k = 1,2,\dots$, we deduce that

$$\|\sum_{n=1}^{\infty} f_n\|_p = \lim_{k\to\infty} \|\sum_{n=1}^{k} f_n\|_p \leq \sum_{n=1}^{\infty} \|f_n\|_p.$$

THEOREM 1. The p-capacity cap_p proposed by us for the space $(X, \mathcal{M}, \mu)$ from above is a capacity, i.e. is a nonnegative extended real-valued set function defined on $\mathcal{M}$ and satisfying the conditions (i),(ii),(iii) from the introduction.

Condition (i) is trivial since $u \equiv 0$ is an admissible function for $E = \emptyset$. It is easy to see that also (ii) holds since if $E_1 \subset E_2$ and $\mathcal{U}_1, \mathcal{U}_2$ are the two corresponding classes of admissible functions, then $\mathcal{U}_2 \subset \mathcal{U}_1$; hence

$$\operatorname{cap}_p E_1 = \inf_{\mathcal{U}_1} (\int |\nabla u|^p d\mu)^{\frac{1}{p}} \leq \inf_{\mathcal{U}_2} (\int |\nabla u|^p d\mu)^{\frac{1}{p}} = \operatorname{cap}_p E_2.$$

In order to establish (iii), let us consider $E \subset \bigcup_{m=1}^{\infty} E_m$, where E, E_m are $\mathcal{M}$-measurable sets $(m = 1,2,\ldots)$, let $\mathcal{U}_m$ be the classes of admissible functions corresponding to $E_m (m = 1,2,\ldots)$ and let $u(x) = \sup\{u_1(x), u_2(x), \ldots\}$, where $u_m \in \mathcal{U}_m (m=1,2,\ldots)$. Now, let us prove that u is Lipschitzian too with the same constant $K > 0$. Indeed, given 2 points $x, y \in X$, we may assume, without loss of generality, that $u(x) > u(y)$. Then, from the definition of u, given $0 < \varepsilon < u(x) - u(y)$, there is an $n \in N$ such that $u_n(x) > u(x) - \varepsilon$; hence

$$|u(x)-u(y)| < |u_n(x)-u(y)+\varepsilon| \leq |u_n(x)-u_n(y)|+\varepsilon \leq K|x-y|+\varepsilon$$

and letting $\varepsilon \longrightarrow 0$, we obtain

$$|u(x)-u(y)| \leq K|x-y|.$$

Now, it is easy to see that $u \in \mathcal{U}$, i.e. it is admissible for $\operatorname{cap}_p E$ because $0 \leq u(x) \leq 1$, as it follows from the definition, u is Lipschitzian with the constant K, as it was proved above, and clearly, $S_u \subset B(R)$, $u|_E = 1$.

Next, let us show that

$$(24) \qquad |\nabla u(x)| \leq \sum_{m=1}^{\infty} |\nabla u_m(x)|.$$

Let us consider a unit vector e_s of direction s and suppose first that

$$(25) \qquad u(x) > u(x+|\Delta x|e_s).$$

Then, given $0 < \varepsilon|\Delta x| < u(x)-u(x+|\Delta x|e_s)$, there is an $m \in N$ such that

$$\frac{|u(x)-u(x+|\Delta x|e_s)|}{|\Delta x|} < \frac{|u_m(x)-u(x+|\Delta x|e_s)+\varepsilon|\Delta x||}{|\Delta x|} \leq \frac{|u_m(x)-u_m(x+|\Delta x|e_s)|}{|\Delta x|} + \varepsilon$$

$$\leq \sum_{m=1}^{\infty} \frac{|u_m(x)-u_m(x+|\Delta x|e_s)|}{|\Delta x|} + \varepsilon;$$

hence, letting $\varepsilon \longrightarrow 0$,

$$\frac{|u(x)-u(x+|\Delta x|e_s)|}{|\Delta x|} \leq \sum_{m=1}^{\infty} \frac{|u_m(x)-u_m(x+|\Delta x|e_s)|}{|\Delta x|}.$$

But, since u and $u_m (m = 1,2,\ldots)$ are Lipschtzian with the Lipschitz constant K in X, they are so also in the axis X_s passing through x

and having the direction s, so that u and u_m (considered as functions of a real variable), have a directional derivative $(\partial/\partial s)u(x)$, $(\partial/\partial s)u_m(x)$ a.e. in X_s. Suppose that the point x from above is such a point where $(\partial/\partial s)u(x)$ and $(\partial/\partial s)u_m(x)$ exists (according to Corollary 2 of Lemma 5, this happens μ-a.e.). Then, letting $|\Delta x| \to 0$ in the preceding inequality, we obtain

$$\left|\frac{\partial u(x)}{\partial s}\right| = \lim_{|\Delta x|\to 0}\frac{|u(x)-u(x+|\Delta x|e_s)|}{|\Delta x|} \leq \overline{\lim_{|\Delta x|\to 0}} \sum_{m=1}^{\infty}\frac{|u_m(x)-u_m(x+|\Delta x|e_s)|}{|\Delta x|}$$

$$\leq \sum_{m=1}^{\infty} \overline{\lim_{|\Delta x|\to 0}} \frac{|u_m(x)-u_m(x+|\Delta x|e_s)|}{|\Delta x|} = \sum_{m=1}^{\infty}\left|\frac{\partial u_m(x)}{\partial s}\right|;$$

hence, on account of Lemma 6, taking the supremum with respect to s,

$$|\nabla u(x)| = \sup_s\left|\frac{\partial u(x)}{\partial s}\right| \leq \sup_s \sum_{m=1}^{\infty}\left|\frac{\partial u_m(x)}{\partial s}\right| \leq \sum_{m=1}^{\infty}\sup_s\left|\frac{\partial u_m(x)}{\partial s}\right| = \sum_{m=1}^{\infty}|\nabla u_m(x)|.$$

Thus, we established (24) in the hypothesis (25).

Now, suppose that the converse inequality holds, i.e. that $u(x) < u(x+|\Delta x|e_s)$. Then, given $0<\varepsilon<u(x+|\Delta x|e_s)-u(x)$, there is an $m \in N$ such that

$$\frac{|u(x)-u(x+|\Delta x|e_s)|}{|\Delta x|} < \frac{|u(x)-u_m(x+|\Delta x|e_s)-\varepsilon|\Delta x||}{|\Delta x|} \leq \frac{|u_m(x)-u_m(x+|\Delta x|e_s)|}{|\Delta x|} + \varepsilon$$

$$\leq \sum_{m=1}^{\infty}\frac{|u_m(x)-u_m(x+|\Delta x|e_s)|}{|\Delta x|} + \varepsilon$$

and arguing as above, in the hypothesis (25), we obtain (24) also in this case.

Finally, from (24) and Corollary 3 of Minkowski inequality, we conclude that

$$\operatorname{cap}_p E \leq \left(\int|\nabla u|^p d\mu\right)^{\frac{1}{p}} \leq \left\{\int\left[\sum_{m=1}^{\infty}|\nabla u_m(x)|\right]^p d\mu\right\}^{\frac{1}{p}} \leq \sum_{m=1}^{\infty}\left(\int|\nabla u_m|^p d\mu\right)^{\frac{1}{p}};$$

hence, since $\forall m = 1,2,\ldots$, u_m was an arbitrary function of $\mathcal{U}_m$,

$$\operatorname{cap}_p E \leq \sum_{m=1}^{\infty}\inf_{\mathcal{U}_m}\left(\int|\nabla u_m|^p d\mu\right)^{\frac{1}{p}} = \sum_{m=1}^{\infty}\operatorname{cap}_p E_m,$$

as desired.

2. Conformal capacity in a product space

Now, in order to generalize also the conformal capacity in the measure space $(X,\mathcal{M},\mu)$ from above, let us recall first some concepts and establish a few preliminary results.

An $\mathcal{M}$-measurable function f is said to be <u>essentialy bounded</u> if there exists a constant $\alpha \geq 0$ such that $|f(x)| \leq \alpha$ μ-a.e.. The infimum of all these α is called the <u>essential supremum</u> of $|f(x)|$ and is denoted

$$\mathrm{ess}_X \sup |f(x)| \quad \text{or} \quad \mathrm{vrai}_X \max |f(x)|.$$

The space $L^\infty(X,\mathcal{M},\mu)$ or simply $L^\infty(X)$ is the set of all $\mathcal{M}$-mesurable essentially bounded functions given μ-a.e. in X. In this space one uses the norm $\|f\|_\infty = \mathrm{ess}_X \sup |f(x)|$.

PROPOSITION 8. *If* $(X,\mathcal{M},\mu)$ *is a measure space*, $\mu(X) < \infty$ *and* $f \in L^\infty(X)$, then

$$(26) \qquad \lim_{p \to \infty} \left(\int |f|^p d\mu\right)^{\frac{1}{p}} = \mathrm{ess}_X \sup |f(x)|.$$

(For the proof, see for instance K.Yosida [9],Chap.3, Theorem 1, p.34).

The preceding proposition is true also in the following more general conditions:

LEMMA 8. *If* $(X,\mathcal{M},\mu)$ *is a mesure space*, $\mu(X) < \infty$ *and* f *is* $\mathcal{M}$*-measurable, then* (26) *still holds*.

If

$$\mathrm{ess}_X \sup |f(x)| < \infty, \text{i.e.} f \in L^\infty(X),$$

(26) is satisfied on account of the preceding proposition. Thus, we have to consider only the case

$$\mathrm{ess}_X \sup |f(x)| = \infty.$$

But, by the definition of the ess sup, it follows that $\forall M < \infty$, there exists a set E with $\mu(E) > 0$ such that $|f(x)| > M$ $\forall x \in E$, but then

$$\left(\int |f|^p d\mu\right)^{\frac{1}{p}} \geq \left(\int_E |f|^p d\mu\right)^{\frac{1}{p}} > M\mu(E)^{\frac{1}{p}};$$

hence, letting $p \to \infty$,

$$\lim_{p \to \infty} \left(\int |f|^p d\mu\right)^{\frac{1}{p}} \geq M$$

and, since M may be chosen as large as one wishes , we get also

$$\lim_{p\to\infty} (\int |f|^p d\mu)^{\frac{1}{p}} = \infty,$$

as desired.

LEMMA 9. *If* $\{X_n, \mathcal{M}_n, \mu_n\}$ *is a sequence of measure spaces and* $(X, \mathcal{M}, \mu)$ *is the corresponding measure space from above, then, for every* $\mathcal{M}$*-measurable function* f, *we have*

$$(27)\qquad \lim_{p\to\infty}\lim_{n\to\infty} [\int |f(x_{J_n}, \tilde{x}_{J'_n})|^p d\mu_{J_n}(x_{J_n})]^{\frac{1}{p}}$$

$$= \lim_{n\to\infty}\lim_{p\to\infty} [\int |f(x_{J_n}, \tilde{x}_{J'_n})|^p d\mu_{J_n}(x_{J_n})]^{\frac{1}{p}} = \mathrm{ess}_X\sup|f(x)|$$

μ*-a.e.*

First, the preceding lemma and corollary of Lemma 1 yield

$$(28)\qquad \mathrm{ess}_X\sup|f(x)| = \lim_{p\to\infty} (\int |f|^p d\mu)^{\frac{1}{p}} = \lim_{p\to\infty}\lim_{n\to\infty} [\int |f(x_{J_n}, \tilde{x}_{J'_n})|^p d\mu_{J_n}(x_{J_n})]^{\frac{1}{p}}$$

μ-a.e. Indeed, $\forall p \in N$, relation (7) holds in a set CE_p with $\mu(E_p) = 0$, so that it holds simultaneously for all p in a set CE_o, where $E_o = \bigcup_p E_p$ and $\mu(E_o) = 0$, allowing us to conclude that (28) holds μ-a.e.

On the other side, clearly,

$$(29)\qquad \lim_{n\to\infty}\lim_{p\to\infty} [\int |f(x_{J_n}, \tilde{x}_{J'_n})|^p d\mu_{J_n}(x_{J_n})]^{\frac{1}{p}} \le \mathrm{ess}_X\sup|f(x)| \lim_{n\to\infty}\lim_{p\to\infty} [\mu_{J_n}(X_{J_n})]^{\frac{1}{p}}$$

$$= \mathrm{ess}_X\sup|f(x)|.$$

Now, in order to prove the opposite inequality, suppose for the moment that $\mathrm{ess}_X\sup|f(x)| < \infty$. Then, on account of the preceding proposition, $\forall\varepsilon > 0$, there is a $p_o \in N$ such that $(\int |f|^p d\mu)^{\frac{1}{p}} > \mathrm{ess}_X\sup|f| - \varepsilon, \forall p \ge p_o$ and by corollary of Lemma 1, there exists an $n_p \in N$ such that, $\forall n > n_p$,

$$\lim_{p\to\infty} [\int |f(x_{J_n}, \tilde{x}_{J'_n})|^p d\mu_{J_n}(x_{J_n})]^{\frac{1}{p}} = \mathrm{ess}_{X_{J_n}}\sup|f(x_{J_n}, \tilde{x}_{J'_n})|$$

$$\ge [\int |f(x_{J_n}, \tilde{x}_{J_n})|^p d\mu_{J_n}(x_{J_n})]^{\frac{1}{p}} > \mathrm{ess}_X\sup|f(x)| - \varepsilon$$

μ-a.e., i.e. this inequality holds in a set CE'_p with $\mu(E'_p) = 0$; hence

$$\lim_{n\to\infty}\lim_{p\to\infty}[\int|f(x_{J_n},\tilde{x}_{J'_n})|^p d\mu_{J_n}(n_{J_n})]^{\frac{1}{p}} \geq \mathrm{ess}_X\sup|f(x)| - \varepsilon$$

in CE'_p. Let us denote $E'_o = \bigcup_p E'_p$. Clearly, $\mu(E'_o) = 0$ and the preceding inequality holds in $CE'_o \forall\ \varepsilon > 0$. Then, letting $\varepsilon \to 0$, we obtain

$$\lim_{n\to\infty}\lim_{p\to\infty}[\int|f(x_{J_n},\tilde{x}_{J'_n})|^p d\mu_{J_n}(x_{J_n})]^{\frac{1}{p}} \geq \mathrm{ess}_X\sup|f(x)|$$

in E'_o (i.e. μ-a.e.) and this inequality, together with (29), implies that

$$\lim_{n\to\infty}\lim_{p\to\infty}[\int|f(x_{J_n},\tilde{x}_{J'_n})|^p d\mu_{J_n})]^{\frac{1}{p}} = \mathrm{ess}_X\sup|f(x)|$$

μ-a.e., whence, on account of (28), we obtain (27) in the hypothesis $\mathrm{ess}_X\sup f(x)| < \infty$.

Next, let us assume that $\mathrm{ess}_X\sup|f(x)| = \infty$. Then, $\forall M < \infty$, there exists a $p_o \in N$ such that

$$(\int|f|^p d\mu)^{\frac{1}{p}} > M \quad \text{for} \quad p \geq p_o$$

and, on account of the corollary of Lemma 1, $\forall p \geq p_o$, there is an n_p so that, $\forall n \geq n_p$,

$$\lim_{p\to\infty}[\int|f(x_{J_n},\tilde{x}_{J'_n})|^p d\mu_{J_n}(x_{J_n})]^{\frac{1}{p}} = \mathrm{ess}_{X_{J_n}}\sup|f(x_{J_n},\tilde{x}_{J'_n})|$$

$$\geq [\int|f(x_{J_n},\tilde{x}_{J'_n})|^p d\mu_{J_n}(x_{J_n})]^{\frac{1}{p}} > M$$

μ-a.e., i.e. exept for $x \in CE''_p$ with $\mu(E''_p) = 0$; hence

$$\lim_{n\to\infty}\lim_{p\to\infty}[\int|f(x_{J_n},\tilde{x}_{J'_n})|^p d\mu_{J_n}(x_{J_n})]^{\frac{1}{p}} > M$$

in CE''_p. If $E''_o = \bigcup_p E''_p$, then the preceding inequality holds in CE''_o for M as large as one wishes, whence

$$\lim_{n\to\infty}\lim_{p\to\infty}[\int|f(x_{J_n},\tilde{x}_{J'_n})|^p d\mu_{J_n}(x_{J_n})]^{\frac{1}{p}} = \infty = \mathrm{ess}_X\sup|f(x)|$$

in CE''_o, i.e. μ-a.e., which, together with (28) gives (27) also in this case.

COROLLARY 1. *In the hypotheses of the preceding lemma,*

$$\lim_{n,p\to\infty} [\int |f(x_{J_n},\tilde{x}_{J'_n})|^p d\mu_{J_n}(x_{J_n})]^{\frac{1}{p}} = \text{ess}_X\sup|f(x)|$$

μ-a.e.

COROLLARY 2. *In the hypotheses of the preceding lemma,*

$$\lim_{n\to\infty} [\int|f(x_{J_n},\tilde{x}_{J'_n})|^n d\mu_{J_n}(x_{J_n})]^{\frac{1}{n}} = \text{ess}_X\sup|f(x)|$$

μ-a.e.

COROLLARY 3. *In the hypotheses of the preceding lemma,*

$$(30)\qquad \lim_{n\to\infty} \text{ess}_X\sup[\int|f(x_{J_n},\tilde{x}_{J'_n})|^n d\mu_{J_n}(x_{J_n})]^{\frac{1}{n}} = \text{ess}_X\sup|f(x)|.$$

Indeed, on account of the preceding corollary, we have

$$\lim_{n\to\infty} [\int|f(x_{J_n},\tilde{x}_{J'_n})|^n d\mu_{J_n}(x_{J_n})]^{\frac{1}{n}} \le \lim_{n\to\infty} \text{ess}_X\sup[\int|f(x_{J_n},\tilde{x}_{J'_n})|^n d\mu_{J_n}(x_{J_n})]^{\frac{1}{n}}$$

$$\le \text{ess}_X\sup|f(x)| = \lim_{n\to\infty} [\int|f(x_{J_n},\tilde{x}_{J'_n})|^n d\mu_{J_n}(x_{J_n})]^{\frac{1}{n}}$$

μ-a.e.; hence (30) holds, as desidered.

LEMMA 10. *In the hypotheses of the preceding corollary, if* $\mathcal{C}$ *is a class of* $\mathcal{M}$*-measurable functions, then*

$$\lim_{n\to\infty} \inf_{\mathcal{C}} \text{ess}_X\sup[\int|f(x_{J_n},\tilde{x}_{J'_n})|^n d\mu_{J_n}(x_{J_n})]^{\frac{1}{n}} = \inf_{\mathcal{C}} \text{ess}_X\sup|f(x)|.$$

The argument is similar to that of Lemma 2, so that we shall point out only the part involving some differences in the proof. Thus, in this case, the relation

$$(31)\qquad \lim_{n\to\infty} [\int|f(x_{J_n},\tilde{x}_{J'_n})|^n d\mu_{J_n}(x_{J_n})]^{\frac{1}{n}} = \sup_n[\int|f(x_{J_n},\tilde{x}_{J'_n})|^n d\mu_{J_n}(x_{J_n})]^{\frac{1}{n}}$$

may be established μ-a.e. in a more simple way. Indeed, clearly, arguing as for the inequality (10), we obtain

$$(32)\qquad \lim_{n\to\infty} [\int|f(x_{J_n},\tilde{x}_{J'_n})|^n d\mu_{J_n}(x_{J_n})]^{\frac{1}{n}} \le \sup_n[\int|f(x_{J_n},\tilde{x}_{J'_n})|^n d\mu_{J_n}(x_{J_n})]^{\frac{1}{n}}.$$

In order to prove also the opposite innequality, we observe that, on account of Corollary 2 of the preceding lemma,

(33) $$[\int|f(x_{J_n},\tilde{x}_{J'_n})|^n d\mu_{J_n}(x_{J_n})]^{\frac{1}{n}} \le \mathrm{ess}_X\sup|f(x)|\mu_{J_n}(X_{J_n})^{\frac{1}{n}}$$

$$=\mathrm{ess}_X\sup|f(x)| = \lim_{n\to\infty}[\int|f(x_{J_n},\tilde{x}_{J'_n})|^n d\mu_{J_n}(x_{J_n})]^{\frac{1}{n}}$$

μ-a.e.; hence, taking the supremum,

$$\sup_n[\int|f(x_{J_n},\tilde{x}_{J'_n})|^n d\mu_{J_n}(x_{J_n})]^{\frac{1}{n}} \le \lim_{n\to\infty}[\int|f(x_{J_n},\tilde{x}_{J'_n})|^n d\mu_{J_n}(x_{J_n})]^{\frac{1}{n}}$$

μ-a.e., which, together with (32), yields (31), as desired. But also in this case, we may obtain even more, i.e.

(34) $$\lim_k[\int|f(x_{J_{n_k}},\tilde{x}_{J_{n_k}})|^{n_k} d\mu_{J_{n_k}}(x_{J_{n_k}})]^{\frac{1}{n_k}}$$

$$=\sup_k[\int|f(x_{J_{n_k}},x_{J_{n_k}})|^{n_k} d\mu_{J_{n_k}}(x_{J_{n_k}})]^{\frac{1}{n_k}}$$

μ-a.e. Indeed, arguing as for (32), we have

(35) $$\lim_k[\int|f(x_{J_{n_k}},\tilde{x}_{J_{n_k}})|^{n_k} d\mu_{J_{n_k}}(x_{J_{n_k}})]^{\frac{1}{n_k}}$$

$$\le\sup_k[\int|f(x_{J_{n_k}},\tilde{x}_{J_{n_k}})|^{n_k} d\mu_{J_{n_k}}(x_{J_{n_k}})]^{\frac{1}{n_k}}$$

and analogously as for (33), and taking into account that if a sequence has a limit, then any of its subsequences has the same limit, we get

$$[\int|f(x_{J_{n_k}},\tilde{x}_{J'_{n_k}})|^{n_k} d\mu_{J_{n_k}}(x_{J_{n_k}})]^{\frac{1}{n_k}} \le \mathrm{ess}_X\sup|f(x)|[\mu_{J_{n_k}}(X_{J_{n_k}})]^{\frac{1}{n_k}}$$

$$=\mathrm{ess}_X\sup|f(x)| = \lim_{n\to\infty}[\int|f(x_{J_n},\tilde{x}_{J'_n})|^n d\mu_{J_n}(x_{J_n})]^{\frac{1}{n}}$$

$$=\lim_k[\int|f(x_{J_{n_k}},\tilde{x}_{J'_{n_k}})|^{n_k} d\mu_{J_{n_k}}(x_{J_{n_k}})]^{\frac{1}{n_k}}$$

μ-a.e., hence, taking the supremum with respect to k and on account of (35),

$$\lim_k [\int |f(x_{J_{n_k}}, \tilde{x}_{J'_{n_k}})|^{n_k} d\mu_{J_{n_k}}(x_{J_{n_k}})]^{\frac{1}{n_k}}$$

$$\le \sup_k [\int |f(x_{J_{n_k}}, \tilde{x}_{J'_{n_k}})|^{n_k} d\mu_{J_{n_k}}(x_{J_{n_k}})]^{\frac{1}{n_k}}$$

$$\le \lim_k [\int |f(x_{J_{n_k}}, \tilde{x}_{J'_{n_k}})|^{n_k} d\mu_{J_{n_k}}(x_{J_{n_k}})]^{\frac{1}{n_k}}$$

μ-a.e., yielding (34), as desired.

And now, from the preceding two corollaries, we deduce

$$\text{ess}_X\sup|f(x)| = \lim_n \text{ess}_X\sup[\int |f(x_{J_n}, \tilde{x}_{J'_n})|^n d\mu_{J_n}(x_{J_n})]^{\frac{1}{n}}$$

$$= \lim_k \text{ess}_X\sup[\int |f(x_{J_{n_k}}, \tilde{x}_{J'_{n_k}})|^{n_k} d\mu_{J_{n_k}}(x_{J_{n_k}})]^{\frac{1}{n_k}}$$

$$= \lim_n [\int |f(x_{J_n}, \tilde{x}_{J'_n})|^n d\mu_{J_n}(x_{J_n})]^{\frac{1}{n}}$$

$$= \lim_k [\int |f(x_{J_{n_k}}, \tilde{x}_{J'_{n_k}})|^{n_k} d\mu_{J_{n_k}}(x_{J_{n_k}})]^{\frac{1}{n_k}}$$

$$= \sup_k [\int |f(x_{J_{n_k}}, \tilde{x}_{J'_{n_k}})|^{n_k} d\mu_{J_{n_k}}(x_{J_{n_k}})]^{\frac{1}{n_k}}$$

$$\le \sup_X \text{ess} \sup_k [\int |f(x_{J_{n_k}}, \tilde{x}_{J'_{n_k}})|^{n_k} d\mu_{J_{n_k}}(x_{J_{n_k}})]^{\frac{1}{n_k}} \le \text{ess}_X\sup|f(x)|$$

μ-a.e., whence

$$\lim_k \text{ess}_X\sup[\int |f(x_{J_{n_k}}, \tilde{x}_{J'_{n_k}})|^{n_k} d\mu_{J_{n_k}}(x_{J_{n_k}})]^{\frac{1}{n_k}}$$

$$= \sup_k \text{ess}_X\sup[\int |f(x_{J_{n_k}}, \tilde{x}_{J'_{n_k}})|^{n_k} d\mu_{J_{n_k}}(x_{J_{n_k}})]^{\frac{1}{n_k}}$$

everywhere since the two parts of this relation are constans.

Next, from the preceding corollary, it follows also that

$$\inf_{\mathscr{C}} \lim_{n} \operatorname{ess}_X\sup[\int|f(x_{J_n},\tilde{x}_{J'_n})|^n d\mu_{J_n}(x_{J_n})]^{\frac{1}{n}} = \inf_{\mathscr{C}} \operatorname{ess}_X\sup|f(x)|.$$

The rest of the proof follows the argument of Lemma 2; we have only to change $\operatorname{ess}_X\inf$ by $\operatorname{ess}_X\sup$ and $(\int|f|^p d\mu)^{\frac{1}{p}}$ by $\operatorname{ess}_X\sup|f(x)|$.

Arguing as in Lemma 7, we have the following

<u>COROLLARY</u>. <u>Let</u> $\{X_n, \mathcal{M}_n, \mu_n\}$ <u>be a sequence of measure spaces with</u> $X_n = R$ <u>and</u> μ_n <u>the Gauss measure</u>, $(X, \mathcal{M}, \mu)$ <u>the corresponding measure space from above</u>, $E \subset B(R)$ <u>an</u> $\mathcal{M}$<u>-measurable set and</u> $\mathcal{U}$ <u>the class of admissible functions for</u> E; <u>then</u>

$$\lim_{n\to\infty} \inf_{\mathcal{U}} \operatorname{ess}_X\sup[\int|\nabla u(x_{J_n},\tilde{x}_{J'_n})|^n d\mu_{J_n}(x_{J_n})]^{\frac{1}{n}} = \inf_{\mathcal{U}} \operatorname{ess}_X\sup|\nabla u(x)|.$$

The following definition of the <u>conformal capacity of an</u> $\mathcal{M}$<u>-measurable set</u> $E \subset B(R)$ <u>of the measure space</u> $(X, \mathcal{M}, \mu)$ from above may be deduced from the conformal capacity in R^n:

$$\operatorname{cap}E = \inf_{\mathcal{U}} \operatorname{ess}_X\sup|\nabla u(x)|, \tag{36}$$

where $\mathcal{U}$ is the class of admissible functions for E.

In order to justify this definition, arguing as in the case of the p-capacity, let $E_{J_n} \subset B(R)$ be a Borel set of R^n, $\operatorname{cap}^{(n)}$ the conformal capacity corresponding to R^n and $\mathcal{U}_n$ the class of admissible functions for E_{J_n} and defined in R^n; then

$$\inf_{\mathcal{U}_n} \operatorname{ess}_X\sup[\int|\nabla u(x_{J_n},\tilde{x}_{J'_n})|^n d\mu_{J_n}(x_{J_n})]^{\frac{1}{n}} = [\inf_{\mathcal{U}_n}\int|\nabla u(x_{J_n})|^n d\mu_{J_n}(x_{J_n})]^{\frac{1}{n}} = [\operatorname{cap}^{(n)}E_{J_n}]^{\frac{1}{n}},$$

so that the conformal capacity proposed by us may be obtained by a limiting process with respect to the dimension n from the power $\frac{1}{n}$ of the conformal capacity in R^n. The fact that the conformal capacity in $(X, \mathcal{M}, \mu)$ does not derive exactly from the conformal capacity in R^n does not matter so much since, in the different theoretical problems, it is not so important to have a precise value of the conformal capacity of a set as it is to be able to make a distinction between the class of exeptional sets (i.e. with conformal capacity zero) and the class of all the other sets, or a set remains in the same class if instead of $\operatorname{cap}^{(n)}$ we use $[\operatorname{cap}^{(n)}]^{1/n}$. But even more, the definition of the quasiconformal mappings given by F.Gehring by means of the rings may be obtained using $[\mathrm{cap}^{(n)}]^{1/n-1}$.

It is true that if $E_{J_n} \subset B(R)$ and if $\mathcal{U}$ is the class of admissible functions for E_{J_n} defined in R^n, then the conformal capacity in $(X, \mathcal{M}, \mu)$ i.e. $\inf_{\mathcal{U}} \mathrm{ess}_X \sup |\nabla u(x_{J_n})|$ does not come to the conformal capacity in R^n, but it is natural to expect such a thing because in any case the conformal capacity in R^{n+1} for $E_{J_n} \subset R^n$ does not coincide with the conformal capacity in R^n even if we consider the class $\mathcal{U}_n$ of admissible functions defined in R^n instead of the class $\mathcal{U}_{n+1}$ of admissible functions defined in R^{n+1}, since

$$\inf_{\mathcal{U}_n} \int |\nabla u(x_{J_{n+1}})|^{n+1} d\mu_{J_{n+1}}(x_{J_{n+1}})$$

$$= \inf_{\mathcal{U}_n} \int\int_{RR^n} |\nabla u(x_{J_n})|^{n+1} d\mu_{J_n}(x_{J_n}) d\mu_{n+1}(x_{n+1})$$

$$= \inf_{\mathcal{U}_n} \int_{R^n} |\nabla u(x_{J_n})|^{n+1} d\mu_{J_n}(x_{J_n}) \mu_n(X_n) = \inf_{\mathcal{U}_n} \int_{R^n} |\nabla u(x_{J_n})|^{n+1} d\mu_{J_n}(x_{J_n}),$$

which, in general, is different from $\inf_{\mathcal{U}_n} \int_{R^n} |\nabla u(x_{J_n})|^n d\mu_{J_n}(x_{J_n})$.

THEOREM 2. *The conformal capacity proposed by us in the space* $(X, \mathcal{M}, \mu)$ *from above is a capacity, i.e. satisfies the conditions* (i), (ii), (iii) *from the introduction*.

The algebra $\mathcal{N}$ involved in the conditions (i),(ii),(iii) will be in our case the σ-algebra $\mathcal{M}$.

Condition (i) is trivial. For condition (ii), we see that if $E_1 \subset E_2$ and $\mathcal{U}_1, \mathcal{U}_2$ are the corresponding two classes of admissible functions, then $\mathcal{U}_2 \subset \mathcal{U}_1$, so that

$$\inf_{\mathcal{U}_1} \mathrm{ess}_X \sup |\nabla u(x)| \leq \inf_{\mathcal{U}_2} \mathrm{ess}_X \sup |\nabla u(x)|.$$

And now, arguing as in the preceding theorem, we shall prove that also (iii) is verified. Indeed, if

$$E = \bigcup_m E_m,$$

$\mathcal{U}_m$ are the classes of admissible functions corresponding to E_m and $u(x) = \sup\{u_1(x), u_2(x), \ldots\}$, where $u_m \in \mathcal{U}_m (m = 1,2,\ldots)$, then, by the same argument as in the preceding theorem, we establish (24); hence

$$\mathrm{cap}E \leq \mathrm{ess}_X\sup|\nabla u(x)| \leq \mathrm{ess}_X\sup \sum_m |\nabla u_m(x)| \leq \sum_m \mathrm{ess}_X\sup|\nabla u_m(x)|,$$

hence, since $\forall m \in N, u_m$ is an arbitrary function of $\mathcal{U}_m$,

$$\mathrm{cap}E \leq \sum_m \inf_{\mathcal{U}_m} \mathrm{ess}_X\sup|\nabla u_m(x)| \leq \sum_m \mathrm{cap}E_m,$$

as desired.

3. p-capacity and conformal capacity in abstract Wiener spaces

In the preceding two paragraphs we proposed concepts for p-capacity and conformal capacity with the analytical expressions (22) and (36), respectively, in a product space $(X,\mathcal{M},\mu)$ obtained by a limiting process with respect to the dimension n. In the present paragraph, we justify the above definitions of the p-capacity and of the conformal capacity also in the case of an abstract Wiener space if the norm of the Banach space involved in its definition is obtained by means of a self-adjoint, strictly positive definite, Hilbert-Schmidt operator.

Now, let us start with some notions and preliminary results needed in this paragraph. Let us give first a constructive characterization of an abstract Wiener space. This has been done in detail in H.H.Kuo's [8] and L.Gross' [6] monographs, but in order to be self-contained, we remind here some basic concepts necessary for the understanding of the idea of an abstract Wiener space.

Let H be a real separable Hilbert space with the scalar product $\langle\cdot,\cdot\rangle$ and the norm $|\cdot|$. Then, one defines the Gauss measure μ_H by (21) on the ring R_H of the cylinder sets $Z=\{x\in H;(\langle x,y_1\rangle,\ldots,\langle x,y_m\rangle)\in E_{J_n}\}$, where $E_{J_n}\in\mathcal{M}_{J_n}$ is a Borel set and the span $[y_1,\ldots,y_m]$ is called the basis of Z.

We remind that a set function $\mu\geq 0$ on R_H is a <u>cylinder set measure</u> if 1) $\mu(H)=1$ and 2) for each finite dimensional subspace $K\subset H$, μ is σ-additive when restricted to the σ-ring R_K of all cylinders based on K.

Since μ_H has not a σ-additive extension to the σ-ring generated by R_H (see H.H.Kuo [8], Proposition 4.1,p.54), then one considers the Banach space B, obtained from H by completion with respect to a measurable norm.

As you know, a norm $\|\cdot\|$ on H is said to be <u>measurable</u> (with respect to a measure μ) if $\forall\varepsilon>0$ there is a a finite dimensional orthogonal projection P_0 on H such that

(37) $\mu\{x\in H; \|Px\|>\varepsilon\}<\varepsilon,$

whenever P is a finite dimensional projection orthogonal to P_o, i.e. such that $(PH) \cap (P_oH) = \{0\}$. Since $\{x \in H; \| Px \| > \varepsilon\}$ is a cylinder set based on the range of P, (37) makes sense.

We deduce a(σ-additive) measure m on the ring R_B of all cylinders

$$(38) \qquad Z = \{x \in B; (\langle x,y_1\rangle, \ldots, \langle x,y_m\rangle) \in E_{J_n}\},$$

where, this time, $y_k \in B^*$ (the topological dual of B) and $\langle x,y_k\rangle = y_k(x)$ $(k = 1,\ldots,m)$, by the relation

$$(39) \qquad m\{x \in B; (\langle x,y_1\rangle, \ldots, \langle x,y_m\rangle) \in E_{J_n}\}$$
$$= \mu_H\{x \in H; (\langle x,y_1\rangle, \ldots, \langle x,y_m\rangle) \in E_{J_n}\},$$

where $E_{J_n} \in \mathcal{M}_{J_n}$, since, regarding $y \in B^*$ as an element of $H^* \equiv H$ by restriction, we can embed B^* in H. Clearly, m is well-defined and satisfies the following properties:

PROPOSITION 9. *The cylinder set measure* m *on the ring* R_B *of all cylinders* (38) *of* B *is* σ*-additive* (L.Gross [6], Theorem 2,p.101).

PROPOSITION 10. m *is* σ*-additive also on the* σ*-algebra* $\mathcal{M}_B$ *generated by* R_B (H.H.Kuo [8], Theorem 4.1,p.63).

This extension of m to the σ-algebra $\mathcal{M}_B$ is called an *abstract Wiener measure on* B.

PROPOSITION 11. *The* σ*-algebra* $\mathcal{M}_B$ *generated by* R_B *is the Borel algebra of* B (H.H.Kuo [8], Theorem 4.2,p.74).

A triple (i,H,B), where $i: H \longrightarrow B$ is a continuous injection of the real separable Hilbert space H into the real Banach space B is called an *abstract Wiener space* if the B-norm pulled back to H is a measurable norm.

PROPOSITION 12. *The completion* B *of a real separable Hilbert space* H *with respect to a measurable norm is separable* (L.Gross [5]).

PROPOSITION 13. *Let* B *be a real separable Banach space. Then there exists a Hilbert space* H *densely embedded in* B *such that the* B*-norm is measurable over* H*, i.e.* (i,H,B) *is an abstract Wiener space, where* i *is the inclusion map from* H *into* B (H.H.Kuo [8], Theorem 4.4,p.79).

PROPOSITION 14. *If* $\dim H = \infty$*, then* $|\cdot| = \sqrt{\langle .,.\rangle}$ *is not a measurable norm* (H.H.Kuo [8],p.59).

There are different ways of obtaining measurable norms. Thus, for instance, an injective Hilbert-Schmidt operator A yields a mesurable norm.

We remind that a linear operator A of H is called a <u>Hilbert-Schmidt operator</u> if, for same orthonormal basis $\{e_n\}$ of H,

$$\sum_{n=1}^{\infty} |Ae_n|^2 < \infty.$$

Its norm is

$$||A||_2 = (\sum_n |Ae_n|^2)^{\frac{1}{2}}$$

and does not depend on the choice of $\{e_n\}$ (H.H.Kuo [8],p.3).

An operator of H is called <u>compact</u> if it takes any bounded subset of H into a set whose closure is compact.

A self-adjoint operator may be characterized as a linear continuous operator $T:H\longrightarrow H$ satisfying the condition $<Tx,y>=<x,Ty>$ $\forall x,y\in H$.

PROPOSITION 15. *If A is a self-adjoint, compact operator of H, then there exists an orthonormal basis $\{e_n\}$ of H such that*

$$Ax = \sum_n \lambda_n < x, e_n > e_n,$$

where λ_n's are real numbers and $\lambda_n \longrightarrow 0$ as $n \longrightarrow \infty$. λ_n are called eigenvalues and e_n eigenvectors $(n=1,2,\dots)$ (H.H.Kuo [8], Theorem 1.5,pp.8,9).

We remind that a self-adjoint operator A is said to be <u>positive definite</u> if $<Ax,x>\geq 0$ $\forall x\in H$ and <u>strictly positive definite</u> if $<Ax,x>>0$ $\forall x\in H$.

We observe that a Hilbert-Schmidt operator is compact and furthemore, A is a Hilbert-Schmidt operator iff $\sum_n \lambda_n^2 < \infty$, where λ_n's are the eigenvalues of $(A^*A)^{\frac{1}{2}}$.

PROPOSITION 16. *Let A be a Hilbert-Schmidt operator. Then the semi-norm $||x|| = |Ax|$ $\forall x \in H$ is measurable* (H.H.Kuo [8],p.59).

COROLLARY. *If A is a self-adjoint, strictly positive definite Hilbert-Schmidt operator, then $||\cdot|| = |A.|$ is a measurable norm.*

LEMMA 11. *If*

$$X = \prod_n X_n,$$

$\{e_n\}$ is an orthonormal basis on X such that

$$\forall x \in X, x=(x_1 x_2,\ldots)=\sum_n x_n e_n, H=\{x \in X; \sum_n x_n^2 < \infty\}$$

(with the scalar product $<x,y>=\sum_n x_n y_n$) is the corresponding Hilbert space, A is a self-adjoint, strictly positive definite, Hilbert-Schmidt operator corresponding to the basis $\{e_n\}$ and B is the Banach space obtained as a completion of H with respect to the measurable norm $||\cdot|| = |A.|$, then $B \subset X$.

Suppose X is provided with Tihonov's topology and cosider a sequence $\{x_m\}$, with $x_m \in H$. Assume $\{x_m\}$ is a Cauchy sequence with respect to the norm $||\cdot||$. According to the completion process of H with respect to this norm,

$$\lim_m x_m = x_\infty \in B.$$

Let us show that the sequence $\{x_m\}$ coverges to x_∞ also with respect to Tihonov's topology.

Let us prove first that any sequence $\{y^m\}, y^m \in H$, which converges to 0 with respect to $||\cdot||$, converges to 0 also with respect to Tihonov's topology. According to Proposition 15 and since a Hilbert-Schmidt operator is compact, we may write

$$Ay^m = \sum_k \lambda_k < y^m, e_k > e_k = \sum_k \lambda_k y_k^m e_k,$$

where

$$y^m = \sum_k y_k^m e_k.$$

But, the convergence to 0 with respect to $||\cdot||$ of the sequence $\{y^m\}$ yields

$$||y^m|| = |Ay^m| = \sum_k \lambda_k^2 (y_k^m)^2 \longrightarrow 0 \quad \text{as} \quad m \longrightarrow \infty,$$

hence, for any natural p,

$$\lim_m \sum_{k=1}^p \lambda_k^2 (y_k^m)^2 \le \lim_m \sum_{k=1}^\infty \lambda_k^2 (y_k^m)^2 = 0$$

and then a fortiori

$$\lim_m y_k^m = 0 \quad (k=1,\ldots,p),$$

i.e. $\{y^m\}$ converges to 0 with respect to Tihonov's topology.

Next, let us consider an arbitrary sequence $\{x_m\}$ with $x_m \in H$ and convergent to a point $x_\infty \in H$ with respect to $||\cdot||$. Then, the sequence $\{y^m\}$, with $y^m = x^m - x_\infty$, will converge to 0 with respect to $||\cdot||$. But, it was just proved that this implies the convergence of y^m to 0 also with respect to Tihonov's topology; hence we derive the convergence of x^m to x_∞ in Tihonov's topology, allowing us to conclude that the topology induced by the norm $||\cdot||$ is stronger than Tihonov's topology, so that any sequence $\{x^m\}, x^m \in H$, convergent to a point $x \in B$ with respect to the B-norm $||\cdot||$ will be convergent to x also with respect to Tihonov's topology, and since the product space X is closed with respect to Tihonov's topology, it follows that $x \in X$, allowing us to deduce that the completion B of H with respect to the norm $||\cdot||$, considered as a set, is contained in X, as desidered.

LEMMA 12. *Let $(X, \mathcal{M}, \mu)$ be the measure space from above (with $X_n = R$ and μ_n the Gauss measure in X_n), $H \subset X$ the corresponding Hilbert space, A a self-adjoint, strictly positive definite Hilbert-Schmidt operator, B the Banach space obtained from H by completion with respect to the measurable norm $||\cdot|| = |A.|$ and $\mathcal{M}, \mathcal{M}_B$ the σ-algebras of Borel sets of X and B, respectively. Then $\mathcal{M}_B \subset \mathcal{M}$.*

Let us show first that the unit ball $\bar{B} = \{x \in B; ||x|| \le 1\} \in \mathcal{M}$. If the operator A is of the form $Ax = \sum_n \lambda_n < x, e_n > e_n$ (where $\lambda_n > 0$ are its eigenvalues) and we consider the cylinders $Z_n = \{x \in X; \sum_{k=1}^{n} \lambda_k^2 x_k^2 \le 1\}$, then, clearly, $\bar{B} \subset \bigcap_n Z_n$. Next, arguing as in Lemma 4, suppose there is an $n \in N$, such that $\sum_{k=1}^{n} \lambda_k^2 x_k^2 > 1$, so that $x \notin \bigcap_n Z_n$; this contradiction implying that $\bar{B} = \bigcap_n Z_n \in \mathcal{M}$ and that $\mathcal{M}_B \subset \mathcal{M}$, as desired.

LEMMA 13. *In the hypotheses of the preceding lemma, μ is AC relatively to $\mathcal{A}\{e_n\} = \{E \in \mathcal{M}_B; E = \bigcup_m E_m, E_m \in \mathcal{A}(e_n)\}$.*

Suppose $E \in \mathcal{A}\{e_n\}$, then, arguing as in Lemma 5, we have, on account of the preceding lemma and of Lebesgue and Fubini theorems, that

$$\mu(E) = \int_X \chi_E d\mu = \int_B \chi_E d\mu \le \int_B \sum_n \chi_{E_n} d\mu = \sum_n \int_B \chi_{E_n} d\mu$$

$$= \sum_n \int_{B_{n'}} \int_{X_n} \chi_{E_n}(x_n, x_{n'}) d\mu_n(x_n) d\mu_{n'}(x_{n'})$$

$$= \sum_{n=1}^{\infty} \int_{B_{n'}} \mu_n[(x_{n'} + Re_n) \cap E_n] d\mu_{n'}(x_{n'}) = 0,$$

where $B_{n'} = \{x_{n'} \in \prod_{k\neq n} X_k;\ \sum_{k\neq n} \lambda_k^2 x_k^2 < \infty\}$, as desired.

LEMMA 14. *In the hypotheses of Lemma 12, if* $E \subset B(R)$ *is a Borel set of* B *and* $\mathcal{U}$ *is the class of admissible functions for* E, *then*

$$\inf_{\mathcal{U}}[\int_B \|\nabla u(x)\|^p dm(x)]^{\frac{1}{p}} = \lim_n \inf_{\mathcal{U}} \operatorname{ess}_B \inf[\int_{R^n} \|\nabla u(x_{J_n}, \tilde{x}_{J_n'})\|^p dm_{J_n}(x_{J_n})]^{\frac{1}{p}}.$$

First of all, on account of Lemma 11, $B \subset X$. Next, let us show that $\|\nabla u\|$ exists μ-a.e. Indeed, according to Proposition 7, u is differentiable (in the sense of Aronszajn) in B except in a set $E_o \in \mathcal{A} \equiv \mathcal{A}_B \subset \mathcal{A}\{e_{n'}\}$ and, since by the preceding lemma $\mu(E_o) = 0$, u is differentiable also μ-a.e. in B. But, according to the definition of admissible functions, the suport of u is contained in a ball $B(R)$ of radius $R > 0$, so that we may extend u as a Lipschitzian function in X, by taking $u(x) = 0$ for $x \in X - B$, and this extended u will be differentiable μ-a.e. in X. But then, $Du(x;e_s)$ exists μ-a.e., whence $\|\nabla u(x)\| = \sup_s |Du(x,e_s)|$ exists μ-a.e. too.

Next, since u is Lipschitzian (with respect to the B-norm), it is clearly $\mathcal{M}_B$-measurable and, arguing as in Lemma 7, we observe that $(1/\rho)[u(x+\rho e_s) - u(x)]$, $Du(x;e_s)$ and $\|\nabla u(x)\| = \sup|Du(x;e_s)|$ are $\mathcal{M}_B$-measurable and then, Lemma 12 implies that u is also $\mathcal{M}$-measurable in B. Finally, considering the above extension of u to X:

$$\tilde{u}(x) = \begin{cases} u(x) & \text{for } x \in B, \\ 0 & \text{for } x \subset X - B, \end{cases}$$

we have

$$\inf_{\mathcal{U}}[\int_B \|\nabla u(x)\|^p dm(x)]^{\frac{1}{p}} = \inf_{\mathcal{U}}[\int_B \|\nabla \tilde{u}(x)\|^p d\mu(x)]^{\frac{1}{p}}$$

$$= \inf_{\mathcal{U}}[\int_X \|\nabla \tilde{u}(x)\|^p d\mu(x)]^{\frac{1}{p}}$$

$$= \lim_n \inf_{\mathcal{U}} \operatorname{ess}_X \inf[\int \|\nabla \tilde{u}(x_{J_n}, \tilde{x}_{J_n'})\|^p d\mu_{J_n}(x_{J_n})]^{\frac{1}{p}}$$

$$= \lim_n \inf_{\mathcal{U}} \operatorname{ess}_B \inf[\int_{R^n} \|\nabla u(x_{J_n}, \tilde{x}_{J_n'})\|^p d\mu_{J_n}(x_{J_n})]^{\frac{1}{p}},$$

as desired.

The preceding lemma allows us to deduce the following definition of the p-*capacity of an* $\mathcal{M}_B$-*measurable set* $E \subset B(R)$ *of a real separable Banach space* B (obtained as above by means of a self-adjoint, strictly positive definite, Hilbert-Schmidt operator) from the corresponding definition in R^n by a limiting process with respect to the dimension:

$$\mathrm{cap}_p E = \inf_{\mathcal{U}} [\int_B \| \nabla u(x) \|^p dm(x)]^{\frac{1}{p}},$$

where m is the probability measure in the measure space $(B, \mathcal{M}_B, m)$ and $\mathcal{U}$ is the class of admissible functions of E.

THEOREM 3. *The p-capacity (as defined above) is a capacity.*

The same proof as for Theorem 1.

Arguing as in the preceding lemma, we have also

LEMMA 15. *In the hypotheses of the preceding lemma,*

$$\lim_n \inf_{\mathcal{U}} \mathrm{ess}_B \sup [\int_{R^n} \| \nabla u(x_{J_n}, x_{J'_n}) \|^n d\mu_{J_n}(x_{J_n})]^{\frac{1}{n}} = \inf_{\mathcal{U}} \mathrm{ess}_B \sup \| u(x) \|.$$

And then, we give the following definition of the *conformal capacity of an $\mathcal{M}_B$-measurable set* $E \subset B(R)$ *of a Banach space* B (of the type considered above):

$$\mathrm{cap}\, E = \inf_{\mathcal{U}} \mathrm{ess}_B \sup \| \nabla u(x) \|,$$

where $\mathcal{U}$ is the class of admissible functions of E.

THEOREM 4. *The conformal capacity (as defined above) is a capacity.*

In the rest of the paper we shall indicate how to obtain an abstract Wiener space (i, H_o, H), with H_o, H real separable Hilbert spaces so that the concepts of p-capacity and conformal capacity from above may be obtained again by a limiting process with respect to the dimension n.

We remind that a compact operator A of a real separable Hilbert space H is called *nuclear (trace class) operator* if

$$\sum_n \lambda_n > \infty$$

where λ_n's are the eigenvalues of $(A^*A)^{\frac{1}{2}}$. Let $L_{(1)}(H)$ denote the collection of nuclear operators of H and $\| A \|_1 = \sum_n \lambda_n$ its norm.

The *covariance operator* S_μ of a Borel measure μ in H is defined by

$$< S_\mu, y > = \int_H < x, z > < z, y > d\mu(z) \qquad (x, y \in H).$$

If S_μ exsists, it is positive definite and self-adjoint.

An operator is called an *S-operator* of H if it is in

$L_{(1)}(H)$, positive definite and self-adjoint. $\mathcal{S}$ denotes the collection of S-operators of H.

Let μ_H be the Gaussian measure in $H, S = S_{\mu_H}$ its covariance operator and $H_o = \sqrt{S}H$ the isometrical image of H. $\sqrt{S}$ is a Hilbert-Schmidt, strictly positive definite operator and H_o is a Hilbert space with the scalar product $<.,.>_o$ defined by $<\sqrt{S}x, \sqrt{S}y>_o = <x,y>, x,y \in H$. If

$$Sx = \sum_n \alpha_n <x, e_n> e_n, \alpha_n > 0 \quad (n = 1,2,\ldots), \text{ then } H_o = \{x = \sum_n x_n e_n ; \sum_n \frac{x_n^2}{\alpha_n} < \infty\}.$$

PROPOSITION 17. *Let* H, H_o *be as given above and* i *the inclusion map of* H_o *in* H. *Then* (i, H_o, H) *is an abstract Wiener space. Conversely, if* B *is a Hilbert space, then* (i,H,B) *arises in the way indicated above* (H.H.Kuo [8], p.63).

Finally, in order to point out the importance of the nuclear operators, we quote the following results :

PROPOSITION 18. *A necessary and sufficient condition that a cylinder set measure* μ *be* σ-*additive is that given* $\varepsilon > 0$, *we can find a bounded set* K *such that* $\bar{\mu}(K) \geq 1-\varepsilon$, *where* $\bar{\mu}$ *is the corresponding outer measure* (A.V.Balakrishnan [2], Theorem 4.1, p.128).

By means of this theorem, he obtains

PROPOSITION 19. *Let* A *be a nonnegative self-adjoint operator mapping* H *into* H. *In order that the cylinder measure induced by* A *be* σ-*additive, it is necessary and sufficient that* A *be nuclear* (A.V. Balakrishnan [2], Theorem 4.2, p.130).

Let us remind now what is a measure induced by an operator.

We recall first that the characteristic functional $\hat{\mu}$ of a Borel measure μ in H is defined by

$$\hat{\mu}(x) = \int_H e^{i<x,y>} d\mu(y), \quad x \in H.$$

And now we have

PROHOROV'S THEOREM. *If* $x_o \in H$ *and* $S \in \mathcal{S}$, *then*

$$\Phi(x) = e^{i<x_o,x> - \frac{1}{2}<Sx,x>}$$

is the characteristic functional of a (Gaussian) measure in H.

The (Gaussian) measure in H given by this theorem is called the *measure induced by the operator* S.

References

[1] ARONSZAJN, N.: Differentiability of Lipschitzian mappings, Studia Math. 57 (1976) 147-190.

[2] BALAKRISHNAN, A.V.: Introduction to optimization theory in a Hilbert spce (Lecture Notes in Operations Research and Math. Systems 42), Springer-Vellag, Berlin-Heidelberg-New York 1971, 154 pp.

[3] BARBU, V. and PRECUPANU,T. Convexity and optimization in Banach spaces, Edit. Acad. Bucureşti România and Sijthoff & Noordhoff, International Publishers 1978, 316 pp.

[4] FUGLEDE, B.: Extremal lengh and functional completion, Acta Math. 98 (1957), 171-219.

[5] GROSS, L.: Potential theory in Hilbert space, J. Functional Anal. 1 (1967), 123-181.

[6] ——: Abstract Wiener measure and infinite dimensional potential theory, in Lecture Notes in Modern Analysis and Applications II by J. Glinn, L.Gross, Harish-Chandra, R.V.Kadison, D.Ruella, I.Segal (Lecture Notes in Math. 140), Springer-Verlag, Berlin-Heidelberg-New York 1970 pp. 84-116.

[7] HEWITT, E. and STROMBERG,K.: Real and abstract analysis. A modern treatment of the theory of functions of a real variable, Springer-Verlag, Berlin-Heidelberg-New York 1965, 476 pp.

[8] KUO, HUI HSIUNG: Gaussian measures in Banach spaces (Lecture Notes in Math. 463) Springer- Verlag , Berlin-Heidelberg-New York 1975, 224 pp.

[9] YOSIDA, K.: Functional analysis, 3. ed., Springer-Verlag, Berlin-Heidelberg-New York 1971, 475 pp.

[10] ZIEMER, P. W.: Extremal lengh and p-capacity, Michigan Math. J. 16 (1969), 43-51.

Institute of Mathematics
University "Al. I. Cuza"
Iaşi, Romania

MÉTHODES DE CONTRÔLE OPTIMAL EN ANALYSE COMPLEXE.
IV. APPLICATIONS AUX ALGÈBRES DE FONCTIONS ANALYTIQUES

Amédée Debiard et Bernard Gaveau (Paris)

Table des matières

Résumé. En utilisant la classe des fonctions plurisousharmoniques de Rickart nous posons le problème de Bremermann pour une algèbre de fonctions générale et nous en déduisons des propriétés des mesures de Jensen. Nous définissons le potentiel capacitaire extrêmal et le relions à l'enveloppe polynomialement convexe et au contrôle optimal stochastique.

Ces résultats ont été annoncés dans deux Notes aux Comptes Rendus de l'Académie des Sciences de Paris [4] et [8] et exposés à la Conférence d'Analyse complexe et harmonique de la Garde-Freinet, juin 1977, au séminaire de Monsieur Leray (Collège de France, juin 1977), à la conférence de Partial differential equations (Park City, Utah, février 1977) et aux Universités de Maryland, Michigan, Princeton (novembre 1977). Nous avons eu connaissance en mars 1978, de l'article géneral de T. Gamelin et N. Sibony.

Amédée Debiard et Bernard Gaveau

Introduction

Dans la théorie des fonctions de plusieurs variables complexes, la réduction à des problèmes de théorie du potentiel peut se faire d'une infinité de façons distinctes, alors qu'en une variable complexe, il n'existe qu'une façon d'associer une théorie du potentiel à transformation conforme près [8,I] . Ce phénomène peut être interprété de deux façons : soit en disant que la classe des fonctions plurisousharmoniques est plus petite qu'une classe de fonctions sousharmoniques, soit en remarquant qu'en plusieurs variables les mesures de Jensen des algèbres de fonctions sont non uniques. Le premier point de vue conduit naturellement au problème de Bremermann [2] et peut être développé dans le cadre d'une algèbre de fonctions abstraite et nous montrons dans la première partie comment on peut résoudre ce problème par une formule de balayage par les mesures de Jensen. Nous utilisons ici la classe de fonctions plurisousharmoniques introduites par Rickart [10] , alors que T. Gamelin utilise une classe plus restreinte. Nous montrons qu'en une variable complexe, les fonctions plurisousharmoniques de Rickart coïncident avec les fonctions finement sousharmoniques de Fuglede [5] (et aussi [3]). En plusieurs variables, la formule de balayage abstraite, jointe à la méthode de contrôle optimal [8], montrent que dans le cas de la boule de $\mathbb{C}^n$, les combinaisons convexes de mesures harmoniques kählériennes sont denses dans toutes les mesures de Jensen ce qui confirme a posteriori, que l'abondance de mesures de Jensen est liée à l'abondance de théories du potentiel adaptées à l'algèbre envisagée.

Dans une seconde partie, nous introduisons le potentiel capacitaire extrêmal d'un compact de $\mathbb{C}^n$. C'est essentiellement le supremum de tous les potentiels capacitaires kählériens et nous montrons comment il fournit le calcul de l'enveloppe polynomialement convexe du compact envisagé lorsque celui-ci est assez gros. Ce potentiel capacitaire est analogue à celui de Bedford-Taylor [1] et de Siciak [11].

Nous introduirons, dans une publication ultérieure des notions d'énergie et de capacité.

I. FONCTIONS PLURISOUSHARMONIQUES ET MESURES DE JENSEN POUR CERTAINES ALGEBRES DE FONCTIONS

1. Fonctions plurisousharmoniques dans une algèbre abstraite de fonctions.

Soit A une algèbre abstraite de fonctions de spectre M, de frontière de Silov X. Nous allons utiliser les définitions dûes à Rickart [10] pour définir les fonctions holomorphes et plurisousharmoniques sur un ouvert $U \subset M$. Notons $H_A(U)$ la classe des limites localement uniformes sur U des fonctions de A : c'est la classe des fonctions holomorphes sur U relativement à A. Notons

$L_A = \left\{ \frac{1}{n} \log |f| / f \in A,\ n \text{ entier} > 0 \right\}$. Notons $L_A^s(U)$ l'ensemble de fonctions définies sur U à valeurs $[-\infty, +\infty[$, scs, qui sont localement le supremum d'une famille de fonctions de L_A. Notons enfin $S_A(U)$ l'ensemble des fonctions sur U qui sont localement limite simple d'une suite décroissante de fonctions de L_A^s ; $S_A(U)$ est appelée classe des fonctions plurisousharmoniques sur U relativement à l'algèbre A.

exemple fondamental [9] : si U est un ouvert de $\mathbb{C}^n$, il y a égalité entre la classe des fonctions plurisousharmoniques sur U au sens usuel et la classe des fonctions plurisousharmoniques sur U relativement à l'algèbre des polynômes sur $\mathbb{C}^n$.

En effet, toute fonctions plurisousharmonique sur U relativement à l'algèbre des polynômes sur $\mathbb{C}^n$ est une fonction plurisousharmonique au sens usuel est limite d'une suite décroissante de fonctions plurisousharmoniques continues localement. De plus si G, est domaine d'holomorphie et si $V \subset \bar{V} \subset G$ et si f est plurisousharmonique continue sur G, f est le supremum V des fonctions $\frac{1}{n} \log |h| \leq f$ avec n entier > 0 et h holomorphe sur $\bar{V}$; par suite cela achève le réciproque.

Nous avons le lemme suivant :

<u>Lemme 1</u> : <u>Soit</u> $(X_t)_t$ <u>une diffusion sur</u> M <u>adaptée à</u> A. <u>Alors pour</u> $f \in H_A(U)$ <u>et</u> T_U <u>le premier temp de sortie</u> $f(X_{t \wedge T_U})$ <u>est martingale locale et si</u> $f \in S_A(U)$, $f(X_{t \wedge T_U})$ <u>est sous-martingale locale</u>.

Preuve : rappelons que X_t est adaptée à A, si pour $f \in A$, $f(X_t)$ est martingale. La propriété de martingale locale, étant tautologiquement locale, il est clair que le lemme est vrai.

Corollaire : si μ est la mesure loi de X_T où T est temps d'arrêt $< T_U$ de la diffusion X_t issue de m, alors μ est une mesure de Jensen et

$$f(m) \leq \int f(q)\, d\mu(q)$$

pour tout $f \in S_A(U)$.

La preuve est évidente par le lemme 1.

Remarque : dans [6] , Gamelin a introduit une classe analogue de fonctions plurisousharmoniques pour certaines algèbres de fonctions. (voir le 3.)

Nous appellerons fonction plurisurharmonique relativement à A, toute fonction f telle que f soit plurisousharmonique relativement à A et fonction pluriharmonique relativement à A une fonction continue sur U qui est limite localement uniforme de fonctions de la classe L_A^{-1} où

$$L_A^{-1} = \left\{ \frac{1}{n} \log |f| / n \text{ entier} > 0, f \in A \text{ inversible} \right\} .$$

On notera $P_A(U)$ la classe des fonctions pluriharmoniques sur U pour l'algèbre A ; une fonction $f \in P_A(U)$ est à la fois plurisousharmonique et plurisurharmonique.

Exemple : dans $\mathbb{C}^n$, si f est pluriharmonique, elle est localement du type $\operatorname{Re} u$ où u est holomorphe (et même dans une algèbre de polynômes locales !) donc du type $\log |\exp u|$.

2. Formule de balayage abstraite et problème de Bremermann abstrait.

Soit A, M, X comme au 1. Notons $J(x_o)$ l'ensemble des mesures de Jensen de $x_o \in M$ pour l'algèbre A de masse 1, i.e. les mesures positives sur M de masse 1 avec

$$\log |f(x_o)| \leq \int \log |f|\, d\mu \qquad \forall f \in A.$$

Soit de même $J_X(x_o)$ le sous ensemble de $J(x_o)$ portées par X.

<u>Théorème 1</u> : ① <u>Soit</u> φ <u>continue réelle sur</u> X.

<u>Alors on a</u>

$$\inf_{\substack{u \geq \varphi \\ u = -\frac{1}{n}\log|f| \\ f\in A,\ n>0 \text{ entier}}} u(x_o) = \sup_{\mu\in J_X(x_o)} \int \varphi\, d\mu$$

② <u>Si</u> φ <u>est continue réelle sur</u> M, <u>alors</u>

$$\inf_{\substack{u \geq \varphi \\ u = -\frac{1}{n}\log|f| \\ f\in A,\ n \text{ entier} > 0}} u(x_o) = \sup_{\mu\in J(x_o)} \int \varphi\, d\mu$$

Preuve : ① Soit $u = -\frac{1}{n}\log|f| \geq \varphi$ et si $\mu \in J_X(x_o)$, on a

$$u(x_o) \geq \int u\, d\mu \geq \int \varphi\, d\mu .$$

d'où
$$\inf_{\substack{u \geq \varphi \\ u = -\frac{1}{n}\log|f| \ldots}} u(x_o) \geq \sup_{\mu\in J_X(x_o)} \int \varphi\, d\mu .$$

Montrons l'inégalité opposée ; la fonctionnelle

$$\varphi \in \mathcal{C}(X) \longrightarrow \inf_{\substack{u \geq \varphi \\ u = -\frac{1}{n}\log|f| \ldots}} u(x_o)$$

est une semi-norme sur $\mathcal{C}(X)$ car facilement

$$\inf_{\substack{u \geq \varphi+\psi \\ u = -\frac{1}{n}\log|f|}} u(x_o) \leq \inf_{\substack{v \geq \varphi,\ \omega \geq \psi \\ v = -\frac{1}{n}\log|f| \\ \omega = -\frac{1}{p}\log|g|}} (v(x_o) + \omega(x_o))$$

et si $\lambda = \frac{p}{q} \in Q^+$, on a

$$\inf_{\substack{u \geq \frac{p}{q}\varphi \\ u = -\frac{1}{p}\log|f|}} u(x_o) = \inf_{\substack{v \geq \varphi \\ v = -\frac{1}{m}\log|f|}} \frac{p}{q} v(x_o)$$

D'après le théorème de Hahn Banach, cette semi norme est le supremum des formes linéaires sur $\mathcal{C}(X)$ qui la minorent, si μ est une telle forme linéaire et si $\varphi \leqslant 0$, comme $1 \in A$,

$$0 \geqslant \inf_{\substack{u = -\frac{1}{n}\log|f| \\ u \geqslant \varphi}} u(x_o) \geqslant \langle \mu, \varphi \rangle$$

donc μ est positive sur les fonctions $\geqslant 0$ et donc μ est mesure $\geqslant 0$ sur X

Prenons pour φ la fonction $\varphi = 1_X$. Alors $\mu(1) \leqslant 1$, car la fonction $f = e$ (constante de Néper) donne $u = 1 \geqslant \varphi$.

Soit alors $\varphi = \min(-\log|f|, p)$ où p est entier > 0 et $f \in A$. et soit $u = -\log|f| \geqslant \varphi$ sur X. Alors par définition de μ

$$\int \min(-\log|f|, p)\, d\mu \leqslant -\log|f(x_o)|$$

et faisant tendre $p \longrightarrow +\infty$, il vient

$$-\int \log|f|\, d\mu \leqslant -\log|f(x_o)|$$

donc μ est mesure de Jensen de x_o portée par X et masse $\leqslant 1$. Ainsi

$$\inf_{\substack{u \geqslant \varphi \\ u = -\frac{1}{n}\log|f|}} u(x_o) = \sup_{\mu \in J_X(x_o)} \int \varphi\, d\mu$$

② se démontre pareillement que ① avec l'espace $\mathcal{C}(M)$ au lieu de $\mathcal{C}(X)$.

<u>Corollaire 1</u> : <u>supposons que pour toute mesure de Jensen</u> $\mu \in J(x_o)$, <u>toute fonction</u> $u \in S_A(M)$, <u>on ait</u>

$$u(x_o) \leqslant \int u\, d\mu.$$

<u>Alors avec les notations du théorème 1, on a égalité des expressions qui figurent avec la quantité suivante</u>

$$\inf_{\substack{u \geqslant \varphi \\ -u \in S_A(M)}} u(x_o)$$

Preuve : on reprend la preuve du théorème 1. L'inégalité $u(x_o) \leq \int u\, d\mu$ pour $u \in S_A(M)$ implique que

$$\operatorname*{Inf}_{\substack{u \geq \varphi \\ -u \in S_A(M)}} u(x_o) \geq \sup_{\mu \in J_X(x_o)} \int \varphi\, d\mu$$

Pour démontrer l'inégalité opposé on introduit la semi-norme

$$\varphi \in \mathcal{C}(X) \longrightarrow \operatorname*{Inf}_{\substack{u \geq \varphi \\ -u \in S_A(M)}} u(x_o)$$

et on conclut domme au théorème 1.

<u>notations</u> : nous appellerons $\mathcal{P}(x_o)$ l'ensemble des mesures ≥ 0 de masse 1, sur M telles que $u(x_o) \leq \int u\, d\mu$ $\forall u \in S_A(M)$. C'est une partie de $J(x_o)$; on introduit ainsi $\mathcal{P}_X(x_o)$, $\mathcal{P}^1(x_o)$. On a alors de même qu'au théorème 1.

<u>Corollaire 2</u> : <u>si φ est continue sur</u> X (resp M)

$$\operatorname*{Inf}_{\substack{u \geq \varphi \\ -u \in S_A(M)}} u(x_o) = \sup_{\substack{\mu \in \mathcal{P}_X(x_o) \\ (\text{resp } \mu \in \mathcal{P}(x_o))}} \int \varphi\, d\mu$$

<u>exemple fondamental</u> : prenons pour D un domaine d'holomorphie borné, pour A l'algèbre $A(\bar{D})$ des fonctions continues sur $\bar{D}$, holomorphies sur D et supposons D tel que le spectre de $A(\bar{D})$ sur $\bar{D}$.

<u>Lemme 2</u> : <u>la classe des fonctions sur</u> D <u>plurisousharmoniques au sens usuel coïncide avec la classe des fonctions</u> $S_A(D)$.

<u>Preuve</u> : Ceci est généralisation immédiate du raisonnement de Rickart du 1 ; car il est clair qu'une fonction de $S_A(D)$ est par construction

plurisousharmonique usuelle et inversement car toute fonction plurisousharmonique usuelle est limite locale d'une suite décroissante de plurisousharmoniques continues ; si B est une boule et si $\bar{V} \subset B$ et si f est plurisousharmonique continue sur B, alors

$$f|_V = \sup_{\frac{1}{n} \log|h| \leq f \text{ sur } V} \left(\frac{1}{n} \log|h|\right)$$

et h holomorphe sur B.

On peut évidemment se borner à prendre h polynôme, donc a fortiori dans $A(\bar{D})$.

<u>Corollaire 3</u> : <u>si</u> u <u>est une fonction plurisousharmonique dans</u> B (<u>boule ou polydisque de</u> $\mathbb{C}^n$) <u>et si</u> μ <u>est une mesure de Jensen de</u> x_0 <u>pour</u> $A(\bar{B})$ <u>à support compact dans</u> D, <u>alors</u>

$$u(x_0) \leq \int u \, d\mu \quad .$$

Preuve : elle découle immédiatement de la construction du lemme 2.

<u>Donc les mesures de Jensen à support compact dans</u> B <u>sont dans</u> $\mathcal{P}(x_0)$. <u>De même si</u> X_t <u>est un processus qui rend sousharmoniques les fonctions psh, la loi de</u> X_T (T <u>temps d'arrêt</u>) <u>sachant</u> $X_0 = x_0$ <u>est dans</u> $\mathcal{P}(x_0)$

Le problème de Bremermann est alors le calcul de

$$v_\varphi(x_0) = \sup_{\substack{u \in S_A(D) \\ \limsup u \leq \varphi \text{ sur Sil } D}} u(x_0)$$

où φ est fonction continue donnée sur Sil D.

<u>Corollaire 4</u> : <u>dans le cas de</u> $A(\bar{D})$ <u>comme ci-dessus, si</u> φ <u>est réelle sur</u> Sil D = X, <u>la solution</u> v_φ <u>du problème de Bremermann est</u>

$$v_\varphi(x_0) = \inf_{\mu \in \mathcal{P}_X(x_0)} \int \varphi \, d\mu$$

Preuve : elle généralise le corollaire 2.

3. Fonctions de Hartogs

Dans [6], Gamelin introduit une classe plus restreinte que $S_A(M)$ qui est la classe $\mathcal{H}$ des fonctions scs sur M, ayant les propriétés suivantes

(i) $\frac{1}{n}\log|f| \in \mathcal{H}$ si $f \in A$, $n > 0$ entier

(ii) si $(w_n)_n$ est suite de $\mathcal{H}$ bornée supérieurement alors $w = \lim\sup w_n$ est dans $\mathcal{H}$.

$\mathcal{H}$ est appelé classe des fonctions de Hartogs.

Alors si $\mu \in J(x_o)$, on a évidemment pour tout $h \in \mathcal{H}$

$$h(x_o) \leq \int h \; d\mu.$$

L'analoguedu théorème 1 qui est démontré dans [ ] est alors :

<u>Théorème 1 bis</u> : <u>si</u> φ <u>est scs sur</u> X, <u>on a</u>

$$\inf_{\substack{u \geq \varphi \\ u \in \mathcal{H}}} u(x_o) = \inf_{\substack{u \geq \varphi \\ u = \frac{1}{n}\log|f|}} u(x_o) = \sup_{\mu \in J_X(x_o)} \int \varphi \, d\mu$$

<u>Exemple 1</u> : Soit X compact de $\mathbb{C}^n$, $A = P(X)$ et u une fonction sur X qui se prolonge de façon plurisousharmonique au voisinage de l'enveloppe polynomialement convexe $\hat{X}$ de X. Alors u est une fonction de Hartogs.

<u>Exemple 2</u> : Si D est domaine d'holomorphie de $\mathbb{C}^n$ et u une fonction plurisousharmonique continue au voisinage de $\bar{D}$, alors $u|_{\bar{D}}$ est fonction de Hartogs pour $A(\bar{D})$ toujours par le raisonnement du type de celui utilisé au lemme 2.

4. Caractérisation des fonctions plurisousharmoniques par les propriétés de sous-moyennes.

Théorème 2 : Soit A algèbre de fonctions de spectre M, u une fonction continue sur M. Supposons que pour tout $x \in M$, toute mesure de Jensen $\mu \in J(x)$ on ait $u(x) \leq \int u \, d\mu$. Alors u est plurisousharmonique sur M relativement à A.

Preuve : comme $\delta_x \in J(x)$, on a donc

$$u(x) = \inf_{\mu \in J(x)} \int u \, d\mu$$

l'infimum étant précisément atteint sur δx. Par le théorème 1 ③, on déduit que

$$u(x) = \sup_{\substack{v \leq u \\ v = \frac{1}{n} \log |f|}} v(x)$$

Mais d'après Rickart [10] p.5, un supremum scs plurisousharmonique par rapport à A est encore plurisousharmonique par rapport à A, donc ici comme u est continu par hypothèse, $u \in S_A(M)$.

5. Une formule de balayage pour des fonctions pluriharmoniques

Soit toujours le contexte abstrait précédent et soit $x_0 \in M$ et $AS(x_0)$ (resp $AS_X(x_0)$) les mesures de Arens-Singer de x_0 pour A (resp celles qui sont portées par X) i.e. les mesures $\mu \geq 0$ de masse 1 telles que

$$\log |f(x_0)| = \int \log |f| \, d\mu$$

pour tout f inversible dans A.

Théorème 3 : ① Soit φ scs ≥ 0 sur M. Alors on a

$$\inf_{\substack{u \geq \varphi \\ u=\frac{1}{n}\log|f| \\ f \text{ inversible de } A}} u(x_0) = \sup_{\mu \in AS(x_0)} \int \varphi \, d\mu$$

② Si φ est scs ≥ 0 sur X, on a la même formule en remplaçant $AS(x_0)$ par $AS_X(x_0)$.

Preuve : ① Posons

$$a = \inf_{\substack{u \geq \varphi \\ u=\frac{1}{n}\log|f| \\ f \text{ inversible de } A}} u(x_0) \qquad b = \sup_{\mu \in AS(x_0)} \int \varphi \, d\mu$$

On a clairement $a \geq b$ par définition de $AS(x_0)$. Montrons que $a \leq b$.

Soit $\alpha < a$. Montrons que $\alpha < b$. Introduisons les ensembles suivants de fonctions

$$C_1 = \{v \in \mathcal{C}_{\mathbb{R}}(M) / v > \varphi - \alpha\}$$

C_2 = l'espace vectoriel engendré par les fonctions $\frac{1}{n}\log|f|$ n entier > 0, f inversible de A et $|f(x_0)| = 1$.

Montrons le

Lemme 3 : $\bar{C}_1 \cap \bar{C}_2 = \emptyset$

Preuve : en effet, sinon soit v_k suite de C_1, $\frac{1}{n_k}\log|f_k|$ suite de C_2 avec

$$\left\| \frac{1}{n_k}\log|f_k| - v_k \right\|_M \longrightarrow 0$$

Comme v_k $\varphi - \alpha$, pour $\varepsilon > 0$ il existe k assez grand avec $\frac{1}{n_k}\log|f_k| + \alpha + \varepsilon > \varphi$. Donc $\frac{1}{n_k}\log|f_k| + \alpha + \varepsilon$ est dans la famille des pluriharmoniques relativement à A majorant φ d'où

$$a \leq \frac{1}{n_k} \log|f_k(x_o)| + \alpha + \varepsilon = \alpha + \varepsilon$$

et ceci est vrai pour tout $\varepsilon > 0$, d'où $a \leq \alpha$ ce qui contredit le choix $\alpha < a$.

Comme maintenant $\bar{C}_1 \cap \bar{C}_2 = \emptyset$, soit μ une mesure nulle sur C_2, positive sur C_1 de masse 1. Soit ψ une fonction continue positive. Comme φ est scs elle a un maximum, et il existe t avec $t\psi \geq \varphi - \alpha$ d'où $\int t\psi > \mu \geq 0$ et donc $\mu \geq 0$. Comme μ annule C_2, et est une probabilité, $\mu \in AS(x_o)$. Mais alors $\int(\varphi - \alpha)\, d\mu = \inf_{v \in C_1} \int v \, d\mu \geq 0$, d'où $\int \varphi \, d\mu \leq \alpha$ et donc $b \geq \alpha$ et $b \geq a$.

② se démontre de la même façon.

<u>Corollaire 1</u> : si u est pluriharmonique relativement à A, sur tout M, alors pour tout $x_o \in A$, toute mesure μ de $AS(x_o)$, on a

$$u(x_o) = \int u \, d\mu.$$

Preuve : appliquons le théorème 3 avec $\varphi = u$ qu'on peut supposer ≥ 0, car u est continue donc minorer et ajouter une constante ne change rien. Alors

$$u(x_o) \leq \int u \, d\mu$$

Considérons maintenant $\varphi = -u$; elle est donc dans $P_A(M)$ et on a la même inégalité d'où l'égalité du corollaire 1.

6. Applications aux algèbres R(K) en une variable complexe.

Dans ce paragraphe, K désigne un compact de $\mathbb{C}$, R(K), l'algèbre des fonctions continues sur K qui s'approchent uniformément sur K par des fonctions holomorphes au voisinage de K. Rappelons encore que si U est ouvert fin de $\mathbb{C}$, une fonction $u : U \longrightarrow [-\infty, +\infty[$ est dite finement sous harmonique si

(i) u est finement scs.

(ii) pour tout $x_o \in U$, tout voisinage fin V de x_o tel que $\bar{V}^f \subset U$ ($\bar{V}^f$ désigne l'adhérence fine de V), on a

$$u(x_o) \leq \int u \cdot d\mathcal{B}^V_{\delta_{x_o}}$$

où $d\mathcal{B}^V_{\delta_{x_o}}$ est la balayée de δ_{x_o} sur la frontière fine de V, $\partial_f V$ (propriété de sous moyenne fine) (voir [3] pour les notations).

En particulier on note K' l'intérieur fin de K et si $x_o \in K'$, on note dV_{x_o} la balayée de δ_{x_o} sur $\partial_f K$. C'est la mesure de Keldych de x_o relativement à K.

Enfin b_t désignera le mouvement brownien de $\mathbb{C}$, T_K le temps de sortie de K.

<u>Lemme 4</u> : <u>Soit</u> K <u>un compact de</u> $\mathbb{C}$, u <u>une fonction continue sur</u> K, <u>finement sousharmonique sur</u> K', $x_o \in K$, μ <u>une mesure de Jensen de</u> x_o <u>pour</u> R(K) (i.e. $\mu \in J^1(x_o)$). <u>Alors</u>

$$u(x_o) \leq \int u \, d\mu.$$

Preuve : La fonction u étant continue sur K finement sousharmonique sur K', elle a une décomposition de Riesz $u = \hat{u} + P_{K'}\alpha$ où $P_{K'}\alpha$ est le potentiel de Green fin relativement à K' d'une mesure $\alpha \geq 0$ portée par K' et $\hat{u}$ est la fonction finement harmonique sur K' solution du problème de Dirichlet fin posé par $u\,|_{\partial_f K}$

$$\hat{u}(x) = E_x\,(u(b_{T_K})) = \int u \, dv_x \quad .$$

En effet dans [3], il est démontré que la fonction $\hat{u}$ définie ci-dessus a bien pour valeur frontière $u|_{\partial_f K}$. D'autre part $P_{K'}$ est donné par le noyau de Green fin $g_{K'}(x,y)$ (voir [3]) que nous prendront finement sousharmonique, i.e. négatif. On va régler le sort de chaque morceau de la décomposition de Riesz séparément. Pour cela soit U_n une suite décroissante d'ouverts réguliers de $\mathbb{C}$ avec $\cap U_n = K$.

1) étude de $\hat{u}$:

Soit $d\lambda_x^{U_n}$ la mesure harmonique de $x \in K$ relativement à U_n. Il a été démontré dans [3] corollaire du lemme 1 que $d\lambda_x^{U_n}$ converge faiblement vers $d\nu_x$. Soit $\tilde{u}$ un prolongement continu quelconque de u au voisinage de K. Pour n assez grand la fonction $\hat{u}_n(x) = \int \tilde{u}\, d\lambda_x^{U_n}$ existe (il suffit que ∂U_n soit contenu dans le voisinage de K où $\tilde{u}$ est prolongée) et converge vers $\hat{u}(x)$. De plus $\hat{u}_n(x)$ est harmonique au voisinage de K et par suite $\hat{u}_n(x_o) = \int \hat{u}_n\, d\mu$ si $x_o \in K$ et si $\mu \in J^1(x_o)$ (voir [3] où est démontré que toute mesure de Arens-Singer de x_o pour R(K) et donc a fortiori toute mesure de Jensen, représente x_o dans l'espace H(K) des fonctions continues sur K qui s'approche uniformément sur K par des fonctions harmoniques au voisinage de K). Comme $\hat{u}_n \longrightarrow \hat{u}$ partout en restant dominée par $\sup|\tilde{u}|$, on déduit donc que $\hat{u}(x_o) = \int \hat{u}\, d\mu$

2) étude de $P_{K'}\alpha$

Par définition de [9], $g_{K'}(x,y) = \lim g_{U_n}(x,y)$ où les $g_{U_n}(x,y)$ sont fonctions de Green sousharmonique de U_n et où la limite est croissante. Si P_{U_n} est le potentiel de Green, comme α est mesure ≥ 0, on a

$$P_{K'}\alpha = \lim P_{U_n}\alpha \quad \text{(limite croissante)}.$$

Maintenant $P_{U_n}\alpha$ est sous harmonique au voisinage U_n de K et donc

$$(P_{U_n}\alpha)(x_o) \leq \int (P_{U_n}\alpha)\, d\mu$$

à cause de la preuve du théorème 4 de [3] d'où puisque $P_{U_n}\alpha$ croît vers $P_{K'}\alpha$

$$(P_{K'}\alpha)(x_o) \leq \int (P_{K'}\alpha)\, d\mu$$

d'où $$u(x_o) \leq \int u\, d\mu .$$

Nous pouvons alors démontrer.

Théorème 3 : Soit u fonction continue sur K. Il y a équivalence entre les trois prorpiétés suivantes.

(i) u est plurisousharmonique relativement à $R(K)$.

(ii) u est finement sous harmonique sur K'.

(iii) pour toute mesure de Jensen $\mu \in J^1(x_o)$ tout $x_o \in K$, on a $u(x_o) \leq \int u \, d\mu$.

Preuve : (i) $\Longrightarrow$ (ii) : utilisant le mouvement brownien, on voit que si u est plurisousharmonique pour $R(K)$, alors $u(b_t)$ est une sous-martingale locale, par le corollaire du lemme 1, on déduit que u est finement localement, finement sousharmonique ; donc finement sousharmonique car la propriété de fine sousharmonicité est finement locale

(ii) $\Longrightarrow$ (iii) : à cause du lemme 4.

(iii) $\Longrightarrow$ (i) : par le théorème 2.

Remarque : ici il y a coïncidence avec la classe des fonctions de Hartogs bien sûr.

On a de même

Théroème 3 bis : si K est compact de $\mathbb{C}$ et si u est continue sur K, il y a équivalence entre

(i) u est pluriharmonique par rapport à $R(K)$.

(ii) u est finement harmonique sur l'intérieur fin K' de K.

(iii) pour tout $x_o \in K$ et toute mesure $\mu \in AS(x_o)$, $u(x_o) = \int u \, d\mu$.

Preuve : (i) $\Longrightarrow$ (ii) car $u(b_t)$ est martingale locale, donc u est finement localement, finement harmonique donc est finement harmonique sur K'.

(ii) $\longrightarrow$ (iii) par [3]

(iii) $\Longrightarrow$ (ii) par [3]

(ii) $\Longrightarrow$ (i) parce que si u est continue sur K, finement harmonique sur K', alors u s'approche uniformément sur K par des fonctions harmoniques au voisinage de K. (th.1 de [3]) On peut donc supposer u harmonique au voisinage U de K. Alors soit $x_o \in K$, $\Delta(x_o)$ petit disque de centre x_o contenu dans U, alors $u = \mathrm{Re}\, f$ où f est holomorphe sur $\Delta(x_o)$, d'où $u = \log|\exp f|$ et s'approche donc localement uniformément près de x_o par des $\log|\exp f_n|$ où f_n sont des polynômes, donc a fortiori dans R(K).

7. Applications aux mesures de Jensen en plusieurs variables complexes.

Soit maintenant D un domaine d'holomorphie borné de $\mathbb{C}^n$ ayant la propriété suivante : pour toute fonction φ continue sur Sil(D) (frontière de Silov relative à $A(\bar{D})$), la solution du problème de Bremermann posé par φ est continue sur $D \cup \mathrm{Sil}\, D$.

Pour un tel domaine, on a introduit dans [8] la classe $\mathcal{K}(D)$ des contrôles kählériens adaptés à D, i.e. des matrices σ^{ij} hermitienne $n \times n, \geq 0$, non anticipantes par rapport au brownien standard de $\mathbb{C}^n$ tel que $\mathrm{Tr}\,\sigma\sigma^* \geq 1$ et telle que le processus $X_s^{(\sigma,z)}(\omega)$ réponse de σ

$$X_s^{(\sigma,z)^i}(\omega) = z^i + \int_0^s \sigma^{ij}(s,\omega)\, db_j(s).$$

converge au bout d'un temps $\zeta_{(\sigma,z)}$ fini ou non vers Sil D. La solution du problème de Bremermann est alors

$$v_\varphi(z_o) = \inf_{\sigma \in \mathcal{K}(D)} E(\varphi(X^{(\sigma,z)}_{\zeta_{(\sigma,z)}}))$$

lorsque évidemment $\mathcal{K}(D)$ n'est pas vide.

Notons $\mu_{(\sigma,z)}$ la loi de la variable aléatoire $X^{(\sigma,z)}_{\zeta_{(\sigma,z)}}$. Il est clair que c'est une mesure de Jensen de l'algèbre $A(\bar{D})$ portée par Sil D par définition de masse 1, pour le point z

Théorème 4 : Supposons D domaine d'holomorphie avec $\mathcal{K}(D)$ non vide satisfaisant l'hypothèse du début de ce paragraphe.

① <u>l'ensemble des combinaisons linéaires convexes des mesures $\mu_{(\sigma,z)}$ pour $z \in D$ fixé et $\sigma \in \mathcal{K}(D)$ est dense dans l'ensemble des mesures $\mathcal{P}_{\mathrm{Sil}\, D}(z)$.</u>

② <u>en particulier si φ est continue sur Sil D, et si v_φ est la solution du problème de Bremermann posé par φ sur Sil D.</u>

$$v_\varphi(z) = \inf_{\sigma \in \mathcal{K}(D)} E(\varphi(X^{(\sigma,z)}_{\xi(\sigma,z)})) = \inf_{\mu \in \mathcal{P}_{\mathrm{Sil}}(z)} \int \varphi\, d\mu = \inf_{\mu \in \mathcal{P}^1_{\mathrm{Sil}}(z)} \int \varphi\, d\mu$$

Preuve : ① par la formule de contrôle précédente et le corollaire 4 du théorème 1 et 2. on a

$$v_\varphi(z) = \operatorname{Inf} E(\varphi(X^{(\sigma,z)}_{\xi(\sigma,z)})) = \inf_{\mu \in \mathcal{P}_{\mathrm{Sil}}(z)} \int \varphi\, d\mu \leq \inf_{\mu \in \mathcal{P}^1_{\mathrm{Sil}}(z)} \int \varphi\, d\mu$$

Si l'assertion du ① n'est pas vraie,il existe φ continue avec $\int \varphi\, d\mu_{(\sigma,z)} = 0 \quad \forall \sigma \in \mathcal{K}(D)$ et $\mu_o \in \mathcal{P}^1_{\mathrm{Sil}}(z)$ avec $\int \varphi d\mu_o < 0$ par exemple. Alors $v_\varphi(z) = 0 \leq \inf_{\mu \in \mathcal{P}^1_{\mathrm{Sil}}(z)} \int \varphi\, d\mu$ d'après le calcul précédent, ce qui contredit le fait que $\int \varphi d\mu_o < 0$.

② les deux premières égalités sont déjà vues. Par le ①, on déduit l'égalité du 2nd et du $4^{\mathrm{ième}}$ terme. Ce n'est pas évident a priori car l'analyse fonctionnelle donne simplement une inégalité dans le corollaire 2.

De façon plus générale, si $\mathcal{K}'(D) \subset \mathcal{K}(D)$ est une sous-classe de contrôles kählériens donnant la solution du problème de Bremermann, alors les $\mu_{(\sigma,z)}$ pour $\sigma \in \mathcal{K}'(D)$ verront leur combinaisons convexes engendrer $\mathcal{P}^1_{\mathrm{Sil}}(z)$.

L'hypothèse $\mathcal{K}(D) \neq \emptyset$ est réalisée pour les types suivants de domaines :

(i) les domaines $D = \{z/p(z) < 0\}$ où p est fonction C^3 au voisinage de $\bar{D}$, plurisousharmonique sur $\bar{D}$ (voir [8]).

(ii) en particulier tous les domaines strictement pseudoconvexes à bord C^2.

(iii) les polyèdres analytiques (voir [8]).

(iv) les domaines symétriques bornés.

Exemple : prenons D strictement pseudoconvexe et $\mathcal{K}'(D)$ la classe des contrôles kählériens dont les réponses vivent localement par droites complexes, i.e. on se donne un processus à saut $\delta_\omega(t)$ dans le projectif $\mathbb{P}^{n-1}(\mathbb{C})$ de $\mathbb{C}^n$ défini par une suite de temps d'arrêt

$$T_0 = 0 < T_1 < T_2 < \ldots < T_n < \ldots$$

continu à droite, constant sur $[T_i, T_{i+1}[$ et on construit sur $\delta_\omega(t)$ $t \in [T_0, T_1[$ le brownien complexe de cette droite issue de z_0 de façon indépendante $b_s(\omega')$, qu'on stoppe à $T_1(\omega)$. Puis arrivé à $b_{T_1(\omega)}(\omega')$, on change de droite en prenant la droite $\delta_\omega(T_1(\omega))$ issue de ce point, et on commence un mouvement brownien indépendant sur $\delta_\omega(T_1(\omega))$ etc...
<u>Alors les mesures</u> $\mu_{(\sigma,z)} (\sigma \in \mathcal{K}'(D))$ <u>ont leur combinaisons convexes denses dans</u> $\mathcal{P}^1_{\partial D}(z)$.
Cela est à rapprocher du résultat suivant.

<u>Théorème 6</u> : <u>Soit</u> D <u>domaine strictement pseudoconvexe</u>, $z \in D$, Vz <u>une variété complexe de dimension</u> 1 <u>passant par</u> z <u>coupant transversalement</u> ∂D. <u>Alors la balayée de</u> δ_z <u>sur</u> $\partial D \cap Vz$ <u>pour la théorème du potentiel le long de</u> Vz <u>est une mesure de Jensen extrêmale de</u> z <u>pour</u> $A(D)$.

Ce théorème résulte du travail de Henkin [12].

Dans le cas des domaines D où agit transsitivement un groupe de Lie d'automorphismes complexes, en particulier les domaines symétriques bornés, on peut construire une fonction de synthèse pour le contrôle (voir [8 I] pour le cas de la boule et [8,II] pour le cas général). On en déduit alors le théorème suivant.

Théorème 7 : Si D est la boule de $\mathbb{C}^n$, les mesures harmoniques des ds^2 kählériens dans D pour z, sur ∂D ont leur combinaison linéaire convexes denses dans l'ensemble des mesures de Jensen de z pour $A(D)$ portées par ∂D.

Preuve : dans ce cas, par convexité polynomiale et l'exemple 1 du 3 il y a identité des fonctions plurisousharmoniques et de Hartogs. Le problème de Bremermann est alors donné par le théorème 1 bis i.e.

$$v_\varphi(x_o) = \operatorname*{Inf}_{\mu\in J_X(x_o)} \int \varphi \, d\mu$$

et aussi par le formule de contrôle kählérien. D'autre part, prenons pour φ une fonction $C^2(\partial D)$ Notant $\mathfrak{M}$ la classe des métriques kählériennes sur D, on a alors

Lemme : $v_\varphi(z) = \operatorname*{Inf}_{ds^2\in\mathfrak{M}_D} E_z(\varphi(X^{(ds^2)}_{T_D}))$

Admettant ce lemme pour un moment, on voit donc que si $\varphi\in C^2(\partial D)$

$$\operatorname*{Inf}_{\mu\in J_X(z_o)} \int \varphi \, d\mu = \operatorname*{Inf}_{ds^2\in\mathfrak{M}} E_{z_o}(\varphi(X^{(ds^2)}_{T_D}))$$

$$\leq \operatorname*{Inf}_{\mu\in J^1_X(z_o)} \int \varphi \, d\mu .$$

Cela est donc vrai pour tout $\varphi\in C^o(\partial D)$ par approximation. Donc si les mesures harmoniques des métriques de $\mathfrak{M}$ pour z_o n'engendrent pas $J^1_X(z_o)$, on conclut exactement comme avant.

Preuve du lemme : (Par transformation conforme, il suffit de le voir si $z = 0$) posons $u = v_\varphi$. D'après [8,I] pour la boule, $u \in L^\infty_{2,loc}$ et $\partial\bar\partial u$ est matrice à coéfficients L^∞ et l'opérateurs L dont les coefficients sont ceux de la matrice $(\partial\bar\partial u)^{\wedge^{n-1}}$ satisfait $Lu = 0$ pp car Lu n'est autre que $(\partial\bar\partial u)^{\wedge^n}$ qui est nul puisque u est solution de $(\partial\bar\partial u)^{\wedge^n} = 0$. D'après Krylov [13], il existe une diffusion Y_s de générateurs L. On a alors

$$u(0) = E_0(\varphi(Y_{\tau_D}(\omega))).$$

Mais si nous régularisons u en u_ε sur $D_\varepsilon \subset D$ avec $u_\varepsilon \in C^\infty$ strictement plurisousharmonique et si nous remarquons que $(\partial\bar\partial u_\varepsilon)^{\wedge^{n-1}}$ tend vers $(\partial\bar\partial u)^{\wedge^{n-1}}$ presque partout, l'opérateur L et la diffusion $X^{(\varepsilon)}$ associée tendent vers L et Y, donc la mesure $\mu^{(\varepsilon)}$ harmonique de 0 pour L_ε sur ∂D_ε tend faiblement vers la loi μ_0 de Y_{T_D} ; comme $X^{(\varepsilon)}$ est à un changement de temps près, la diffusion $X^{(ds^2_\varepsilon)}$ de la métrique $ds^2_\varepsilon = \partial\bar\partial u_\varepsilon$ dans D_ε (puisque son laplacien $\Delta_{ds^2_\varepsilon}$ coïncide avec L_ε au facteur $(\partial\bar\partial u_\varepsilon)^{\wedge^n}$ près) ; donc $\mu^{(\varepsilon)}$ est encore mesure harmonique de ds^2_ε pour le point 0 sur ∂D_ε.

On peut évidemment supposer que D_ε est boule concentrique à D de rayon p_ε, alors on fait une homothétie de centre 0 et de rapport $\frac{1}{p_\varepsilon}$, on déduit que $\mu^{(\varepsilon)}$ est la mesure harmonique sur ∂D d'une métrique $d\tilde{s}^2_\varepsilon$ pour le point 0 et donc

$$E_0(\varphi(X_{T_D}^{(d\tilde{s}^2_\varepsilon)})) = \int \varphi \, d\mu^{(\varepsilon)} \longrightarrow \int \varphi \, d\mu_0 = v_\varphi(0) .$$

Mais comme on a toujours par plurisousharmonicité de u

$$u(0) \leq E_0(\varphi(X_{T_D}^{(d\tilde{s}^2_\varepsilon)})) ,$$

on déduit donc le lemme pour $z = 0$.

Méthodes de contrôle optimal en analyse complexe

II. POTENTIEL CAPACITAIRE EXTREMAL ET ENVELOPPES POLYNOMIALEMENT CONVEXES

1. Le potentiel capacitaire extrêmal et ses propriétés au bord.

Considérons la situation suivante : $D \subset \mathbb{C}^n$ est un domaine ouvert et $K \subset D$ est un compact. Notons toujours $\mathcal{K}$ l'ensemble des contrôles kählériens : cela signifie que l'on s'est donné a priori l'espace de probabilités Ω du mouvement brownien n dimensionnel complexe $(b^1 \ldots b^n)$ de $\mathbb{C}^n$ et que $\sigma \in \mathcal{K}$ est une matrice hermitienne ≥ 0, $n \times n$, non anticipante avec $t \longrightarrow \sigma(t,\omega)$ continu en t ps en ω et dét $\sigma\sigma^* \geq 1$. Contrairement à [8] nous ne supposerons pas $\sigma \leq N$ Id. Alors si $z \in \mathbb{C}^n$, on note par $X_t^{(\sigma,t)}$ l'intégrale stochastique

$$X_t^{(\sigma,z)^i} = z^i + \int_0^t \sigma^{ij}(s,\omega)\, db_\omega^j(s)$$

qui existe car $t \longrightarrow \sigma(t,\omega)$ est continu ps, donc localement borné ps. Soit maintenant $z \in D \setminus K$ et soit $\zeta_{(\sigma,z)}$ le 1er temps de rencontre de $\partial D \cup K$ pour $X^{(\sigma,z)}$

$$\zeta_{(\sigma,z)} = \mathrm{Inf}\left\{ t / X_t^{(\sigma,z)} \in \partial D \cup K \right\} .$$

Appelons <u>potentiel capacitaire extrêmal</u> de K relativement à D, la fonction u^* régularisée scs de la fonction u suivante :

$$u(z) = \inf_{\sigma \in \mathcal{K}} P(X_{\zeta_{(\sigma,z)}}^{(\sigma,z)} \in \partial D) \tag{1}$$

<u>Lemme 1</u> : <u>Supposons</u> D <u>strictement pseudoconvexe. Alors</u> $u(z) \longrightarrow 1$ <u>si</u> z <u>tend vers</u> ∂D <u>et par suite</u> $u^*|_{\partial D} = 1$

Preuve : soit p fonction d'exhaustion de D strictement plurisousharmonique de classe C^2 près de $\bar{D}$ avec $D = \{z/p(z) < 0\}$ et soit $a > 0$ tel que $K \subset \{z/p(z) \leq -a\}$. Clairement, on a donc

$$(2)\qquad E(p(X^{(\sigma,z)}_{\zeta_{(\sigma,z)}})) \leq -a\,P(X^{(\sigma,z)}_{\zeta_{(\sigma,z)}} \in \partial K)$$

Or par le lemme de Itô et une intégration

$$(3)\quad E(p(X^{(\sigma,z)}_{\zeta_{(\sigma,z)}})) = p(z) + E\left(\int_0^{\zeta_{(\sigma,z)}} \sum_{i,j} (\sigma\sigma^*)^{ij}(s,\omega)\,\frac{\partial^2 p}{\partial z_i\,\partial \bar{z}_j}\,(X^{(\sigma,z)}_s)ds\right)$$

et donc par (2)

$$0 \leq aP(X^{(\sigma,z)}_{\zeta_{(\sigma,z)}} \in \partial K) \leq -p(z) - E\left(\int_0^{\zeta_{(\sigma,z)}} \sum_{i,j} (\sigma\sigma^*)^{ij}\,(s,\omega)\,\frac{\partial^2 p}{\partial z_i\,\partial \bar{z}_j}\,(X^{(\sigma,z)}_s)ds\right).$$

Maintenant comme p est plurisousharmonique et σ hermitienne ≥ 0, on a que le 2ème terme est ≤ 0, d'où

$$(4)\qquad 0 \leq P(X^{(\sigma,z)}_{\zeta_{(\sigma,z)}} \in \partial K) \leq \frac{-p(z)}{a}$$

et par suite tend vers 0 si $z \longrightarrow \partial D$. D'autre part, le temps de sortie $\eta_{(\sigma,z)}$ de D est fini car par la formule de Itô, on a

$$(5)\quad E\left(\int_0^{\eta_{(\sigma,z)}} \sum_{i,j} (\sigma\sigma^*)^{ij}(s,\omega)\,\frac{\delta_p}{\partial z_i\,\partial \bar{z}_j}\,(X^{(\sigma,z)}_s)ds\right) \leq -p(z) + E(p(X^{(\sigma,z)}_{\eta_{(\sigma,z)}}))$$

et par l'inégalité de la moyenne géométrique arithmétique on déduit comme dans [8] que

$$E(\eta_{(\sigma,z)}) < +\infty.$$

donc $P(X^{(\sigma,z)}_{\zeta_{(\sigma,z)}} \in \partial D) = 1 - P(X^{(\sigma,z)}_{\zeta_{(\sigma,z)}} \in \partial D) \longrightarrow 1$ si $z \longmapsto \partial D$.

<u>Lemme 2</u> : <u>Supposons que en</u> $z_o \in K$, <u>l'ensemble</u> $\complement K$ <u>ne soit pas effilé pour la théorie du potentiel usuelle (ou ce qui est le même, pour une théorie du potentiel au voisinage de</u> D). <u>Alors</u> $u(z) \longrightarrow 0$ <u>si</u> $z \longrightarrow z_o$ <u>et donc</u> $u^*(z_o) = 0$.

Preuve : soit $B(z_o,\rho)$ une boule de centre z_o et de rayon ρ contenue dans D et

$$W(z_o) = B(z_o,\rho) \cap \complement K.$$

soit $b^{(z)}$ le mouvement brownien standart issu de z ; il correspond au contrôle kählérien $\sigma = Id$ et on a $u(z) \leq P(b^{(z)}_{\zeta(Id,z)} \in \partial D)$.

Mais par conséquent si $z \in W(z_o)$

$$P(b^{(z)}_{\zeta(Id,z)} \in \partial D) \leq P_z(b^{(z)}_{T_{W(z_o)}} \in \partial W(z_o) - \partial K)$$

où $T_{W(z_o)}$ est le premier temps de sortie de $W(z_o)$ et ceci peut être rendu $< \varepsilon$ pourvu que $z \in V_{(z_o)}$ voisinage assez petit de z_o à cause de la continuité de la mesure harmonique en z_o puisque $\complement K$ n'est pas effilé en z_o. D'où $u(z) \leq \varepsilon$ si $z \in V_{(z_o)}$ et le lemme 2 est achevé.

2. Plurisousharmonicité du potentiel capacitaire extrêmal.

Lemme 3 : pour tout $z_o \in D - K$, tout voisinage $B \subset \bar{B} \subset D$ de z_o, toute métrique kählérienne ds^2, on a

$$u(z_o) \leq \int_{\partial B} u(z)\, d\mu_{z_o}(z) \tag{6}$$

où $d\mu_{z_o}$ est la mesure harmonique de z_o relativement à B pour le ds^2 considéré.

Preuve : elle est un corollaire du théorème suivant appliqué à l'ouvert $D - K$ et à la fonction $\varphi = 1_{\partial D}$.

Théorème 1 : Soit D un domaine borné quelconque de $\mathbb{C}^n$, φ une fonction mesurable bornée sur ∂D et

$$u(z) = \operatorname*{Inf}_{\sigma \in \mathcal{K}} E(\varphi(X^{(\sigma,z)}_{\zeta(\sigma,z)})) \tag{7}$$

où $\zeta_{(\sigma,z)}$ est temps de sortie de D. Pour tout $z_o \in D$, tout voisinage $B \subset \bar{B} \subset D$ de z_o, toute métrique kählérienne ds^2, on a

$$u(z_o) \leq \int_{\partial B} u(z)\, d\mu_{z_o}(z)$$

Preuve du théorème 1 : soit $\varepsilon > 0$ fixé et pour tout $y \in \partial D$, soit $\sigma^y \in \mathcal{K}$ un contrôle kählérien de la classe $\mathcal{K}$ avec

$$(8) \qquad u(y) \leq E(\varphi(X^{(y,\sigma^y)}_{\zeta(y,\sigma^y)})) \leq u(y) + \varepsilon \quad .$$

Définissons un nouveau contrôle kählérien ainsi : considérons la métrique ds^2 et soit ds'^2 la métrique conforme à ds^2 dont le déterminant de la matrice g_{ij} est égal à 1, ie $ds'^2 = \frac{ds^2}{\text{dét}\,(g_{ij})}$. Considérons la diffusion $X^{(ds'^2)}$ associée à ds'^2 ; elle définit un contrôle kählérien

$$\Sigma^{ij}(t,\omega) = \sigma'^{ij}(X^{(ds'^2)}_t(\omega))$$

où $\sigma' \sigma'^* = g^{-1}$. Prenons $\Sigma^{ij}(t,\omega)$ jusqu'au temps de sortie $T_B(\omega)$ de B et pour $t \leq T_B(\omega)$, notons $X^{(z_0,\Sigma)}_t(\omega) = X^{(ds'^2)}_t(\omega)$.

Donnons nous un hasard ω' indépendant de ω (donc une autre copie Ω' de Ω) et posons

$$X^{(z_0,\Sigma)}_{t+T_B(\omega)}(\omega') = X^{(z_0,\Sigma)}_{T_B(\omega)}(\omega) + \int_{T_B(\omega)}^{t+T_B(\omega)} \sigma^{(y)ij}(s-T_B(\omega),\omega')\, db^j_s(\omega')$$

où dans l'intégrale stochastique $y = X^{(z_0,\Sigma)}_{T_B(\omega)}(\omega)$ et cette intégrale stochastique est prise sur l'espace de probabilités Ω'. Ceci définit ainsi un contrôle kählérien sur $\Omega \times \Omega'$, par

$$(9) \qquad \begin{aligned} \Sigma(t,\bar{\omega}) &= \sigma'(X^{(ds'^2)}_t(\omega)) & t \leq T_B(\omega) \\ \Sigma(t,\bar{\omega}) &= \sigma^{X^{(ds'^2)}_{T_B(\omega)}(\omega)}(t-T_B(\omega),\omega') & t \geq T_B(\omega) \end{aligned}$$

où $\bar{\omega} = (\omega,\omega') \in \Omega \times \Omega'$ et par construction, $t \longrightarrow \Sigma(t,\bar{\omega})$ est continu ps.

Maintenant, remarquons que puisque T_B est temps d'arrêt du brownien standard de $\mathbb{C}^n$, on peut tout réaliser sur le même Ω en recollant $(\omega,\omega') \in \Omega \times \Omega'$ en $\omega'' \in \Omega$ puisque conditionnellement par rapport à la tribu du temps d'arrêt T_B, le passé et le futur de b sont indépendants. On obtient ainsi un contrôle $\Sigma(t,\bar{\omega})$ $(\bar{\omega}\in\Omega)$ pour $t \leq \zeta_{(z_o,\Sigma)}$ où

$$\zeta_{(z_o,\Sigma)}(\bar{\omega}) = T_B(\omega) + \zeta_{(y,\sigma^y)}(\omega')$$

où $y = X^{(ds'^2)}_{T_B(\omega)}(\omega)$ et $\bar{\omega} = (\omega,\omega')$.

Montrons alors que le processus $X_s^{(z_o,\Sigma)}(\bar{\omega})$ a la propriété de Markov relativement à T_B, i.e. montrons que

$$(10)\quad E_{z_o}\left(f(X^{(z_o,\Sigma)}_{\zeta(z_o,\Sigma)})\,\middle|\,B_{T_B}\right) = E'_y(f(X^{(y,\sigma^y)}_{\zeta(y,\sigma^y)}(\omega')))$$

où $y = X^{(ds'^2)}_{T_B(\omega)}(\omega)$ et E' est l'espérance pour ω'.

On a par construction

$$E_{z_o}(f(X^{(z_o,\Sigma)}_{\zeta(z_o,\Sigma)})\,\Pi_A(X^{(ds'^2)}_{T_B})) = E_{z_o}(\Pi_A(X^{ds^2}_{T_B})\,f(X^{(ds^2)}_{T_B} +$$

$$+\int_{T_B}^{\zeta(\Sigma)} \Sigma^{(y)^{i,j}}(s-T_B(\omega),\omega')\,db_s^j(\omega')))$$

Conditionnant par rapport à B_{T_B}, on voit que le passé de T_B n'intervient que par $X^{(ds'^2)}_{T_B}$ (puisque le futur de T_B pour $X^{(\Sigma)}$ est indépendant du passé par construction même) et donc on a

$$= E_{z_o}(\Pi_A(X^{(ds^2)}_{T_B}(\omega))\,E'(f(y + \int_{T_B}^{\zeta(z_o,\Sigma)} \sigma^{(y)^{ij}}(s-T_B(\omega),\omega')\,db_s^j(\omega'))\;\Big|\;X^{(ds'^2)}_{T_B}(\omega) = y))$$

et puisque $\zeta_{(y,\sigma_y)} + T_B = \zeta_{(z_o,\Sigma)}$

$$= E_{z_o}(\Pi_A(X^{(ds^2)}_{T_B})\,E'_{X^{(ds'^2)}_{T_B}(\omega)}\,(f(X^{(y,\sigma^y)}_{\zeta(y,\sigma^y)(\omega')}(\omega'))))$$

où on a posé $y = X^{(ds'^2)}_{T_B}(\omega)$ pour alléger l'écriture. Ceci démontre (10).

Mais alors revenons à u : on a évidemment

$$u(z_0) \leq E_{z_0}\left(\varphi\left(X^{(z_0,\Sigma)}_{\varsigma(z_0,\Sigma)}\right)\right)$$

et par la propriété de Markov précédente (10)

$$= E_{z_0}\left(E_{z_0}\left(\varphi\left(X^{(z_0,\Sigma)}_{\varsigma(z_0,\Sigma)}\right) \mid B_{T_B}\right)\right) =$$

$$= E_{z_0}\left(E'_{X^{(ds'^2)}_{T_B}(\omega)}\left(\varphi\left(X^{(y,\sigma^y)}_{\varsigma(y,\sigma^y)}\right)\right)\right)$$

$$\leq E_{z_0}\left(u\left(X^{(ds'^2)}_{T_B}(\omega)\right) + \varepsilon\right)$$

à cause du choix de σ^y (8).

Maintenant ds'^2 étant conforme à ds^2, $X^{(ds'^2)}_{T_B}$ a même loi que $X^{(ds^2)}_{T_B}$, i.e. la mesure harmonique $d\mu_{z_0}$ d'où

$$u(z_0) \leq \int_{\partial B} u(z)\, d\mu_{z_0}(z) + \varepsilon$$

et ceci est vrai pour tout ε, d'où (6)

Remarque : dans [𝔊, I] nous avons démontré un téhroème analogue pour les équations de Monge Ampère avec 2nd membre (théorème 3 de [𝔊, I]) et nous avions utilisé la continuité de u ; ici nous ne savons pas que u est continue (ni même scs !) et d'autre part ce théorème est démontré avec une classe plus large de contrôles (puisque nous n'exigeons pas que $\sigma \leq N\, Id$, mais seulement que $t \longrightarrow \sigma(t,\omega)$ soit continue pour presque tout ω).

Théorème 2 : *La fonction* u^* *est plurisousharmonique dans* D-K.

Preuve : on va montrer que pour tout opérateur kählérien

$$\Delta_a = \sum a^{ij} \frac{\partial^2}{\partial z_i \partial \bar{z}_j}$$

à coefficients constants u^* est sous harmonique. Cela achèvera le théorème 2 en vertu du lemme évident suivant.

Lemme 4 : si u^* est sousharmonique par rapport à tout opérateur kählérien Δ_u à coefficients constants, elle est plurisousharmonique (car elle implique que le courant $i\partial\bar\partial u^*$ est ≥ 0).

Or, d'abord u^* est scs par définition. Soit ensuite $z_o \in D-K$; comme $u^* = u$ presque partout, on a pour presque tout p que $u^*|_{\partial B(z_o,p)} = u|_{\partial B(z_o,p)}$ $d\sigma$-presque partout (où $d\sigma$ est l'aire euclidienne des sphères). Fixons un tel p. Mais la matrice a étant fixée une fois pour toute et $d\mu_B^{(z)}$ désignant la mesure harmonique de $z \in B$ relativement à B pour Δ_a, on a évidemment $d\mu_B^{(z)} \ll \sigma$. Soit alors $z_n \longrightarrow z_o$ avec $\lim u(z_n) = u^*(z_o)$. Par le lemme 3, on a

$$u(z_n) \leq \int_{\partial B(z_o,p)} u(\zeta)\, d\mu_{B(z_o,p)}^{(z_n)}(\zeta).$$

Mais $z \longrightarrow \int_{\partial B(z_o,p)} u(\zeta)\, d\mu_{B(z_o,p)}^{(z)}(\zeta)$ est harmonique pour Δ_a, donc continue dans $B(z_o, p)$, donc

$$u^*(z_o) \leq \int_{\partial B(z_o,p)} u(\zeta)\, d\mu_{B(z_o,p)}^{(z_o)}(\zeta) = \int_{\partial B(z_o,p)} u^*(\zeta)\, d\mu_{B(z_o,p)}^{(z_o)}(\zeta)$$

la dernière égalité à cause de la propriété du p choisi d'où la propriété de sous moyenne pour Δ_a pour tout z_o et presque tout p., donc la propriété de sous harmonicité.

3. Calcul de l'enveloppe polynomiale.

Lemme 5 : Soit K compact de $\mathbb{C}^n$, $D=B$ une boule contenant K dans son intérieur et u la fonction définie au 1. Soit $z_o \in D$ tel que $u(z_o) = 0$. Alors z_o est dans l'enveloppe polynomiale $\hat{K}$ de K.

Preuve : admettons pour un moment le lemme suivant.

Lemme 6 : Soit $\sigma \in \mathcal{K}$, T un temps d'arrêt avec $T \leq \zeta_{(z,\sigma_z)}$. Alors $X_T^{(z,\sigma)}$ a pour loi une mesure de Jensen de z pour l'algèbre des fonctions holomorphes dans D.

Finissons le lemme 5 : soit $z_0 \in D-K$ avec $u(z_0) = 0$. Il existe une suite $\sigma^{(n)} \in \mathcal{K}$ de contrôles tels que

$$P(X^{(z_0,\sigma^{(n)})}_{\zeta(z_0,\sigma^{(n)})} \in \partial D) \longrightarrow 0$$

Soit μ_n la loi de $X^{(z_0,\sigma^{(n)})}_{\zeta(z_0,\sigma^{(n)})}$. C'est une mesure de Jensen de z_0 pour $P(\bar{D})$ sur $\partial D \cup K$; on peut extraire une sous suite μ_m convergente faiblement sur $\partial D \cup K$ vers μ qui est encore mesure représentative de z_0 pour $P(\bar{D})$; de plus comme $1\!\!1_{\partial D}$ est fonction continue sur $\partial D \cup K$, il s'ensuit que $\mu(\partial D) = 0$; donc μ est portée par K et par conséquent pour tout $f \in P(\bar{D})$

$$|f(z_0)| \leq \max_K |f|$$

Preuve du lemme 6 : généralisation immédiate du lemme 1 de [3].

Lemme 7 : Supposons que u soit continu sur D-K et soit $z_0 \in D-K$ avec $u(z_0) > 0$. Alors z_0 n'est pas dans l'enveloppe polynomiale $\hat{K}$ de K.

Preuve : soit $D' \subset \bar{D}' \subset D$ une boule contenant $K \cup \{z_0\}$. D'après les lemmes 3 et 4, u est alors plurisousharmonique continue et elle se prolonge évidemment à tout D par $u \equiv 0$ sur K de façon plurisousharmonique continue. D'après un théorème de Bremermann sur les fonctions psh continue [2], ε étant donné il existe $f_1 \dots f_p$ holomorphes sur $\bar{D}'$, et des $c_i > 0$ tels que

$$u(z) - \varepsilon \leq \sup_{1 \leq i \leq p} (c_i \log |f_i| (z)) \leq u(z).$$

Alors pour ε tel que $u(z_0) > \varepsilon$, il existe i tel que $|f_i(z_0)| > 1$ et d'autre part $|f_i|\big|_K \leq 1$. puisque $u \equiv 0$ sur K comme f_i est holomorphe au voisinage d'une boule, d'après le théorème d'Oka-Weil, on peut supposer que f_i est un polynôme, d'où le lemme 7.

Ceci nous donne en particulier.

<u>Théorème 3</u> : <u>Soit</u> $K \subset D$ (D <u>boule de</u> $\mathbb{C}^n$) <u>et</u> u <u>la fonction définie au 1</u>. <u>Supposons que</u> u <u>soit continu sur</u> $\overline{D-K}$. <u>Alors</u> u <u>est le potentiel capacitaire extrêmal et on a</u>

$$\hat{K} = \{z \in D/u(z) = 0\}$$

(*)

Donnons un exemple où la continuité de u se produit toujours : c'est le cas où K est l'adhérence $\overline{U}$ d'un domaine borné strictement pseudoconvexe de $\mathbb{C}^n$. On a alors :

<u>Théorème 4 : Soit</u> U <u>un domaine strictement pseudoconvexe borné de</u> $\mathbb{C}^n$, B <u>une boule contenant</u> $\overline{U}$. <u>Alors la fonction</u> u <u>est plurisousharmonique continue sur</u> $\overline{B}$, <u>vaut</u> 1 <u>sur</u> ∂B, 0 <u>sur</u> $\overline{U}$ <u>et l'enveloppe polynomiale</u> $\hat{U}$ <u>de</u> U <u>est</u>

$$\hat{U} = \{z \in B/u(z) = 0\} .$$

Preuve : utilisons la proposition 8-1 de [1]:

il existe v psh continue avec

$$v=1 \text{ sur } \partial B$$

$$v=0 \text{ sur } \partial U$$

$$(i\partial\bar{\partial} v)^{\wedge n} = 0 \text{ sur } B-\overline{U} .$$

et v est le supremum de la classe $\mathcal{G}$ des fonctions plurisousharmonique w

(*) <u>remarque</u> : si u est continue, il est clair que $\{u=0\}$ est holomorphiquement convexe contenant K, donc $\hat{K}$. L'expression du contrôle optimal dit que c'est exactement $\hat{K}$.

sur $B-\bar{U}$ avec

$$\limsup w \leq 1 \quad \text{sur} \quad \partial B$$

$$\limsup w \leq 0 \quad \text{sur} \quad \partial U.$$

Soit u la fonction définie au 1 et u* ; par le théorème 2 u* est psh continue et par les lemmes 1 et 2 $u^* = 1$ sur ∂B et 0 sur ∂U., donc $u^* \in \mathcal{E}$.

D'autre part, soit $w \in \mathcal{E}$; alors $0 \leq w \leq 1$ sur $B-U$ et par suite comme $w(X_t^{(z,\sigma)})$ est une sous martingale bornée, par le théorème de Doob

$$w(z) \leq E(w_{\zeta_{(z,\sigma)}}^{(z,\sigma)})) \leq P(X_{\zeta_{(z,\sigma)}}^{(z,\sigma)} \in \partial B)$$

d'où $w(z) \leq u(z) \leq u^*(z)$, donc $u^* = \sup_{w \in \mathcal{E}} w$ d'où $u^* = v$. Or $v \in \mathcal{E}$, donc $v \leq u \leq u^*$ d'où $u = v = u^*$ et u est continue une application des lemmes 5 ey 7 achève la preuve.

Remarque : nous aurions pu aussi utiliser l'espace des contrôles dont les processus associés vivant localement par droites complexes comme dans [8].

Exemple 1 : Voici un exemple très simple de domaines strictement pseudoconvexe circulaire non polynomialement convexe qui nous a été communiqué par J. Chaumat et A.M. Chollet : dans $\mathbb{C}^2$ on considère

$$U : (|z|^2 - r_1^2)(|z|^2 - |z|_2^2) + |W|^2 \leq 0 \qquad r_1 < r_2$$

pour r_1 et r_2 convenable on montre que ce domaine est strictement pseudoconvexe ; il est clair qu'il n'est pas polynomialement convexe.

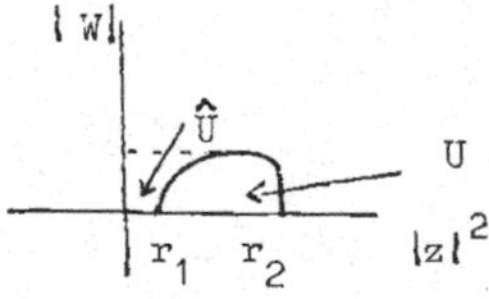

Exemple 2 : Dans le théorème 3, la propriété que u est continue ne peut pas être supprimée : en effet soit par exemple une variété analytique complexe V de dimension 1 de $\mathbb{C}^2$ et K une courbe fermée dessinée sur V ;

alors $\hat{K}$ est le compact et V qu'elle borde. Maintenant si B est une boule contenant K, alors la fonction u est $u = 0$ sur K

$u = 1$ sur $B-K$.

En effet, clairemant $u|_K = 0$; soit $z \notin K$ et $\sigma \in \mathcal{K}$ un contrôle, f une fonction holomorphe nulle sur V. Alors $f(X_t^{(z,\sigma)})$ est brownien complexe changé de temps, donc est constamment nul si il s'annule avec probabilité > 0. Donc si $P(X^{(\sigma,z)}_{\zeta(\sigma,z)} \in \partial D) < 1$, cela signifie que $f(X_t^{(z,\sigma)}) \equiv 0$ pour $t \leq \zeta_{(\sigma,z)}$. Mais alors $X_t^{(z,\sigma)}$ vit sur la variété V et par conséquent $\text{dét}\,\sigma\sigma^* = 0$.

4. Un autre espace de contrôle et sa relation avec l'enveloppe polynomiale

Introduisons l'espace $\mathcal{K}'$ des contrôles kählériens σ qui sont des matrices hermitiennes ≥ 0, $n \times n$, non anticipantes par rapport au brownien, tels que $t \longrightarrow \sigma(t,\omega)$ soit continu ps et enfin tels que $\mathrm{Tr}\,\sigma\sigma^* \geq 1$. Alors $\mathcal{K}' \supset \mathcal{K}$ certainement. On construit alors l'analogue u' de u en prenant dans (1) l'infimum sur le $\sigma \in \mathcal{K}'$. Les lemmes 1, 2, 3 subsistent. (Le seul point est de voir que l'inégalité (5) implique $E(\eta_{(\sigma,z)}) < +\infty$ mais cela est évident car

$$\sum_{i,j} (\sigma\sigma^*)^{ij} \frac{\partial^2 p}{\partial z_i \partial \bar{z}_j} \geq \min(\lambda_k) > \varepsilon > 0$$

où $\min(\lambda_k)$ est la plus petite valeur propre de la forme de Lévi de p ; le lemme 5 subsite encore de sorte que

$$\hat{K} \supset \{ z \in \mathbb{C}^n / u'(z) \neq 0 \} .$$

Si nous reprenons l'exemple 2. du 3. nous voyons que $\{u' = 0\}$ donne $\hat{K}$; car si $z_0 \in \hat{K}$, en utilisant la diffusion qui reste constamment dans V, cela fournit un contrôle qui ne sort pas de B avant de toucher K, donc $u'(z_0) = 0$ et si $z_0 \notin V$, il existe f holomorphe avec $f(z_0) \neq 0$ et $f|_V \equiv 0$. Alors pour tout $\sigma \in \mathcal{K}'$ $f(X_t^{(z_0,\sigma)})$ est brownien complexe changé de temps qui ne peut pas s'annuler, donc

$u'(z_0) = 1$. Enfin si $z_0 \in V \cap \complement \hat{K}$, le seul processus qui peut toucher K doit rester sur V d'après le raisonnement du 3, auquel cas il a une probabilité > 0 de sortir de B avant de toucher K.

Cependant si u' est continu, alors $u' = u = v$ de la même façon qu'au théorème 4. L'espace des contrôles $\mathcal{K}'$ contient tous les processus qui vivent localement (dans le temps) sur des ensembles analytiques complexes de dimension 1. Mais, a priori il contient beaucoup plus, car il contient toutes les limites possibles de ces processus.

Adresse: Mathématiques
Université Pierre et Marie Curie
4, Place Jussieu
F-75230 Paris Cedex 05, France

Références

[1] BEDFORD, E. and B. A. TAYLOR: Variational properties of the complex Monge-Ampère equatin. I. Dirichlet principle, Duke Math. J. 45 ('978), 375-403.

[2] BREMERMANN, H.: On a generalized Dirichlet problem for plurisubharmonic functions and pseudoconvex domains. Characterization of Šilov boundaries, Trans. Amer. Math. Soc. 91 (1959), 246-276.

[3] DEBIARD, A. et B. GAVEAU: Potentiel fin et algèbres de fonctions analytiques I-III, J. Functional Analysis 16 (1974), 289-304, 17 (1974), 296-310, et 21 (1976), 448-468.

[4] —— et ——: Balayage des fonctions plurisousharmoniques de Rickart et applications à certaines algèbres de fonctions, CRAS Paris 286 (1978), 1117-1120.

[5] FUGLEDE, B.: Finely harmonic functions (Lecture Notes in Mathematics 289), Springer-Verlag, Berlin - Heidelberg - New York 1972.

[6] GAMELIN, T.: Uniform algebras spanned by Hartogs series, Pacific J. Math. 62 (1976), 401-417.

[7] —— et N. SIBONY: preprint d'Orsay (à paraitre).

[8] GAVEAU, B.: Méthodes de contrôle optimal en analyse complexe et en topologie, CRAS Paris 284 (1977), 24-26; Méthodes de contrôle optimal en analyse complexe I-II, J. Functional Analysis 25 (1977), 391-411 et Bull. Sci. Mathématiques 102 (1978), 101-128.

[9] ——: Potentiel de Green fin et approximation des fonctions finement harmoniques, CRAS Paris 280 (1975), 410-412.

[10] RICKART, C.: Plurisubharmonic functions and convexity properties for general function algebras,,Trans. Amer. Math. Soc. 169 (1972), 1-24.

[11] SICIAK, J.: Extremal plurisubharmonic functions in $\mathbb{C}^n$, Proc. of the First Finnish-Polish Summer School in Complex Analysis at Podlesice. Part I, ed. by J. Ławrynowicz and O. Lehto, Uniwersytet Łódzki, Łódź 1978, pp. 115-152.

[12] HENKIN, G. M.: à paraitre.

[13] KRYLOV, N.: Theor. Prob. Appl. 14 (1969), 330-336 et 16 (1971), 438-448.

ON AN INTEGRAL TRANSFORMATION DUE TO N. OBRECHKOFF

Ivan Hristov Dimovski and Virjinia Stoineva Kirjakova (Sofia)

In [1] N.Obrechkoff had introduced the following generalization of the integral transforms of Laplace and Meijer:

$$(1)\qquad F(x) = \int_0^\infty \Phi(xt)f(t)dt$$

with the kernel function

$$(2)\qquad \Phi(x) = \int_0^\infty \cdots \int_0^\infty u_1^{\beta_1} u_2^{\beta_2} \cdots u_p^{\beta_p} \exp(-u_1 - \cdots - u_p - \frac{x}{u_1 \cdots u_p}) du_1 \cdots du_p$$

with arbitrary real $\beta_1, \beta_2, \ldots, \beta_p$. In the same paper a real inversion formula of Post-Widder type is found. Now we propose a complex inversion formula and establish some operational properties for a modification of the Obrechkoff transform (1). This modification has been introduced in [2].

Definition 1. Let $\gamma_1 \le \gamma_2 \le \cdots \le \gamma_m$ be a sequence of m real numbers and

$$(3)\qquad K(z) = \underbrace{\int_0^\infty \cdots \int_0^\infty}_{(m-1)} \exp(-u_1 - \cdots - u_{m-1} - \frac{z}{u_1 \cdots u_{m-1}}) \prod_{k=1}^{m-1} u_k^{\gamma_k - \gamma_m - 1} \times du_1 \cdots du_{m-1}.$$

The integral transform

$$(4)\qquad \mathcal{K}_\gamma\{f(t);z\} = \beta \int_0^\infty K[(zt)^\beta] t^{\beta(\gamma_m+1)-1} f(t)dt,$$

defined for the functions on $(0,\infty)$ with $\beta > 0$, is said to be the *modified Obrechkoff transform*.

For the sake of brevity we name (4) *Obrechkoff's transform*. It is determined by the non-decreasing sequence

$$\gamma = (\gamma_1, \gamma_2, \ldots, \gamma_m) \text{ and by } \beta > 0.$$

Definition 2. Ω is the space of local integrable functions on $(0,\infty)$ which are $O(t^p)$ with arbitrary $p > \alpha = -\beta(\gamma_1 + 1)$ for $f \to +0$

and $O(\exp \lambda t^{\beta/m})$ for $t \to +\infty$ with an arbitrary real λ.

In [2] it is shown that each $f(t) \in \Omega$ has well defined Obrechkoff transform $\mathcal{K}_\gamma\{f(t);z\}$ as an analytic function in the truncated angle domain

(5) $$D_f = \{z : \operatorname{Re} z > \lambda\} \cap \{z : |\arg z| < \frac{\pi m}{\beta}\}$$

In the same paper a convolution theorem for the Obrechkoff transform (4) is proved.

<u>THEOREM 1</u>. <u>Let</u> $f(t), g(t) \in \Omega$ <u>and</u>

(6) $$f * g = T(f \circ g),$$

<u>where</u>

$$(f \circ g)(t) = t^\beta \int_0^1 \cdots \int_0^1 f[t(t_1 \ldots t_m)^{1/\beta}] g[t((1-t_1)(1-t_2)\ldots \times (1-t_m))^{1/\beta}] \prod_{k=1}^m [t_k(1-t_k)]^{\gamma_k} dt_1 \ldots dt_m.$$

<u>and</u>

$$Tf(t) = [t^{\gamma_m \beta} / \prod_{k=1}^s \Gamma(\gamma_m - \gamma_k)] \underbrace{\int_0^1 \cdots \int_0^1}_{s} f[t(t_1 \ldots t_s)^{1/\beta}] \prod_{k=1}^s t_k^{2\gamma_k} \times (1-t_k)^{\gamma_m - \gamma_k - 1} dt_1 \ldots dt_s$$

<u>with</u>

$$s : \gamma_1 \leqq \gamma_2 \leqq \ldots \leqq \gamma_s < \gamma_{s+1} = \ldots = \gamma_m.$$

<u>Then</u>

(7) $$\mathcal{K}_\gamma\{f * g; z\} = \mathcal{K}_\gamma\{f; z\} \mathcal{K}_\gamma\{g; z\}.$$

For a proof see [2] and [3].

The Oberchkoff transform (4) is closely related to a general differential operator of Bessel-type:

(8) $$B = t^{\alpha_0} \frac{d}{dt} t^{\alpha_1} \ldots \frac{d}{dt} t^{\alpha_{m-1}} \frac{d}{dt} t^{\alpha_m},$$

where

$$\alpha_0 = -\beta - \beta\gamma_1' + 1; \ \alpha_k = \beta\gamma_k' - \beta\gamma_{k+1}' + 1, \quad k = 1,2,\ldots,m-1; \ \alpha_m = \beta\gamma_m',$$

and $\gamma' = (\gamma_1', \ldots, \gamma_m')$ is a rearrangement of the sequence $\gamma = (\gamma_1, \ldots, \gamma_m)$

The Obrechkoff transform can be used as a transform basis of an operational calculus for each of the differential operators B, just as the usual Laplace transform is used in the classical operational calculus for the operator d/dt (see [3]).

Indeed, the operator B is right-invertable and its initial right inverse operator L for the point $t = 0$ on Ω has the form:

$$(10)\qquad Lf(t) = (t^{\beta}/\beta^{m}) \int_0^1 \cdots \int_0^1 f[t(t_1 \dots t_m)^{1/\beta}] \prod_{k=1}^{m} t_k^{\gamma_k} dt_1 \dots dt_m .$$

It satisfies the initial conditions

$$\lim_{t\to+0} t^{\alpha_k}\frac{d}{dt} t^{\alpha_{k+1}} \dots \frac{d}{dt} t^{\alpha_m} Lf(t) = 0, \quad k = 1,2\dots,m.$$

As it is shown in [3], the identity

$$(11)\qquad \mathcal{K}_\gamma\{Lf(t);z\} = \frac{1}{\beta^m z^\beta}\mathcal{K}_\gamma\{f(t);z\}$$

holds in Ω. It is a basic operational property of (4).

The objective of this note is to establish a corresponding differential law for the considered transform.

Definition 3. The operator

$$(12)\qquad F = I - LB,$$

where I denotes the identity operator, is said to be the "*initial value operator*" of the right inverse operator L of B.

It is easy to prove the following

LEMMA 1. *If* $f \in \Omega$ *and* $f \in C^m[0,+\infty)$, *then*

$$(13)\qquad Ff(t) = \sum_{i=1}^{m} [\beta^{i-m} \prod_{k=i+1}^{m} (\gamma_k - \gamma_i)^{-1} \lim_{t\to+0} t^{\alpha_i}\frac{d}{dt}\dots t^{\alpha_m} f(t)] t^{-\beta\gamma_i},$$

provided

$$\gamma_1 < \gamma_2 < \dots < \gamma_m < \gamma_1 + 1.$$

An analogue of the well known differential property of the Laplace transform:

$$\mathcal{L}\{\frac{d}{dt} f(t);z\} = z\mathcal{L}\{f(t);z\} - f(+0)$$

holds.

THEOREM 2. *Let* $f(t)$ *be a m-times continuously differentiable function on* $[0,+\infty)$ *and let* $f(t), Bf(t) \in \Omega$. *If*

$$(14)\qquad \gamma_1 < \gamma_2 < \dots < \gamma_m < \gamma_1 + 1,$$

then

$$(15)\qquad \mathcal{K}_\gamma\{Bf(t);z\} = \beta^m z^\beta K_\gamma\{f(t);z\} - \sum_{i=1}^{m} [\beta^i z^{\beta(\gamma_i - \gamma_m)} \prod_{j=1}^{i-1} \Gamma(\gamma_j - \gamma_i + 1) \times \prod_{j=i+1}^{m} \Gamma(\gamma_j - \gamma_i)] \lim_{t\to+0} t^{\alpha_i}\frac{d}{dt}\dots t^{\alpha_m} f(t).$$

P r o o f. According to (11), for each $f_1(t) \in \Omega$:

$$\mathcal{K}_\gamma\{f_1(t);z\} = \beta^m z^\beta \mathcal{K}_\gamma\{Lf_1(t);z\}.$$

Since $f_1(t) = Bf(t) \in \Omega$, then

$$\mathcal{K}_\gamma\{Bf(t);z\} = \beta^m z^\beta \mathcal{K}_\gamma\{LBf(t);z\}$$

$$= \beta^m z^\beta \mathcal{K}_\gamma\{f(t);z\} - \beta^m z^\beta \mathcal{K}_\gamma\{Ff(t);z\}.$$

Replacing $Ff(t)$ by (13) we receive (15).

COROLLARY 1. *If $f(t)$ is of the form $f(t) = t^p \tilde{f}(t)$ with $p > \alpha+\beta$ and with m-times continuously differentiable function $\tilde{f}(t)$ on $[0,+\infty)$, where $\tilde{f}^{(m)}(t) = O(\exp \lambda t^{\beta/m})$ for $t \to +\infty$, then the following simple relation*

$$(16) \quad \mathcal{K}_\gamma\{Bf(t);z\} = \beta^m z^\beta \mathcal{K}_\gamma\{f(t);z\}$$

holds, provided that (14) is assumed.

P r o o f. Under the assumptions made for $f(t)$ it is easy to verify that $Bf(t) \in \Omega$ and then

$$\mathcal{K}_\gamma\{Bf;z\} = \beta^m z^\beta \mathcal{K}_\gamma\{f;z\} - \beta^m z^\beta \mathcal{K}_\gamma\{Ff;z\}.$$

On the other hand, if (14) holds, then the operator $Ff(t) \equiv 0$, since

$$\lim_{t \to +0} t^{\alpha_i} \frac{d}{dt} t^{\,i+1} \dots \frac{d}{dt} t^{\,m} f(t) = 0$$

Thus the relation (16) is proved.

N. Obrechkoff had obtained a real inversion formula for the transform (1). Now we extend this result for the modification (14) of this transform. In a manner similar to that used in [1], it can be proved the following analogue of the Post-Widder real inversion formula ([4] p. 288):

$$f(t) = \lim_{k \to \infty} \frac{(-1)^k}{k!} \left(\frac{k}{t}\right)^{k+1} F^{(k)}\left(\frac{k}{t}\right),$$

where $F(x) = \mathcal{L}\{f(t);x\}$.

THEOREM 3. *Let the function $f(t)$ be Lebesgue integrable on every finite interval $(0,T)$, $T > 0$ and let the integral*

$$\mathcal{F}(x) = \int_0^\infty K[(xt)^\beta]\, t^{\beta(\gamma_m+1)-1} f(t)\,dt$$

converge for some $x = x_0 > 0$. If the modified Obrechkoff transform

$\mathfrak{F}(x)$ has derivatives of arbitrary order, then we form the sequence of functions

(17) $\mathfrak{F}_o(x) = \mathfrak{F}(x),\ \mathfrak{F}_1(x) = B^*\mathfrak{F}(x), \ldots, \mathfrak{F}_k(x) = B^{*k}\mathfrak{F}(x), \ldots$

where

(18) $B^* = x^{1-\beta}\frac{d}{dx}x^{\alpha_m - 1}\frac{d}{dx}x^{\alpha_m - 2}\ldots\frac{d}{dx}x^{\alpha_1}\frac{d}{dx}x^{\alpha_o+\alpha_m+\beta-1}.$

Then at every point of continuity $t = t_o$ of $f(t)$ we have

(19) $f(t_o) = \lim\limits_{k\to\infty}[(-1)^k/\beta^k k!]^m k^{-\sum_{i=1}^m \gamma_i}\left(\frac{k^{m/\beta}}{t_o}\right)^{\beta(\gamma_m+k+1)}\mathfrak{F}_k\left(\frac{k^{m/\beta}}{t_o}\right).$

A specialization of this formula leads to real inversion formulas for the Meijer transform, for the integral transform, investigated by Krätzel [9], etc.

In order to obtain a complex inversion formula, we use the following relation between (4) and m-dimensional Laplace transform, established in [3]:

LEMMA 2. If $f(t) \in \Omega$, then

(20) $$\mathfrak{K}_\gamma\{f(t);(z_1z_2\ldots z_m)^{1/\beta}\} = \left(\prod_{k=1}^m z_k^{\gamma_k-\gamma_m}\right)\mathcal{L}_m\{f[(t_1t_2\ldots t_m)^{1/\beta}] \times \prod_{k=1}^m t_k^{\gamma_k}; z_1, z_2, \ldots, z_m\},$$

where

$$\mathcal{L}_m\{f(t_1,\ldots,t_m);z_1,z_2,\ldots,z_m\} = \int_0^\infty\ldots\int_0^\infty \exp(-z_1t_1 - \ldots - z_mt_m) \times f(t_1,\ldots,t_m)\,dt_1\ldots dt_m.$$

Now using the relation (20) we obtain a complex inversion formula for the Obrechkoff transform (4).

THEOREM 4. If $f(t) \in \Omega$ is of the form $f(t) = t^p\tilde{f}(t)$ with $p > \alpha = -\beta(\gamma_1 + 1)$ and with $\tilde{f}^{(m)}(t) = O(\exp\lambda t^{\beta/m})$ for $t \to +\infty$, then

(21) $$f(t) = \frac{1}{(2\pi i)^m}t^{-\frac{\beta}{m}(\gamma_1+\ldots+\gamma_m)}\int_{\sigma-i\infty}^{\sigma+i\infty}\ldots\int_{\sigma-i\infty}^{\sigma+i\infty}\exp[t^{\frac{\beta}{m}}(z_1+\ldots+z_m)] \times \prod_{k=1}^m z_k^{\gamma_m-\gamma_k}\,\mathfrak{F}[(z_1\ldots z_m)^{1/\beta}]\,dz_1\ldots dz_m,$$

where

$\mathfrak{F}(z) = \mathfrak{K}_\gamma\{f(t);z\}$

and σ is a real constant.

Proof. Under the hypothesis of the theorem, the function $f[(t_1\ldots t_m)^{1/\beta}]\prod_{k=1}^{m} t_k^{\gamma_k}$ satisfies the conditions for validity of the complex inversion formula of the m-dimensional Laplace transformation ([5], p. 319), since

$$f[(t_1\ldots t_m)^{1/\beta}] \leq M\exp[\lambda(t_1\ldots t_m)^{\frac{1}{m}}] \leq M\exp(\tfrac{\lambda}{m}t_1+\ldots+\tfrac{\lambda}{m}t_m),$$

then

$$f[(t_1\ldots t_m)^{\frac{1}{\beta}}] = \frac{1}{(2\pi i)^m}\prod_{k=1}^{m} t_k^{-\gamma_k}\int_{\sigma-i\infty}^{\sigma+i\infty}\ldots\int_{\sigma-i\infty}^{\sigma+i\infty}\exp(z_1t_1+\ldots+z_mt_m) \times\prod_{k=1}^{m} z_k^{\gamma_m-\gamma_k}\mathcal{F}[(z_1\ldots z_m)^{1/\beta}]dz_1\ldots dz_m, \tag{22}$$

provided $\sigma > \lambda/m$. We put $t_1 = t_2 = \ldots = t_m = t^{\beta/m}$ and get the inversion formula (21).

COROLLARY 2. If $\gamma_1 = -\frac{1}{2}\nu$, $\gamma_2 = \frac{1}{2}\nu$, $\nu \geq 0$ and $\beta = 2$, then the Obrechkoff transform (4) coincides to a constant multiplier with a modification of the Meijer transform ([6]). In this case the formula (21) is identical to the well known complex inversion formula ([7], p. 81).

The transform (4), considered here, is a generalization of the classical Laplace and Meijer transforms. Also it generalizes some integral transformations of Bessel-type, proposed by Ditkin and Prudnikov [7], Botashev [8] and Krätzel [9], as a transform basis of more special type than differential operators (8). The operational properties and inversion formulas for these transforms could be received by specilization of the corresponding results for the modified Obrechkoff transform.

References

[1] Н. ОБРЕШКОВ: Върху някои интегрални представяния на реални функции върху реалната полуос, Известия на Мат. институт при БАН 3,1 (1958), 3-28.

[2] I. H. DIMOVSKI: A transform approach to operational calculus for the general Bessel-type differential operator, Comptes rendus Acad. Bulg. Sci. 27,2 (1974), 155-158.

[3] ——: Foundations of operational calculi for the Bessel-type differential operators, Serdica, Bulgaricae math. publ. 1,1 (1975), 51-63.

[4] D. V. WIDDER: The Laplace transform, Princeton Univ. Press, Princeton 1946.

[5] В. А. ДИТКИН, А. П. ПРУДНИКОВ: Операционное исчисление, Изд. "Высшая школа", Москва 1975.

[6] Н. ОБРЕШКОВ: Върху някои представяния с интеграли на реални функции, Известия на Мат. институт при БАН 1,1 (1953), 83-110.

[7] В. А. ДИТКИН, А. П. ПРУДНИКОВ: Интегральные преобразования и операционное исчисление, Изд. "Наука", Москва 1974.

[8] А. И. БОТАШЕВ: К теории операторного исчисления, Исследов. по интегро-дифф. уравн. в Киргизии 2 (1962), 297-304.

[9] E. KRÄTZEL: Eine Verallgemeinerung der Laplace- und Meijer-Transformations, Wiss. Zeitschrift der FSU Jena, Math.-Naturwiss. Reihe 14,5 (1965), 369-381.

Institute of Mathematics of the
Bulgarian Academy of Sciences
BG-1090 Sofia, P.O.Box 373, Bulgaria

EIN NEUER EXISTENZBEWEIS FÜR QUASIKONFORME ABBILDUNGEN MIT VORGEGEBENER KOMPLEXER DILATATION

Bodo Dittmar (Halle an der Saale)

Inhaltsverzeichnis

1. Einleitung

In [2] a.)(vgl.auch [2] b.)) wurden im Anschluß an eine bekannte Abhandlung von Schiffer und Hawley [8] gewisse quasikonforme Abbildungen eines n-fach zusammenhängenden Gebietes durch ein reelles nichtlineares Funktional charakterisiert. Hier soll nun gezeigt werden, wie dieses Funktional einen neuen Existenzbeweis für quasikonforme Abbildungen ermöglicht. Liegen nämlich keine Randkomponenten vor, das heißt, betrachten wir quasikonforme Abbildungen der Vollebene E mit vorgegebener komplexer Dilatation $\mu(z)$, dann geht dieses Funktional - wegen der fehlenden Randintegrale - in ein lineares Funktional über. Dieses Funktional kann auch erhalten werden, indem man die zu dem Differentialgleichungssystem

$$(1) \qquad w_{\bar{z}} = \mu(z) \cdot w_z + \mu_z$$

gehörende Eulersche Differentialgleichung für $\operatorname{Re} w(z)$ benutzt, um ein entsprechendes quadratisches Funktional für den Realteil der Lösung von (1) zu gewinnen. Verwendet man dieses Funktional, so können zunächst mit der sogenannten Hilbertraummethode zu hinreichend glatten komplexen Dilatationen $\mu(z)$ Lösungen der zu (1) gehörenden inhomogenen reellen Beltramischen Differentialgleichung gefunden werden. Hiermit erhält man dann leicht schlichte Lösungen der Beltramischen Differentialgleichung

(2) $w_{\bar z} = \mu(z) \cdot w_z$.

Durch bekannte Grenzübergänge kann dann auf den allgemeinen Fall quasikonformer Abbildungen mit in E summierbarer komplexer Dilatation geschlossen werden. Das wird in der vorliegenden Arbeit durchgeführt, und damit wird den zahlreichen Existenzbeweisen zur Beltramischen Differentialgleichung (s. etwa die Übersicht in [7], S. 49) ein weiterer neuartiger hinzugefügt.

Auch hier möchte ich mich wieder herzlich bei Herrn Reiner Kühnau bedanken, durch dessen ständige Anregung und Anleitung diese Arbeit ermöglicht wurde.

§ 2. Hilfsmittel

Bevor wir zwei Hilfssätze beweisen, sollen einige im folgenden Paragraphen benötigte Größen definiert werden. Zu gegebenem $\mu(z) = \mu_1 + i\mu_2 \in C_0^3(E)$ definieren wir:

(3) $$E = |1+\mu|^2, \quad F = 2\mu_2, \quad G = |1-\mu|^2, \quad W = \sqrt{EG-F^2} = 1-|\mu|^2$$

und

(4) $$H = \frac{1}{16W}(2F(E_x - G_x + 2F_y) + (2F_x - E_y + G_y)(G - E - 4)),$$

$$I = \frac{1}{16W}(2F(G_y - E_y + 2F_x) + (2F_y + E_x - G_x)(E - G - 4)).$$

Weiterhin werden auch hier wieder die Beltramischen Differentialoperatoren benötigt [1]

(5) $$\nabla(\varphi,\psi) = \frac{E}{W^2}\cdot\varphi_y\psi_y - \frac{F}{W^2}\cdot(\varphi_x\psi_y + \varphi_y\psi_x) + \frac{G}{W^2}\cdot\varphi_x\psi_x ,$$

(6) $$\nabla(\varphi) = \nabla(\varphi,\varphi),$$

(7) $$\Delta(\varphi) = \frac{1}{W}\left(\frac{E}{W}\varphi_y - \frac{F}{W}\varphi_x\right)_y + \frac{1}{W}\left(\frac{G}{W}\varphi_x - \frac{F}{W}\varphi_y\right)_x .$$

Ist $|\mu(z)| \le k < 1$, dann haben wir für beliebig reelle x_1, x_2

(8) $$\frac{1-k}{1+k}(x_1^2 + x_2^2) \le \frac{G}{W}\cdot x_1^2 + \frac{E}{W}\cdot x_2^2 - \frac{2F}{W}\cdot x_1x_2 \le \frac{1+k}{1-k}(x_1^2 + x_2^2).$$

Wir betrachten die lineare Mannigfaltigkeit H der in der (endlichen) Ebene E definierten Funktionen, die in jedem Kompaktum quadratisch summierbar sind. Diese Funktionen mögen verallgemeinerte 1. Ableitungen nach x und y besitzen [9], die ihrerseits in der ganzen Ebene quadratisch summierbar sind. In dieser Funktionenmenge wird das folgende Skalarprodukt eingeführt:

$$(9) \qquad (u,v) = \iint_E W \cdot \nabla(u,v)\,dxdy, \quad u,v \in H.$$

Weiterhin wird gefordert:

$$(10) \qquad \int_{|z|=R_o} u\,ds = 0, \quad R_o > 0 \text{ fest.}$$

Dann kann bewiesen werden das

<u>LEMMA 1</u>. <u>Die lineare Mannigfaltigkeit</u> H <u>bildet mit</u> (9) <u>und</u> (10) <u>einen Hilbertraum, falls</u> $\mu \in C_o^3(E)$ <u>und</u> $|\mu| \leq k < 1$ <u>in</u> E <u>gilt</u>.

Es genügt offenbar, den V o l l s t ä n d i g k e i t s b e w e i s zu führen. Dazu betrachten wir in $W_2^1(K_R = \{z : |z| < R\})$ [9], $R \geq R_o$, die Norm

$$(11) \qquad \|u\|_{W_2^1(K_R)} = \sqrt{\iint_{|z|<R} u_x^2} + \sqrt{\iint_{|z|<R} u_y^2} + \left|\int_{|z|=R_o} u\,ds\right|.$$

(11) ist nach [9], S. 307 f., eine äquivalente Norm in $W_2^1(K_R)$. Denn die linke Seite von (10) stellt ein lineares und beschränktes Funktional in $W_2^1(K_R)$, $R \geq R_o$, nach den Sobolevschen Einbettungssätzen dar [9], S. 318. Es sei also $\{u_n\}$ eine Fundamentalfolge in H, dann existieren wegen (8) und der Vollständigkeit von $L^2(E)$ zwei Funktionen $v_1, v_2 \in L^2(E)$, mit

$$(12) \qquad \|u_{n_x} - v_1\|_{L^2(E)} \longrightarrow 0, \quad \|u_{n_y} - v_2\|_{L^2(E)} \longrightarrow 0 \quad \text{für } n \longrightarrow \infty.$$

Ist nun ein beliebiges Kompaktum K gegeben, so existiert ein $R' \geq R_o$ mit $K \subset K_{R'}$, und nach den Einbettungssätzen ist $\{u_n\}$ auch Fundamentalfolge in $L^2(K_{R'})$. Somit wird in der ganzen (endlichen) Ebene f. ü. eindeutig eine Funktion u definiert mit

(13) $\qquad \| u - u_n \|_{L^2(K)} \longrightarrow 0, \quad n \longrightarrow \infty.$

Nach Definition der verallgemeinerten Ableitungen gilt

(14) $\qquad \iint_E u_n \cdot \varphi_x \, dxdy = - \iint_E u_{n_x} \cdot \varphi \, dxdy, \qquad \varphi \in C_0^1(E).$

Für jedes beliebige $\varphi \in C_0^1(E)$ kann in (14) wegen (12) und (13) der Grenzwert genommen werden, so daß folgt $u_x = v_1$, $u_y = v_2$. Schließlich folgt aus den Einbettungssätzen noch die Normierungsbedingung (10) für u, womit der Vollständigkeitsbeweis erbracht ist.

Mit Lemma 1 folgt leicht

LEMMA 2. *Ist* $\mu(z) \in C_0^3(E)$ *und gilt*:

$$|\mu(z)| \leq k < 1 \quad \text{in} \quad E,$$

dann existiert ein $u_0 \in C^2(E)$ *mit endlichem Dirichletschen Integral und*

$$\Delta u_0 = - (H_y + I_x)/W.$$

Zum B e w e i s betrachten wir das Funktional

(15) $\qquad L[u] = \iint_E u(H_y + I_x) dxdy, \qquad u \in H.$

Dieses Funktional ist offenbar linear und da $H, \; I \in C_0^1(E)$ nach Voraussetzung gilt, folgt nach Definition der verallgemeinerten Ableitungen

$$L[u] = - \iint_E (H \cdot u_y + I \cdot u_x) dxdy.$$

Mit der Cauchy-Schwarzschen Ungleichung folgt also die Beschränktheit von (15) in H. Der Satz von Riesz bringt somit:

(16) $\qquad L[u] = (v_0, u), \quad \forall u \in H, \quad v_0 \in H.$

(16) gilt auch für $u = \varphi \in C_0^\infty(E)$, da die wegen der Normierungsvorschrift (10) eventuell nötige Addition einer Konstanten am Bestehen von (16) offenbar nichts ändert. Es kann also partiell integriert werden, und wir erhalten

$$(17)\qquad \iint_E v_o \cdot W \cdot \Delta(\varphi)\,dxdy = - \iint_E \varphi(H_y + I_x)\,dxdy, \qquad \varphi \in C_o^\infty(E).$$

Mit $\mu(z) \in C_o^3(E)$ sind die Voraussetzungen des Weylschen Lemmas [4], S. 189, erfüllt, und damit existiert ein $u_o \in C^2(E)$ mit $u_o = v_o$ f. ü. in E. Offenbar genügt u_o auch (17), und zweimalige partielle Integration bringt die Aussage des Lemmas.

§ 3. Existenzbeweis

Der folgende Existenzsatz für quasikonforme Abbildungen wird bewiesen:

SATZ 1. In einem Gebiet D der komplexen Vollebene sei eine summierbare Funktion $\mu(z)$ gegeben mit

$$(18)\qquad \sup_{z\in D} |\mu(z)| < 1.$$

Dann gibt es eine quasikonforme Abbildung von D, deren komplexe Dilatation f. ü. in D mit $\mu(z)$ übereinstimmt.

B e w e i s. Ohne Einschränkung der Allgemeinheit kann das Gebiet D als die gesamte endliche Ebene angenommen werden, wobei $\mu(z) \equiv 0$ für z im Komplement von D gesetzt wird. Zu $\mu(z) \in L(E)$ mit (18) kann eine Folge $\{\mu_k(z)\} \subset C_o^\infty(E)$, mit $\mu_k(z) \longrightarrow \mu(z)$ in L(E) folgendermaßen konstruiert werden [5], S.149:

$$\mu_k(z) = \iint_E \varphi_k(z-\zeta)\,\theta_{n_k}(\zeta)\,d\tau_\zeta,$$

$$\varphi_k(z) = \begin{cases} \mu(z), & |z| < k, \\ 0 & \text{sonst,} \end{cases}$$

$$\theta_{n_k}(z) = \begin{cases} a_{n_k} \exp(1/|z|^2 - \frac{1}{n_k^2}), & |z| < \frac{1}{n_k}, \\ 0 & \text{sonst,} \end{cases}$$

wobei die a_{n_k} so zu wählen sind, daß

$$\iint_E \theta_{n_k}(z)\,d\tau_z = 1$$

gilt. Weiterhin gilt, wie man leicht nachrechnet:

$$(19)\qquad \sup_{z\in E} |\mu_k(z)| \le \sup_{z\in E} |\mu(z)|.$$

Zu jedem $\mu_k(z)$ gehört nach Lemma 2 ein u_k (in Lemma 2 u_o genannt), und hierzu definieren wir die bis auf eine additive Konstante eindeutig bestimmte Funktion

$$v_k = \int_{z_o}^{z} \left(\frac{F_k}{W_k} \cdot u_{k_x} - \frac{E_k}{W_k} \cdot u_{k_y} - H_k\right)dx + \left(\frac{G_k}{W_k} \cdot u_{k_x} - \frac{F_k}{W_k} \cdot u_{k_y} + I_k\right)dy,$$

die offenbar aus $C^2(E)$ ist (E_k, F_k, G_k und W_k sind hierbei die zu $\mu_k(z)$ durch (3) definierten Größen).

Die Funktion $f_k(z) = u_k + iv_k$ erfüllt damit (1), mit $\mu_k(z)$ anstelle von $\mu(z)$. Folglich genügt die eindeutige Funktion

$$(20) \qquad w_k(z) = \int_{z_o}^{z} e^{f_k(z)} dz + \mu_k(z) \cdot e^{\overline{f_k(z)}}$$

der mit $\mu_k(z)$ gebildeten Beltramischen Differentialgleichung (2). Die Funktionen u_k, v_k haben nach Lemma 2 ein endliches Dirichletsches Integral, so daß die in einer Umgebung von $z = \infty$ analytische Funktion $f_k(z)$ in $z = \infty$ auch noch analytisch ist. Wir können annehmen, daß $f_k(\infty) = 0$ gilt. Setzen wir die Reihenentwicklung von $f_k(z)$ in (20) ein, so folgt wegen der Eindeutigkeit von $w_k(z)$ die hydrodynamische Normiertheit dieser Funktionen in $z = \infty$. Nach einem allgemeinen topologischen Satz von Hadamard [3] folgt nun aus der offensichtlichen lokalen Schlichtheit der $w_k(z)$ die Schlichtheit in der gesamten Vollebene. Es kann auch leicht unter Verwendung des Argumentprinzips für (2) wie in [10], II, § 5, auf die Schlichtheit der $w_k(z)$ geschlossen werden. Die Folge quasikonformer Abbildungen $\{w_k(z)\}$ erlaubt nun, wie in [5], S. 203, zu schließen: Wir wählen aus $\{\mu_k(z)\}$ eine f.ü. gegen $\mu(z)$ konvergente Teilfolge $\{\mu_{k'}(z)\}$ aus. Die entsprechende Folge $\{w_{k'}(z)\}$ kann durch Ähnlichkeitstransformationen so normiert werden, daß $w_{k'}(\infty) = \infty$, $w_{k'}(1) = 1$ und $w_{k'}(0) = 0$ gilt. Damit kann aus dieser Folge eine in jedem kompakten Teil von E gleichmäßig konvergente Teilfoge ausgewählt werden [5], S. 76, Satz 5. 1, die gegen eine quasikonforme Abbildung $w(z)$ strebt [5], S. 77, Satz 5. 3, deren komplexe Dilatation f.ü. mit $\mu(z)$ übereinstimmt [5], S. 197, Satz 5. 2.

Literatur

[1] BLASCHKE,W.: Differentialgeometrie I, Berlin 1945.

[2] a.) DITTMAR,B.: Übertragung eines Extremalproblems von M.Schiffer und N.S.Hawley auf quasikonforme Abbildungen. Math.Nachr., im Druck.

[2] b.) ——: Bemerkungen zu einem Funktional von M.Schiffer und N.S.Hawley in der Theorie der Konformen Abbildungen. Math.Nachr., im Druck.

[3] HADAMARD,J.: Sur les transformations pounctuelles, Bull.Soc.Math. France 34, 71-84 (1906).

[4] HELLWIG,G.: Partielle Differentialgleichungen, Stuttgart 1960.

[5] LEHTO,O. und K.I.VIRTANEN: Quasikonforme Abbildungen, Berlin-Heidelberg-New York 1965.

[6] MICHLIN,S.G.: Numerische Realisierung von Variationsmethoden, Berlin 1969 (Übers. aus dem Russ.).

[7] NITSCHE,J.C.C.: Vorlesungen über Minimalflächen, Berlin-Heidelberg-New York 1975.

[8] SCHIFFER,M. and N.S.HAWLEY: Connections and conformal mapping, Acta math. 107, 175-274 (1962).

[9] SMIRNOV,V.: Lehrgang der höheren Mathematik V, 7. Auflage, Berlin 1976 (Übers. aus dem Russ.).

[10] VEKUA,I.N.: Verallgemeinerte analytische Funktionen, Berlin 1963 (Übers. aus dem Russ.).

Sektion-Mathematik der Martin-
Luther-Universität Halle-Wittenberg
Universitätsplatz 8/9
DDR-402 Halle an der Saale, DDR

EXAMPLES OF HARMONIC AND HOLOMORPHIC MAPS

James Eells (Coventry)

Contents

In this expository lecture my main object is to present certain geometric examples of harmonic and holomorphic maps between surfaces - and above all, to illustrate interrelationships between those concepts. For background and unexplained notation, consult [4,6].

Harmonic and holomorphic maps

First of all, let M and N be domains in the complex plane $\mathbb{C}$. A smooth map $\phi:M \to N$ is said to be holomorphic (resp., antiholomorphic; we shall combine these by writing $\pm$ holomorphic) if the differential $d\phi(x)$ is $\mathbb{C}$-linear (resp., $\mathbb{C}$-antilinear) for every point $x \in M$. Equivalently, if the complex differential $\bar{\partial}\phi \equiv 0$ on M (resp., $\partial\phi \equiv 0$ on M). A map $\phi:M \to N$ is harmonic if $\mathrm{div}(d\phi) \equiv 0$ on M. Of course, every $\pm$ holomorphic map is harmonic.

The first of these notions depends only on the complex structures of M and N. By way of contrast, the second notion really involves their Hermitian metrics (induced from that of $\mathbb{C}$). To be more precise in these last statements, I emphasise that I wish geometric concepts of holomorphicity and harmonicity; e.g., ones which are meaningful for maps between Riemann surfaces.
For that generality, the very notion of the divergence operator "div" requires Hermitian structures on both M and N. Once again,

<u>every</u> $\pm$ <u>holomorphic map is harmonic</u>.

There is a general theory of $\pm$ holomorphic and harmonic maps - the natural context being that of Kähler geometry. In this lecture, however, I shall restrict my attention to the case of Riemann surfaces.

Surfaces in 3-space

Now let M be a surface in $\mathbb{R}^3$. Then M has an induced orientation, so that in particular we may speak of the outward drawn normal to the oriented tangent $T_x(M)$ at any point $x \in M$. Thus if S denotes the unit sphere of $\mathbb{R}^3$ centred at the origin, we have

Gauss's map $\gamma : M \longrightarrow S$,

which assigns the outward drawn unit normal vector $\gamma(x)$ to each $x \in M$.

If g_o denotes the Euclidean metric of $\mathbb{R}^3$ and $i : M \longrightarrow \mathbb{R}^3$ the embedding, then $g = i{*}g_o$ is a Riemannian metric on M; i.e., for each $x \in M$, g_x is an inner product on the tangent space $T_x(M)$. g is called the <u>first fundamental form of</u> M.

That metric (rather, that embedding) induces a complex structure on M; indeed, we introduce a complex structure J_x on each vector space $T_x(M)$ by defining $J_x(v)$ to be rotation of v through $+\pi/2$, for every $v \neq 0$. With that, g_x determines a Hermitian structure on $T_x(M)$. It is a standard fact that such an almost complex structure J) i.e., tensor field $x \longrightarrow J_x$) determines a complex structure on the surface M itself.

The extent to which M is curved in $\mathbb{R}^3$ is measured by the <u>second fundamental form</u> β of the embedding. In fact, β_x is a quadratic form on $T_x(M)$, which is just the differential $d\gamma(x)$. (Indeed, the induced bundle $\gamma^{-1}(T(S)) = T^*(M) \otimes V(\mathbb{R}^3,M)$, where $T^*(M)$ denotes the contagent bundle of M and $V(\mathbb{R}^3,M)$ the normal bundle of M in $\mathbb{R}^3$; and $d\gamma(x) \in T^*_x(M) \otimes T_{\gamma(x)}(S)$. Otherwise said, $d\gamma$ is a section of

(1) $\quad T^*(M) \otimes \gamma^{-1}(T(S)) = \odot^2 T^*(M) \otimes V(\mathbb{R}^3,M)$.

But $V(\mathbb{R}^3,M)$ is a trivialised line bundle over M, so that $d\gamma$ can be interpreted as a section of $\odot^2 T^*(M)$.); see [12,15].

Gauss maps of surfaces

Let us now consider special classes of Gauss maps:

Example. γ is a constant map. In that case, $\beta \equiv 0$; thus i is totally geodesic, carrying M into an affine plane of $\mathbb{R}^3$.

Example. γ is $\pm$ holomorphic. It is a classical fact that M is minimal in $\mathbb{R}^3$ (i.e., M locally minimises surface area) iff γ is $-$ holomorphic. I.e., the conjugate map $\bar{\gamma} : M \longrightarrow S$ is a meromorphic function on M. That can be described by saying that

(2) $\quad \mathrm{Trace}\ \beta \equiv 0.$

I.e., that M has 0 mean curvature in $\mathbb{R}^3$.

Example. γ is harmonic. That could be described by saying that the covariant differential

(3) $\quad \nabla\ \mathrm{Trace}\ \beta \equiv 0.$

And is interpreted by the geometric condition that M has constant mean curvature in $\mathbb{R}^3$ [12].

If we express the Hermitian metric g of M in an isothermal chart as $g = \rho^2 dz d\bar{z}$, and that of S as $h = \sigma^2 dw d\bar{w}$, then, using subscripts to denote complex differentiation,

γ is $-$ holomorphic iff $\gamma_z \equiv 0$.

γ is harmonic iff

(4) $$\gamma_{z\bar{z}} + \frac{2\sigma_w}{\sigma}\gamma_z\gamma_{\bar{z}} \equiv 0$$

It should be noted that harmonicity of γ uses strongly the metric structure of S; but only the conformal structure of M.

There is another characterisation of harmonicity:

$\gamma^* h$ is a 2-covariant tensor field on M, the third fundamental form of M in $\mathbb{R}^3$; we let $(\gamma^* h)^{2,0}$ denote the component involving dz^2. Then

<u>γ is harmonic iff $(\gamma^* h)^{2,0}$ is a holomorphic quadratic differential.</u>

Remarks on these constructions

I wish to give specific examples of such surfaces in $\mathbb{R}^3$, but first let me make the following remarks:

(a) There are no compact minimal surfaces in $\mathbb{R}^3$, by the maximum principle. The only compact surfaces of constant mean curvature in $\mathbb{R}^3$ are the Euclidean spheres, by a theorem of Alexandrov.

We can generalise our constructions to include immersions of M in $\mathbb{R}^3$; however, that does not broaden the class of compact M, in the case of minimal submanifolds - and may well not broaden the class of compact surfaces of constant mean curvature.

(b) In another direction, we can make our constructions for embeddings of M in the torus $T^3 = \mathbb{R}^3/\mathbb{Z}^3$, and obtain substantially greater generality; see [9,10,11]. For instance there are many minimal immersions of compact surfaces M of genus p into T^3, obtainded by generalising a construction of H.A. Schwarz; their Gauss maps $\gamma : M \longrightarrow S$ are - holomorphic maps of Brouwer degree 1-p. That is especially interesting, for we do not have sufficient information about the existence of meromorphic functions of degree < p on compact Riemann surfaces.

(c) Generally, <u>a harmonic map $\phi : M \longrightarrow S$ of a compact orientable surface of genus ≥ 0 is $\pm$ holomorphic</u> [7]. Thus in our present examples we are confronted with the first case of maps γ which are harmonic but not $\pm$ holomorphic.

(d) The third example above can be generalised as follows: <u>A branched immersion $i : M \longrightarrow \mathbb{R}^3$ has constant mean curvature iff its Gauss map $\gamma : M \longrightarrow S$ is harmonic</u> [5]. The nonsingular case was established by Ruh-Vilms [12]; the presence of branch points requires special knownledge of the properties of solutions of the harmonic map equation (3).

Delaunay's surfaces

Let C be the curve traced by a focus as a conic rolls in a plane along an axis L. Consider the surface of revolution M by revolving C around L.

If the conic is a parabola, then C is a catenary; thus M is the catenoid in $\mathbb{R}^3$. <u>Its Gauss map $\gamma : M \longrightarrow S$ is - holomorphic</u>.

If the conic is a parabola, then C is a catenary; thus M is the catenoid in $\mathbb{R}^3$. Its Gauss map $\gamma : M \longrightarrow S$ is - holomorphic. It is well known that the catenoid is the only complete minimal surface of revolution in $\mathbb{R}^3$.

If the conic is an ellipse, then M is the unduloid. It is a complete surface of revolution embedded in $\mathbb{R}^3$, Its Gauss map $\gamma : M \longrightarrow S$ is harmonic, and covers a band around an equator of S. An analytic version of that fact was first discovered by R.T. Smith [13].

If the conic is an hyperbola, then M is the nodoid. It is a complete surface immersed in $\mathbb{R}^3$. Its Gauss map $\gamma : M \longrightarrow S$ is harmonic, and is surjective.

Delaunay proved that the complete immersed surfaces of revolution in $\mathbb{R}^3$ with constant mean curvature are precisely those obtained by revolving about their axes L the roulettes C of conics.

These constructions are applied in [5] to produce significant harmonic maps of tori and Klein bottles into S.

References

[1] DELAUNAY, C.: Sur la surface de révolution dont la courbure moyenne est constante. J.Math. pures et appl. Sér.1(6) (1841), 309-320; with a note appended by M.STURM.

[2] EELLS,J.: Harmonic maps of surfaces, Proc. Eighth Annual National Mathematics Conference, Tehran. Bull. Iranian Math. Soc. (1978).

[3] ——: On the surfaces of Delaunay and their Gauss maps, Fourth Coloq. Inter. Geo. Dif. (In honour of Professor E.Vidal), Santiago de Compostela.

[4] —— and L.LEMAIRE: A report on harmonic maps, Bull. London Math. Soc. 10 (1978), 1-68.

[5] —— and ——: On the construction of harmonic and holomorphic maps between surfaces, to appear.

[6] —— and J.H. SAMPSON: Harmonic mapping of Riemannian manifolds, Ann. J. Math. 86 (1964), 109-160.

[7] —— and J.C. WOOD: Restrictions on harmonic maps of surfaces, Topology 15 (1976), 263-266.

[8] KENMOTSU,K.: Weierstrass formula for surfaces of prescribed mean curvature, to appear.

[9] MEEKS,W.H.: The conformal structure and geometry of triply periodic minimal surfaces in $\mathbb{R}^3$, Bull. Am. Math. Soc. 83 (1977), 134-6.

[10] NAGANO,T. and B. SMYTH: Sur les surfaces minimales hyperelliptique dans un tore, C.R. Paris A280 (1975), 1527-1529.

[11] ——— and ———: Periodic minimal sufaces, Comm. Math. Helv. (1978).

[12] RUH,E.A. and J.VILMS:, The tension field of the Gauss map. Trans. Am. Math. Soc. 149 (1970), 509-513.

[13] SMITH,R.T.: Harmonic mapping of spheres, Warwick Thesis (1972).

[14] THOMPSON,D'A.W.: Growth and Form, Cambridge (1917).

[15] VILMS,J.: Submanifolds of Euclidean space with parallel second fundamental form, Proc. Ann. Math. Soc. 32 (1972), 263-267.

Mathematics Institute
University of Warwick
Coventry CV4 7AL, U.K.

THE MODULUS AND THE HYPERBOLIC MEASURE

Dorin Ghişa (Timişoara)

Studying the possibility of extending to Riemann surfaces the notion of modulus for a family of measures (cf. [1] and [4]), we considered in [2] the measures compatible in a precise way with conformal structure of the surface. So, with every real function f, defined on a Riemann surface W and having a.e. partial derivatives which belong locally to $\mathcal{L}^2$, we may associate a Borel measure μ_f on W so that for every local map (V,h) and for every Borel set $B\subset V$, the measure $\mu_f(B)$ coincide with

$$\int_{h(B)} \|\operatorname{grad} f\circ h^{-1}\|^2 dxdy .$$

If φ is a K-quasiconformal mapping of the Riemann surface W onto a Riemann surface W' and $f'=f\circ\varphi^{-1}$, we proved (cf. [3]) that for every Borel set $B\subset W$ we have

$$(1)\qquad (1/K)\mu_f(B)\le \mu_{f'}(\varphi(B))\le K\mu_f(B).$$

The measures μ_f and $\mu_{f'}$ are called <u>associated measures</u> by means of the quasiconformal mapping φ.

Now, if W is a hyperbolic Riemann surface, we may define a measure on it in a natural way. Let $\tilde{W}$ be the universal covering surface of W and let Π be its canonical projection on W. It has the property that every point $\tilde{p}\in\tilde{W}$ has a neighbourhood $\tilde{V}$ which is homeomorphic by means of Π with a neighbourhood V of $p=\Pi(\tilde{p})$. There is a countable family ϑ of neighbourhoods V with this property, so that $W=\bigcup_{V\in\vartheta} V$.

On the other hand $\tilde{W}$ is conformally equivalent to the unit disk $E=\{z;\ |z|<1\}$ and the surface element $(1-|z|^2)^{-2}dxdy$ is invariant under conformal mapping of E onto itself. Hence, if $\tilde{B}\subset\tilde{W}$, then

$$\tilde{m}(\tilde{B}) = \int_{\psi^{-1}(B)} (1-|z|^2)^{-2}dxdy$$

is the same for every conformal mapping ψ of E onto $\tilde{W}$.

If B is a Borel set on W so that there is a V in ϑ which contains B, we put $\Pi_{\tilde{V}}$ for the restriction of Π to $\tilde{V}$ and

$$m(B) = \tilde{m}(\Pi_{\tilde{V}}^{-1}(B)).$$

It may be easily verified that $m(B)$ does not depend on the neighbourhood V, nor on $\tilde{V}$ which projects on V. In a similar way as in [2], we pass from this case to the case where B is included in a compact set of W and finally to the case where B is an arbitrary Borel set. It follows that m is a Borel measure on W, which can be named the <u>hyperbolic measure</u> of W.

We will use the measure m as the basic measure in the definition of the module of a family of measures on W, in the sense of M. Ohtsuka [4]. Namely, let $\mathcal{M}$ be a family of measures on W and ϱ a positive m-measurable function defined on W so that $\int_W \varrho\, d\mu \geq 1$ for any measure μ of $\mathcal{M}$. (ϱ is called an $\mathcal{M}$-admissible function; $\mathcal{M}$-ad. ϱ). If Φ is an increasing continuous function on $\bar{R}$ such that $\Phi(0) = 0$, then

$$M_{\Phi,m}(\mathcal{M}) = \inf_{\mathcal{M}\text{-ad}.\varrho} \int_W \Phi(\varrho)dm$$

is called the (Φ,m)-module of $\mathcal{M}$.

Let φ be a quasiconformal mapping of W onto a Riemann surface W'. Then, for any measure μ on W there is a measure μ' on W' with the property that for every measurable function f and for every Borel set B on W we have

$$\int_B f\, d\mu = \int_{\varphi(B)} f \circ \varphi^{-1}\, d\mu'.$$

The measure μ' is called the <u>carried measure</u> of μ by φ. The family $\mathcal{M}'$ of carried measures μ of $\mathcal{M}$ is called the <u>carried family</u> of $\mathcal{M}$. With this notion we may prove the following

<u>THEOREM</u>. <u>Let W be a hyperbolic Riemann surface and let φ be a K-quasiconformal mapping of W onto a Riemann surface W'. If $\mathcal{M}$ is a family of measures on W and $\mathcal{M}'$ is its carried family on W', m is the hyperbolic measure of W and m' is its associated measure by means of φ, then</u>

$$(2) \qquad (1/K)M_{\Phi,m}(\mathcal{M}) \leq M_{\Phi,m'}(\mathcal{M}') \leq K\, M_{\Phi,m}(\mathcal{M}).$$

P r o o f. At first let ϱ be the characteristic function of a Borel set B contained in a parametric neighbourhood V. Then

$$\int_W \Phi(\varrho)\,dm = \Phi(1)m(B) = \Phi(1)\tilde{m}(\Pi_{\tilde{V}}^{-1}(B)) = \int_{\psi^{-1}\circ\Pi_{\tilde{V}}^{-1}(B)} (1-|z|^2)^{-2}\,dxdy,$$

where ψ is a conformal mapping of E onto $\tilde{W}$ and $\Pi_{\tilde{V}}$ is the restriction of the canonical projection Π to $\tilde{V}$.

Then $\theta = \psi^{-1}\circ\Pi_{\tilde{V}}^{-1}$ is a homeomorphism of V onto an open set of the complex plane and h, defined by $h(p) = \log[(1-|\theta(p)|)/(1+|\theta(p)|)]^{\frac{1}{2}}$ is a real function defined on V. Hence $h\circ\theta^{-1}(z) = \log[(1-|z|)/(1+|z|)]^{\frac{1}{2}}$ and

$$\mu_h(B) = \int_{\theta(B)} \|\operatorname{grad} h\circ\theta^{-1}\|^2 dxdy = \int_{\theta(B)} \|\operatorname{grad}\log[(1-|z|)/(1+|z|)]^{\frac{1}{2}}\|^2 dxdy$$

$$= \int_{\psi^{-1}\circ\Pi_{\tilde{V}}^{-1}(B)} (1-|z|^2)^{-2}\,dxdy = \int_W \Phi(\varrho)\,dm\,.$$

Let $h' = h\circ\varphi^{-1}$. In an analogous way we find

$$\mu_{h'}(\varphi(B)) = \int_{W'} \Phi(\chi_{\varphi(B)})\,dm' = \int_{W'} \Phi(\chi_B\circ\varphi^{-1})\,dm',$$

where χ_B is the characteristic function of B. By (1) we have

$$(1/K)\mu_h(B) \le \mu_{h'}(\varphi(B)) \le K\mu_h(B)$$

and therefore we obtain the relation

$$(1/K)\int_W \Phi(\varrho)\,dm \le \int_{W'} \Phi(\varrho\circ\varphi^{-1})\,dm' \le K\int_W \Phi(\varrho)\,dm$$

valid for the characteristic function ϱ of every Borel set B contained in a parametric neighbourhood. This result is sufficient for the general case: in view of the limit theorems for integrals of positive functions, the above inequalities are satisfied for every $\mathcal{M}$-admissible function ϱ.

On the other hand, it can be easily seen that if ϱ is $\mathcal{M}$-admissible then $\varrho\circ\varphi^{-1}$ is $\mathcal{M}'$-admissible and, conversely, if ϱ' is $\mathcal{M}'$-admissible, then $\varrho'\circ\varphi$ is $\mathcal{M}$-admissible, so finally we arrive at (2).

Bibliography

[1] FUGLEDE, B.: Extremal length and functional complection, Acta Math. 98 (1957), 171-219.

[2] GHIŞA, D.: Mesures sur les surfaces de Riemann, Actas del V Congreso de la Agrupación de Matemáticos de Expresión Latina, Madrid 1978.

[3] ——: Associated measures and the quasiconformality, Proceedings of the Colloquium on Complex Analysis, Joensuu 1978.

[4] OHTSUKA, M.: Extremal length and precise functions in 3-space, Hiroshima University, Hiroshima 1973.

Institute of Mathematics
University of Timişoara
Timişoara, Romania

ON THE ORDER OF GROWTH FOR THE NORM OF THE HOLOMORPHIC COMPONENT OF A MEROMORPHIC FUNCTION

Levon Derenikovič Grigorjan (Erevan)

Suppose that D is a domain in the complex plane $\mathbb{C}$, Γ-the boundary of D, and n - an arbitrary positive integer. Denote by $M_n(D)$ the class of all functions f meromorphic in D and such that

a) the number of poles (with their multiplicity) of the function f in D is not greater than n,

b) $\|f\|_\Gamma = \sup_{z\in\Gamma} \lim_{\zeta\to z,\zeta\in D} |f(\zeta)| \leq 1$.

Let R_f be the sum of principal parts of the functions $f\in M_n(D)$ with respect to all the poles in D and let $f^* = f - R_f$ be the regular part of f. We set

$$\Lambda_n(d) = \sup\{\|f^*\|_\Gamma : f\in M_n(D)\}.$$

We shall show below that for an arbitrary domain D whose boundary consists of a finite number of pairwise disjoint piecewise smooth curves we have $\Lambda_n(D) \asymp n$ (in the estimate from above with a multiplicative constant depending on D).

We have the following

<u>THEOREM</u>. <u>Let</u> D <u>be a domain in</u> $\mathbb{C}$ <u>whose boundary</u> Γ <u>consists of a finite number of pairwise disjoint piecewise smooth Jordan curves. Then, for any</u> $n \geq 1$,

$$n \leq \Lambda_n(d) \leq C(D)\cdot n$$

<u>where</u> $C(D)$ <u>depends only on the domain</u> D $(0 < C(D) < +\infty)$.

To prove the estimate from below it is enough to see that the example given in [2] is, in fact, of a local character. Therefore, if the boundary Γ of the domain D has the tangent at any point of Γ, we can already conclude that $\Lambda_n(D) \geq n$.

We proceed now to prove the estimates of $\Lambda_n(D)$ from above. In [3] it was shown that for any Jordan domain with smooth boundary there exist constants $r_o > 0$ and $A > 0$ depending on the domain only, such that for any point z_o of the boundary, every circle $|z - z_o| = r < r_o$ intersects the boundary exactly at two points and $|d(z - z_o)| \leq A|d|z - z_o||$ for all points of the boundary lying in the disc $|z - z_o| \leq r_o$, d being the symbol of the differential. It can be seen easily that for every domain D, whose boundary consists of a finite number of pairwise disjoint piecewise smooth Jordan curves, there exist constants $r_o = r_o(D) > 0$ and $A = A(D) > 0$ with the above mentioned properties.

Let us consider an arbitrary function $f \in M_n(D)$, whose poles (in D) are denoted by $\zeta_1, \ldots, \zeta_n$ (including their multiplicity; we assume that f has exactly n poles in D). Let us fix the point $z_o \in \Gamma$. We are going to estimate $|R_f(z_o)|$; without any loss of generality we may put $z_o = 0$.

Denote by $g_D(z, \zeta)$ the Green function of D with singularity at ζ. Since the function

$$U(z) = \log|f(z)| - \sum_{\nu=1}^{n} g_D(z, \zeta_\nu)$$

is subharmonic in D and $\|U\|_\Gamma \leq 0$, then by the maximum principle we arrive at

$$(1) \qquad \log|f(z)| \leq \sum_{\nu=1}^{n} g_D(z, \zeta_\nu), \quad z \in D.$$

Let

$$E_n = \{z \in D : \sum_{\nu=1}^{n} g_D(z, \zeta_\nu) \geq 1\}.$$

Then (1) yields

$$(2) \qquad |f(z)| \leq e, \quad z \in D \setminus E_n.$$

Let us set $D^- = \overline{\mathbb{C}} \setminus D$ and consider the condenser (D^-, E_n). Using Lemma 1 of [1] we obtain

$$(3) \qquad C(D^-, E_n) = n,$$

$C(D^-, E_n)$ being the capacity of condenser (D^-, E_n); for the capacity of condenser we refer to [1].

Let $E^*_{n,r_o} = \{|z| : z \in E_n\} \cap [0, r_o]$ be the intersection of the circular projection of the set E_n on the positive half-axis with the segment $[0, r_o]$. Since we assume that the point $z_o = 0$ belongs to the boundary of D, the choice of the constant $r_o = r_o(D) > 0$ implies that $[-r_o, 0]$ is contained in the circular projection of the set D^- on the negative half-axis. By circular projection the capacity of a condenser does not increase. By monotoneity of the capacity and by (3) we obtain

$$(4) \qquad C([-r_o, 0], E_n^*) \leq n.$$

We set $E_{n,r_o}^- = \{-|z| : z \in E_{n,r_o}^*\}$; since $E_{n,r_o}^- \subset [-r_o, 0]$, from (4) we deduce

$$C(E_{n,r_o}^-, E_{n,r_o}^*) \leq n.$$

From the symmetry of the condenser $(E_{n,r_o}^-, E_{n,r_o}^*)$ with respect to the imaginary axis it follows that $C(\Pi^-, E_{n,r_o}^*) = 2C(E_{n,r_o}^-, E_{n,r_o}^*)$; Π^- being the closed left half-plane. Consequently,

$$(5) \qquad C(\Pi^-, E_{n,r_o}^*) \leq 2n.$$

Thus, in the same way as in [2], we obtain the inequality

$$(6) \qquad \int_{E_{n,r_o}^*} \frac{dr}{r} \leq \pi^2 \cdot n.$$

Suppose that the boundary Γ of D consists of p pairwise disjoint piecewise smooth Jordan curves (the domain D is p-connected). In this case the set E_n consists of no more than $n+p-1$ connected components (each such component has to contain either the pole of f or a bounded component of the complement of D). Consequently, the set E_{n,r_o}^* is the union of no more than $n+p-1$ disjoint segments lying on the positive half-axis. Let us set

$$F_n = \{z \in \overline{D} : |z| \in E_{n,r_o}^*\}, \quad D_o = \{z \in D : |z| > r_o\}.$$

From the definition of the number $r_o = r_o(D)$ it follows that the set F_n consists of $q_o \leq n+p-1$ "half-rings", and that each of these half-rings intersects at least one component of E_n. All the poles of f belong to $D_o \cup F_n$ (and are its interior points). By the Cauchy integral formula we obtain

$$R_f(0) = \frac{1}{2\pi i} \int_{\partial(D_o \cup F_n)} \frac{f(\zeta)}{\zeta} d\zeta.$$

From the inequality (2) and the construction of D_o and F_n it follows that at the points of $\partial(D_o \cup F_n)$ the values of f (or its limit values along nontangent paths) do not exceed the number e. Thus

(7) $$|R_f(0)| \leq \frac{e}{2\pi} \int_{\partial(D_o \cup F_n)} (1/|\zeta|)|d\zeta| .$$

The set $\partial(D_o \cup F_n)$ is contained in $\partial D_o \cup \partial F_n$; on the other hand ∂F_n is the union of $(\partial F_n)^* = \Gamma \cap \partial F_n$ and $\partial F_n \setminus (\partial F_n)^*$. The set (∂F_n) is the union of two sets and each of them is projected in E^*_{n,r_o}. Finally, $\partial F_n \setminus (\partial F_n)^*$ is the union of no more than $2q_o \leq 2(n+p-1)$ arcs of circles centred at $z=0$. We have

(8) $$\int_{\partial D_o} (1/|\zeta|)|d\zeta| \leq (1/r_o)|\Gamma| + 2\pi ,$$

where $|\Gamma|$ is the sum of lengths of Jordan curves creating Γ. Also

(9) $$\int_{(\partial F_n)^*} \frac{|d\zeta|}{|\zeta|} \leq A \int_{(\partial F_n)^*} \frac{|d|\zeta||}{|\zeta|} \leq 2A \int_{E^*_{n,r_o}} \frac{dr}{r} \leq 2A\pi^2 \cdot n$$

(here we have used the definition of the number $A = A(D)$ and the estimate (6)) and

(10) $$\int_{\partial F_n \setminus (\partial F_n)^*} \frac{|d\zeta|}{|\zeta|} \leq 4\pi(n+p-1) .$$

Combining the estimates (8) - (10) we obtain, by (7),

$$|R_f(0)| \leq \frac{e}{2\pi} \int_{\partial D_o \cup \partial F_n} (1/|\zeta|)|d\zeta| \leq C_1(D) \cdot n ,$$

where $C_1(D)$ depends on D only. Thus $\|R_f\|_\Gamma \leq C_1(D) \cdot n$ and $\|f^*\|_\Gamma \leq \|R_f\|_\Gamma + 1 \leq C(D) \cdot n$. Thus the theorem is proved.

References

[1] GONČAR, A. A.: The rate of rational approximation and the property of univalence of an analytic function in the neighborhood of an isolated singular point [in Russian], Mat. Sb. (N. S.) 94 (136) (1974), 265-282 and 336.

[2] GRIGORJAN, L. D.: Estimates of the norm of holomorphic components of meromorphic functions in domains with a smooth boundary [in Russian], Mat. Sb. (N. S.) 100 (142) (1976), 156-164 and 166.

[3] MUSHELIŠVILI, N. I.: Singular integral equations [in Russian], Izdat. "Nauka", Moscow 1968.

Chair of Higher Mathematics
Polytechnical Institute
105, Teryan; Erevan 09, USSR

ANALYTIC REPRESENTATION FOR CR-FUNCTIONS ON SUBMANIFOLDS OF CODIMENSION 2 IN $\mathbb{C}^n$

Gennadiĭ Markovič Henkin (Moskva)

Contents

Introduction

Let M be a smooth manifold in $\mathbb{C}^n$; $T_\xi(M)$ - real tangent space to M at $\xi \in M$; $T^c_\xi(M)$ - the biggest complex subspace of $T_\xi(M)$.

M is called a CR-manifold if $\dim_{\mathbb{C}} T^c_\xi(M)$ does not depend on $\xi \in M$. A function f defined on M is called a CR-function if $\bar\partial_M f = 0$, i.e. if

$$\int_M f\,\bar\partial\beta = 0$$

for every smooth finite differential form β of type $(k - 1, l - 1)$, where $k = \dim_{\mathbb{R}} T_\xi(M)$ and $l = \dim_{\mathbb{C}} T^c(M)$.

The theory of CR-functions initiated by the classical works of S. Bochner and H. Lewy contain already a number of important results, applications and unsolved problems (cf. e.g. [15]).

In the papers [1], [2], [19], [10] and [5] it was noticed that for solving some problems concerning the theory of CR-function (the problems of approximation, uniqueness and analytic continuation...) it is useful to have an analytic representation of CR-functions. It is essential to assume that a CR-function f on M admits an analytic representation if there exist domains D_ν, $\nu = 1,2,\ldots,N$, in $\mathbb{C}^n$ and functions f_ν holomorphic in these domains, such that $M \subset \partial D_\nu$, $\nu = 1,2,\ldots,N$; f_ν have (in a proper sense) boundary values on M, and

$$f = \sum_{\nu=1}^{N} f_\nu \quad \text{on } M.$$

However the possibility of analytic representation (though locally) of CR-functions is proved only in very special cases. We shall mention here the two best known results about the analytic representation of CR-functions.

THEOREM I (Martineau [7]). Let the manifold M have a form $M = \mathbb{R}^n \cap \Omega$, where $\mathbb{R}^n$ is the real part of the space $\mathbb{C}^n$ and Ω is some domain of holomorphy in $\mathbb{C}^n$. Besides, let $\{\Omega_\nu\}$ be a finite covering of $\Omega \setminus M$ by domains of holomorphy Ω_ν, $\nu = 1,2,\ldots,N > n$. Then for any (generalized) function f on M we have a representation

$$f = \sum_{1 \le \nu_1 < \ldots < \nu_n \le N} f_{\nu_1,\ldots,\nu_n},$$

where every function $f_{\nu_1,\ldots,\nu_n}$ is holomorphic in the domain $\Omega_{\nu_1} \cap \Omega_{\nu_2} \cap \ldots \cap \Omega_{\nu_n}$ and has on M a slice in the space of L. Schwartz's distributions .

In the case where the covering of the domain $\Omega \setminus M$ has a special form

$$\Omega_\nu^{\pm} = \{z : \operatorname{Im} z_\nu \gtrless 0\}, \quad \nu = 1,2,\ldots,n,$$

the analytic representation (by Cauchy integrals) for (generalized) functions on M was obtained earlier by Tillmann [14].

Theorem I plays an important role in the theory of Sato-Martineau hyperfunctions (cf. [13]).

THEOREM II (Andreotti-Hill [1], Kashiwara-Kawai [6]). Suppose that $M = S_{2n-1} \cap \Omega$, where $S_{2n-1} = \{z \in \mathbb{C}^n : \varrho(z) = 0\}$, is a smooth closed hypersurface in $\mathbb{C}^n$ and Ω is a domain of holomorphy in $\mathbb{C}^n$. Then any (generalized) CR-function f on M has a representation of the form $f = f_+ + f_-$ where f_+ and f_- are holomorphic in $\Omega_+ = \{z \in \Omega : \varrho(z) > 0\}$ and $\Omega_- = \{z \in \Omega : \varrho(z) < 0\}$, respectively, having a slice on M in the space of L. Schwartz's distributions.

The useful variants of this result, referring to other classes of functions, are to be found in [19] and [10].

In this paper, using the integrals of Cauchy-Leray type, we shall prove the possibility of (local) analytic representation of CR-functions, defined on an arbitrary CR-manifold of codimension 2 in $\mathbb{C}^n$ which contains a hypersurface with the nondegenerated Levi form.

We say that a CR-manifold M in $\mathbb{C}^n$ satisfies at ξ, $\xi \in M$, the R. Nirenberg condition (cf. [8] and [15]) if it contains a hypersurface

$\{z : \varrho(z) = 0\}$, such that the restriction of its Levi form $\partial\bar\partial\varrho(\xi)$ on the complex tangent plane $T^c_\xi(M)$ either has all the eigenvalues of the same sign or one of them is positive and one negative.

The main result of the present paper is

THEOREM III. *Suppose that M is a smooth CR-manifold of codimension 2 in $\mathbb{C}^n$, satisfying R. Nirenberg's condition at $z_0 \in M$ and f is a (generalized) CR-function on M. Then there exist a domain $\Omega \ni z_0$, a covering of $\Omega \setminus M$ by the domains of the form $\Omega_i = \{z \in \Omega : \varrho_i(z) < 0, \ \varrho_i \in C^2(\Omega)\}$, $1 \le i \le 3$, and the functions $f_{i,j}$ holomorphic in the domains $\Omega_i \cap \Omega_j$, such that*

$$\varrho_i(z) = 0, \quad d\varrho_i(z) \wedge d\varrho_j(z) \neq 0 \quad \text{for every } z \in M, \quad 1 \le i < j \le 3,$$

the functions $f_{i,j}$ have generalized boundary values on the manifold $M \cap \Omega \subset \partial\Omega_i \cap \partial\Omega_j$, and on $M \cap \Omega$ we have the equality

$$f = f_{12} + f_{23} + f_{13}.$$

It is unknown how far the assumption in Theorem III that the CR-manifold M satisfies R. Nirenberg's condition is important. It is unknown either how to prove the analogue of Theorem III for CR-manifolds of arbitrary codimension in $\mathbb{C}^n$.

1. Preliminary results

At first, following [9], [18], [16] and [11] we shall write an integral formula for differential forms defined in domains contained in $\mathbb{C}^n$, with a piecewise smooth boundary, in the form useful for further investigations. Let D be a domain in $\mathbb{C}^n$ of a form

$$(1.1) \quad D = \{z \in \Omega : \ \varrho_j(z) < 0, \quad j = 1,2,\ldots,N\},$$

where $\{\varrho_j\}$ are real-valued smooth functions in $\Omega \supset \bar D$ such that for every $1 \le k \le N$ and $1 \le j_1 < j_2 < \ldots < j_k \le N$ we have $d\varrho_{j_1} \wedge \ldots \wedge d\varrho_{j_k} \neq 0$ on the set

$$\Gamma_J = \Gamma_{j_1\ldots j_k} = \{z \in \partial D : \ \varrho_{j_2} = \ldots = \varrho_{j_k} = 0\}.$$

The orientation of $\mathbb{C}^n$ is chosen so that the form $(-i)^n\omega(\bar z) \wedge \omega(z)$ is positive. The hypersurface

$$\partial D = \sum_{j=1}^{n} \Gamma_j$$

is endowed with the orientation induced from D and, further, for every J the hypersurface

$$\partial\Gamma_J = \bigcup_{j=1}^{N} \Gamma_{Jj}$$

is endowed with the orientation induced from Γ_J.

Let $\eta^0(\xi,z) = (\bar{\xi} - \bar{z})/|\xi - z|^2$; besides let $\eta^\nu : \Gamma_\nu \times D \to \mathbb{C}^n$, $\nu = 1,\dots,N$, be smooth mappings from the manifolds $\Gamma_\nu \times D$ to $\mathbb{C}^n$, such that for every $\nu = 1,2,\dots,N$, $\xi \in \Gamma_\nu$ and $z \in D$ we have

$$\sum_{k=1}^{n} \eta_k^\nu(\xi,z)(\xi_k - z_k) \equiv 1.$$

Let further

$$\Delta = \{\lambda = (\lambda_0,\dots,\lambda_{N-1}) \in \mathbb{R}^n : \lambda_j \geq 0, \sum_{j=0}^{N-1} \lambda_j = 1\}$$

be an N-simplex in $\mathbb{R}^n$ endowed with the standard orientation. Let further

$$\Delta_{\nu_0\dots\nu_k} = \{\lambda \in \Delta : \sum_{j=0}^{k} \lambda_{\nu_j} = 1\}.$$

We shall endow the simplexes $\Delta_{\nu_0\dots\nu_k}$ $(1 \leq k \leq N-1)$ with orientations induced by the orientation of Δ.

$$\eta^{\nu_0\dots\nu_k} = \sum_{j=0}^{k} \lambda_{\nu_j}\eta^{\nu_j}, \quad \text{where} \quad \lambda \in \Delta_{\nu_0\dots\nu_k}.$$

$C_{p,q}(\bar{D})$ is the space of (p,q)-forms with continuous coefficients on $\bar{D}$.

In the space $\mathbb{C}^{3n}$ of complex variables

$$(\eta,\xi,z) = (\eta_1,\dots,\eta_n, \xi_1,\dots,\xi_n, z_1,\dots,z_n)$$

on the analytic hypersurface

$$\{(\eta,\xi,z) : \sum_{k=1}^{n} \eta_k(\xi_k - z_k) = 1\}$$

we shall consider a holomorphic form $\omega'(\eta) \wedge \omega(\xi) \wedge \omega(z)$ of degree $3n-1$, where

$$\omega'(\eta) = \sum_{k=1}^{n} (-1)^{k-1} \eta_k \bigwedge_{j\neq k} d\eta_j,$$

$$\omega(\xi) = \bigwedge_{j=1}^{n} d\xi_j, \quad \omega(z) = \bigwedge_{j=1}^{n} dz_j.$$

If $\eta = \eta(\xi,z,\lambda)$ is a smooth function with respect to ξ, z and the parameter λ, then we have the equality

$$\omega'(\eta)\wedge\omega(\xi)\wedge\omega(z)=\sum_{q=0}^{n-1}\omega'_q(\eta)\wedge\omega(\xi)\wedge\omega(z),$$

where $\omega'_q(\eta)$ is a form of degree q with respect to $d\bar{z}$ and of degree $n-q-1$ with respect to $d\bar{\xi}$ and $d\lambda$. We shall need the following important formulae (cf. Lemma 2.2 in [11]):

$$(1.2)\quad d_\lambda\omega'_q(\eta)+\bar{\partial}_\xi\omega'_q(\eta)=-\bar{\partial}_z\omega'_{q-1}(\eta)\qquad (q=0,1,2,\dots,n).$$

In the papers [9], [18], [16] and [11] the following result, though not explicitly formulated, was obtained:

PROPOSITION 1. *Let* $f\in C_{n,q}(\bar{D})$ *and* $\bar{\partial}f\in C_{n,q+1}(D)$, *where* D *is a domain of type* $(1,1)$, $q=0,1,2,\dots,n-1$. *Then for every* $z\in D$ *the following equality holds*:

$$(-1)^q\frac{(2\pi i)^n}{(n-1)!}f(z)$$

$$=\sum_{k=1}^{n-q}\sum_{\substack{1\le\nu_0<\dots\\ \dots<\nu_{k-1}\le N}}\int_{(\xi,\lambda)\in\Gamma_{\nu_0\dots\nu_{k-1}}\times\Delta_{\nu_0\dots\nu_{k-1}}} f(\xi)\omega'_q(\eta^{\nu_0\dots\nu_{k-1}}(\xi,z,\lambda))\wedge\omega(\xi)$$

$$-\sum_{k=1}^{n-q-1}\sum_{\substack{1\le\nu_0<\dots\\ \dots<\nu_{k-1}\le N}}\int_{\Gamma_{\nu_0\dots\nu_{k-1}}\times\Delta_{0,\nu_0\dots\nu_{k-1}}}\bar{\partial}f(\xi)\wedge\omega'_q(\eta^{0,\nu_0\dots\nu_{k-1}}(\xi,z,\lambda))\wedge\omega(\xi)$$

$$-\int_{\xi\in D}\bar{\partial}f\wedge\omega'_q(\eta^0(\xi,z))\wedge\omega(\xi)$$

$$+\bar{\partial}_z\sum_{k=1}^{n-q}\sum_{\substack{1\le\nu_0<\dots\\ \dots<\nu_{k-1}\le N}}\int_{\Gamma_{\nu_0\dots\nu_{k-1}}\times\Delta_{0,\nu_0\dots\nu_{k-1}}} f(\xi)\omega'_{q-1}(\eta^{0,\nu_0\dots\nu_{k-1}}(\xi,z,\lambda))\wedge\omega(\xi)$$

$$-\bar{\partial}_z\int_D f(\xi)\omega'_{q-1}(\eta^0)\wedge\omega(\xi).$$

The case $q=0$ is essentially contained in [9] and [18]. Extenfor $q>0$ was obtained in [16] and [11].

Suppose that M is an arbitrary CR-manifold being the whole intersection of codimension S in a domain $\Omega\in\mathbb{C}^n$, that is

$$M = \{z \in \Omega : \ \varrho_1 = \varrho_2 = \dots = \varrho_s = 0\},$$

where $\varrho_1,\dots,\varrho_s$ are smooth functions in Ω, such that $\bar\partial\varrho_1\wedge\bar\partial\varrho_2\wedge\dots\wedge\bar\partial\varrho_s \neq 0$ on M. We have $\dim_{\mathbb{R}} M = 2n-s$, $\mathrm{CR}\dim M = n-s$. Let $D_\nu = \{z \in \Omega : \varrho_\nu < 0\}$.

Let further ϱ_{s+1} be another smooth function in Ω such that

$$(1.3) \quad \{z\in\Omega : \varrho_1 = \varrho_2 = \dots = \varrho_{\nu-1} = \varrho_{\nu+1} = \dots = \varrho_{s+1}\} = (-1)^{s+1+\nu} M,$$
$$\nu = 1,2,\dots,s+1;$$

$$\bar\partial\varrho_{\nu_1}\wedge\dots\wedge\bar\partial\varrho_{\nu_s} \neq 0 \quad \text{on } M, \quad 1\le\nu_1<\dots<\nu_s\le s+1;$$

$$\bigcup_{\nu=1}^{s+1} D_\nu = \Omega\setminus M.$$

Let $P^{(\nu)}(\xi,z) = \{P_1^{(\nu)}(\xi,z),\dots,P_n^{(\nu)}(\xi,z)\}$ and $\Phi^{(\nu)}(\xi,z)$ be arbitrary smooth functions with respect to $\xi\in\partial D_\nu$ and $z\in\bar D_\nu$, which fulfil the conditions

$$\eta^\nu(\xi,z) = \frac{P^{(\nu)}(\xi,z)}{\Phi^{(\nu)}(\xi,z)} ; \quad \Phi^{(\nu)}(\xi,z) = \sum_{i=1}^n P_i^{(\nu)}(\xi,z)(\xi_i - z_i);$$

$$(1.4) \quad 2\mathrm{Re}[\Phi^{(\nu)}(\xi,z)] \ge \varrho_\nu(\xi) - \varrho_\nu(z) + \gamma|\xi-z|^2,$$

where γ is a nonnegative constant, $\nu = 1,2,\dots,s+1$. Let $z_0\in M$ and the domain $D = \{z\in\Omega : \varrho_0(z)<0\}$, where ϱ_0 is a strictly plurisubharmonic function in Ω, is a neighbourhood of z_0 in Ω. Let $P(\xi,z) = \{P_1(\xi,z),\dots,P_n(\xi,z)\}$ and $\Phi(\xi,z)$ be barrier functions of D constructed in [17]. We set $\eta^*(\xi,z) = P(\xi,z)/\Phi(\xi,z)$. Let $\varepsilon>0$. We set $D^\varepsilon = \{z\in\Omega : \varrho_0(z)<0, \ \varrho_1(z)<\varepsilon,\dots,\varrho_{s+1}(z)<\varepsilon\}$;

$$\Gamma_\nu^\varepsilon = \{z\in\Omega : \varrho_\nu(z)=\varepsilon; \ \varrho_0(z)\le 0, \ \varrho_\mu(z)\le\varepsilon \text{ for } \mu\neq\nu\},$$
$$\nu = 1,\dots,s+1;$$

$$\Gamma_0^\varepsilon = \{z\in\Omega : \ \varrho_0(z)=0, \ \varrho_\nu(z)\le\varepsilon, \ 1\le\nu\le s+1\};$$

$$\Gamma^\varepsilon_{\nu_0\dots\nu_{k-1}} = \bigcap_{i=0}^{k-1}\Gamma^\varepsilon_{\nu_i}, \quad \text{where } \{\nu_0,\dots,\nu_{k-1}\}\subset\{*,1,2,\dots,s+1\}.$$

For a differential form

$$f = \sum_\alpha f_{I,J}\, dz^I\wedge d\bar z^J$$

we set $|f| = \sum |f_{I,J}|$. We denote by $C^\alpha_{p,q}(M)$ the space of (p,q)-forms on M whose coefficients fulfil the Hölder condition with the exponent α.

PROPOSITION 2. Let $\beta \in C^{\alpha}_{0,n-s}(\Omega)$, $0<\alpha\leq 1$, and $|d\beta(z)| = O(|\varrho(z,M)|^{\alpha-1})$. Then for every $z\in M\cap D$ the following relations hold:

(i) $$\lim_{\varepsilon\to 0}\int_{\Gamma^{\varepsilon}_{\nu_0\dots\nu_{k-1}}\times\Delta_{0,\nu_0\dots\nu_{k-1}}} \bar\partial\beta\wedge\omega'_{n-s}(\eta^{0,\nu_0\dots\nu_{k-1}}(\xi,z,\lambda))\wedge\omega(\xi)=0,\quad k=1,\dots,s-1;$$

(ii) $$\lim_{\varepsilon\to 0}\int_{\Gamma^{\varepsilon}_{\nu_0\dots\nu_{k-1}}\times\Delta_{\nu_0\dots\nu_{k-1}}} \beta(\xi)\wedge\omega'_{n-s}(\eta^{\nu_0\dots\nu_{k-1}}(\xi,z,\lambda))\wedge\omega(\xi)=0,\quad k=1,\dots,s-1;$$

(iii) $$\lim_{\varepsilon\to 0}\int_{\Gamma^{\varepsilon}_{\nu_0\dots\nu_{k-1}}\times\Delta_{0,\nu_0\dots\nu_{k-1}}} \beta(\xi)\wedge\omega'_{n-s-1}(\eta^{0,\nu_0\dots\nu_{k-1}}(\xi,z,\lambda))\wedge\omega(\xi)=0,\quad k=1,\dots,s-1,$$

where $\{\nu_0,\dots,\nu_{k-1}\}\subset\{*,1,2,\dots,s+1\}$.

Proof. At first we shall prove (ii). For the sake of simplicity, let $(\nu_0,\dots,\nu_{k-1})=(1,2,\dots,k)$. Then we have

$$\mathrm{II}^{\varepsilon}=\int_{\Gamma^{\varepsilon}_{12\dots k}\times\Delta_{12\dots k}}\beta(\xi)\wedge\omega'_{n-s}(\eta^{12\dots k}(\xi,z,\lambda))\wedge\omega(\xi)=\sum_{\alpha,\beta}C_{\alpha,\beta}\times\mathrm{II}^{\varepsilon}_{\alpha,\beta},$$

where

$$\mathrm{II}^{\varepsilon}_{\alpha,\beta}=\sum_{I,J}\int_{\Gamma^{\varepsilon}_{12\dots k}}\beta(\xi)\Big[\prod_{l=1}^{k}\Phi^{(l)}\prod_{i=1}^{s-k}\Phi^{(\alpha_i)}\prod_{j=1}^{n-s}\Phi^{(\beta_j)}\Big]^{-1}$$

$$\times\det[P^{(1)}\dots P^{(k)},(\partial/\partial\xi_{i_1})P^{(\alpha_1)}\dots(\partial/\partial\xi_{i_{s-k}})P^{(\alpha_{s-k})}\cdot(\partial/\partial\bar z_{j_1})P^{(\beta_1)}\dots$$

$$\cdot(\partial/\partial\bar z_{j_{n-s}})P^{(\beta_{n-s})}]\,d\xi^{I}\wedge d\bar z^{J}\wedge\omega(\xi),$$

where $C_{\alpha,\beta}$ is an absolute constant, $I=(i_1,\dots,i_{s-k})$, $J=(j_1,\dots,j_{n-s})$, $\alpha=(\alpha_1,\dots,\alpha_{s-k})$, $\beta=(\beta_1,\dots,\beta_{n-s})$ with $1\leq i_1<\dots<i_{s-k}\leq n$; $\alpha_1,\alpha_2,\dots,\alpha_{s-k}=1,2,\dots,k$; $\beta_1,\beta_2,\dots,\beta_{n-s}=1,2,\dots,k$. We shall prove that

$$\lim_{\varepsilon\to 0}\mathrm{II}^{\varepsilon}_{\alpha,\beta}=0.$$

For simplicity we assume that $\alpha_1=\dots=\alpha_{s-k}=1$, $\beta_1=\dots=\beta_{n-s}=1$. In the remaining cases the proof is analogous.

We have

$$|\Pi_{1,1}^{\varepsilon}| \le O(1)\int_{\Gamma_{12\dots k}^{\varepsilon}} \sum_{I}\{|\Phi^{(1)}|^{n-k+1}|\Phi^{(2)}|\dots|\Phi^{(k)}|\}^{-1}|d\bar{\xi}^{I}\wedge\omega(\xi)|,$$

where $|I| = n-k$. Introducing the coordinates $t_{2i-1} = \mathrm{Re}\,\Phi^{(i)}$, $t_{2i} = \mathrm{Im}\,\Phi^{(i)}$, $i = 1,2,\dots,k$, and applying the inequality (1.6) we have

$$|\Phi^{(i)}| \ge (t_{2i-1} + t_1^2 + \dots + t_{2n}^2 + t_{2i});$$

$$\Gamma_{12\dots k}^{\varepsilon} = \{(t_1\dots t_{2n}): t_{2i-1} = \varepsilon,\ 0 \le t_{2i} \le \delta\ \ (i = 1,\dots,k),$$

$$0 \le t_{2k+j} \le \gamma\varepsilon\ \ (j = 1,\dots,k),\ 0 \le t_{3k+j} \le \delta;\ j = 1,\dots,2n-3k\}.$$

Hence

$$|\Pi_{1,1}^{\varepsilon}|$$

$$\le O(1)\int_0^{\delta}\dots\int_0^{\delta}\int_0^{\gamma\varepsilon}\dots\int_0^{\gamma\varepsilon}\int_0^{\delta}\dots\int_0^{\delta}\frac{dt_2\dots dt_{2k}\,dt_{2k+1}\dots dt_{3k}\,dt_{3k+1}\dots dt_{2n}}{(t_1 + \sum_{j=1}^{2n} t_j^2 + t_2)^{n-k+1}\cdot\prod_{i=2}^{k}(t_{2i-1} + \sum_{j=1}^{2n} t_j^2 + t_{2i})}$$

$$\le O(1)\int_0^{\delta}\dots\int_0^{\delta}\int_0^{\gamma\varepsilon}\dots\int_0^{\gamma\varepsilon}\int_0^{\delta}\dots\int_0^{\delta}\frac{dt_2\dots dt_{2k}\,dt_{2k+1}\dots dt_{3k}\,dt_{3k+1}\dots dt_{2n}}{(\varepsilon + \sum_{j=2k+1}^{2n} t_j^2 + t_2)^{n-k+1}\prod_{i=2}^{k}(\varepsilon + t_{2i})}$$

$$\le O(1)\int_0^{\delta}\dots\int_0^{\delta}\int_0^{\gamma\varepsilon}\dots\int_0^{\gamma\varepsilon}\int_0^{\delta}\dots\int_0^{\delta}\frac{dt_4\dots dt_{2k}\,dt_{2k+1}\dots dt_{3k}\,dt_{3k+1}\dots dt_{2n}}{(\sqrt{\varepsilon} + \sum_{j=2k+1}^{2n} t_j)^{2n-2k}\cdot\prod_{i=2}^{k}(\varepsilon + t_{2i})}$$

$$\le O(1)\ln^{k-1}(1/\varepsilon)\cdot\ln(1+\sqrt{\varepsilon}) \le O(\sqrt{\varepsilon}\cdot\ln^{k-1}(1/\varepsilon)).$$

We shall prove next (i). We have (for $(\nu_0,\dots,\nu_{k-1}) = (1,2,\dots,k)$) the equality

$$\Pi^{\varepsilon} = \int_{\Gamma_{12\dots k}^{\varepsilon}\times\Delta_{012\dots k}} \bar{\partial}\beta(\xi)\wedge\omega'_{n-s}(\eta^{012\dots k}(\xi,z,\lambda))\wedge\omega(\xi) = \sum_{\alpha,\beta} c_{\alpha,\beta}\times\Pi_{\alpha,\beta}^{\varepsilon},$$

where

$$\Pi_{\alpha,\beta}^{\varepsilon} = \sum_{I,J}\int_{\Gamma_{12\dots k}^{\varepsilon}} \bar{\partial}\beta\Big[\prod_{l=0}^{k}\Phi^{(l)}\prod_{i=1}^{s-k-1}\Phi^{(\alpha_i)}\prod_{j=1}^{n-s}\Phi^{(\beta_j)}\Big]^{-1}$$

$$\times\det[P^{(0)},\dots,P^{(k)},(\partial/\partial\bar{\xi}_{i_1})P^{(\alpha_1)},\dots,(\partial/\partial\bar{\xi}_{i_{s-k-1}})P^{(\alpha_{s-k-1})},$$

$$(\partial/\partial\bar z_{j_1})P^{(\beta_1)},\dots,(\partial/\partial\bar z_{j_{n-s}})P^{(\beta_{n-s})}]d\bar\xi^I\wedge d\bar z^J\wedge\omega(\xi);$$

$$\Phi^{(0)}=|\xi-z|^2;\ P^{(0)}=\bar\xi-\bar z;\ \alpha_1,\dots,\alpha_{s-k-1},\beta_1,\dots,\beta_{n-s}=0,1,\dots,k.$$

Since in a neighbourhood of z we have $|\Phi^{(j)}|\geq|\xi-z|^2$, $j\geq 0$, then

$$|\mathrm{II}^\varepsilon|\leq O(1)\int_{\Gamma^\varepsilon_{12\dots k}}\sum_I\{|\Phi^{(1)}|\dots|\Phi^{(k)}|\cdot|\xi-z|^{2n-2k-1}\}^{-1}|d\bar\xi^I\wedge\omega(\xi)|,$$

where $|I|=n-k$.

Introducing the coordinates $t_{2i-1}=\mathrm{Re}\,\Phi^{(i)}$, $t_{2i}=\mathrm{Im}\,\Phi^{(i)}$, $i=1,\dots,k$, we get

$$|\mathrm{II}^\varepsilon|$$

$$\leq O(1)\int_0^\delta\dots\int_0^\delta\int_0^{\gamma\varepsilon}\dots\int_0^{\gamma\varepsilon}\int_0^\delta\dots\int_0^\delta\frac{dt_2\dots dt_{2k}\,dt_{2k+1}\dots dt_{3k}\,dt_{3k+1}\dots dt_{2n}}{\prod_{i=1}^k(t_{2i-1}+\sum_{j=1}^{2n}t_j^2+t_{2i})(t_1+\dots+t_{2n})^{2n-2k-1}}$$

$$\leq O(1)\int_0^\delta\dots\int_0^\delta\int_0^{\gamma\varepsilon}\dots\int_0^{\gamma\varepsilon}\int_0^\delta\dots\int_0^\delta\frac{dt_2\dots dt_{2k}\,dt_{2k+1}\dots dt_{3k}\,dt_{3k+1}\dots dt_{2n}}{\prod_{i=1}^k(\varepsilon+t_{2i})(\varepsilon+\sum_{j=2k+1}^{2n}t_j)^{2n-2k-1}}$$

$$\leq O(1)\ln^k(\tfrac{1}{\varepsilon})\int_0^{\gamma\varepsilon}\dots\int_0^{\gamma\varepsilon}\int_0^\delta\dots\int_0^\delta\frac{dt_{2k+1}\dots dt_{3k}\,dt_{3k+1}\dots dt_{2n}}{(\varepsilon+\sum_{j=2k+1}^{2n}t_j)^{2n-2k-1}}\leq O(1)\varepsilon\cdot\ln^k(\tfrac{1}{\varepsilon}).$$

The relation (iii) may be proved analogously. Thus Proposition 2 is proved.

2. General integral representations for CR-functions

In this section we shall derive integral representations (generally speaking with a non-holomorphic kernel) for CR-functions on CR-manifolds of an arbitrary codimension in $\mathbb{C}^n$.

Let $L(M)$ denote the space of the measure-type generalized functions on M and let $f\in L(M)$. By (1.4) for $z\in D\cap D_{\nu_0}\cap\dots\cap D_{\nu_{s-1}}$, where $1\leq\nu_0<\nu_1<\dots<\nu_{s-1}\leq s+1$, the function

$$(2.1)\quad f_{\nu_0,\dots,\nu_{s-1}}=\int_{(\xi,\lambda)\in\Gamma_{\nu_0,\dots,\nu_{s-1}}\times\Delta_{\nu_0,\dots,\nu_{s-1}}}f(\xi)\,\omega'(\eta^{\nu_0,\dots,\nu_{s-1}}(\xi,z,\lambda))\wedge\omega(\xi)$$

$$- \int\limits_{\substack{(\xi,\lambda)\in\partial\Gamma_{\nu_0,\dots,\nu_{s-1}} \\ \times\Delta_{*,\nu_0,\dots,\nu_{s-1}}}} f(\xi)\,\omega'(\eta^{*,\nu_0,\dots,\nu_{s-1}}(\xi,z,\lambda))\wedge\omega(\xi).$$

is well-defined.

PROPOSITION 3. Let $\bar\partial_M f = 0$ on M. Then

$$\bar\partial_z f_{\nu_0,\dots,\nu_{s-1}}(z) = - \int\limits_{\substack{(\xi,\lambda)\in\Gamma_{\nu_0,\dots,\nu_{s-1}} \\ \times\partial\Delta_{\nu_0,\dots,\nu_{s-1}}}} f(\xi)\,\omega_1'(\eta^{\nu_0,\dots,\nu_{s-1}}(\xi,z,\lambda))\wedge\omega(\xi)$$

$$+ \sum_{i=0}^{s-1}(-1)^{i+1} \int\limits_{\substack{(\xi,\lambda)\in\partial\Gamma_{\nu_0,\dots,\nu_{s-1}} \\ *,\nu_0,\dots,\hat\nu_i,\dots,\nu_{s-1}}} f(\xi)\,\omega_1'(\eta^{*,\nu_0,\dots,\hat\nu_i,\dots,\nu_{s-1}}(\xi,z,\lambda))\wedge\omega(\xi),$$

where $1\le\nu_0\le\nu_1<\dots<\nu_{s-1}\le s+1$, $z\in D\cap D_{\nu_0}\cap\dots\cap D_{\nu_{s-1}}$.

Proof. Using the relation (1.2), from (2.1) we have

$$\bar\partial_z f_{\nu_0,\dots,\nu_{s-1}}(z)$$

$$= \int\limits_{\substack{(\xi,\lambda)\in\Gamma_{\nu_0,\dots,\nu_{s-1}} \\ \times\Delta_{\nu_0,\dots,\nu_{s-1}}}} f(\xi)(\bar\partial_\xi + d_\lambda)\,\omega_1'(\eta^{\nu_0,\dots,\nu_{s-1}}(\xi,z,\lambda))\wedge\omega(\xi)$$

$$+ \int\limits_{\substack{(\xi,\eta)\in\partial\Gamma_{\nu_0,\dots,\nu_{s-1}} \\ \times\Delta_{*,\nu_0,\dots,\nu_{s-1}}}} f(\xi)(\bar\partial_\xi + d_\lambda)\,\omega_1'(\eta^{*,\nu_0,\dots,\nu_{s-1}}(\xi,z,\lambda))\wedge\omega(\xi).$$

If we use the equality $\bar\partial_M f = 0$, integrating by parts with respect to ξ and further applying the Stokes theorem with respect to λ, we arrive at the conclusion of Proposition 3, as desired.

PROPOSITION 4. Let $f\in L(M)$. Then every function $f_{\nu_0,\dots,\nu_{s-1}}$ of the form (2.1) in the domain $D\cap D_{\nu_0}\cap\dots\cap D_{\nu_{s-1}}$ has the generalized boundary values on M. Moreover, for any form $\beta\in C^\alpha_{0,n-s}(\Omega)$, $0<\alpha<1$, finite in D, and having there the property $|d\beta(z)| = \sigma(|\varrho(z,M)|^{\alpha-1})$, the following equality holds:

$$\lim_{\varepsilon\to 0}\int_{z\in\Gamma^{\varepsilon}_{\nu_0,\dots,\nu_{s-1}}} f_{\nu_0,\dots,\nu_{s-1}}(z)\beta(z)\wedge\omega(z) = \int_{z\in\Gamma^{0}_{\nu_0,\dots,\nu_{s-1}}} f_{\nu_0,\dots,\nu_{s-1}}(z)\beta(z)\wedge\omega(z).$$

P r o o f. Let

$$(2.2)\quad \tilde f_{\nu_0,\dots,\nu_{s-1}}(z) = \int_{\substack{(\xi,\lambda)\in\Gamma_{\nu_0,\dots,\nu_{s-1}}\\ \times\Delta_{\nu_0,\dots,\nu_{s-1}}}} f(\xi)\,\omega'(\eta^{\nu_0,\dots,\nu_{s-1}}(\xi,z,\lambda))\wedge\omega(\xi).$$

By (2.1) and (2.2) we have

$$(2.3)\quad \int_{z\in\Gamma^{\varepsilon}_{\nu_0,\dots,\nu_{s-1}}} f_{\nu_0,\dots,\nu_{s-1}}(z)\beta(z)\wedge\omega(z)$$

$$= \int_{z\in\Gamma^{\varepsilon}_{\nu_0,\dots,\nu_{s-1}}} \tilde f_{\nu_0,\dots,\nu_{s-1}}(z)\wedge\beta(z)\wedge\omega(z)$$

$$- \int_{z\in\Gamma^{\varepsilon}_{\nu_0,\dots,\nu_{s-1}}}\Big\{\int_{\substack{(\xi,\lambda)\in\partial\Gamma_{\nu_0,\dots,\nu_{s-1}}\\ \times\Delta_{*,\nu_0,\dots,\nu_{s-1}}}} f(\xi)\omega'(\eta^{*,\nu_0,\dots,\nu_{s-1}}(\xi,z,\lambda)\wedge\omega(\xi)\Big\}\wedge\beta(z)\wedge\omega(z).$$

It is obvious that the second component on the right-hand side of (2.3) converges as $\varepsilon\to 0$. We shall show that

$$\lim_{\varepsilon\to 0}\int_{\Gamma^{\varepsilon}_{\nu_0,\dots,\nu_{s-1}}} \tilde f_{\nu_0,\dots,\nu_{s-1}}(z)\beta(z)\wedge\omega(z) = \int_{\Gamma^{0}_{\nu_0,\dots,\nu_{s-1}}} \tilde f_{\nu_0,\dots,\nu_{s-1}}(z)\beta(z)\wedge\omega(z).$$

Let

$$\tilde\Gamma^{\varepsilon}_{\nu_0\dots\nu_{s-1}} = \bigcup_{\varepsilon+\mu\le t\le\varepsilon}\Gamma^{t}_{\nu_0,\dots,\nu_{s-1}},\quad \mu<0.$$

By Stokes' formula we have

$$(2.4)\quad \int_{\tilde\Gamma^{\varepsilon}_{\nu_0\dots\nu_{s-1}}} d(\tilde f_{\nu_0\dots\nu_{s-1}}(z)\beta(z)\wedge\omega(z))$$

$$= \int_{\Gamma^{\varepsilon+\mu}_{\nu_0\dots\nu_{s-1}}} \tilde f_{\nu_0\dots\nu_{s-1}}(z)\beta(z)\wedge\omega(z) - \int_{\Gamma^{\varepsilon}_{\nu_0\dots\nu_{s-1}}} \tilde f_{\nu_0\dots\nu_{s-1}}(z)\beta(z)\wedge\omega(z).$$

Since

$$\lim_{\varepsilon\to 0}\int_{\Gamma^{\varepsilon+\mu}_{\nu_0\ldots\nu_{s-1}}} \tilde f_{\nu_0\ldots\nu_{s-1}}(z)\beta(z)\wedge\omega(z) = \int_{\Gamma^{\mu}_{\nu_0\ldots\nu_{s-1}}} \tilde f_{\nu_0\ldots\nu_{s-1}}(z)\beta(z)\wedge\omega(z),$$

it is enough to show that on the left-hand side of (2.4) there exists a finite limit. We have

$$(2.5)\qquad \int_{\tilde\Gamma^{\varepsilon}_{\nu_0\ldots\nu_{s-1}}} d(\tilde f_{\nu_0\ldots\nu_{s-1}}(z)\wedge\beta(z)\wedge\omega(z))$$

$$= \int_{\tilde\Gamma^{\varepsilon}_{\nu_0\ldots\nu_{s-1}}} \bar\partial\tilde f_{\nu_0\ldots\nu_{s-1}}(z)\wedge\beta(z)\omega(z) + \int_{\tilde\Gamma^{\varepsilon}_{\nu_0\ldots\nu_{s-1}}} \bar\partial\beta(z)\tilde f_{\nu_0\ldots\nu_{s-1}}(z)\wedge\omega(z).$$

Let

$$z_0\in\tilde\Gamma^{\varepsilon}_{\nu_0\ldots\nu_{s-1}} \quad\text{and}\quad \varrho(z_0,\Gamma_{\nu_0\ldots\nu_{s-1}}) = \delta.$$

From (2.2) and Proposition 3, in analogy to the proof of Proposition 2, Part (ii), we obtain $|\tilde f_{\nu_0\ldots\nu_{s-1}}(z_0)| \le \mathcal{O}(\ln^s\delta)$, $|\bar\partial\tilde f_{\nu_0\ldots\nu_{s-1}}(z_0)| \le \mathcal{O}(\ln^s\delta)$. Since

$$|d\beta(z_0)| = \mathcal{O}(\delta^{\alpha-1}),$$

then

$$\Big|\int_{\tilde\Gamma^{\varepsilon}_{\nu_0\ldots\nu_{s-1}}} \bar\partial\tilde f_{\nu_0\ldots\nu_{s-1}}(z)\wedge\beta(z)\wedge\omega(z)\Big| \le \int_{|\varepsilon|}^{|\varepsilon+\mu|}\ln^s t\,dt \int_{\Gamma^0_{\nu_0\ldots\nu_{s-1}}} |\beta(z)\wedge\omega(z)| < +\infty;$$

$$\Big|\int_{\tilde\Gamma^{\varepsilon}_{\nu_0\ldots\nu_{s-1}}} \tilde f_{\nu_0\ldots\nu_{s-1}}(z)\,\bar\partial\beta(z)\wedge\omega(z)\Big| \le \sum_I \int_{|\varepsilon|}^{|\varepsilon+\mu|}\ln^s t\cdot t^{\alpha-1}\,dt \int_{\Gamma^0_{\nu_0\ldots\nu_{s-1}}} |d\bar z^I\wedge\omega(z)|,$$

where $I=(i_1,\ldots,i_{n-s})$; $1\le i_1<i_2<\ldots<i_{n-s}\le n$. This proves our statement.

THEOREM 0. *Let* $f\in L(M)$ *and* $\bar\partial_M f = 0$. *Then on* M *we have*

$$f = (-1)^{n-s}\frac{(n-1)!}{(2\pi i)^n}\sum_{\nu_0<\ldots<\nu_{s-1}} f_{\nu_0,\ldots,\nu_{s-1}}.$$

<u>Proof</u>. For any form $\beta \in C^{\alpha}_{0,n-s}(\Omega)$, by Proposition 1, we have

$$(-1)^{n-s}\frac{(2\pi i)^n}{(n-1)!}\int_M f(z)\beta(z)\wedge\omega(z)$$

$$= \int_M f(z)\{\sum_{k=1}^{s}\sum_{\nu_0<\dots<\nu_{k-1}}\int_{\substack{\Gamma_{\nu_0\dots\nu_{k-1}}\\ \times\Delta_{\nu_0\dots\nu_{k-1}}}}\beta(\xi)\omega'_{n-s}(\eta^{\nu_0\dots\nu_{k-1}}(\xi,z,\lambda))\wedge\omega(\xi)$$

$$-\sum_{k=1}^{s-1}\sum_{\nu_0<\dots<\nu_{k-1}}\int_{\substack{\Gamma^{\varepsilon}_{\nu_0\dots\nu_{k-1}}\\ \times\Delta_{0,\nu_0\dots\nu_{k-1}}}}\bar\partial\beta\wedge\omega'_{n-s}(\eta^{0,\nu_0\dots\nu_{k-1}}(\xi,z,\lambda))\wedge\omega(\xi)$$

$$-\int_{\xi\in D^{\varepsilon}}\bar\partial\beta\wedge\omega'_{n-s}(\eta^0)\wedge\omega(\xi)$$

$$+\bar\partial_z\sum_{k=1}^{s}\sum_{\nu_0<\dots<\nu_{k-1}}\int_{\substack{\Gamma^{\varepsilon}_{\nu_0\dots\nu_{k-1}}\\ \times\Delta_{0,\nu_0\dots\nu_{k-1}}}}\beta(\xi)\omega'_{n-s-1}(\eta^{0,\nu_0\dots\nu_{k-1}}(\xi,z,\lambda))\wedge\omega(\xi)$$

$$-\bar\partial_z\int_{\xi\in D^{\varepsilon}}\beta(\xi)\omega'_{n-s-1}(\eta^0(\xi,z))\wedge\omega(\xi)\}\wedge\omega(z),$$

where $\{\nu_0\dots\nu_{k-1}\}\subset\{*,1,2,\dots,s+1\}$. By Propositions 2 and 4 we get

$$(-1)^{n-s}\frac{(2\pi i)^n}{(n-1)!}\int_M f(z)\beta(z)\wedge\omega(z)$$

$$= \int_M f(z)\{\sum_{1\le\nu_0<\dots<\nu_{s-1}\le s+1}\int_{\substack{\Gamma^0_{\nu_0\dots\nu_{s-1}}\\ \nu_0\dots\nu_{s-1}}}\beta(\xi)\omega'_{n-s}(\eta^{\nu_0\dots\nu_{s-1}}(\xi,z,\lambda))\wedge\omega(\xi)\}\omega(z)$$

$$+\int_M f(z)\{\bar\partial_z\sum_{1\le\nu_0<\dots<\nu_{s-1}\le s+1}\int_{\substack{\Gamma^0_{\nu_0\dots\nu_{s-1}}\\ \times\Delta_{0,\nu_0\dots\nu_{s-1}}}}\beta(\xi)\omega'_{n-s-1}(\eta^{0,\nu_0\dots\nu_{s-1}}(\xi,z,\lambda))$$

$$\wedge\omega(\xi)\}\wedge\omega(z).$$

Hence, by the theorem of Fubini and Stokes' formula,

$$(2.6)\quad (-1)^{n-s}\frac{(2\pi i)^n}{(n-1)!}\int_M f(z)\beta(z)\wedge\omega(z)$$

$$= \sum_{1\le\nu_0<\dots<\nu_{s-1}\le s+1}\int_M \beta(z)\Big\{\int_{\Gamma^0_{\nu_0\dots\nu_{s-1}}\times\Delta_{\nu_0,\nu_1\dots\nu_{s-1}}} f(\xi)\omega'(\eta^{\nu_0\dots\nu_{s-1}}(\xi,z,\lambda))\wedge\omega(\xi)\Big\}\wedge\omega(z)$$

$$- \sum_{1\le\nu_0<\dots<\nu_{s-1}\le s+1}\int_M \beta(z)\Big\{\int_{\partial\Gamma^0_{\nu_0\dots\nu_{s-1}}\times\Delta_{0,\nu_0\dots\nu_{s-1}}} f(\xi)\omega'(\eta^{0,\nu_0\dots\nu_{s-1}}(\xi,z,\lambda))\wedge\omega(\xi)\Big\}\omega(z).$$

Besides, by Stokes' formula,

$$0 = \int_{\partial\Gamma_{\nu_0\dots\nu_{s-1}}\times\Delta_{*,0,\nu_0\dots\nu_{s-1}}} f(\xi)(\bar\partial_\xi + d_\lambda)\omega'(\eta^{*,0,\nu_0\dots\nu_{s-1}}(\xi,\lambda,z))\wedge\omega(\xi)$$

$$= \int_{\partial\Gamma_{\nu_0\dots\nu_{s-1}}\times\Delta_{0,\nu_0\dots\nu_{s-1}}} f(\xi)\omega'(\eta^{0,\nu_0\dots\nu_{s-1}}(\xi,\lambda,z))\wedge\omega(\xi)$$

$$- \int_{\partial\Gamma_{\nu_0\dots\nu_{s-1}}\times\Delta_{*,\nu_0\dots\nu_{s-1}}} f(\xi)\omega'(\eta^{*,\nu_0\dots\nu_{s-1}}(\xi,\lambda,z))\wedge\omega(\xi)$$

$$+ \sum_{i=0}^{s-1}(-1)^{i+1}\int_{\partial\Gamma_{\nu_0\dots\nu_{s-1}}\times\Delta_{*,0,\nu_0\dots\hat\nu_i\dots\nu_{s-1}}} f(\xi)\omega'(\eta^{*,0,\nu_0\dots\hat\nu_i\dots\nu_{s-1}}(\xi,z,\lambda)\wedge\omega(\xi).$$

Therefore we obtain

$$(2.7)\quad \int_M \beta(z)\Big\{\int_{\partial\Gamma_{\nu_0\dots\nu_{s-1}}\times\Delta_{0,\nu_0\dots\nu_{s-1}}} f(\xi)\omega'(\eta^{0,\nu_0\dots\nu_{s-1}}(\xi,z,\lambda))\wedge\omega(\xi)\Big\}\wedge\omega(z)$$

$$= \int_M \beta(z)\{ \int_{\substack{\partial\Gamma_{\nu_0\dots\nu_{s-1}} \\ \times\Delta_{*,\nu_0\dots\nu_{s-1}}}} f(\xi)\omega'(\eta^{*,\nu_0\dots\nu_{s-1}}(\xi,z,\lambda))\wedge\omega(\xi)\}\wedge\omega(z)$$

$$- \int_M \beta(z)\{ \sum_{i=0}^{s-1}(-1)^{i+1} \int_{\substack{\partial\Gamma_{\nu_0\dots\nu_{s-1}} \\ \times\Delta_{*,0,\nu_0\dots\hat{\nu}_i\dots\nu_{s-1}}}} f(\xi)\omega'(\eta^{*,0,\nu_0\dots\hat{\nu}_i\dots\nu_{s-1}}(\xi,z,\lambda)) \wedge\omega(\xi)\}\wedge\omega(z).$$

Since $\partial\Gamma_{1,2\dots i\dots s+1} = (-1)^{s+1+i}\partial M$, then

$$\sum_{1\leq\nu_0<\dots<\nu_{s-1}\leq s+1} \int_M \beta(z)\{ \sum_{i=0}^{s-1}(-1)^{i+1} \int_{\substack{\partial\Gamma_{\nu_0\dots\nu_{s-1}} \\ \times\Delta_{*,0,\nu_0\dots\hat{\nu}_i\dots\nu_{s-1}}}} f(\xi)$$

$$\times\,\omega'(\eta^{*,0,\nu_0\dots\hat{\nu}_i\dots\nu_{s-1}}(\xi,z,\lambda))\wedge\omega(\xi)\}\omega(z) = 0.$$

Thus, by (2.7),

$$(2.8)\qquad \sum_{\nu_0<\dots<\nu_{s-1}} \int_M \beta(z)\{ \int_{\substack{\partial\Gamma_{\nu_0\dots\nu_{s-1}} \\ \times\Delta_{0,\nu_0\dots\nu_{s-1}}}} f(\xi)\omega'(\eta^{0,\nu_0\dots\nu_{s-1}}(\xi,z,\lambda))\wedge\omega(\xi)\}\wedge\omega(z)$$

$$= \sum_{\nu_0<\dots<\nu_{s-1}} \int_M \beta(z)\{ \int_{\substack{\partial\Gamma_{\nu_0\dots\nu_{s-1}} \\ \times\Delta_{*,\nu_0\dots\nu_{s-1}}}} f(\xi)\omega'(\eta^{*,\nu_0\dots\nu_{s-1}}(\xi,z,\lambda))\wedge\omega(\xi)\}\wedge\omega(z).$$

Finally we insert the right-hand side of (2.8) into (2.6) concluding that

$$(-1)^{n-s}\frac{(2\pi i)^n}{(n-1)!}\int_M f(z)\beta(z)\wedge\omega(z)$$

$$= \sum_{1\leq\nu_0<\dots<\nu_{s-1}\leq s+1} \int_M \beta(z)\{ \int_{\substack{\Gamma_{\nu_0\dots\nu_{s-1}} \\ \times\Delta_{\nu_0\dots\nu_{s-1}}}} f(\xi)\omega'(\eta^{\nu_0\dots\nu_{s-1}}(\xi,z,\lambda))\wedge\omega(\xi)$$

$$- \int\limits_{\substack{\partial\Gamma_{\nu_0\ldots\nu_{s-1}} \\ \times\Delta_{*,\nu_0\ldots\nu_{s-1}}}} f(\xi)\omega'(\eta^{*,\nu_0\ldots\nu_{s-1}}(\xi,z,\lambda))\wedge\omega(\xi)\}\wedge\omega(z)$$

and Theorem 0 is proved.

3. Analytic integral representations for CR-functions

In this section we are going to show that the general integral representations for CR-functions, derived in Section 2, may be "brought" in some cases to analytic integral representations of CR-functions, which constitutes the principal aim of the present paper.

At first suppose that the CR-manifold M of general form is full intersection of codimension $s=n$ in a certain domain $\Omega\subset\mathbb{C}^n$. Then, as we know (cf. [8] and [15]), M is a completely real manifold and may be represented in the form (1.3), where the functions $\varrho_1, \varrho_2, \ldots, \varrho_{n+1}$ are strictly plurisubharmonic in a strictly pseudoconvex domain $D\subset\Omega$.

By virtue of a result in [17] the functions $\{P^{(\nu)}(\xi,z)\}$ and $\{\Phi^{(\nu)}(\xi,z)\}$ may be chosen as functions which not only satisfy (1.4), but are also holomorphic with respect to $z\in\bar{D}_\nu$, $\nu=1,2,\ldots,n+1$. For any function $f\in L(M)$, we introduce, according to the definition (2.1), the functions

$$(3.1)\quad f_{\nu_0\ldots\nu_{n-1}}(z) = \int\limits_{\substack{(\xi,\lambda)\in\Gamma_{\nu_0\ldots\nu_{n-1}} \\ \times\Delta_{\nu_0\ldots\nu_{n-1}}}} f(\xi)\omega'(\eta^{\nu_0\ldots\nu_{n-1}}(\xi,z,\lambda))\wedge\omega(\xi).$$

By (1.6) these functions are well-defined and holomorphic for $z\in D\cap D_{\nu_0}\cap\ldots\cap D_{\nu_{s-1}}$.

Using now Theorem 0 and Proposition 4 we obtain the following more precise form of a result of Martineau [7] (cf. Theorem 1):

THEOREM 1. *Let $\tilde{M}$ be a smooth completely real submanifold of the form (1.3) in the domain Ω, where ϱ_ν are strictly plurisubharmonic functions in Ω, D is a strictly pseudoconvex domain contained in Ω, and $M=\tilde{M}\cap D$. Then every function $f\in L(M)$ has analytic representation of the form*

$$f = \frac{(n-1)!}{(2\pi i)^n}\sum_{\nu_0<\ldots<\nu_{n-1}} f_{\nu_0\ldots\nu_{n-1}}.$$

Besides, for every finite function $\beta \in C^{\alpha}(M)$, $\alpha > 0$, we have the equality

$$\sum_{\nu_0<\dots<\nu_{n-1}} \int_{z\in M} f_{\nu_0\dots\nu_{n-1}}(z)\beta(z)\omega(z) = \int_{z\in M} f(z)\beta(z)\omega(z).$$

Remark. If $M = R^n \cap D$, $D^{\pm}_{\nu} = \{z\in D:\ \operatorname{Im} z_{\nu} \gtrless 0\}$, and if in the definition (3.1) we set $\Phi^{(\nu)}(\xi,z) = \xi_{\nu} - z_{\nu}$, then the statement of Theorem 1 is equivalent to a Tillmann's result [14].

Let now $s = 1$, i.e. $\tilde{M}$ is a hypersurface in Ω. The hypersurface $M = \tilde{M} \cap D$, where D is a strictly pseudoconvex domain in Ω, "divides" D into an "upper" part D^+ and a "lower" part D^-. According to the definition (2.1), for a function $f \in L(M)$, we set

$$(3.2)\quad \int_{M^{\pm}} f(\xi)\omega'(\eta^{\pm}(\xi,z))\wedge\omega(\xi) - \int_{\partial M^{\pm}\times[0,1]} f(\xi)\omega'((1-\lambda)\eta^{*} + \lambda\eta^{\pm})\wedge\omega(\xi) = \begin{cases} f^{+}(z) & \text{for } z\in D^{+}, \\ f^{-}(z) & \text{for } z\in D^{-}, \end{cases}$$

where M^+ (resp. M^-) is the manifold M endowed with the orientation induced by the orientation of D^+ (resp. D^-), $\eta^{\pm}$ are any functions which fulfil the conditions (1.4) in the domains D^+ and D^-, whereas $\eta^{*}(\xi,z) = P(\xi,z)/\Phi(\xi,z)$ is a barrier function [17] of the domain D.

From Propositions 3, 4, and Theorem 0 it results immediately the following more precise version of the results of Andreotti-Hill [1], Kashiwara-Kawai [6], and Čirka [19].

THEOREM 2. Suppose that $\tilde{M}$ is a closed hypersurface in the domain Ω, D is a strictly pseudoconvex domain, and $M = D\cap\tilde{M}$. Then for every CR-function $f\in L(M)$ the functions f^+ and f^-, which are holomorphic in the domains D^+ and D^-, respectively, have generalized boundary values on M and

$$f = f^+ + f^- \quad \text{on } M.$$

Moreover, for any finite form $\beta\in C^{\alpha}_{0,n-1}(M)$, $\alpha > 0$, we have the equality

$$\int_{z\in M} f^{+}\beta\wedge\omega + \int_{z\, M} f^{-}\beta\wedge\omega = \int_{z\in M} f\beta\wedge\omega.$$

Remark. If in the definition (3.2) we set $\eta^{\pm} = (\bar{\xi} - \bar{z})/|\xi - z|^2$, then the statement of Theorem 2 is equivalent to the basic result obtained by Čirka in [19].

We will prove now as our basic result that the general scheme of Section 2 may be adapted to obtain local analytic representations for CR-functions on CR-manifolds of codimension 2 in $\mathbb{C}^n$.

Let further $\hat{M}$ be a CR-manifold of codimension 2 in $\mathbb{C}^n$. Then, either M is a complex manifold or $\operatorname{CR}\dim M = \dim_{\mathbb{C}} T^c_{\xi}(M) = n-2$.

The following proposition may be found in the paper of R. Nirenberg [8] (cf. Lemma 2.4.4 in [8]), though it is not explicitly formulated there:

PROPOSITION 5. *Let* M *be a CR-manifold of codimension* 2 *in* $\mathbb{C}^n$, *and assume that, at a point* z_0, M *fulfils the R. Nirenberg condition (cf. Introduction), and* $\operatorname{CR}\dim M = n-2$. *Then there exist a domain* $\Omega_0 \ni z_0$ *and functions* $\varrho_\nu \in C^2(\Omega_0)$, $\nu = 1,2,3$, *such that:*

(i) $\Omega_0 \setminus M = \bigcup\limits_{\nu=1}^{3} \{z \in \Omega_0 : \varrho_\nu(z) < 0\}$;

(ii) $\varrho_\nu = 0$; $\bar\partial\varrho_{\nu_1} \wedge \bar\partial\varrho_{\nu_2} \neq 0$ *on* $M \cap \Omega_0$, $1 \le \nu_1 < \nu_2 \le 3$, $\nu = 1,2,3$;

(iii) *either the Levi form* $\partial\bar\partial\varrho_\nu(z)$ *of every function* ϱ_ν *has for all* $z \in M \cap \Omega_0$ *one positive and three negative eigenvalues or the Levi form of every* ϱ_ν *has for all* $z \in M$ *all eigenvalues of the same sign.*

The following proposition arises as a simple corollary of the papers by Fischer-Lieb [3] and Range-Siu [12].

PROPOSITION 6. *Suppose that* $z_0 \in \Omega_0 \subset \mathbb{C}^n$, ϱ *is a real-valued function in* $C^2(\Omega_0)$, $\varrho(z_0) = 0$, $\operatorname{grad}\varrho(z_0) \neq 0$, *and the Levi form* $\partial\bar\partial\varrho(z_0)$ *has* s *negative eigenvalues. Then there exist a neighbourhood* Ω *of* z_0 *and a function* $\Phi \in C^2(\Omega\times\Omega)$ *such that::*

(i) $\Phi(\xi,\xi) = 0, \quad \xi \in \Omega$;

$\quad\operatorname{Re}\Phi(\xi,z) \ge \varrho(\xi) - \varrho(z) + \gamma|\xi - z|^2, \quad \xi, z \in \Omega$;

(ii) $\Phi(\xi,z) = \sum\limits_{\nu=1}^{n} P_\nu(\xi,z)(\xi_\nu - z_\nu), \quad \xi, z \in \Omega$;

(iii) $P_\nu(\cdot,z) \in C^\infty(\Omega)$ *for all* $z \in \Omega$, $\nu = 1,2,\dots,n$;

$\quad\bar\partial_\xi P_{\nu_1} \wedge \bar\partial_\xi P_{\nu_2} \wedge \dots \wedge \bar\partial_\xi P_{\nu_{n-s+1}}$ *in* $\Omega \times \Omega$ *for all* $\nu_1, \nu_2, \dots, \nu_{n-s+1}$.

THEOREM 3. *Let* M *be a smooth (of the class* C^2*) CR-manifold of codimension* 2 *in* $\mathbb{C}^n$. *Suppose that* M *fulfils the R. Nirenberg*

<u>condition at</u> $z_0 \in M$. <u>Besides, let</u> $f \in L(M)$ <u>and</u> $\bar\partial_M f = 0$ <u>on</u> M. <u>Then there exist a domain</u> $\Omega \ni z_0$, <u>a covering of the domain</u> $\Omega \setminus M$ <u>by domains of the form</u> $\Omega_i = \{z \in \Omega : \varrho_i(z) < 0, \varrho_i \in C^2(\Omega)\}$, $i = 1,2,3$, <u>and functions</u> f_{ij}, <u>holomorphic in</u> $\Omega_i \cap \Omega_j$, $1 \le i < j \le 3$, <u>such that</u>

$$\varrho_i(z) = 0;\ \bar\partial\varrho_i(z) \wedge \bar\partial\varrho_j(z) \neq 0 \quad \text{for every} \quad z \in M \cap \Omega,\ 1 \le i < j \le 3;$$

<u>the functions</u> f_{ij} <u>have generalized boundary values on the manifold</u> $M \cap \Omega \subset \partial\Omega_i \cap \partial\Omega_j$ <u>and on</u> $M \cap \Omega$ <u>we have the equality</u>

$$f = f_{12} + f_{23} + f_{13}.$$

<u>Moreover, for every finite form</u> $\beta \in C_{0,n-2}(M)$, $\alpha > 0$, <u>we have</u>

$$\int_{M\cap\Omega} f_{12}\beta\wedge\omega + \int_{M\cap\Omega} f_{23}\beta\wedge\omega + \int_{M\cap\Omega} f_{13}\beta\wedge\omega = \int_{M\cap\Omega} f\cdot\beta\wedge\omega.$$

<u>Proof</u>. Assume at the beginning that such a hypersurface passes through M and that the restriction of its Levi form $\partial\bar\partial\varrho(z_0)$ to $T^c_{z_0}(M)$ has one positive and one negative eigenvalue. Then, by Proposition 5, there exist a domain $\Omega_0 \ni z_0$ and functions $\varrho_\nu \in C^2(\Omega_0)$, $\nu = 1,2,3$, with the properties (i), (ii) of Proposition 5 and with the following property (iii)′: for all $z \in M \cap \Omega_0$ the Levi form $\partial\bar\partial\varrho_\nu(z)$ of every function ϱ_ν has one positive and three negative eigenvalues.

By Proposition 6 we can find a strictly pseudoconvex neighbourhood D of z_0 and functions $\Phi^{(\nu)}(\xi, z)$, $\nu = 1,2,3$, with the properties (i)-(iii) of Proposition 6, where $s = 3$.

Let $D_\nu = \{z \in D : \varrho_\nu(z) < 0\}$; $D_{\nu_1\nu_2} = D_{\nu_1} \cap D_{\nu_2}$; $\Gamma_\nu = \{z \in D : \varrho_\nu(z) = 0\}$; $\Gamma_{\nu_1,\nu_2} = \Gamma_{\nu_1} \cap \Gamma_{\nu_2}$; $\eta^\nu = P^\nu/\Phi^\nu$, $\nu = 1,2,3$; $\eta^* = P/\Phi$, where P and Φ are barrier functions of D.

By Theorem 0 on $M \cap D$ we have

$$(3.3)\qquad f = (-1)^{n-2}\frac{(n-1)!}{(2\pi i)^n}\sum_{\nu_1<\nu_2} \tilde f_{\nu_1,\nu_2},$$

where $\tilde f_{\nu_1,\nu_2}$ is a slice on $M \cap D$ of the function of $z \in D_{\nu_1,\nu_2}$, having the form

$$\tilde f_{\nu_1,\nu_2}(z) = \int_{(\xi,\lambda)\in\Gamma_{\nu_1\nu_2}\times\Delta_{\nu_1\nu_2}} f(\xi)\omega'(\eta^{\nu_1\nu_2}(\xi,z,\lambda))\wedge\omega(\xi)$$
$$- \int_{(\xi,\lambda)\in\partial\Gamma_{\nu_1\nu_2}\times\Delta_{\nu_1\nu_2}} f(\xi)\omega'(\eta^{*,\nu_1,\nu_2}(\xi,z,\lambda))\wedge\omega(\xi).$$

Using Proposition 3, we have

$$\bar\partial_z \tilde f_{\nu_1\nu_2}(z) = -\int_{\xi\in\Gamma_{\nu_1\nu_2}} f(\xi)\omega_1'(\eta^{\nu_2}(\xi,z))\wedge\omega(\xi)$$

$$+\int_{\xi\in\Gamma_{\nu_1\nu_2}} f(\xi)\omega_1'(\eta^{\nu_1}(\xi,z))\wedge\omega(\xi)$$

$$-\int_{(\xi,\lambda)\in\partial\Gamma_{\nu_1\nu_2}\times\Delta_{*,\nu_1}} f(\xi)\omega_1'(\eta^{*,\nu_1}(\xi,z,\lambda))\wedge\omega(\xi)$$

$$+\int_{(\xi,\lambda)\in\partial\Gamma_{\nu_1\nu_2}\times\Delta_{*,\nu_2}} f(\xi)\omega_1'(\eta^{*,\nu_2}(\xi,z,\lambda))\wedge\omega(\xi),$$

where $z\in D_{\nu_1\nu_2}$. By Proposition 6 (the property (iii)) we get

$$\int_{\Gamma_{\nu_1\nu_2}} f(\xi)\omega_1'(\eta^{\nu_i}(\xi,z))\wedge\omega(\xi) = 0,\quad 1\le\nu_1<\nu_2\le 3.$$

Thus we obtain

$$\bar\partial_z \tilde f_{\nu_1\nu_2}(z) = \int_{\partial\Gamma_{\nu_1\nu_2}\times\Delta_{*,\nu_2}} f(\xi)\omega_1'(\eta^{*,\nu_2}(\xi,z,\lambda))\wedge\omega(\xi) - \int_{\partial\Gamma_{\nu_1\nu_2}\times\Delta_{*,\nu_1}} f(\xi)\omega_1'(\eta^{*,\nu_1}(\xi,z,\lambda))\wedge\omega(\xi). \tag{3.4}$$

Let further Ω be a pseudoconvex neighbourhood of $z_0\in M$, such that $\Omega\subset D$ and for every $z\in\Omega$ we have $\{\xi : \operatorname{Re}\Phi^{(\nu)}(\xi,z)=0\}\cap\partial M=\emptyset$, $\nu=1,2,3$. The existence of such a neighbourhood follows from the properties (i) and (ii) of Proposition 6 for the functions $\Phi^{(\nu)}$. Therefore the forms

$$F_{\nu_i} = \int_{\partial\Gamma_{12}\times\Delta_{*,\nu_i}} f(\xi)\omega_1'(\eta^{*,\nu_i}(\xi,z,\lambda))\wedge\omega(\xi),\quad i=1,2,$$

appearing on the right-hand side of (3.4), are well-defined in Ω. We shall prove that these forms are $\bar\partial_z$-closed in Ω. By (1.2), integration by parts with respect to ξ and the use of Stokes' formula with respect to λ, gives for every $z\in\Omega$:

$$\bar\partial_z F_{\nu_i} = - \int_{(\xi,\lambda)\in\partial\Gamma_{12}\times\Delta_{*,\nu_i}} f(\xi)(\bar\partial_\xi + d_\lambda)\omega_2'(\eta^{*,\nu_i}(\xi,z,\lambda))\wedge\omega(\xi)$$

$$= - \int_{\xi\in\partial\Gamma_{12}} f(\xi)\omega_2'(\eta^{\nu_i}(\xi,z))\wedge\omega(\xi).$$

Let us notice now that by the property (iii) of Proposition 6 for the functions $\Phi^{(\nu_i)}$ and every $z\in\Omega$ the form $\omega_2'(\eta^{\nu_i}(\xi,z))\wedge\omega(\xi)$ is a $\bar\partial_\xi$-closed $(n, n-3)$-form with respect to ξ in the domain $\Omega\setminus\{\xi : \Phi^{(\nu_i)}(\xi,z)=0\}$, containing the manifold ∂M. Using the approximation theorem of Andreotti-Grauert-Hörmander (cf. [4]), we choose for a fixed $z\in\Omega$ a sequence $\{\chi_k(\xi,z)\}$ of $(n, n-3)$-forms so that they be well-defined and $\bar\partial_\xi$-closed on $\bar M$, and so that $\chi_k(\xi,z)\to\omega_2'(\eta^{\nu_i}(\xi,z))\wedge\omega(\xi)$ on M as $k\to\infty$. Consequently, by Stokes' formula, we have

$$\int_{\xi\in\partial\Gamma_{12}} f(\xi)\omega_2'(\eta^{\nu_i}(\xi,z))\wedge\omega(\xi) = \lim_{k\to\infty}\int_{\xi\in\partial\Gamma_{12}} f(\xi)\chi_k(\xi,z) = 0.$$

Therefore $\bar\partial_z F_{\nu_i} = 0$.

By the Oka theorem we may choose functions $u_{\nu_i}(z)$, $z\in\Omega$, so that

(3.5) $\bar\partial_z u_{\nu_i}(z) = F_{\nu_i}$.

Finally, let us introduce functions of the form

$f_{12} = \tilde f_{12} + u_1 - u_2$ in the domain $D_{12}\cap\Omega$,

$f_{13} = \tilde f_{13} - u_1 + u_3$ in the domain $D_{13}\cap\Omega$,

$f_{23} = \tilde f_{23} + u_2 - u_3$ in the domain $D_{23}\cap\Omega$.

Directly from the definition and from (3.3) there follows the equality

$$f = (-1)^{n-2}\frac{(n-1)!}{(2\pi i)^n}\sum_{\nu_1<\nu_2} f_{\nu_1\nu_2} \quad \text{on } M\cap\Omega.$$

From (3.4), (3.5), and from the fact that $\partial M\cap D = \partial\Gamma_{12} = \partial\Gamma_{23} = -\partial\Gamma_{13}$ it follows that the functions $f_{\nu_1\nu_2}$ are holomorphic in the domains $D_{\nu_1\nu_2}$.

In this way in one of two considered cases (one positive and one negative eigenvalue of the Levi form $\partial\bar\partial\varrho$ on $T_z^c(M)$) the theorem is completely proved.

In the remaining case (all the eigenvalues of the Levi form $\partial\bar\partial\varrho$

on $T^c_{z_0}(M)$ are of the same sign) the proof remains the same if during the construction of the functions $\Phi^{(\nu)}$ we change the reference to Proposition 6 into the reference [17] concerning the existence of the corresponding barrier functions of strictly pseudoconvex (and strictly pseudoconcave) domains.

References

[1] ANDREOTTI, A. and C. DENSON-HILL: E. E. Levi convexity and the Hans Lewy problem, Ann. Scuola Norm. Sup. Pisa 26 (1972), 325-363.

[2] ——, ——, S. ŁOJASIEWICZ, and B. MACKICHAN: Complexes of differential operators, Invent. Math. 35 (1976), 43-86.

[3] FISCHER, W. und I. LIEB: Lokale Kerne und beschränkte Lösungen für den $\bar\partial$-Operator auf q-konvexen Gebieten, Math. Ann. 208 (1974), 249-265.

[4] HÖRMANDER, L.: L^2-estimates and existence theorems for the $\bar\partial$-operator, Acta Math. 113 (1965), 89-152.

[5] HUNT, L. R., J. C. POLKING, and J. STRAUSS: Unique continuation for solutions to the induced Cauchy-Riemann equations, J. Diff. Equations 23 (1977), 436-447.

[6] KASHIWARA, M. and T. KAWAI: On the boundary value problem for elliptic systems of linear differential equations, Proc. Japan Acad. 48 (1972), 712-715.

[7] MARTINEAU, A.: Distributions et valeurs au bord des fonctions holomorphes, Proc. of the Internat. Summer Institute, Lisbon 1964.

[8] NIRENBERG, R.: On the H. Lewy extension phenomenon, Trans. Amer. Math. Soc. 168 (1972), 337-356.

[9] NORGUET, F.: Problèmes sur les formes différentielles des courants, Ann. Inst. Fourier 11 (1960), 1-88.

[10] POLKING, J. C. and R. O. WELLS: Hyperfunction boundary values and a generalized Bochner-Hartogs theorem, Proc. Symp. Pure Math. 30 (1977), 187-193.

[11] RANGE, R. M. and Y. T. SIU: Uniform estimates for the $\bar\partial$-equation on domains with piecewise smooth strictly pseudoconvex boundaries, Math. Ann. 206 (1974), 325-354.

[12] —— and ——: C^k-approximation by holomorphic functions and $\bar\partial$-closed forms on C^k-submanifolds of a complex manifold, Math. Ann. 210 (1974), 105-122.

[13] SATO, M., T. KAWAI, and M. KASHIWARA: Microfunctions and pseudo-differential equations, Springer-Verlag, Berlin - Heidelberg - New York 1973.

[14] TILLMANN, H. G.: Randverteilungen analytischer Funktionen und Distributionen, Math. Z. 59 (1953), 61-83.

[15] WELLS, R. O.: Function theory on differentiable submanifolds, in: Contributions to Analysis, Academic Press, New York 1974, pp. 407-441.

[16] POLJAKOV, P. L.: Banach cohomology on piecewise strictly pseudoconvex domains [in Russian], Mat. Sb. (N. S.) 88 (130) (1972), 239-255.

[17] HENKIN, G. M.: Integral representation of functions which are holomorphic in strictly pseudoconvex regions, and some applica-

tions [in Russian], Mat. Sb. (N. S.) 78 (120) (1969), 611-632.

[18] ——: A uniform estimate for the solution of the $\bar{\partial}$-problem in a Weil region [in Russian], Uspehi Mat. Nauk 26 (1971), no. 3 (159), 211-212.

[19] ČIRKA, E. M.: Analytic representations of CR-functions [in Russian], Mat. Sb. (N. S.) 98 (140) (1975), 591-623 and 640.

Центральный экономико-математический
институт Академии наук СССР
ул. Вавилова 44-2, SU-117 333 Москва В-333, СССР

ON THE STABILITY OF HOLOMORPHIC FOLIATIONS

Harald Holmann (Fribourg)

Contents

Abstract.

In 1978 D.B.A. Epstein and E. Vogt succeeded in constructing an unstable real-analytic periodic flow on a 4-dimensional compact real-analytic manifold. This cannot be generalized to the complex-analytic case:

Proposition 1: A periodic holomorphic flow of a compact complex variety is always stable (H. Holmann, 1977).

This year Th. Müller found the first example of an unstable compact holomorphic foliation of a non-compact complex manifold in form of a periodic holomorphic flow all orbits being equivalent complex tori. The underlying 3-dimensional complex manifold of this example cannot carry a Kähler structure because of the following proposition proved in this paper:

Proposition 2: On a (not necessarily compact) Kähler manifold all compact holomorphic foliations are stable.

Its proof uses that Kähler-manifolds are characterized by the fact that local-analytic submanifolds are minimal surfaces with respect to the Kähler metric. Proposition 2 is a special case of a more general result obtained with different methods by H. Rummler (1978):

Proposition 3: A compact differentiable foliation of a differentiable manifold is stable iff it carries a Riemannian-metric such that all leaves are minimal surfaces with respect to this metric.

Harald Holmann

1. Introduction.

About 25 years ago Reeb and Haefliger posed the problem: are compact differentiable foliations of compact differentiable manifolds stable? In1952 already Reeb showed that 1-codimensional compact differentiable foliations are always stable and gave examples of unstable 2-codimensional compact foliations of non-compact manifolds (compare [9]). There was no progress until Epstein proved in 1972 that periodic differentiable flows on 3-dimensional compact differentiable manifolds are stable (compare [2]). In 1975 the Reeb-Haefliger conjecture was confirmed for all 2-codimensional compact foliations of compact differentiable manifolds by Edwards, Millett, Sullivan and Vogt (compare [1] and [14]). At the same time Sullivan and Thurston produced counterexamples in codimension 4 (compare [12] and [13]) and in 1978 Epstein and Vogt succeeded in constructing an unstable real-analytic periodic flow on a 4-dimensional compact real-analytic manifold (compare [4]). These examples cannot be generalized to the complex-analytic case because of the following proposition:

Proposition 1: A periodic holomorphic flow on a compact complex variety is always stable (compare [6]).

By a periodic holomorphic flow we understand a holomorphic action of the additive group $\mathbb{C}$ of complex numbers such that all orbits are complex tori.

Up to now no counterexamples to the Reeb-Haefliger conjecture are known in the complex analytic case. Only this year Th. Müller found the first example of an unstable compact holomorphic foliation of a non-compact complex manifold in form of a periodic holomorphic flow all orbits being biholomorphically equivalent complex tori (compare [8]). The underlying 3-dimensional complex manifold of this example cannot carry a Kähler-structure because of the following proposition which will be proved in section 3 and which is well known for compact Kähler-manifolds.

<u>Proposition 2:</u> On a (not necessarily compact) Kähler-manifold all compact holomorphic foliations are stable.

Its proof uses that Kähler-manifolds are characterized by the fact that local-analytic submanifolds are minimal surfaces with respect to the Kähler-metric (compare [10]). Proposition 2 is a special case of a more general result obtained with different methods by H. Rummler (compare [11]):

<u>Proposition 3:</u> A compact differentiable foliation of a differentiable manifold is stable iff it carries a Riemannian-metric such that all leaves are minimal surfaces with respect to this metric.

2. Some definitions and results on holomorphic foliations.

In this section we recollect some definitions and properties of holomorphic foliations. Differentiable and real-analytic foliations are defined analogously (compare [1], [3], [5] and [7]).

Definition 2.1: By a p-dimensional [(n-p)-codimensional] holomorphic foliation of an n-dimensional complex manifold M we understand a decomposition $F = (L_i)_{i\in I}$ of M into pairwise disjoint connected subsets L_i - which are called leaves of the foliation F - with the following property:

Each point $x \in M$ has a neighbourhood U carrying a complex chart $\varphi: U \longrightarrow Q_r^n := \{z = (z_1,\ldots,z_n) \in \mathbb{C}^n;\ |z_\nu| < r,\ \nu = 1,\ldots,n\},\ r > 0$, such that for each leaf L_i the connected components of $\varphi(L_i \cap U)$ are of the form $\{z \in Q_r^n;\ z_{p+1} = \text{const},\ldots,z_n = \text{const}\}$, that means they are fibres of the projection $p: Q_r^n \longrightarrow Q_r^{n-p}$, $p(z_1,\ldots,z_n) := (z_{p+1},\ldots,z_n)$.

The set $M/F := \{L_i;\ i \in I\}$ of all leaves of F equipped with the quotient topology is called the leaf-space of F.

The foliation F is called compact if all leaves L_i are compact.

The local behaviour of a foliation is quite well described by the holonomy groups which we are going to define now.

Definition 2.2: (We use the notations of definition 2.1 above). Let $h: V \longrightarrow V'$ be a biholomorphic mapping between two open neighbourhoods V, V' of the origin 0 in Q_r^{n-p} with $h(0) = 0$. We say that h is compatible with the foliation F if $\varphi^{-1}(p^{-1}(h(v)))$ and $\varphi^{-1}(p^{-1}(v))$ lie in the same leaf for each $v \in V$.

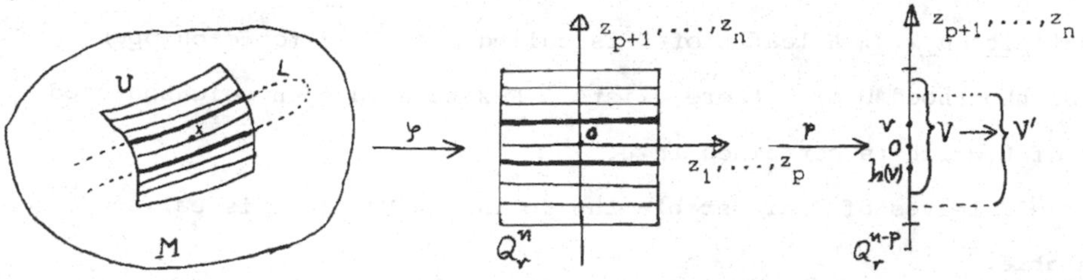

The germs at the origin of such F-compatible mappings form a group which is called the holonomy group H(x) of F at $x \in M$.

One sees easily that up to isomorphy this definition of H(x) does not depend on the special chart (U,φ). Actually it depends only on the leaf L_x of F that passes through x (compare [7]). We therefore wright also $H(L_x)$ instead of H(x).

Later on we shall need also the following relative version of holonomy groups: Let X be a closed F-saturated subset of M (that means: for each $y \in X$ the whole leaf L_y passing through y belongs also to X). We obtain the so called relative holonomy group H(x,X) for $x \in X$, if in the above definition of F-compatibility of a biholomorphic mapping $h: V \longrightarrow V'$ we request only for all $v \in V \cap p(\varphi(U \cap X))$ that the sets $\varphi^{-1}(p^{-1}(h(v)))$ and $\varphi^{-1}(p^{-1}(v))$ lie in the same leaf of F.

There is a canonical group homomorphism (restriction) $H(x) \rightarrow H(x,X)$ for each $x \in X$.

For a compact holomorphic foliation F of a complex manifold M we define the stability of a leaf as follows:

<u>Definition 2.3:</u> A leaf L of F is called stable if for each open neighbourhood U of L there exists a F-saturated open neighbourhood V of L which is contained in U.
If all leaves of F are stable the foliation F itself is called stable.

If X is closed F-saturated subset of M, a leaf $L \subset X$ is called stable relative to X if for each neighbourhood U of L in X there exists a F-saturated neighbourhood V of L in X with $V \subset U$.

We have the following criteria for stability (compare [1] and [7]):

Proposition 1.4: For a leaf L of a compact holomorphic foliation F of a complex manifold M the following conditions are equivalent:

(1) L is stable.

(2) The holonomy group H(L) is finite.

(3) Vol: $M \longrightarrow \mathbb{R}$ is bounded in a neighbourhood of L.
(Vol(x) can be defined as the volume of the leaf L_x passing through x with respect to any fixed Riemannian-metric on M)

Remark: The relative version of this proposition is also true (compare [1]).

Examples of holomorphic foliations can be obtained by complex Lie group actions on a complex manifold when the dimension of all orbits is constant. We shall discuss a special example of this kind in the last section.

3. Stability of compact holomorphic foliations on Kähler-manifolds.

In this section the following proposition about holomorphic foliations of not necessarily compact Kähler-manifolds will be proved.

Proposition 3.1: A compact holomorphic foliation $F = (L_i)_{i \in I}$ of a Kähler-manifold M is always stable.

<u>Proof:</u> I. If $I_o := \{i \in I;\ L_i$ is unstable$\}$ then $X_o := \bigcup_{i \in I_o} L_i$ is called the bad set of F. X_o is F-saturated, closed and nowhere dense in M (compare [1]). Suppose X_o is not empty. Let X_1 denote the union of all leaves of F in X_o which are unstable relative to X_o. Again X_1 is closed and nowhere dense in X_o (compare [1]). Without any loss of generality we can assume that $X_1 = \emptyset$. (Otherwise we take the Kähler-manifold $M' := M - X_1$. F induces a compact holomorphic foliation F' on M', such that the bad set X_o' of F' is the non-empty set $X_o - X_1$.)

Then X_o decomposes into two classes:

$X_o' := \{x \in X_o\ ;\ H(x,X_o) = \{Id\}\}$

$X_o'' := \{x \in X_o\ ;\ H(x,X_o)$ finite, but $\neq \{Id\}\}$

X_o'' is closed and nowhere dense in X_o. Again we can assume that $X_o'' = \emptyset$, that means the relative holonomy groups $H(x,X_o)$ are trivial for all $x \in X_o$.

II. Let x be a point of X_o (we use the notations of definition 2.1 in the following). The holonomy group H(x) of F at x is not finite but finitely generated, that means there exist biholomorphic mappings $h_i : V_i \longrightarrow V_i'$, $i = 1,\dots,N$ $(N \geqslant 1)$, with the following properties:

1) V_i, V_i' are open neighbourhoods of 0 in Q_r^{n-p} and $h_i(0) = 0$,

2) $\varphi^{-1}(p^{-1}(h_i(v)))$ and $\varphi^{-1}(p^{-1}(v))$ lie in the same leaf of F for each $v \in V_i$,

such that the germs $[h_i]_o$, $i = 1,\dots,N$, generate H(x).

$Y^o := p(\varphi(U \cap X_o))$ is closed and nowhere dense in Q_r^{n-p}, since $U \cap X_o$ is closed and nowhere dense in U. Since $H(x,X_o)$ is trivial, we can assume that $h_i | Q_\varepsilon^{n-p} \cap Y^o = \mathrm{Id}$ for a sufficiently small polycylinder Q_ε^{n-p} in Q_r^{n-p} which lies inside all V_i, and $h_i | Q_\varepsilon^{n-p} \neq \mathrm{Id}$ for all $i = 1,\dots,N$. The points of Q_ε^{n-p} which remain fixed under all mappings h_i form a proper analytic subset A of Q_ε^{n-p}. $Q_\varepsilon^{n-p} \cap Y_o$ is contained in A. $U' := \varphi^{-1}(p^{-1}(Q_\varepsilon^{n-p}))$ is an open neighbourhood of x in M such that $U' \cap X_o$ is contained in the proper analytic subset $\varphi^{-1}(p^{-1}(A))$ of U'. This shows that X_o is not only closed and nowhere dense in X, it also does not decompose M locally. Here we have used explicetely the complex analytic structure of M. In the real-analytic or differentiable case things are more complicated, here the manifold can be decomposed by the bad set locally. Because of this the proof of the more general theorem of Rummler is completely different.

III. We can assume, that M is connected. As we just showed $M-X_o$ is connected too. Since F is stable on $M-X_o$ all holonomy groups $H(x)$, $x \in M-X_o$, are finite. $S := \{x \in M-X_o;\ H(x) \neq \{\mathrm{Id}\}\}$ is a proper analytic subset of $M-X_o$ (compare [1] and [7]). $M_{reg} := M-(X_o \cup S)$ is therefore a dense, open, connected subset of M.
Now we use that M and consequently M_{reg} are Kähler-manifolds. All leaves of F in M_{reg} are minimal surfaces and since M_{reg} is connected $\mathrm{Vol}(x) = C = \mathrm{const}$ for all $x \in M_{reg}$. (One can say even more: each compact complex submanifold of a Kähler-manifold has minimal volume in its homology class with respect to the Kähler-metric; the volume of a non-analytic member of the same homology class is strictly smaller.) C is also an upper bound for the volume of all

leaves in S (compare [1]). More precisely the following holds (compare [11]): $\mathrm{Vol}(x) = \frac{C}{\mathrm{Ord}\ H(x)}$ for all $x \in M-X_o$. Since all the leaves of F contained in X_o are stable relative to X_o. The volume fonction $\mathrm{Vol}|X_o \longrightarrow \mathbb{R}$ is locally bounded on X_o (compare the remark following Proposition 2.4). It follows that $\mathrm{Vol}: M \longrightarrow \mathbb{R}$ is locally bounded on M. Consequently all leaves of F are stable which contradicts the assumption that X_o is not empty.

The leaf-space M/F of a stable compact holomorphic foliation F of a complex manifold M is hausdorff and consequently carries a canonical complex structure (compare [5]), in general with singularities. Therefore the leaves of F are nothing else but the connected fibres of the canonical holomorphic projection $\pi: M \longrightarrow M/F$. As a corollary of Proposition 3.1 we obtain:

Corollary 3.2: All compact holomorphic foliations of Kähler-manifolds are holomorphic fibrations with smooth connected fibres.

4. Example of an unstable foliation.

Recently Th. Müller (compare [8]) found the first example of an unstable compact holomorphic foliation of a non-compact complex manifoldd. I want to describe here its real-analytic analogue which is also very similar to an example given by D.B.A. Epstein in [2]. I shall indicate how by certain replacements this can be transformed into Müller's example.

We shall construct a 3-dimensional real-analytic manifold M as a quotient of the open set $X := X_1 \cup X_2$ of $\mathbb{R}^3$, where $X_1 := \mathbb{R} \times \mathbb{R} \times \mathbb{R}^*$

and $X_2 := I \times \mathbb{R} \times \mathbb{R}$ ($\mathbb{R}^* := \mathbb{R} - \{0\}$, $I := \{x \in \mathbb{R};\ 0 < x < 1\}$).
On X we have the following real-analytic flow $\varphi: \mathbb{R} \times X \longrightarrow X$,
$\varphi(t; z_1, z_2, z_3) := (z_1 + z_3 \cdot t,\ z_2 + t,\ z_3)$ which will induce an unstable real-analytic flow on M with compact orbits all analytically equivalent to the circle S^1.

For the construction of M we need the following real-analytic actions of the additive group $\mathbb{Z}$ of integers:

1) $\tilde{\varphi}: \mathbb{Z} \times X_1 \longrightarrow X_1$, $\tilde{\varphi}(g; z_1, z_2, z_3) := \varphi(\frac{g}{z_3}; z_1, z_2, z_3) = (z_1 + g, z_2 + \frac{g}{z_3}, z_3)$.

2) $\psi: \mathbb{Z} \times X \longrightarrow X$, $\psi(g; z_1, z_2, z_3) := (z_1, z_2 + g, z_3)$.

We use the symbols $\mathbb{Z}_{\tilde{\varphi}}$ and $\mathbb{Z}_\psi$ to indicate by which operations $\mathbb{Z}$ acts on X_1 and X respectively.

We observe that the action of $\mathbb{Z}_{\tilde{\varphi}}$ on X_1 is free and properly discontinous. The same is true for the action of $\mathbb{Z}_\psi$ on X. X_1 and X_2 are invariant under the action of $\mathbb{Z}_\psi$. We can form the group $H := \mathbb{Z}_\psi \bullet \mathbb{Z}_{\tilde{\varphi}} = \mathbb{Z}_{\tilde{\varphi}} \bullet \mathbb{Z}_\psi$ which also acts freely and properly discontinously on X_1. Consequently $\hat{X}_1 := X_1/H$ and $\hat{X}_2 := X_2/\mathbb{Z}_\psi$ are 3-dimensional real-analytic manifolds in a canonical way.

$D := X_1 \cap X_2 = I \times \mathbb{R} \times \mathbb{R}^*$ is invariant under $\mathbb{Z}_\psi$ but not under H, but the equivalence relation induced by H on D coincides with the one defined by $\mathbb{Z}_\psi$. Therefore we can regard $\hat{D} := D/\mathbb{Z}_\psi$ as an open subset of $\hat{X}_2$ as well as of $\hat{X}_1$; to be precise: $\hat{D} = \hat{X}_1 \cap \hat{X}_2$. Since $M := \hat{X}_1 \cup \hat{X}_2$ is hausdorff it is a real-analytic manifold in a canonical way.

One proves easily that $\varphi: \mathbb{R} \times X \longrightarrow X$ induces a real-analytic flow $\hat{\varphi}: \mathbb{R} \times M \longrightarrow M$ with compact orbits all real-analytically equivalent to S^1. The flow is stable on $\hat{X}_1$; $M-\hat{X}_1$ consists only of unstable orbits.

One obtains the unstable compact holomorphic flow found by Th. Müller by making the following substitutions:

$\mathbb{R} \rightsquigarrow \mathbb{C}$, $\mathbb{R}^* \rightsquigarrow \mathbb{C}^*$, $Z \rightsquigarrow Z+iZ$, $I \rightsquigarrow I+iI$.

References

[1] EDWARDS, R., MILLETT, K., SULLIVAN, D.: Foliations with all leaves compact. Topology 16 (1977) 13-32.

[2] EPSTEIN, D.B.A.: Periodic flows on three-manifolds. Ann. of Math. 95 (1972), 68-82.

[3] EPSTEIN, D.B.A.: Foliations with all leaves compact. Ann. Inst. Fourier, Grenoble, 26,1 (1976), 265-282.

[4] EPSTEIN, D.B.A., VOGT, E.: A counterexample to the periodic orbit conjecture in codimension 3. Ann. of Math. 108 (1978), 539-552.

[5] HOLMANN, H.: Holomorphe Blätterungen komplexer Räume. Comment. Math. Helv. 47 (1972), 185-204.

[6] HOLMANN, H.: Analytische periodische Strömungen auf kompakten komplexen Räumen, Comment. Math. Helv. 52 (1977), 251-257.

[7] HOLMANN, H.: On the stability of holomorphic foliations with all leaves compact. 683, Springer Lecture Notes in Mathematics: Variétés Analytiques Compactes, Colloque, Nice (1977), 217-248.

[8] MÜLLER, Th.: Beispiel einer periodischen instabilen holomorphen Strömung. Erscheint in L'Enseignment mathématique.

[9] REEB, G.: Sur certaines propriétés topologiques des variétés feuilletées. Act.Sci. et Ind. No 1183, Hermann, Paris (1952).

[10] RUMMLER, H.: Métriques Kähleriennes et surfaces minimales. L'Enseignement mathématique T.XXIV, fasc. 3-4 (1978), 305-310.

[11] RUMMLER, H.: Kompakte Blätterungen durch Minimalflächen. Habilitationsschrift, Universität Freiburg i. Ue. (1978).

[12] SULLIVAN, D.: A counterexample to the periodic orbit conjecture. Publ. I.H.E.S. No 46 (1976).

[13] SULLIVAN, D.: A new flow. Bulletin Am.Math.Soc. 82 (1976),331-332.

[14] VOGT, E.: Foliations of codimension 2 with all leaves compact. Manuscripta math. 18 (1976), 187-212.

Université de Fribourg
Institut de Mathématiques
CH-1700 Fribourg/Suisse

STABILITY IN THE DIFFERENTIAL EQUATIONS FOR QUASIREGULAR MAPPINGS

Tadeusz Iwaniec and Ryszard Kopiecki (Warszawa)

1. Notations. We will work with the space $\mathbb{R}^{n\times n}$ of all square matrices $A = (a_{ij})$, with transposed matrices $A^* = (a_{ji})$, and the norm

$$|A| = (\sum_{i,j=1}^{n} |a_{ij}|^2)^{1/2} = (\operatorname{Tr} A^*A)^{1/2}.$$

The symbols $L_p(D,\mathbb{R})$, $L_p(D,\mathbb{R}^n)$, $L_p(D,\mathbb{R}^{n\times n})$ will denote L_p-spaces of scalar-, vector-, and matrix-valued functions on a subset $D\subset \mathbb{R}^n$. On the space $L_p(D,\mathbb{R}^{n\times n})$ we will consider the norm

$$\|A\|_{p,D} = (\int_D (\operatorname{Tr} A^*(x)A(x))^{p/2}dx)^{1/p}.$$

When D is open in $\mathbb{R}^n$, then $W_p^1(D,\mathbb{R})$, $W_p^1(D,\mathbb{R}^n)$ will denote the corresponding Sobolev spaces of functions in $L_p(D,\cdot)$, whose distributional first derivatives belong also to the space $L_p(D,\cdot)$. Similarly, $W^1_{p,loc}(D,\cdot)$ will be the spaces of functions, which lie in $W_p^1(F,\cdot)$ for every compact subset $F\subset D$. The Jacobi matrix Df and the Jacobian $Jf = \det Df$ are well defined for any function $f\in W^1_{p,loc}(D,\mathbb{R}^n)$ and, by definition, $Df\in L_{p,loc}(D,\mathbb{R}^{n\times n})$.

2. First of all we define quasiregular maps.

Definition. A map $f : D \longrightarrow D'$ of two domains in $\mathbb{R}^n$ is said to be K-quasiregular, $1\le K<\infty$, if

(α) $f\in W^1_{n,loc}(D,\mathbb{R}^n)$,

(β) $|Df(x)|^n \le n^{n/2}K^n Jf(x)$ almost everywhere in D;

when f is homeomorphic on D, then we call it K-quasiconformal. In other words, f is K-quasiregular on D if the function

$$K_f(x) = \frac{|Df(x)|}{n^{1/2} Jf(x)^{1/n}}$$

is bounded on D by the constant K. The number

$$K_f = \| K_f(x) \|_{\infty,D} \leq K$$

is called the maximal dilatation of f.

In the case of $n=2$ differential inequality (β) implies the Beltrami equation

$$(1) \qquad f_{\bar{z}} = q\, f_z ,$$

where q is a measurable function on D, such that $\| q \|_{\infty,D} \leq q_o < 1$; the function $K_f(z) = (1 - |q(z)|)^{-1}(1 + |q(z)|)$. This equation may be considered as a particular case of the elliptic sysetm

$$(2) \qquad f_{\bar{z}}(z) = q_1(z,f)\, f_z(z) + q_2(z,f)\, \overline{f_z(z)}$$

with measurable coefficients and the ellipticity condition

$$|q_1(z,f)| + |q_2(z,f)| \leq q_o < 1$$

In this paper we consider the following generalization of (2) for many veriables:

$$(3) \qquad Df^*(x)\, H(x,f)\, Df(x) = Jf(x)^{2/n} G(x,f),$$

where $H, G \in L_\infty(D, \mathbb{R}^{n\times n})$ are given symetric and positive-defined matrices; i.e.

$$(4) \qquad \begin{array}{l} c|v|^2 \leq \langle H(x,f)v, v\rangle, \\ c|w|^2 \leq \langle G(x,f)w, w\rangle, \end{array} \qquad c > 0, \quad v, w \in \mathbb{R}^n,$$

for almost all $(x,f) \in D \times D'$, and

$$(4') \qquad \det H = \det G = 1 \text{ almost everywhere in } D \times D'.$$

By generalized solution of (3) we mean any vector-valued function $f \in W^1_{n,loc}(D, \mathbb{R}^n)$ which satisfies system (3) almost everywhere.

Conditions (4) quarantee quasiregularity of the generalized solutions of (3). The maximal dilatation can be estimated by a constant, which depends only on c and n. When $H = G = E$ (-the identity matrix) almost everywhere, then the sysetm defines conformal maps. The lecture Bojarski and Iwaniec [7] contains a survey of the basic properties of (3) in the general form.

When $n=2$, then we get from (3) quasilinear system (2). Actually, the first one was obtained from a geometrical interpretation of (2). In the planar case we know much more than in the general one. The investigation of (2) has been developed from many different points of view by several authors, among of them L.Ahlfors [1], L.Bers [5], B.Bojarski [6], L.Nirenberg, and I.N.Vekua [21]. We recall here only three fundamental results.

(i) (The Riemann mapping theorem). For every measurable q_1, q_2 there is a homeomorphic solution $f : D \longrightarrow D'$.

(ii) (The regularity theorem). When q_1, q_2 are of the class $C^{k,\alpha}(D\times D')$, with $k \geq 1$ and $0<\alpha<1$, then the solutions are of the class $C^{k+1,\alpha}(D)$.

(iii) (The stability theorem). If the sequences q_1^j, q_2^j uniformly satisfy the inequality

$$|q_1^j(z,y)| + |q_2^j(z,y)| \leq q_o < 1$$

and converge to q_1, q_2 almost everywhere in $D \times D'$, then there exists a subsequence $f_{j_k} : D \longrightarrow D'$ such that f_{j_k} go to a quasiregular map f locally uniformly and also in $W^1_{p,loc}(D,\mathbb{R}^2)$, for some $p > 2$, and $f_{\bar{z}} = q_1 f_z + q_2 \bar{f}_z$ almost everywhere in D.

In the general case of $n \geq 2$ the problem of existence of the solutions of (3) is still open and seems to be difficult. System (3) is overdetermined so that the pair (G,H) must satisfy some integrability conditions. When (3) has a nonconstant solution f, then (G,H) - the characteristics of f - we call integrable. In the particular case $H = E$, $G(x,f) = G(x)$, $G \in C^2(D)$, an integrability condition was given by H.Weyl and J.A.Schouten (see Chapter VI, §5 of [20]). As a result of our consideration we give a generalization of this condition. The regularity problem for system (3) was completely solved by T.Iwaniec [9] and Yu.G.Reshetnjak [19]. Namely, when $G(x,f)$ and $H(x,f)$ are of the class $C^{k,\alpha}(D\times D')$, $k \geq 1$, $0<\alpha<1$, then the solutions are of the class $C^{k+1,\alpha}(D)$. Moreover, the solutions are local diffeomorphisms and this fact distinguishes the higher dimensional case from the planar case.

3. The aim of this paper is to prove the following stability

THEOREM. Let $f_j : D \longrightarrow D'$ be a sequence of quasiregular mappings of a domain $D \subset \mathbb{R}^n$ into a bounded doamin $D' \subset \mathbb{R}^n$, $n \geq 2$, and $G_j(x,y)$, $H_j(x,y)$ be their characteristics, i.e.

$$Df_j^* H_j Df_j = Jf_j^{2/n} G_j \quad \text{a.e. in } D.$$

Suppose that G_j and H_j satisfy uniformly conditions (4) and (4'), that $G_j \longrightarrow G$, $H_j \longrightarrow H$ almost everywhere in $D \times D'$. Then there exists a subsequence f_{j_k} such that

A) f_{j_k} tends locally uniformly to some quasiregular mapping $f : D \longrightarrow D'$ which satisfies the limit equation

$$DF^* H\, Df = Jf^{2/n} G \quad \text{a.e. in } D;$$

B) f_{j_k} converges to f with respect to the norm of the Sobolev space $W^1_{p,loc}(D, \mathbb{R}^n)$ for some $p = p(c,n) > n$.

Before we give the proof we point out some interesting implications of the theorem. When $H = E$, $G(x,f) = G(x)$, and $G \in C^2(D)$, then the Weyl - Schouten condition of integrability of sysetm (3) reads

$$(5) \qquad \mathfrak{C}(G) = 0,$$

where $\mathfrak{C}$ is a non-linear differential operator of order 2 when $n > 3$ (the Weyl corvature tensor) and of order 3 when $n = 3$. By weak solution of the last equation we mean any measurable matrix G, which can be approximated almost everywhere by smooth solutions G_j of the Weyl - Schouten equation (5). In this case we say that the pair (G,E) is weakly integrable. As a result of our theorem we have

COROLLARY 1. Any weakly integrable pair (G,E) is integrable.

Proof. Let $G(x) = \lim_{j\to\infty} G_j(x)$ for almost all $x \in D$, and $G_j \in C^2(D)$. Then there exists a sequence f_j of local diffeomorphic, which satisfy the equations

$$Df_j^*(x) Df_j(x) = Jf_j(x)^{2/n} G_j(x) \quad \text{a.e. in } D.$$

We may assume that $f_j(x_1) = y_1 \neq y_2 = f_j(x_2)$ for some $x_1, x_2 \in D$ and for all $j = 1,2,\ldots$ This normalizing condition can be always obtained by composing of f_j with suitable conformal (Möbius) mappings. Therefore we may choose a subsequence f_{j_k}, which locally uniformly converges to a quasiregular map f that satisfies the limit equation

$$Df^*(x)Df(x) = Jf(x)^{2/n}G(x) \quad \text{a.e. in } D.$$

Of course, $f(x_1) = y_1 \neq y_2 = f(x_2)$, this means f is not constant.

Actually we have proved a little more:

COROLLARY 2. *Every quasiregular map f with weakly integrable characteristics (G,E) can be approximated locally uniformly and also in the space $W^1_{p,loc}(D,\mathbb{R}^n)$ by quasiregular local diffeomorphisms.*

However, we do not know whether any integrable pair (G,E) is weakly integrable.

4. In the proof of the stability theorem we use the following well known results.

1° Any family $\{f_j\}$ of K-quasiregular mappings of a domain D into a bounded domain D' is normal; this means we may choose a subsequence f_{j_k} that locally uniformly converges either to a K-quasiregular map f or to a constant. Moreover

$$\int_D Jf_{j_k}(x)\varphi(x)\,dx \longrightarrow \int_D Jf(x)\varphi(x)\,dx$$

for every function $\varphi \in C_0^\infty(D)$. In other words, functions Jf_{j_k} weakly converge in $L_{1,loc}(D,\mathbb{R})$ to the function Jf.

The normality of such a family is obvious. For the proof of the second statement see Reshetnjak [13], Theorem 4.

2° (*Gehring's theorem*). First generalized derivatives of a quasiregular map $f : D \longrightarrow D'$, $D,D' \subset R^n$, belong to the space $L_{p,loc}(D)$ with $p > n$ and p depends only on n and the maxiaml dilatation K_f of f. This implies that the Jacobian Jf is locally integarble in a power grater than 1. Moreover, the uniform estimate holds

$$\int_F |Df|^p \leq C = C(p,n,K_f,F,D,D'),$$

for any compact subset $F \subset D$.

For qasiconformal mappings it was proved by F.W.Gehring [8]. We will need here a generalization, as above, of Gehring's result. The proof is found in the paper [12] by N.G.Meyers and A.Elcrat. Let us only notice here that in 1976 Yu.G.Reshetnjak proved $p \longrightarrow \infty$ when $\sup_x K_f(x) \longrightarrow 1$ (see [17]).

We will use also the following two properties of the introduced norms.

LEMMA 1. *Let A, A_j belong to the space $L_p(F, \mathbb{R}^{n\times n})$ and A be the weak limit of A_j. Then*

$$\| A\|_{p,F} \le \liminf_{j\to\infty} \| A_j\|_{p,F}.$$

P r o o f. By the definition

$$\| A\|_{p,F} = \left(\int_F (\mathrm{Tr}\, A^*A)^{p/1}\right)^{1/p} = \frac{\int_F \mathrm{Tr}\, AB^*}{\left(\int_F (\mathrm{Tr}\, B^*B)^{\frac{p}{2(p-1)}}\right)^{\frac{p-1}{p}}},$$

where $B = A(\mathrm{Tr}\, A^*A)^{\frac{p-2}{2}} \in L_{\frac{p}{p-1}}(F, \mathbb{R}^{n\times n})$. The last ratio is equal to

$$\lim_{j\to\infty} \frac{\int_F \mathrm{Tr}\, A_jB^*}{\| B\|_{\frac{p}{p-1},F}} \le \liminf_{j\to\infty} \frac{\int_F (\mathrm{Tr}\, A_j^*A_j)^{1/2}(\mathrm{Tr}\, B^*B)^{1/2}}{\| B\|_{\frac{p}{p-1},F}}$$

$$\le \liminf_{j\to\infty} \frac{\left(\int_F (\mathrm{Tr}\, A_j^*A_j)^{p/2}\right)^{1/p}\left(\int_F (\mathrm{Tr}\, B^*B)^{\frac{p}{2(p-1)}}\right)^{\frac{p-1}{p}}}{\| B\|_{\frac{p}{p-1},F}}$$

$$= \liminf_{j\to\infty} \| A_j\|_{p,F} \frac{\| B\|_{\frac{p}{p-1},F}}{\| B\|_{\frac{p}{p-1},F}} = \liminf_{j\to\infty} \| A_j\|_{p,F}.$$

LEMMA 2. *If $A, B,$ and C are $n\times n$ matrices, then*

$$|(\mathrm{Tr}\, A^*BA)^{p/2} - (\mathrm{Tr}\, A^*CA)^{p/2}| \le c(p,n)|B-C||A|^p(|B|^{\frac{p-2}{2}} + |C|^{\frac{p-2}{2}}),$$

where $c(p,n)$ is a constant depending only on p and n.

P r o o f. The computation gives:

$$|(\mathrm{Tr}\, A^*BA)^{p/2} - (\mathrm{Tr}\, A^*CA)^{p/2}| = \left|\int_0^1 \frac{\partial}{\partial t}(\mathrm{Tr}\, A^*(tB + (1-t)C)A)^{p/2}dt\right| =$$

$$\left|\int_0^1 \frac{p}{2}(\mathrm{Tr}\, A^*(tB + (1-t)C)A)^{\frac{p-2}{2}}\, \mathrm{Tr}\, A^*(B-C)A\, dt\right| =$$

$$\frac{p}{2}|\mathrm{Tr}\, A^*(B-C)A|\,|\int_0^1(\mathrm{Tr}\, A^*(tB+(1-t)C)A)^{\frac{p-2}{2}}dt|$$

$$\leq c(p,n)|B-C|\,|A|^p(|B|^{\frac{p-2}{2}}+|C|^{\frac{p-2}{2}}).$$

On the basis of Lemma 2 we prove the following

LEMMA 3. Let H, H_j, A_j are $n\times n$ matrix-valued functions on a bounded domain U and such that $|H_j|\leq M$ for every j, H_j tends to H almost everywhere. Assume that A_j are locally integrable in the power $p(1+\varepsilon)$, for some $\varepsilon>0$, and

$$\int_F |A_j|^{p(1+\varepsilon)} \leq c(p,n,F),\quad F\subset U,$$

for a constant $c(p,n,F)$. Then

$$\liminf_{j\to\infty}\int_F(\mathrm{Tr}\, A_j^*H_jA_j)^{p/2} = \liminf_{j\to\infty}\int_F(\mathrm{Tr}\, A_j^*HA_j)^{p/2}.$$

P r o o f.

$$\liminf_{j\to\infty}\int_F(\mathrm{Tr}\, A_j^*H_jA_j)^{p/2} = \liminf_{j\to\infty}\Big[\int_F(\mathrm{Tr}\, A_j^*HA_j)^{p/2}$$

$$+\int_F((\mathrm{Tr}\, A_j^*H_jA_j)^{p/2}-(\mathrm{Tr}\, A_j^*HA_j)^{p/2}\Big].$$

By Hölder's inequality and Lemma 2,

$$|\int_F((\mathrm{Tr}\, A_j^*H_jA_j)^{p/2}-(\mathrm{Tr}\, A_j^*HA_j)^{p/2})|\leq \mathrm{const.}\int_F|H_j-H|\,|A_j|^p$$

$$\leq \mathrm{const.}\Big(\int_F|H_j-H|^{\frac{1+\varepsilon}{\varepsilon}}\Big)^{\frac{\varepsilon}{1+\varepsilon}}\Big(\int_F|A_j|^{p(1+\varepsilon)}\Big)^{\frac{1}{1+\varepsilon}}\leq \mathrm{const.}\Big(\int_F|H_j-H|^{\frac{1+\varepsilon}{\varepsilon}}\Big)^{\frac{\varepsilon}{1+\varepsilon}},$$

which tends to 0 when $H_j\longrightarrow H$ almost everywhere.

5. Now we give the p r o o f of the stability theorem. Step by step we find subsequence of f_j having the expected properties. We will use the same notation f_j for all these sequence. Thus we may assume that f_j converges locally uniformly to a quasiregular mapping $f: D\longrightarrow D'$ and the Jacobians Jf_j tend weakly to Jf in $L_{1,\mathrm{loc}}(D,\mathbb{R})$ (see 1°). Since K_{f_j} are uniformly bounded, then there exists a common power $p>n$ and a number $\varepsilon>0$ such that

$$\int_F |Df_j|^{p+\varepsilon} \le c(F)$$

for any comapct subset $F \subset D$; the constant $c(F)$ depends only on F. In particular

$$\int_F |Jf_j|^{\frac{p+\varepsilon}{n}} \le c(F)$$

The sequence Jf_j weakly converges to Jf in $L_{1,loc}(D,\mathbb{R})$ and is uniformly bounded in $L_{\frac{p+\varepsilon}{n}}(D,\mathbb{R})$. Actually Jf_j weakly converges to Jf in $L_{\frac{p+\varepsilon}{n}}(D,\mathbb{R})$. Of course, we may suppose that also Df_j weakly converges to Df in the space $L_{p+\varepsilon,loc}(D,\mathbb{R}^{n\times n})$. We shaw that a subsequence of f_j and the number p have the mentioned properties A) and B).

Let us denote $G_j(x) = G_j(x,f_j(x))$, $H_j(x) = H_j(x,f_j(x))$, $G(x) = G(x,f(x))$, and $H(x) = H(x,f(x))$. Since f_j locally uniformly go to f, then $H_j(x) \longrightarrow H(x)$ and $G_j(x) \longrightarrow G(x)$ for almost all $x \in D$. Certainly we may assume that f is not constant. We define almost everywhere the matrix-valued function

$$\tilde{G}(x) = \frac{Df^*(x)H(x)Df(x)}{Jf(x)^{2/n}} .$$

The mapping f is quasiregular, then $\tilde{G}(x)$ is bounded by a constant $c(K_f)$, symetric, positive-defined, and $\det \tilde{G}(x) = 1$ a.e. in D. One has to prove that $\tilde{G} = G$ a.e. in D. For any constant matrix A we have

$$\int_F Jf(x)^{p/n}(\mathrm{Tr}\ A^*\tilde{G}(x)A)^{p/2}dx = \int_F (\mathrm{Tr}\ A^*Df^*(x)H(x)Df(x)A)^{p/2}$$

and, similarly,

$$\int_F Jf_j(x)^{p/n}(\mathrm{Tr}\ A^*G_j(x)A)^{p/2} = \int_F (\mathrm{Tr}\ A^*Df_j^*(x)H_j(x)Df_j(x)A)^{p/2};$$

F is a compact subset of D. In the first setp we put $p=n$; $H(x)$ may be expressed $H(x) = P(x)^*P(x)$. Apply Lemma 3 to the matrix Df_jA and Lemma 1 to $P(x)Df_j(x)A$ we get

$$\liminf_{j\to\infty} \int_F Jf_j(\mathrm{Tr}\ A^*G_jA)^{n/2} = \liminf_{j\to\infty} \int_F (\mathrm{Tr}\ A^*Df_jH_jDf_jA)^{n/2} =$$

$$\liminf_{j\to\infty} \int_F (\mathrm{Tr}\ A^*Df_j^*HDf_jA)^{n/2} = \liminf_{j\to\infty} \int_F (\mathrm{Tr}(PDf_jA)^*(PDf_jA))^{n/2} \ge$$

$$\geq \int_F (\mathrm{Tr}(PDfA)^*(PDfA))^{n/2} = \int_F (\mathrm{Tr}\, A^*Df^*H\, Df\, A)^{n/2} =$$

$$\int_f Jf(\mathrm{Tr}\, A^*\tilde{G}\, A)^{n/2}.$$

The first limit

$$\liminf_{j\to\infty} \int_F Jf_j(\mathrm{Tr}\, A^*G_j\, A)^{n/2} = \liminf_{j\to\infty} \int_F Jf_j(\mathrm{Tr}\, A^*G\, A)^{n/2}.$$

Indeed, by Lemma 2 and Hölder's inequality

$$\int_F Jf_j|(\mathrm{Tr}\, A^*G_jA)^{n/2} - (\mathrm{Tr}\, A^*G\, A)^{n/2}| \leq$$

$$\mathrm{const.}(\int_F |Jf_j|^{1+\alpha})^{\frac{1}{1+\alpha}}(\int_F |G_j - G|^{\frac{1+\alpha}{\alpha}})^{\frac{\alpha}{1+\alpha}},$$

where $\alpha = \varepsilon/n$. But the last factor tends to 0 with $j \longrightarrow \infty$. Since the Jacobians Jf_j weakly converge to Jf in $L_{1,loc}(D,\mathbb{R})$, then finally we get

$$\liminf_{j\to\infty} \int_F Jf_j(\mathrm{Tr}\, A^*G_jA)^{n/2} = \int_F Jf(\mathrm{Tr}\, A^*G\, A)^{n/2},$$

and

$$\int_F Jf(\mathrm{Tr}\, A^*\tilde{G}\, A)^{n/2} \leq \int_F Jf(\mathrm{Tr}\, A^*G\, A)^{n/2}.$$

The last inequality implies

(6) $\quad \mathrm{Tr}\, \tilde{G}(x)AA^* \leq \mathrm{Tr}\, G(x)AA^*,$

for almost all $x \in D$ and any constant matrix A. We prove now that $\tilde{G} = G$ almost everywhere in D.

First we decompose $\tilde{G}(x)G^{-1}(x) = O^*(x)L(x)O(x)$, where $O(x)$ is orthogonal and $L(x)$ is diagonal with $\det L(x) = 1$, then we put $AA^* = G^{-1}(x)O^*(x)Q(x)O(x)$, where $Q(x)$ is an arbitrary diagonal and positive-defined matrix. From (6) we get

$$\mathrm{Tr}(\tilde{G}\, G^{-1}O^*Q\, O) \leq \mathrm{Tr}(G\, G^{-1}O^*Q\, O),$$

and

$$\mathrm{Tr}(O^*L\, O\, O^*Q\, O) \leq \mathrm{Tr}(O^*Q\, O).$$

Hence

$$\mathrm{Tr}\ L\ Q \leq \mathrm{Tr}\ Q.$$

Since $\det L = 1$, we have $L = E$. Therefore $\tilde{G}G^{-1} = O^*EO$ and $\tilde{G} = G$. This ends the proof of part A) of the theorem.

Observe that we have proved

$$\liminf_{j\to\infty} \int_F (\mathrm{Tr}\ Df_j^* H Df_j)^{n/2} = \liminf_{j\to\infty} \int_F (\mathrm{Tr}\ Df_j^* H_j Df_j)^{n/2} =$$

$$\liminf_{j\to\infty} \int_F Jf_j (\mathrm{Tr} G_j)^{n/2} = \int_F Jf (\mathrm{Tr}\ \tilde{G})^{n/2} = \int_F Jf (\mathrm{Tr}\ G)^{n/2} =$$

$$\int_F (\mathrm{Tr}\ Df^* H\ Df)^{n/2}.$$

The sequence $X_j = P\ Df_j$, where $H = P^*P$, has the property

$$\liminf_{j\to\infty} \| X_j - X \|_{n,F} = 0 \quad \text{for} \quad X = P\ Df.$$

In fact, we know that $X_j \longrightarrow X$ weakly in $L_n(F, \mathbb{R}^{n\times n})$ and

$$\liminf_{j\to\infty} \| X_j \|_{n,F} = \| X \|_{n,F}.$$

Now we use Lemma 1 and the Clarkson inequality

$$\left\| \frac{X_j + X}{2} \right\|^n_{n,F} + \left\| \frac{X_j - X}{2} \right\|^n_{n,F} \leq \frac{1}{2} \| X_j \|^n_{n,F} + \frac{1}{2} \| X \|^n_{n,F}.$$

Then we get

$$\| X \|^n_{n,F} \leq \liminf_{j\to\infty} \left\| \frac{X_j + X}{2} \right\|_{n,F},$$

and

$$\| X \|^n_{n,F} + \liminf_{j\to\infty} \left\| \frac{X_j - X}{2} \right\|_{n,F} \leq \frac{1}{2} \| X \|^n_{n,F} + \frac{1}{2} \| X \|^n_{n,F}.$$

Hence $\liminf_{j\to\infty} \| X_j - X \|_{n,F} = 0$. Finally, by Hölder's inequality

$$\int_F |X_j - X|^p \leq \left(\int_F |X_j - X|^n\right)^{\frac{\varepsilon}{p+\varepsilon-n}} \left(\int_F |X_j - X|^{p+\varepsilon}\right)^{\frac{p-n}{p+\varepsilon-n}},$$

therefore $\liminf_{j\to\infty} \| X_j - X \|_{p,F} = 0$. This means that there exists a subsequence f_{j_k} which converges to f in $W^1_{p,loc}(D, \mathbb{R}^n)$.

6. H i s t o r i c a l n o t e s. The problem of stability for quasiconfromal mappings in the plane was stated for the first time by M.A.Lavrentiev in the thirties and recently reformulated in several ways. The question was: how close to a conformal map is a quasiconformal map (of a plane domain) with dilatation close (in the sense of the supremum norm) to 1 (= the dilatation of conformal maps). Some results concerning this problem are found in the early papers of Lavrentiev [10] (see also [11]) and Belinskii [2], [3], [4]. The last author found an absolute constant of the stability in the supremum norm which was a kind of final result. Many interesting theorems for arbitrary $n > 2$ have been obtained by Reshetnjak [13] - [18]; the estimates of the stability have been given in the L_p and Sobolev norms. Paper [6] by Bojarski contains a stability theorem for system (2) under the assumption that the coefficients converge almost everywhere.

R e f e r e n c e s

[1] AHLFORS, L.V.: Lectures on Quasiconformal Mappings, Van Nostrand Mathematical Studies 10, Princeton, New Jersey 1966.

[2] BELINSKIĬ, P.P.: On continuity of n-dimensional quasiconformal mappings and the Liouville theorem (Russian), Doklady Akad. Nauk SSSR 147 (1962) N° 5, 1003-1004.

[3] ——: On the order of aproximity of quasiconformal mappings in space to conformal mappings (Russian), Doklady Akad. Nauk SSSR 200 (1971), 759-761; Siberian Math. J. 14 (1973), 475-483.

[4] ——: Stability in the Liouville theorem on conformal mappings in space (Russian),

[5] BERS, L.: Theory of Pseudoanalytic Functions, New York University Press, 1953.

[6] BOJARSKI, B.: Generalized solutions of first order elliptic equation with discontinuous coefficients, Russian Math. Sbornik 43 (85) (1957), 451-503.

[7] —— and T.Iwaniec: Topics in quasiconformal theory in several variables, Proceedings of the First Finnish-Polish Summer School in Complex Analysis at Polesice, Part II, edited by J.Ławrynowicz and O.Lehto, University of Łódź Press, 1978, pp.21-44.

[8] GEHRING, F.W.: The L^p-integrability of the partial derivatives of a quasiconformal mapping, Acta Math. 130 ((1973), 265-277.

[9] IWANIEC, T.: Regularity theorems for the solutions of partial differential equations for quasiconformal mappings in several dimensions, Institute of Mathematics of Polish Academy of Sciences, preprint N° 153, Warsaw, October 1978 (to appear in Dissertationes Mathematicae).

[10] LAVRENTIEV, M.A.: Stability in the Liouville theorem (Russian), Doklady Akad. Nauk SSSR 95, N°5 (1954), 925-926.

[11] ——: On the theory of mappings in space, Siberian Math. J. 3, N°5 (1962), 710-714.

[12] MEYERS, N.G. and A.ELCRAT: Some results on regularity for solutions of non-linear elliptic systems and quasi-regular functions, Duke Math. J. 42 (1975), 121-136.

[13] RESHETNJAK, Yu.G.: On stability of conformal mappings in space, Siberian Math. J. 8, N°4 (1967), 91-114.

[14] ——: Theorems on stability of mappings with bounded distortion, Siberian Math. J. 9, N°3 (1968), 667-684.

[15] ——: An estimate of stability in the Liouville theorem on conformal mappings in space, Siberian Math. J. 11, N°5 (1970), 1121-1139.

[16] ——: Stability in the Liouville theorem on conformal mappings in space for domain with non-smooth boundary, Siberian Math. J. 17, N°2 (1976), 361-369.

[17] ——: Estimates for stability in the Liouville theorem and L_p-integrability of derivatives of quasiconformal mappings, Siberian Math. J. 17, N°4 (1976), 868-896.

[18] ——: Estimates in W_p^1 for stability in the Liouville theorem on conformal mappings for closed domains, Siberian Math. J. 17, N°6 (1976), 1383-1394.

[19] ——: Differentiability properties of quasiconformal mappings and conformal mappings of Riemann spaces, Siberian Math. J. 19, N°5 (1878), 1166-1183.

[20] SCHOUTEN, J.A.: Ricci-Calculus, Springer-Verlag, Berlin, 1954.

[21] VEKUA, I.N.: Generalized Analytic Functions, Pergamon Press, Oxford, 1962.

Institute of Mathematics
University of Warsaw
Pałac Kultury i Nauki, IXp.
PL-00-901 Warszawa, Poland.

DECOMPOSITION PROPERTY OF $A^k(D)$ IN STRICTLY PSEUDOCONVEX DOMAINS

Piotr Jakóbczak (Kraków)

In this note we consider the decomposition property of the algebra $A^k(D)$ of all functions holomorphic in a domain D in $\mathbb{C}^n$ and continuous with all their derivatives of order $\leq k$ in $\bar{D}$. We prove that if D is strictly pseudoconvex with sufficiently smooth boundary (depending on k), then the algebra $A^k(D)$ has a decomposition property. The precise formulation of this result will be given later.

Let us recall some notations and definitions.

We say that a bounded domain D in $\mathbb{C}^n$ is <u>strictly pseudoconvex</u> with $\mathcal{C}^p$ boundary, $p \geq 2$, if there exists a real-valued function ρ of class $\mathcal{C}^p$, defined in some neighborhood U of ∂D such that

(i) $D = (D \setminus U) \cup \{z \in U : \rho(z) < 0\}$,

(ii) $\operatorname{grad} \rho(z) \neq 0$ for each $z \in \partial D$, and

(iii) ρ is strictly plurisubharmonic in U.

If $\alpha = (\alpha_1, \ldots, \alpha_n)$ is an arbitrary n-tuple of non-negative integers we set

$$D^\alpha = \frac{\partial^{|\alpha|}}{\partial z_1^{\alpha_1} \ldots \partial z_n^{\alpha_n}},$$

where $|\alpha| = \alpha_1 + \ldots + \alpha_n$. Similarly,

$$\bar{D}^\beta = \frac{\partial^{|\beta|}}{\partial \bar{z}_1^{\beta_1} \ldots \partial \bar{z}_n^{\beta_n}}.$$

If these operators are applied to functions which depend on two groups of variables, we put a subscript, to indicate the group of variables, with respect to which the differentiation holds.

If α and β are as above, we set $(\alpha, \beta) = (\alpha_1 + \beta_1, \ldots, \alpha_n + \beta_n)$.

Given a bounded domain $D \subset \mathbb{C}^n$ and a function $f \in \mathcal{C}^k(\bar{D})$, denote

$$\|f\|_{D,k} = \sum_{|\alpha|+|\beta|\le k} \sup_D |D^{\alpha}\bar{D}^{\beta} f| .$$

We write $\|f\|_D$ instead of $\|f\|_{D,0}$.

If a function $f(\xi,z)$ is of class $\mathcal{C}^k$ with respect to (ξ,z) in $\bar{D}\times\bar{D}$ and of class $\mathcal{C}^{k+1}$ with respect to z in $\bar{D}$, we put

$$\|f\|_{D,k_{\xi},(k+1)_z} = \Sigma \sup_{D\times D} |D_z^{\beta'}\bar{D}_z^{\beta''} D_{\xi}^{\alpha'}\bar{D}_{\xi}^{\alpha''} f| ,$$

where the summation is extended over all multiindices $\alpha',\alpha'',\beta',\beta''$ such that $|\alpha'|+|\alpha''|+|\beta'|+|\beta''| \le k+1$ and $|\alpha'|+|\alpha''| \le k$.

The algebra $A^k(D)$ will be always considered with the topology induced by the norm $\|\ \|_{D,k}$; with this norm, $A^k(D)$ is a Banach algebra.

We prove here the following theorem on the decomposition property of the algebra $A^k(D)$:

THEOREM. *Let D be a strictly pseudoconvex domain in $\mathbb{C}^n$. Suppose that ∂D is of class $\mathcal{C}^{k+4}$, k being a non-negative integer. Then for every $f\in A^k(D)$ there exist functions $f_1(z,s),\dots,f_n(z,s)$ holomorphic in $D\times D$ and $\mathcal{C}^k$ in the set $(\bar{D}\times\bar{D}) \setminus \{(z,z) : z\in\partial D\}$, such that*

$$f(z) = f(s) + \sum_{i=1}^{n} (z_i - s_i)\cdot f_i(z,s)$$

for every $(z,s)\in(\bar{D}\times\bar{D} \setminus \{(z,z) : z\in\partial D\}$.

Moreover, for every fixed $s\in D$, the mapping

$$(*) \qquad A^k(D)\ni f \longrightarrow (f_1(\cdot,s),\dots,f_n(\cdot,s)) \in (A^k(D))^n$$

is linear and continuous.

In the case $k = 0$ this theorem was proved by Ahern and Schneider in [1], under weaker assumption that ∂D is only of class $\mathcal{C}^3$. Related results were obtained earlier by Henkin [4] and Øvrelid [7] (for the case $k = 0$), but without the holomorphic dependence of the components $f_i(z,s)$ on the second variable. The above theorem without the holomorphic dependence of $f_i(z,s)$ on s, was proved by the author in [6].

Our proof follows that of [1], and we make use of the techniques developed by Siu in [9].

By Henkin's integral formula ([3], p.303) we have for any function $g\in\mathcal{C}^1(\bar{D})$, the representation

$$(1) \qquad g(z) = \int_{\partial D} g(\xi)\cdot C(\xi,z) + \int_{\partial D} \bar{\partial} g(\xi)\wedge K(\xi,z) - \int_D \bar{\partial} g(\xi)\wedge L(\xi,z),$$

for any $z \in D$, where $C(\xi,z)$ and $K(\xi,z)$ are the integral kernels which depend only on the strictly pseudoconvex domain D, the coefficients of $C(\xi,z)$ are holomorphic in $z \in D$, and $L(\xi,z)$ is Bochner-Martinelli kernel (we keep here the notation of [9], p. 165).

LEMMA 1. *Let D be a strictly pseudoconvex doamin in $\mathbb{C}^n$ with $\mathcal{C}^{k+4}$ boundary, and let $f \in A^k(D)$. Suppose that $g(\xi,z) \in \mathcal{C}^{k+1}(\bar{D}\times\bar{D})$ and is of class $\mathcal{C}^{k+2}$ with respect to z in $\bar{D}$. Let f_g be the function defined by*

$$f_g(z) = \int_{\partial D} f(\xi)g(\xi,z)C(\xi,z), \quad z \in D.$$

Then $f_g \in \mathcal{C}^k(\bar{D})$, and

$$(2) \qquad \| f_g\|_{D,k} \le c\cdot \| f\|_{D,k}\cdot \| g(\xi,z)\|_{D,(k+1)_\xi,(k+2)_z},$$

for some constant $c > 0$, independent of f and g.

Replacing $g(\xi)$ by $f(\xi)g(\xi,z)$ in (1), we obtain

$$(3) \qquad f(z)g(z,z) = \int_{\partial D} f(\xi)g(\xi,z)C(\xi,z) + \int_{\partial D} f(\xi)\bar{\partial}_\xi g(\xi,z)\wedge K(\xi,z) - \int_D f(\xi)\bar{\partial}_\xi g(\xi,z)\wedge L(\xi,z), \quad z \in D$$

(since f in holomorphic in D).

In order to simplify notations, we will write $h(\xi,z)$ instead of $f(\xi)\bar{\partial}_\xi g(\xi,z)$ throughout the rest of this paper. Then $h(\xi,z)$ is a $(0,1)$-form with respect to ξ, $h(\xi,z) = \Sigma_{i=1}^n h_i(\xi,z)d\bar{\xi}_i$, and the coefficients $h_i(\xi,z)$ are of class $\mathcal{C}^k$ in $\bar{D}\times\bar{D}$ and of class $\mathcal{C}^{k+1}$ with respect to z in $\bar{D}$.

Given $z,y \in D$, define (following [9])

$$T_D h(\cdot,y)(z) = \int_{\partial D} h(\xi,y)\wedge K(\xi,z) - \int_D h(\xi,y)\wedge L(\xi,z).$$

Define also the mapping $T_D h(\cdot,y)$ by

$$D \ni z \longrightarrow T_D h(\cdot,y)(z),$$

and the mapping $T_D h(\cdot,y)\big|_{y=z}$ by

$$D \ni z \longrightarrow T_D h(\cdot,z)(z).$$

Now Lemma 1 results from (3) and the following

LEMMA 2. *Let* D *be as in Lemma 1, and let* $h(\xi,z) = \Sigma_{i=1}^{n} h_i(\xi,z)\,d\bar{\xi}_i$ *with* $h_i(\xi,z)$ *in* $\mathcal{C}^k(\bar{D}\times\bar{D})$ *and of class* $\mathcal{C}^{k+1}$ *with respect to* z *in* $\bar{D}$. *Then, for any* $\alpha,\beta \in Z_+^n$, *such that* $|\alpha| + |\beta| \le k$, *the function*

$$D^\alpha\bar{D}^\beta (T_D h(\cdot,y)\big|_{y=z})$$

extends continuously to all of $\bar{D}$, *and*

$$(4)\qquad \| D^\alpha\bar{D}^\beta (T_D h(\cdot,y)\big|_{y=z}) \|_D \le c\cdot \| h \|_{D,k_\xi,(k+1)_z}$$

for some constant c *independent of* h. *(Here* $\|h\|_{D,k_\xi,(k+1)_z} = \Sigma_{i=1}^{n} \|h_i\|_{D,k_\xi,(k+1)_z}$*).*

P r o o f. Consider first the derivatives of the function $T_D h(\cdot,y)\big|_{y=z}$ with respect to z_i, $i = 1,\ldots,n$. Let

$$D^\alpha = \frac{\partial^{|\alpha|}}{\partial z_1^{\alpha_1}\ldots\partial z_n^{\alpha_n}}$$

be a differential operator of order $|\alpha| \le k$. Write

$$D^\alpha = T_1^{(1)}\circ\ldots\circ T_{\alpha_1}^{(1)}\circ\ldots\circ T_1^{(n)}\circ\ldots\circ T_{\alpha_n}^{(n)},$$

$$\text{where}\quad T_j^{(i)} = \frac{\partial}{\partial z_i},\quad i = 1,\ldots,n.$$

Consider now all the $T_j^{(i)}$'s as the different operators. Denote by $C(\alpha)$ the set

$$\{T_1^{(1)},\ldots,T_{\alpha_1}^{(1)},\ldots,T_1^{(n)},\ldots,T_{\alpha_n}^{(n)}\}.$$

We claim that

$$(5)\qquad D^\alpha (T_D h(\cdot,y)\big|_{y=z})(z) = \sum_{\alpha'} D^{\alpha'} (T_D(D_z^{\alpha''}h)(\cdot,y))(z)\big|_{y=z},\qquad z\in D,$$

where the sum is extended over all subsets α' of the set $C(\alpha)$. Here $D^{\alpha'}$ denotes the composition of all operators in α', and $D_z^{\alpha''}$ is the composition of all operators of the set

$$\alpha'' = C(\alpha)\setminus\alpha',\quad\text{and}\quad D_z^{\alpha''}h = \sum_{i=1}^{n}(D_z^{\alpha''}h_i)(\xi,z)\cdot d\bar{\xi}_i.$$

We prove the formula (5) by induction on $|\alpha|$. The case $|\alpha| = 0$ is obvious, so we assume that (5) holds for all α where $|\alpha| \le l$, and l in an integer, $l < k$. Now let $|\alpha| = l+1$. Then $D^\alpha = D^j D^\beta$ for some β with $|\beta| = l$ and some j, $1 \le j \le n$. (Here and afterwards we write D^j and $\bar{D}^j$ instead of $\partial/\partial z_j$ and $\partial/\partial \bar{z}_j$ respectively). By inductional assumption,

$$(6) \qquad D^\beta(T_D h(\cdot,y)|_{y=z})(z) = \sum_{\beta'} D^{\beta'}(T_D(D_z^{\beta''} h)(\cdot,y))(z)|_{y=z}, \quad z \in D,$$

where β' varies over all subsets of the set $C(\beta)$. Note that we can apply [9, Prop. 4.2] to each component of the right-hand side of (6). In virtue of this result, for each $\beta' \subset C(\beta)$,

$$D^{\beta'}(T_D(D_z^{\beta''} h)(\cdot,y))(z)|_{y=z} = \sum_{\substack{|\gamma|+|\gamma'|<|\beta'| \\ 1\le i\le n}} a^{\beta'}_{\gamma\gamma' i}(z)(D_\xi^\gamma \bar{D}_\xi^{\gamma'} D_z^{\beta''} h_i)($$

$$(7) \qquad)(z,z) + \sum_{\substack{|\gamma|+|\gamma'|\le|\beta'| \\ 1\le i\le n}} \int_{\partial D} (D_\xi^\gamma \bar{D}_\xi^{\gamma'} D_z^{\beta''} h_i(\cdot,z))^{(|\beta'|-1-|\gamma|-|\gamma'|)}(\xi,z)$$

$$\cdot K^{\beta'}_{\gamma\gamma' i}(\xi,z) + \sum_{|\gamma|+|\gamma'|\le|\beta'|} \int_D (D_\xi^\gamma \bar{D}_\xi^{\gamma'} D_z^{\beta''} h_i(\cdot,z))^{(|\beta'|-1-|\gamma|-|\gamma'|)}($$

$$)(\xi,z) L^{\beta'}_{\gamma\gamma' i}(\xi,z).$$

We use here the following notations (see also [9], p. 173):

a) for each $g(\xi,z) \in \mathcal{C}^p(\bar{D} \times D)$ and each l, $0 \le l \le p$,

$$(g(\cdot,y))^{(l)}(\xi,z) =$$

$$g(\xi,y) - \sum_{|\eta|+|\eta'|\le l} \frac{1}{\eta!} \frac{1}{(\eta')!} (D_\xi^\eta \bar{D}_\xi^{\eta'} g)(z,y)(\xi-z)^\eta (\bar{\xi}-\bar{z})^{\eta'},$$

and for each $g \in l(\bar{D} \times D)$,

$$(g(\cdot,y))^{(-1)}(\xi,z) = g(\xi,y);$$

b) each function $a_{\gamma\gamma' i}(z)$ is 1/2 - Hölder continuous in D;

c) $K^{\beta'}_{\gamma\gamma' i}(\xi,z)$ is an $(n,n-1)$-form in ξ whose coefficients are obtained by applying to the coefficients of $K(\xi,z)$ the operators $D_\xi^s + D_z^s$ and $\bar{D}_\xi^s$, $1 \le s \le n$, a total number of $|\beta'|-|\gamma|-|\gamma'|$ times and then taking linear combinations with coefficients in $\mathcal{C}^{k+2}(\bar{D})$;

d) $L_{\gamma\gamma' i}(\xi,z)$ is an (n,n)-form in ξ whose coefficients are obtained by applying to the coefficients of $L(\xi,z)$ the operators D_ξ^s, $1 \le s \le n$, $|\beta'|-|\gamma|-|\gamma'|$ times and then taking linear combinations with coefficients in $\mathbb{C}$.

(Here $K(\xi,z)$ and $L(\xi,z)$ are the integral kernels occuring in (1)). Differentiating both sides of (7) and applying once more [9, Prop. 4.2] we conclude that

$$D^j(D^{\beta'}(T_D(D_z^{\beta''}h)(\cdot,y))\big|_{y=z})(z) = D^{\beta'}(T_D(D_z^jD_z^{\beta''}h)(\cdot,y))(z)\big|_{y=z}$$

$$+ D^jD^{\beta'}(T_D(D_z^{\beta''}h)(\cdot,y))(z)\big|_{y=z}.$$

Summing over all β', we obtain (5).

Note that to each term of the right-hand side of (5) we can apply the methods of estimation similar to those of [8]. We obatin then the estimates

$$\|D^{\alpha'}(T_D(D_z^{\alpha''}h)(\cdot,y))(z)\big|_{y=z}\|_{D,\varepsilon} \le c\cdot\|h\|_{D,k_\xi,(k+1)_z},$$

where $c>0$ is independent of h, and ε is some positive constant. (Here $\|\ \|_{D,\varepsilon}$ denotes the Hölder norm of order ε, defined by

$$\|f\|_{D,\varepsilon} = \|f\|_D + \sup_{x,y\in D}|f(x)-f(y)|/|x-y|^{\varepsilon}).$$

In fact, the purpose of deriving the formula (5) was to obtain the above estimates.

The details are a bit complicated, though not difficult, and we omit them here.

We have thus proved Lemma 2 for arbitrary α, and $\beta = (0,\ldots,0)$.

Consider now the arbitrary differential operator T of order $\le k$. Since the function $T_Dh(\cdot,y)\big|_{y=z}$ is of class $\mathcal{C}^k$ in D, we may change the order of differentiation, and assume that $T = D^\alpha\bar{D}^\beta$ with $|\alpha| + |\beta| \le k$. Note that

$$\bar{D}^j(T_Dh(\cdot,y)\big|_{y=z})(z) = h_j(z,z) + (T_D(\bar{D}_z^jh)(\cdot,y))(z)\big|_{y=z},$$

since $\bar{\partial}T_Dh(\cdot,y)\big|_{y=z} = h(z,z)$. Similarly,

$$\bar{D}^i\bar{D}^j(T_Dh(\cdot,y)\big|_{y=z})(z) = \bar{D}^i(h_j(z,z)) + (\bar{D}_z^jh_i)(z,z)$$

$$+ (T_D(\bar{D}_z^i\bar{D}_z^jh)(\cdot,y))(z)\big|_{y=z}.$$

Proceeding in this way we conclude that

$$(8)\qquad \bar{D}^\beta(T_Dh(\cdot,y)\big|_{y=z}) = \sum_s p_s + T_D(\bar{D}_z^\beta h)(\cdot,y)\big|_{y=z},$$

where each p_s is a function of class $\mathcal{C}^{k-|\beta|+1}$ in $\bar{D}$. Now apply the operator D^α to both sides of (8). We see that

$$D^\alpha \bar{D}^\beta (T_D h(\cdot,y)\big|_{y=z} = \sum_s r_s + D^\alpha (T_D (\bar{D}_z^{\beta'} h)(\cdot,y)\big|_{y=z}),$$

where each r_s is a continuous function in $\bar{D}$. It is not difficult to prove that, in addition, each r_s satisfies the estimate

(9) $$\| r_s \|_D \le c \cdot \| h \|_{D,k_\xi,k_z}.$$

Moreover, using the results of the first part of the proof, we conclude that

(10) $$\| D^\alpha (T_D (\bar{D}_z^{\beta'} h)(\cdot,y)\big|_{y=z}) \|_{D,\varepsilon} \le c \cdot \| h \|_{D,k_\xi,(k+1)_z}$$

for some positive constants ε and c independent of h. The formulas (9) and (10) together give the desired estimate (4) for all derivatives of the function $T_D h(\cdot,y)\big|_{y=z}$ of order $\le k$. This ends the proof of Lemma 2.

P r o o f of the Theorem. After having proved Lemma 2 (and, consequently, Lemma 1), we proceed in the same way as in [1]. There exist a neighborhood U of ∂D and a neighborhood $\tilde{D}$ of $\bar{D}$ in $\mathbb{C}^n$, and the functions $\Phi(\xi,z)$ and $K(\xi,z)$ defined in $U \times \tilde{D}$, which are of class $\mathcal{C}^{k+2}$ for $(\xi,z) \in U \times \tilde{D}$, and holomorphic with respect to z in $\tilde{D}$, and $\Phi(\xi,z) \ne 0$ on $\partial D \times D$, such that Henkin's integral kernel $C(\xi,z)$ can be written in the form

$$\frac{K(\xi,z)}{\Phi^n(\xi,z)}\, d\sigma(\xi),$$

$d\sigma(\xi)$ being a volume form on ∂D. Therefore for any $z,s \in D$,

$$f(z) - f(s) = \int_{\partial D} f(\xi) \frac{\Phi^n(\xi,s)K(\xi,z) - \Phi^n(\xi,z)K(\xi,s)}{\Phi^n(\xi,z)\Phi^n(\xi,s)}\, d\sigma(\xi)$$

Denote by $L(\xi,z,s)$ the difference in the numerator of the integrand. This can be written in the form

$$L(\xi,z,s) = \sum_{i=1}^n (z_i - s_i)\, L_i(\xi,z,s),$$

where $L_i(\xi,z,s)$ are of class $\mathcal{C}^{k+2}$ with respect to $(\xi,z,s) \in U \times \tilde{D} \times \tilde{D}$, and are holomorphic with respect to (z,s) in $\tilde{D} \times \tilde{D}$ (see [1]). Therefore, for any $z,s \in D$,

$$f(z) - f(s) = \sum_{i=1}^{n} (z_i - s_i)\, f_i(z,s),$$

with

$$(11) \qquad f_i(z,s) = \int_{\partial D} f(\xi)\, \frac{L_i(\xi,z,s)}{\Phi^n(\xi,z)\,\Phi^n(\xi,s)}\, d\sigma(\xi).$$

Fix $(z_o,s_o) \in (\bar{D}\times\bar{D}) \setminus \{(z,z) : z \in \partial D\}$. We must show that the functions $f_i(z,s)$ are continuous with all their derivatives of order $\leq k$ in the set $W = (\bar{D}\times\bar{D}) \cap (U \times V)$, where U and V are suitably chosen small neighborhoods (in $\mathbb{C}^n$) of the points z_o and s_o respectively. We assume that $z_o \in \partial D$ and $s_o \in \partial D$, and $z_o \neq s_o$; the other cases can be treated in a similar way. Choose the $\mathcal{C}^\infty$ function $\varphi : R_+ \longrightarrow R_+$ with the properties listed in [1], p. 13, and write (11) in the form

$$
\begin{aligned}
f_i(z,s) &= \int_{\partial D} f(\xi)\, \frac{L_i(\xi,z,s)\cdot\varphi(|z-\xi|^2)}{K(\xi,z)\,\Phi^n(\xi,s)}\, C(\xi,z) \\
(12) \qquad &+ \int_{\partial D} f(\xi)\, \frac{L_i(\xi,z,s)\cdot\varphi(|s-\xi|^2)}{K(\xi,s)\,\Phi^n(\xi,z)}\, C(\xi,s) \\
&+ \int_{\partial D} f(\xi)\, \frac{L_i(\xi,z,s)}{\Phi^n(\xi,z)\,\Phi^n(\xi,s)}\, [1 - \varphi(|z-\xi|^2) - \varphi(|s-\xi|^2)]\, d\sigma(\xi).
\end{aligned}
$$

The third term of the right-hand side of (12) is infinitely many differentiable in W, provided that U and V are chosen sufficiently small. As to the first term, differentiate first with respect to s under the integral sign, and then apply Lemma 1. We conclude that this term is of class $\mathcal{C}^k$ in W. Similarly, the second term is of class $\mathcal{C}^k$ in W. Since (z_o,s_o) is an arbitrary point of $(\bar{D}\times\bar{D}) \setminus \{(z,z) : z \in \partial D\}$, the functions $f_i(z,s)$ are actually of class $\mathcal{C}^k$ in this set. By the properties of functions $L_i(\xi,z,s)$ and $\Phi(\xi,z)$, we conclude that $f_i(z,s)$ are holomorphic in $D \times D$. It is obvious from the above construction that the mapping (*) is linear, and its continuity follows from the assertion (2) of Lemma 1. This ends the proof of the Theorem.

N o t e. The above theorem (but with stronger assumption on the regularity of ∂D) can also be proved in another way, by the methods of [2]. Consider first the case when D is a strictly convex domain in $\mathbb{C}^n$. It is easy to see that the components $f_i(z,s)$ obtained by Leibenzon's construction (see [4]) satisfy the assertion of our theorem. One can also show that if L is the extension operator from [2, Th. 4], p. 563, then $L(A^k(\psi(D))) \subset A^k(C)$, $k = 1,2,\ldots$, provided that ∂D is sufficiently smooth (depending on k). The proof is then completed by applying Fornaess' imbedding theorem [2, Th.9].

This alternative method of proof gives a weaker version of our theorem, since the aforementioned extension of [2, Th.4] to the case of the algebras $A^k(\psi(D))$ requires greater differentiability of ∂D then we need in Lemma 2. On the other hand, the proof of our theorem for strictly convex domains by Leibenzon's method is valid also for domains with boundaries of class $\mathcal{C}^2$. Therefore, it is reasonable to conjecture that the theorem proved in this note holds for every strictly pseudoconvex domain with $\mathcal{C}^2$ boundary, and for every value of k.

References

[1] AHERN, P. and R.SCHNEIDER: The boundary behavior of Henkin's kernel, Pacific J. Math. 66 (1976), 9-14.

[2] FORNAESS, J.-E.: Embedding strictly pseudoconvex domains in convex domains, Amer. J. Math. 98 (1976), 529-569.

[3] HENKIN, G.M.: Integral representations of functions in strictly pseudoconvex domains and applications to the $\bar\partial$-problem (In Russian), Mat. Sbornik 82 (1970), 300-308.

[4] ——: Approximation of functions in strictly pseudoconvex domains and a theorem of Z.L.Leibenzon (in Russian), Bull. Acad. Polon. Sci. Sér. Sci. Math. Astronom. Phys. 19 (1971), 37-42.

[5] ——: H.Lewy's equation and the analysis on pseudoconvex manifold, (in Russian), Uspehi Mat. Nauk. 32, N° 3 (1977), 57-118.

[6] JAKÓBCZAK, P.: Approximation and decomposition theorems for the algebras of analytic functions in strictly pseudoconvex domains, to appear.

[7] ØVRELID, N.: Generators of the maximal ideals of A(D), Pacific J. Math. 39 (1971), 219-223.

[8] RANGE, M. and Y.-T.SIU: Uniform estimates for the $\bar\partial$-equation on domains with piecewise smooth strictly pseudoconvex boundaries, Math. Ann. 206 (1973), 325-354.

[9] SIU, Y.-T.: The $\bar\partial$-problem with uniform bounds on derivatives, Math. Ann. 207 (1974), 163-176.

Institute of Mathematics
Jagiellonian University
Reymonta 4, PL-30-059 Kraków, Poland

ON SOME BIHOLOMORPHIC INVARIANTS IN THE ANALYSIS ON MANIFOLDS

Jerzy Kalina, Julian Ławrynowicz (Łódź)
and Ewa Ligocka, Maciej Skwarczyński (Warszawa)

Contents

Summary

The paper gives an exposition of recent results connected with applications of the Bergman function and complex capacities (intrinsic semi-norms), in particular in the case of a strictly pseudoconvex domain, smooth extensions of biholomorphic mappings (with the use of $\bar{\partial}$ Neumann problem), nonlinear Dirichlet problems, and foliations related to the generalized complex Monge-Ampère equations. The paper includes recent, unpublished results of the first named author which extend some achievements of E. Bedford and M. Kalka to the case of foliations related to the generalized complex Monge-Ampère equations.

Introduction

The theory of biholomorphic invariants gives important information on holomorphic mappings. It has valuable mathematical applications, e.g. to the problems of boundary behaviour of mappings, theory of removable singularities and, more generally, approximation theory. This exposition is partly a survey article corresponding to the papers by Caraman [14], Debiard and Gaveau [18], Ghişa [24], Holmann [25],

Ławrynowicz [34], Ligocka [35], and Skwarczyński [40], appearing in this volume, but describes rather recent developments in the field, including the authors' own results.

One of central problems treated in complex analysis is to determine whether two given complex manifolds can be mapped onto each other by a biholomorphic mapping. The basic tool, discovered by Stefan Bergman, is to associate with the manifold in question a function (the Bergman function) which is invariant under biholomorphic transformations. This fact was used later by Bergman for constructing numeruous biholomorphic invariants as well as the invariant Kähler metric.
We concentrate in recalling some fundamental results [22] of J. J. Kohn in the theory of the $\bar{\partial}$ problem. These results were used by him to describe the properties of the Neumann operator in strictly pseudoconvex domains. Since the Bergman function can be described in terms of the Neumann operator, the properties of the Bergman function can be studied in greater detail in this case. We apply here results of Kerzman [30] and Boutet de Monvel and Sjöstrand [12]. We also describe the $\bar{\partial}$ Neumann problem in connection with smooth extensions of biholomorphic mappings.

In turn we study nonlinear Dirichlet problems in $\mathbb{C}^n$ and, more generally, on hermitian manifolds. We deal with the variation of "complex" capacities (intrinsic semi-norms) within the admissible class of plurisubharmonic functions, and from basic properties of these biholomorphic invariants due to Chern, Levine and Nirenberg [16] we arrive at the existence problems for the generalized complex Monge-Ampère equations. In particular, using foliation techniques, we present an extension of some results due to Bedford and Kalka [5] to the case of foliations related to the generalized complex Monge-Ampère equations.

1. The Bergman function on a complex manifold

In 1922 Stefan Bergman discovered that with every domain $D \subset \mathbb{C}^n$ one can associate in a very natural way a function of $2n$ complex variables defined in the Cartesian product $D \times D$. One of several possible definitions of this function is based on a theorem on representation of a functional in a Hilbert space. Denote by $L^2H(D)$ the Hilbert space of all square integrable holomorphic functions in D. Then for every $t \in D$ the evaluation functional is continuous and there exists a unique element $\chi_t \in L^2H(D)$, such that for all $f \in L^2H(D)$

$$f(t) = (f, \chi_t).$$

The <u>Bergman</u> <u>function</u> is then defined by

$$K_D(z,t) = (\chi_t, \chi_z).$$

It is holomorphic in variable z and antiholomorphic in variable t. If $\varphi_j \in L^2H(D)$ is a complete orthonormal system, then

$$K_D(z,t) = \sum_j \varphi_j(z)\overline{\varphi_j(t)}$$

and the series converges normally (i.e. the series of absolute values converges locally uniformly in $D \times D$).

The following three theorems for an arbitrary domain D are of fundamental importance since they permit us to relate the Bergman function of D with the Bergman functions of some other domains.

<u>THEOREM 1</u> (Bergman [11]). <u>For every biholomorphic mapping</u> $g: D \to G$ <u>we have</u>

$$K_D(z,t) = K_G(g(z), g(t)) \det g'(z) \overline{\det g'(t)},$$

<u>where</u> g' <u>denotes the complex Jacobi matrix of</u> g.

<u>THEOREM 2</u> (Bremermann [13]). <u>Let</u> $D = D_1 \times D_2$ <u>be the product of domains</u> $D_1 \subset \mathbb{C}^{n_1}$ <u>and</u> $D_2 \subset \mathbb{C}^{n_2}$. <u>If</u> $z = (z_1, z_2)$ <u>and</u> $t = (t_1, t_2)$ <u>are points in</u> D, <u>then</u>

$$K_D(z,t) = K_{D_1}(z_1,t_1)\, K_{D_2}(z_2,t_2).$$

<u>THEOREM 3</u> (Ramadanov [38]). <u>If</u> D_j, $j = 1,2,\dots,$ <u>is an increasing sequence of domains in</u> $\mathbb{C}^n$ <u>and</u>

$$D = \bigcup_{j=1}^{\infty} D_j,$$

<u>then</u>

$$\lim_{j\to\infty} K_{D_j}(z,t) = K_D(z,t)$$

<u>and the convergence is locally uniform in</u> $D \times D$.

For some domains it is possible to obtain an explicit formula for the Bergman function. For example, if $B \subset \mathbb{C}^n$ is the unit ball, then

$$K_B(z,t) = n!/\pi^n[1 - (z_1\bar{t}_1 + \dots + z_n\bar{t}_n)]^n.$$

Analogous formulae for classical bounded homogeneous domains were established by Hua [27]. For other domains the explicit formulae are known in a few more cases. We shall indicate here the one due to Zinov'ev [43]. For a domain

$$D = \{z \in \mathbb{C}^n : |z_1|^{2/p_1} + \dots + |z_n|^{2/p_n} < 1\}$$

where $p_1, p_2, \ldots, p_n$ are positive integers, the Bergman function is given by

$$K_D(z,t) = \frac{1}{\pi^n p_1 \cdots p_n} \frac{\partial^n}{\partial q_1 \cdots \partial q_n} \sum \frac{1}{1-(v_1+\ldots+v_n)}.$$

Here $q_j = z_j \bar{t}_j$, $j = 1,2,\ldots,n$, and each v_j in the sum belongs to the set of all roots of order p_j of q_j. Actually the sum is a rational function of variables $q_1, q_2, \ldots, q_n$. For other examples see [4]. Note that for every $D \subset \mathbb{C}^n$ we have $K_D(t,t) = \|\chi_t\|^2$ and therefore $D_1 \subset D_2$ implies

$$K_{D_2}(t,t) \le K_{D_1}(t,t).$$

In particular $K_D(t,t)$ is positive for every t if D is bounded.

The Bergman function permits us to construct various biholomorphic invariants. For instance, one can introduce a distance function defined explicitly in terms of the Bergman function and invariant under biholomorphic mappings. For simplicitly we shall consider only the case when D is bounded. Define for every $z, t \in D$

$$\rho_D(z,t) = \operatorname{dist}(\{e^{i\alpha} \chi_z/\|\chi_z\|: 0 \le \alpha \le 2\pi\}, \{e^{i\beta} \chi_t/\|\chi_t\|: 0 \le \beta \le 2\pi\}).$$

Then it is not difficult to verify that ρ has all properties of a distance function, and

$$\rho_D(z,t) = [1 - (K_D(z,t)\, K_D(t,z)/ K_D(z,z)\, K_D(t,t))^{\frac{1}{2}}]^{\frac{1}{2}}.$$

It follows that $H_D(z,t) = (1-\rho^2)^2$ is real-analytic in $D \times D$. The _invariant distance_ ρ_D is never greater than 1, and $\rho_D(z,t) = 1$ if and only if $K_D(z,t) = 0$. If the latter possibility does not occur, then D is called a Lu Qi-keng domain. For details see [39]. The following theorem proved in [39] shows that the information necessary to determine whether there exists a biholomorphic mapping $g: D \to G$ is contained in the Bergman functions K_D and K_G.

THEOREM 4. _Assume that_ D _and_ G _are bounded domains in_ $\mathbb{C}^n$, _complete with respect to the invariant distances_ ρ_D _and_ ρ_G, _respectively. Let_ $g: U \to V$ _be a holomorphic mapping from an open subdomain_ U _of_ D _into an open subdomain_ V _of_ G _such that for all_ $z, t \in U$

$$\rho_G(g(z), g(t)) = \rho_D(z,t).$$

Then g _can be extended to a biholomorphic mapping of_ D _onto_ G.

The Bergman function is useful also in a study of boundary behaviour of biholomorphic mappings. By a _compactification_ of a domain $D \subset$

$\mathbb{C}^n$ we shall understand a homeomorphism $q: D \to X$ of D onto an open dense subset in a compact Hausdorff space X. Two compactifications $q_j: D \to X_j$ are equivalent if there exists a homeomorphism $w: X_1 \to X_2$ of X_1 onto X_2 such that $q_2 = w \circ q_1$. By $PH(D)$ we shall denote the Hausdorff space af all equivalence classes in $H(D) \setminus \{0\}$ under proportionality relation. For a bounded $D \subset \mathbb{C}^n$ we have a natural mapping $p: D \to PH(D)$ given by $p(t) = [K_D(\ ,t)]$. Denote by $\hat{D}$ the closure of $p(D)$ in $PH(D)$. For some domains $p: D \to \hat{D}$ is a compactification. If this is the case we say that D _admits the invariant compactification_, and we call p the _invariant compactification_ of D. The compact set $\hat{D} \setminus p(D)$ is called the _ideal boundary_ of D (cf. [41]). The folowing result, based only on elementary topology, gives a sufficient condition for the existence of the invariant compactification.

LEMMA 1. Assume that the compactification $q: D \to X$ has the following property: the mapping $p \circ q^{-1}: q[D] \to p[D]$ extends to the continuous one-to-one mapping from X into $PH(D)$. Then D admits the invariant compactification p, and q is equivalent to p.

If the euclidean compactification $id: D \to \mathrm{cl}\, D$ is equivalent to the invariant compactification, then the domain D is called _regular_. From Lemma 1 the following result follows directly:

COROLLARY 1. Assume that the Bergman function K_D extends to a continuous mapping $K_D: \mathrm{cl}\, D \times \mathrm{cl}\, D \to \mathbb{C} \cup \{\infty\}$ such that $K_D(z,t) = \infty$ if and only if $z = t \in \partial D$. Then D is regular.

The definition of the invariant compactification and the word "invariant" are justified by the following

LEMMA 2. Consider a biholomorphic mapping $g: D \to G$. If D admits the invariant compactification $p_D: D \to \hat{D}$, then G admits the invariant compactification $p_G: G \to \hat{G}$. Furthermore, the mapping

$$p_G \circ g \circ p_D^{-1}: p_D[D] \to p_G[G]$$

extends to a homeomorphism $\hat{g}: \hat{D} \to \hat{G}$ (onto).

It is now easy to verify

THEOREM 5. Let D and G be bounded domains in $\mathbb{C}^n$. Assume that $K_D(z,t)$ extends to a continuous function $K_D: \mathrm{cl}\, D \times \mathrm{cl}\, D \to \mathbb{C} \cup \{\infty\}$ such that $K_D(z,t) = \infty$ if and only if $z = t \in \partial D$. Assume further that $K_G(z,t)$ has analogous properties. Then every biholomorphic mapping $g: D \to G$ extends to the homeomorphism $g: \mathrm{cl}\, D \to \mathrm{cl}\, D$.

In the case of a simply connected domain $D \subset \mathbb{C}$ the invariant compactification is equivalent to the classical Carathéodory compactification. In this case the ideal boundary of D can be identified in a natural way with the set of all prime ends in D. For details see [41].

2. The case of a strictly pseudoconvex domain

In this section we shall recall some fundamental results [22] of J. J. Kohn in the theory of the $\bar\partial$ problem. These results were used by him to describe the properties of the Neumann operator in strictly pseudoconvex domains. Since the Bergman function can be described in terms of the Neumann operator, the properties of the Bergman function can be studied in greater detail in this case. This was done in the papers [30] and [12].

Denote by $L_{p,q}(D)$ the space of all forms of type (p,q) with square integrable coefficients. This is a Hilbert space. It contains as a dense subset the linear space $A_{p,q}(\mathrm{cl}\,D)$ of all forms which are smooth on $\mathrm{cl}\,D$. Consider the operator $\bar\partial : A_{p,q} \to A_{p,q+1}$. Then the operator $S =$ closure of $\bar\partial$ is a closed, densely defined operator on the space $L_{p,q}$:

$$S : L_{p,q} \to L_{p,q+1}.$$

If we define in a similar way the operator T on $L_{p,q-1}$, then

$$L_{p,q-1} \xrightarrow{T = \text{closure of } \bar\partial} L_{p,q} \xrightarrow{S = \text{closure of } \bar\partial} L_{p,q+1}$$

and $ST = 0$. Consider the subspace $Q_{p,q} \subset L_{p,q}$, defined as $Q_{p,q} = \mathrm{Dom}\,T^* \cap \mathrm{Dom}\,S$. Let

$$\mathcal{H}_{p,q} = \{\varphi \in Q_{p,q},\ T^*\varphi = 0,\ S\varphi = 0\}.$$

The elements of $\mathcal{H}_{p,q}$ are called <u>harmonic</u> <u>forms</u>. Since the space $\mathcal{H}_{p,q}$ is closed, we may consider the projector on this space. It will be denoted by $H_{p,q}$. The space $\mathcal{H}_{p,q}$ can also be characterized as the null space of the Laplace operator

$$L = TT^* + S^*S.$$

The operator L is self-adjoint [23]. Therefore we have the weak orthogonal decomposition of the space $L_{p,q}$:

$$\begin{aligned} L_{p,q} &= \text{closure of } R(L) \oplus \mathcal{H}_{p,q} \\ &= \text{closure of } R(TT^*) \oplus \text{closure of } R(S^*S) \oplus \mathcal{H}_{p,q}. \end{aligned}$$

Assume now that the space $R(L)$ is closed. In such a case we have the strong orthogonal decomposition of the space $L_{p,q}$:

$$L_{p,q} = R(L) \oplus \mathcal{H}_{p,q} = R(T\,T^*) \oplus R(S^*\,S) \oplus \mathcal{H}_{p,q}.$$

The Neumann operator $N\colon L_{p,q} \to \mathcal{H}^{\perp}_{p,q}$ is defined on the whole space $L_{p,q}$ by the formula

$$N\alpha = \begin{cases} 0 & \text{if } \alpha \in \mathcal{H}_{p,q}, \\ \varphi & \text{if } \alpha \in \mathcal{H}^{\perp}_{p,q};\ L\varphi = \alpha \text{ and } \varphi \in \mathcal{H}^{\perp}_{p,q}. \end{cases}$$

This is a bounded operator. In fact the assumption that $R(L)$ is closed is equivalent to the estimate

$$(1) \qquad \|\varphi\| \le \text{const}\, \|L\varphi\|$$

for all $\varphi \in \operatorname{Dom} L \cap \mathcal{H}^{\perp}_{p,q}$. The following identities are obvious from the definition:

$$LN = I - H, \quad NL = I - H.$$

From the first one and the strong orthogonal decomposition of $L_{p,q}$ follows that

$$I = T\,T^*\,N + S^*\,S\,N + H.$$

Let us consider now the following equation with an unknown form $\varphi \in L_{p,q-1}$:

$$T\varphi = \alpha.$$

This $\bar\partial$ problem is solvable if and only if the form $\alpha \in L_{p,q}$ is orthogonal both to $\mathcal{H}_{p,q}$ and $R(S^*)$. These conditions mean that $H(\alpha) = 0$ and $S\alpha = 0$. Since $T\varphi \in R(T)$, the conditions are obviously necessary. They are also sufficient since they imply that the decomposition

$$I\alpha = T\,T^*\,N\alpha + S^*\,S\,N\alpha + H\alpha$$

reduces to $\alpha = T\,T^*\,N\alpha$. Therefore $\varphi = T^*\,N\alpha$ solves the $\bar\partial$ problem. Furthermore this is the only solution which is orthogonal to the null space of the operator T. In some following considerations it will be convenient to indicate explicitly the type (p,q) of forms in the space on which the operators H or N are considered. Therefore we shall introduce a more explicit notation $N_{p,q} = N$ and $H_{p,q} = H$.

Assume now that the operators $N_{p,q-1}$, $N_{p,q}$, $N_{p,q+1}$ are well-defined and consider the diagram

$$\xrightarrow{U} L_{p,q-1} \xrightarrow{T} L_{p,q} \xrightarrow{S} L_{p,q+1} \xrightarrow{V}$$

where the operators U and V are defined in an analogous way as T and S. We are going to show that

$$N_{p,q+1}\,S = S\,N_{p,q},\quad N_{p,q-1}\,T^* = T^*\,N_{p,q}.$$

The proof is based on the fact that the Neumann and Laplace operators commute. Consider first the case of operator S. Since $R(S)\subset \mathsf{H}^{\perp}_{p,q+1}$, we have

$$(V^*V + SS^*)N_{p,q+1}\,S\,N_{p,q} = S\,N_{p,q}.$$

Therefore

$$S\,N_{p,q} = N_{p,q+1}\,S\,S^*\,S\,N_{p,q} = N_{p,q+1}\,S(S^*S + TT^*)N_{p,q}$$

$$= N_{p,q+1}\,S(I - H_{p,q}) = N_{p,q+1}\,S.$$

The case of T^* can be proved in a similar way.

We can now express the projector $H_{p,q}$ in terms of Neumann operators $N_{p,q-1}$ and $N_{p,q+1}$, namely

$$H_{p,q} = I - S^*\,N_{p,q+1}\,S - T\,N_{p,q-1}\,T^*.$$

It is easy to see that

$$S^*\,N_{p,q+1}\,S + T\,N_{p,q-1}\,T^*$$

takes values in $\mathsf{H}^{\perp}_{p,q}$. Therefore it will suffice to show that the right-hand side is harmonic.

To prove that the right-hand side is the null space of S note that $V^*V\,N_{p,q+1}\,S = 0$ since V commutes with $N_{p,q+1}$ and $VS = 0$. Therefore

$$S\,I = S^*\,N_{p,q+1}\,S = S - (SS^* + V^*V)N_{p,q+1}\,S = S - S = 0.$$

A similar reasoning applies to the operator T^* instead of S. This proves the desired representation of the projector $H_{p,q}$. The case $q = 0$ is particularly simple, because then $T^* = 0$ and we have

$$H_{p,0} = I - S^*\,N_{p,1}\,S.$$

Note that in the case $p = 0$ the harmonic space $\mathsf{H}_{0,0}$ is equal to $L^2H(D)$. Therefore

$$H_{0,0} = I - S^*\,N_{0,1}\,S$$

is equal to the Bergman projector and for every $\varphi\in L^2(D)$ we have

$$H_{0,0}\,(t) = \int_D \varphi(z)\,\overline{K_D(z,t)}\,dm(z).$$

In order to state sufficient conditions for existence of the Neumann operator $N_{p,q}$ we have to introduce some notation. For any point $P\in\partial D$ there exists a local coordinate neighbourhood G with coordinates $x_1,\dots,x_{2n-1}$, $r=x_{2n}$ such that $G\cap D=\{z\in G:\ r(z)<0\}$. For any $\varphi\in C_0(G\cap \mathrm{cl}\,D)$ the function $\Lambda_b^u\varphi\in C^\infty\{x\in\mathbb{R}^{2n}:\ x_{2n}\le 0\}$ is defined by the formula

$$\widetilde{\Lambda_b^u\varphi}(\xi,r)=(1+|\xi|^2)^{\frac{1}{2}u}\tilde\varphi(\xi,r),$$

where $\tilde\varphi$ denotes the partial Fourier transform of φ:

$$\tilde\varphi(\xi,r)=\int_{\mathbb{R}^{2n-1}} e^{-ix\xi}\,\varphi(x,r)\,dx.$$

The tangential Sobolev norm of a function $\varphi\in C_0^\infty(G\cap\mathrm{cl}\,D)$ is defined in terms of the operator Λ_b^s as follows:

$$|||\varphi|||_s^2=||\Lambda_b^s\varphi||^2=\int_{-\infty}^{0}||\varphi(\ ,r)||_s^2\,dr.$$

For a smooth form φ with compact support in $G\cap\mathrm{cl}\,D$, let

$$|||\varphi|||_s^2=\sum_{I,J}|||\varphi_{I,J}|||_s^2$$

and

$$|||D\varphi|||_s^2=\sum_{j=1}^{2n}|||\frac{\partial}{\partial x_j}\varphi|||_s^2+|||\varphi|||_s^2\sim|||\varphi|||_{s+1}^2+|||\frac{\partial}{\partial r}\varphi|||_s^2.$$

Consider the subspace $Q_{p,q}\subset L_{p,q}$ defined as

$$Q_{p,q}=\mathrm{Dom}\,T^*\cap\mathrm{Dom}\,S$$

and denote by $Q_{p,q}^\infty$ the set $Q_{p,q}\cap A_{p,q}(\mathrm{cl}\,D)$. Assume that $D=\{z\in\mathbb{C}^n:\ r(z)<0\}$. Then $\varphi\in Q_{p,q}$ if and only if $\varphi\in A_{p,q}(\mathrm{cl}\,D)$ and

$$\sigma(\vartheta,dr)\varphi\,|\,\partial D=0.$$

Here ϑ is the formal adjoint operator for $\bar\partial$, and $\sigma(\vartheta,dr)$ denotes the symbol of this operator. It follows that the linear space $Q_{p,q}^\infty$ is closed under multiplications by functions from $A_{p,q}(\mathrm{cl}\,D)$. We say that a domain $D\subset\mathbb{C}^n$ satisfies the e-<u>subelliptic estimate</u> at $P\in\partial D$ if there exists a local coordinate neighbourhood G of P such that for all $\varphi\in Q_{p,q}$ with compact support in G the inequality

$$|||D\varphi|||_{e-1}\lesssim Q(\varphi,\varphi)$$

holds, where $f(\varphi)\lesssim g(\varphi)$ means $f(\varphi)\le c\,g(\varphi)$ for some constant $c>0$. Here $Q(\varphi,\psi)$ is defined by

$$Q(\varphi,\psi) = (T^*\varphi, T^*\psi) + (S\varphi, S\psi) + (\varphi,\psi) \quad \text{for } \varphi,\psi \in Q^{\infty}_{p,q}.$$

Denote by $\tilde{Q}_{p,q}$ the completion of $Q^{\infty}_{p,q}$ with respect to the norm $Q(\varphi) = Q(\varphi,\varphi)^{\frac{1}{2}}$. Of course we may identify $\tilde{Q}_{p,q}$ with a dense subset of $L_{p,q}$. For any $\alpha \in L_{p,q}$ the continuous linear functional on $\tilde{Q}_{p,q}$ given by $\psi \longmapsto (\psi,\alpha)$ can be represented by the unique form $\alpha \in \tilde{Q}_{p,q}$ so that

$$(\psi,\alpha) = Q(\psi,\varphi)$$

for all $\psi \in \tilde{Q}_{p,q}$. The operator $K: L_{p,q} \longrightarrow \tilde{Q}_{p,q}$, given by $K\alpha = \varphi$, is bounded, self-adjoint, and one-to-one. The self-adjoint operator K^{-1} is denoted by F. It was proved by Hörmander that the condition for the Levi form at every point $P \in \partial D$ to have at least $n-q$ positive eigenvalues is equivalent to the following inequality known as the <u>basic estimate</u>:

$$E(\varphi)^2 \le \text{const}\, Q(\varphi,\varphi) \quad \text{for } \varphi \in Q^{\infty}_{p,q},$$

where

$$E(\varphi)^2 = \sum_{I,J,k} \|(\partial/\partial \bar{z}_k)\varphi_{I,J}\|^2 + \int_{\partial D} |\varphi|^2 + \|\varphi\|^2.$$

In particular, the basic estimate is valid in every strictly pseudoconvex domain for all $q > 0$. For every $P \in \partial D$ there exists a local coordinate neighbourhood G such that for all smooth forms φ with compact support in G

$$|||D\varphi|||^2_{-\frac{1}{2}} \lesssim E(\varphi)^2.$$

It follows that for $q > 0$ every strictly pseudoconvex domain D satisfies the $\frac{1}{2}$-subelliptic estimate at every point $P \in \partial D$. From this fact follows the main theorem of Folland and Kohn [22]:

<u>THEOREM 6</u>. <u>Assume that the basic estimate is satisfied in</u> D. <u>If</u> U <u>is a domain in</u> $\operatorname{cl} D$ <u>and</u> $\alpha|U \in A_{p,q}(\operatorname{cl} U)$, <u>then</u> $\varphi = K\alpha$ <u>implies</u> $\varphi|U \in A_{p,q}(\operatorname{cl} U)$. <u>Furthermore, if</u> ξ, ξ_1 <u>are two smooth functions such that</u> $\operatorname{spt} \xi_1 \subset U$ <u>and</u> $\xi_1 \equiv 1$ <u>on</u> $\operatorname{spt} \xi$, <u>then</u>:

<u>1</u>. <u>If</u> $U \cap \partial D = \emptyset$, <u>then for any positive integer</u> s <u>there exists a constant</u> c_s <u>such that</u>

$$(2) \qquad \|\xi\varphi\|^2_{s+2} \le c_s(\|\xi_1\alpha\|^2_s + \|\alpha\|^2).$$

<u>2</u>. <u>If</u> $U \cap \partial D \neq 0$, <u>then for any positive integer</u> s <u>there exists a constant</u> c_s <u>such that</u>

$$(3) \qquad \|\xi\varphi\|^2_{s+1} \le c_s(\|\xi_1\alpha\|^2_s + \|\alpha\|^2).$$

<u>Here</u> $\|\varphi\|_s$ <u>denotes the Sobolev norm of order</u> s <u>of the form</u> $\varphi \in A_{p,q}(\operatorname{cl} D)$.

From this theorem it follows that if $\alpha|U \in H_s$ (the Sobolev space of order s), then $\xi\varphi \in H_{s+1}$ and (3) holds. In particular, by taking $U = D$, we obtain the inequality

$$\|\varphi\|_{s+1} \leq \text{const}\, \|\alpha\|_s.$$

It follows that K is a compact operator by the Rellich lemma. We can now easily prove that $F = L + I$. Since both operators are self-adjoint, it is sufficient to show that $F \subset L + I$. Assume that $F\varphi = \alpha$. Let $\alpha_n \in A_{p,q}(\text{cl}\, D)$ be any sequence which converges to α. Let $\varphi_n = K\alpha_n$. By Theorem 6, $\varphi_n \in A_{p,q}(\text{cl}\, D)$ and, by continuity of K, $\lim \varphi_n = \varphi$. Since $F\varphi_n = L\varphi_n + \varphi_n$, we see that $\lim(L + I)\varphi_n = \alpha$. Since the operator $L + I$ is closed we conclude that $\varphi \in \text{Dom}(L + I)$ and that $(L + I)\varphi = F\varphi$. Therefore $F = L + I$, as claimed.

It can be proved now that, under the hypotheses of Theorem 6, the space $R(L)$ is closed and the Neumann operator $N_{p,q}$ is well-defined. Assume that (1) does not hold. It follows that we can find a normed sequence $\varphi_n \in \text{Dom}\, L \cap \mathcal{H}^{\perp}_{p,q}$, such that $\lim L\varphi_n = 0$. Let $\alpha_n = F\varphi_n$. The sequence $L\varphi_n = \alpha_n - \varphi_n$ is bounded, and therefore the sequence α_n is bounded as well. Since $K = F^{-1}$ is compact, there exists a subsequence φ_{n_k} which converges to some form φ_0. Since $L\varphi_{n_k}$ converges to zero and L is closed, it follows that $\varphi_0 \in \mathcal{H}_{p,q}$ is a normed harmonic form. Such a form cannot be equal to the limit of forms $\varphi_{n_k} \in \mathcal{H}^{\perp}_{p,q}$. Therefore (1) must hold and $N_{p,q}$ is well-defined. Note that for any bounded sequence $\varphi_n \in \mathcal{H}_{p,q}$ we have $F\varphi_n = \varphi_n$ or, equivalently, $K\varphi_n = \varphi_n$. By compactness of K there exists a convergent subsequence φ_{n_k}. This shows that the unit ball in $\mathcal{H}_{p,q}$ is relatively compact and therefore $\dim \mathcal{H}_{p,q} < \infty$.

From Theorem 6 it follows also that all eigenforms of F belong to $A_{p,q}(\text{cl}\, D)$. If s is any positive integer and $F\varphi = \lambda$ for some $\lambda \neq 0$, then

$$\|\varphi\|_s \lesssim \|F\varphi\|_s \lesssim \|\varphi\|_{s-1} \lesssim \|F\varphi\|_{s-1} \lesssim \|\varphi\|_{s-2} \lesssim \cdots \lesssim \|\varphi\|_0.$$

In particular, $\mathcal{H}_{p,q} \subset A_{p,q}(\text{cl}\, D)$. One can also prove the inequalities

$$\|\varphi\|^2_{s+1} \lesssim \|L\varphi\|^2_s + \|\varphi\|^2$$

and

$$\|N\varphi\|^2_{s+1} \lesssim \|\varphi\|^2_s.$$

It follows that φ is smooth whenever $L\varphi$ is smooth and that N is a compact operator.

When $q = 0$ the situation is somewhat different since the basic estimate does not hold in this case. However one can show that if $L_{p,1}$

admits the strong orthogonal decomposition, then so does $L_{p,0}$. Consider the diagram

$$L_{p,0} \xrightarrow{T} L_{p,1} \xrightarrow{S} L_{p,2}.$$

By assumption the space $R(t)$ is closed and this is equivalent to the condition that $R(T^*)$ is closed. The Laplace operator on $L_{p,0}$ reduces to T^*T. Since $R(T^*)$ is closed, we have the strong orthogonal decomposition

$$L_{p,0} = R(T^*) \oplus \mathcal{H}_{p,0}.$$

It follows that $R(T^*T) = R(T^*)$ is closed and the Neumann operator $N_{p,0}$ is well-defined.

The estimates in Sobolev norms for the Neumann operator yield analogous estimates for the solution $\varphi = T^* N\alpha$ of the $\bar{\partial}$ problem. Namely,

$$\|\varphi\|_s \leq \text{const}\ \|\alpha\|_s.$$

This Kohn's inequality and the methods of the theory of Fourier integral operators with complex phase were used by Boutet de Monvel and Sjöstrand [12] to describe the boundary behaviour of the Bergman function in strictly pseudoconvex domains. They proved that for every strictly pseudoconvex domain $D = \{z \in \mathbb{C}^n : r(z) < 0\}$ there exist smooth functions F and G on $\mathrm{cl}\, D \times \mathrm{cl}\, D$ such that

$$K_D(z,t) = F(z,t)[-i\Psi(z,t)]^{-n} + G(z,t)\ \mathrm{Log}[-i\Psi(z,t)].$$

Here Ψ is a smooth function on $\mathbb{C}^n \times \mathbb{C}^n$ such that

i) $\Psi(z,z) = -i\rho(z)$ and the functions $(\partial/\partial\bar{z})\Psi(z,t)$ and $(\partial/\partial t)\Psi(z,t)$ vanish to infinite order on the diagonal $z = t$,

ii) $\Psi(z,t) = -\overline{\Psi(t,z)}$.

Moreover, $F(z,z) \neq 0$ for $z \in \partial D$.

From this deep result it follows in particular that every strictly pseudoconvex domain is regular in the sense of Section 1.

3. The $\bar{\partial}$ Neumann problem and smooth extensions of biholomorphic mappings

The theory of the Bergman kernel function can be applied to the study of smooth extensions of biholomorphic mappings. In 1974 Charles Fefferman proved the following theorem [20]:

THEOREM 7. *If D and G are strictly pseudoconvex domains in $\mathbb{C}^n$, with C^∞-boundaries, and g is a biholomorphic mapping from D onto G, then g can be extended to a C^∞-diffeomorphism between $\mathrm{cl}\, D$ and $\mathrm{cl}\, G$.*

The original proof of this theorem is long and complicated. It is based on a difficult analysis of the singularity of the Bergman function at the boundary and the study of geodesics with respect to the Bergman metric. In 1978 Ligocka [36] and Webster [42] have independently obtained explicit conditions for the Bergman functions K_D and K_G which permit to eliminate the differential geometric considerations from Fefferman's proof. Recently Bell [10] has proved the following interesting fact:

THEOREM 8. *Let D be a bounded domain in $\mathbb{C}^n$ with C^∞-boundary. Denote by $W^s(D)$ the Sobolev space and by $H^s(D)$ its subspace consisting of holomorphic functions. Let $H = H_{0,0}$ be an orthogonal projector from $L^2(D)$ onto the space $L^2H(D)$ of square integrable holomorphic functions. Assume that there exist $s > 2n$ and $m \geq 0$ such that H is a bounded operator from $W^{s+m}(D)$ into $H^s(D)$. Then*

$$H[W^{s+m}(D)] \subset \mathrm{cl}\, H[C_0^\infty(D)] \subset H^s(D),$$

where the closure is taken in the space $H^s(D)$.

From Kohn's inequality discussed in Section 2 it follows that the assumptions of Bell's theorem are satisfied in every strictly pseudoconvex domain with C^∞-boundary.

The explicit conditions for the Bergman functions K_D and K_G were restated by Ligocka [35] in a form which is more general and easier to verify. It involves only the values of the Bergman functions and their derivatives of the first order. It was observed in the same paper that these new conditions are satisfied in every domain with C^∞-boundary for which the assumptions of Bell's theorem are valid.

Thus the results of [10] and [35] permit to obtain Fefferman's theorem as an elementary consequence of the $\bar\partial$ Neumann estimates due to J. J. Kohn (discussed in Section 2).

The conditions for the Bergman function stated in [35] are valid also for other classes of domains such as the class of bounded strictly starlike, complete circular domains, or the class of weakly pseudoconvex domains with real-analytic boundary, and are invariant under forming the Cartesian product. Therefore the analogue of Fefferman's theorem holds also for domains of the above mentioned classes.

4. Nonlinear Dirichlet problems

With the standard notation $d = \partial + \bar\partial$ and $d^c = i(\partial - \bar\partial)$ in $\mathbb{C}^n$ we have $dd^c = 2i\partial\bar\partial$. Write $(dd^c u)^n = dd^c u \wedge \ldots \wedge dd^c u$ (n times). Also, let

$$dV = (\tfrac{1}{2}i)^n \bigwedge_{j=1}^{n} dz_j \wedge d\bar{z}_j$$

denote the Kähler form in $\mathbb{C}^n$. We are interested here in the nonlinear Dirichlet problem

(4) $\quad (dd^c u)^n = f\,dV$ on Ω, u is C^2-plurisubharmonic on Ω, $u|\partial\Omega = \varphi$,

where Ω is a strictly pseudoconvex bounded domain in $\mathbb{C}^n$ and $f \geq 0$. This problem was investigated in detail by Bedford and Taylor [7]. The operator $(dd^c u)^n$ has an invariance property under holomorphic mappings, i.e. if $F:\Omega \to \Omega$ is holomorphic, then $[dd^c(u \circ F)]^n = |F'|^{2n} \times (dd^c u)^n \circ F$. If $f = 0$, (4) is a natural generalization of the Dirichlet problem for harmonic functions in the complex plane.

Now we consider the transformation $v = e^{-u}$ or $-\log v = u$, which yields the identity

$$e^{-(n+1)u}(dd^c u)^n = J(v),$$

where $J(v)$ is defined as the (n,n)-form

(5) $$J(v) = n!\,(-4)^n \det\begin{bmatrix} v & v_{\bar{k}} \\ v_j & v_{j\bar{k}} \end{bmatrix} dV$$

and $v_j = (\partial/\partial z_j)v$, $v_{\bar{j}} = (\partial/\partial \bar{z}_j)v$, and $v_{j\bar{k}} = (\partial^2/\partial z_j \partial \bar{z}_k)v$. The determinant in (5) is related to the Bergman kernel function [21]. Take into account the Dirichlet problem

(6) $$\mathfrak{J}(v) = (-1)^n \det\begin{bmatrix} v & v_{\bar{k}} \\ v_j & v_{j\bar{k}} \end{bmatrix} = 1 \text{ on } \Omega,\ v = 0 \text{ on } \partial\Omega.$$

In the case of the unit ball B we see that the Bergman kernel function is given by $K_B(z,z) = c_n/(1-|z|^2)^{n+1}$, and $v(z) = 1-|z|^2$ satisfies (6). In general the problem arises how closely is the Bergman kernel function $K_\Omega(z,z)$ approximated by $c_n[v(z)]^{n+1}$ with v being a solution of (6). It is an open question whether $\mathfrak{J}(c_n K_\Omega^{-1/(n+1)})$ is close to 1. By a result of Hörmander [26],

$$\mathfrak{J}(c_n K_\Omega^{-1/(n+1)}) = 1 \text{ on } \partial\Omega.$$

More recent works of Christoffers [17] and Diederich [19] show that also the first derivatives of $\mathfrak{J}(c_n K_\Omega^{-1/(n+1)}) - 1$ vanish at the boundary. On the other hand it is known that $\mathfrak{J}(c_n K_\Omega^{-1/(n+1)}) \not\equiv 1$ so that the relationship of (6) to the Bergman kernel function is only asymptotic. We do not know whether the Bergman kernel function differs asymptotically from a negative power of a solution of (6).

The complex Monge-Ampère equation

$$(7) \qquad (dd^c u)^n \equiv 4^n n! \det[u_{j\bar{k}}] dV = 0,$$

involving the complex hessian $[u_{j\bar{k}}] = [(\partial^2/\partial z_j \partial \bar{z}_k)u]$ has a formal similarity to the real Monge-Ampère equation which involves the determinant of the real hessian $[u_{jk}]$. Analogously, we can formulate a Dirichlet problem for the real Monge-Ampère equation:

$$u \text{ convex on } \Omega, \ u|\partial\Omega = \varphi, \ \det[(\partial^2/\partial x_j \partial x_k)u] = 0 \text{ on } \Omega,$$

where Ω is a strictly convex domain in $\mathbb{R}^n$. This problem has been studied extensively in relation to problems of differential geometry, but it seems difficult to solve it in a completely satisfactory way. Aleksandrov [1], using the theory of convex surfaces, proved the existence and uniqueness of convex (generalized) solutions. Interior regularity of the solution was discussed for $n \geq 3$ by Pogorelov [37] and, more generally, by Cheng and Yau [15].

Real (nonparametric) hypersurfaces $x_{n+1} = u(x_1, \ldots, x_n)$ in $\mathbb{R}^{n+1}$ have been studied in this connection. If the hessian of u has constant rank: $\operatorname{rank}[u_{jk}] = p$, then the graph of u can be developed by real planes of dimension $n-p$ and, in fact, if $p = 1$ and $n = 2$, the solution is locally a cylinder. This arises because the matrix $[u_{jk}]$ may be interpreted as the Jacobi matrix of the normal mapping $x \longmapsto (u_1, \ldots, u_n, -1)$ and this geometrical interpretation is a powerful tool for the study of real hessians.

Equations involving the complex hessian appear also in the theory of semi-norms intoduced by Chern, Levine and Nirenberg [16].

Namely, let $\mathbb{M}$ be a complex manifold of complex dimension n. Consider the family F of C^2-plurisubharmonic functions u on $\mathbb{M}$ such that $0 < u(z) < 1$ everywhere. For any homology class γ of $\mathbb{M}$ with real coefficients, following [16], we set

$$(8) \qquad N[\gamma] = \begin{cases} \sup\limits_{u \in F} \inf\limits_{T \in \gamma} |T[d^c u \wedge (dd^c u)^{k-1}]| & \text{if } \dim\gamma = 2k-1, \\ \sup\limits_{u \in F} \inf\limits_{T \in \gamma} |T[du \wedge d^c u \wedge (dd^c u)^{k-1}]| & \text{if } \dim\gamma = 2k, \end{cases}$$

where T runs over all currents of γ (in the sense of de Rham). It is proved in [16] that the mapping $N: \Gamma \longrightarrow \mathbb{R}^+$, where Γ is a family of homology classes on $\mathbb{M}$, is a semi-norm on Γ and is always nonincreasing under holomorphic mappings. If T_1 and T_2 are closed currents of γ such that $T_1[\varphi] \neq T_2[\varphi]$, where $\varphi = d^c u \wedge (dd^c u)^{k-1}$ if $\dim\gamma = 2k-1$ and $\varphi = du \wedge d^c u \wedge (dd^c u)^{k-1}$ if $\dim\gamma = 2k$, then there exists a real number t satisfying the condition $\{tT_1 + (1-t)T_2\}[\varphi] = 0$, while the current $tT_1 + (1-t)T_2$ still belongs to γ. Therefore in the definition (8) we only need to consider functions u for which φ is closed. These functions u satisfy the differential equations

$$(dd^c u)^k = 0 \text{ if } \dim\gamma = 2k-1, \quad du \wedge (dd^c u)^k = 0 \text{ if } \dim\gamma = 2k.$$

Let us take into account the following special case of the definition (8). Suppose that $\mathbb{M}$ is a bounded domain D in $\mathbb{C}^n$ whose boundary consists of two components which are $(2n-1)$-dimensional smooth submanifolds of $\mathbb{C}^n$. By F we denote the class of C^2-smooth real-valued functions on $\mathrm{cl}\,D$, which are supposed to be plurisubharmonic on D and equal to 0 on one component of the boundary and 1 on the other (we assume that $F \neq 0$). Consider the functional

(9) $$F \ni u \longmapsto \int_D du \wedge d^c u \wedge (dd^c u)^{n-1} \in \mathbb{R}$$

and a natural problem of minimizing the functional (9). The equations (7) arise here as the Euler equations for a stationary point of (9) (cf. [28] and [8]).

Now we shall give some remarks on the Dirichlet problem (4), based in general on the paper of Bedford and Taylor [7]. We shall begin with the following lemma [16]:

LEMMA 3. *Let u be a C^2-plurisubharmonic function in a polydisc $\Delta = \{z \in \mathbb{C}^n : |z_j| < r_j,\ j = 1,\dots,n\}$, satisfying the conditions $0 < u(z) < 1$ everywhere. Let Δ_1 be a compact subpolydisc of Δ: $\Delta_1 = \{z \in \mathbb{C}^n : |z_j| \leq \rho_j < r_j,\ j = 1,\dots,n\}$. Then there is a constant B independent of u such that for any $r \times r$-minor of $[u_{j\bar{k}}]$ we have the estimate*

$$\int_{\Delta_1} \left[\mathrm{abs}\begin{vmatrix} u_{j_1\bar{k}_1} & \dots & u_{j_1\bar{k}_r} \\ \dots & \dots & \dots \\ u_{j_r\bar{k}_1} & \dots & u_{j_r\bar{k}_r} \end{vmatrix} + \sum_k |u_k|^2\right] dV \leq B$$

It follows from Lemma 3 that if u is a bounded C^2-plurisubharmonic function on an open set $\Omega \subset \mathbb{C}^n$ and K is a compact subset of Ω, then

$$\int_K (dd^c u)^n \leq C(\|u\|_\Omega)^n,$$

where $\|u\|_\Omega = \sup\{|u(z)| : z \in \Omega\}$. With help of this estimate it can be shown that the operator $u \longmapsto (dd^c u)^n$, considered as a mapping from the collection of C^2-plurisubharmonic functions on Ω into the space of nonnegative Borel measures on Ω, has a continuous extension to the space of continuous plurisubharmonic functions. For such extension of

$(dd^c u)^n$ we have the following theorem [7]:

THEOREM 9. Let Ω be a bounded open set in $\mathbb{C}^n$. If $u, v \in C(\mathrm{cl}\,\Omega)$ are plurisubharmonic and if $(dd^c u)^n \leq (dd^c v)^n$, then

$$\min\{u(z) - v(z)\colon z \in \mathrm{cl}\,\Omega\} = \min\{u(z) - v(z)\colon z \in \partial\Omega\}.$$

An immediate consequence of the theorem is the uniqueness of continuous solutions of (4).

Another problem connected with (4) is to prove the existence of generalized solutions of (4). This can be done (cf. e.g. [7]) with help of the Perron-Bremermann family

$$B(\varphi,f) = \{v \text{ plurisubharmonic}, \Phi(v) \geq f, \limsup_{\zeta \to z} v(\zeta) \leq \varphi(z), z \subset \partial\Omega\},$$

where $\Phi(v)$ is essentially equal to $(\det[v_{j\bar{k}}])^{1/n}$. It is obtained from $dd^c v$ via a general measure-theoretic construction (cf. e.g. [7]) and defined for all plurisubharmonic functions v. The following regularity result holds [7]:

THEOREM 10. Let Ω be the unit ball in $\mathbb{C}^n$ and let $\varphi \in C^2(\partial\Omega)$, $g \in C^2(\mathrm{cl}\,\Omega)$, $g \geq 0$. If $u = \sup\{v\colon v \in B(\varphi,g)\}$ is the upper envelope of $B(\varphi,g)$, then $u \in C(\mathrm{cl}\,\Omega)$ and the second partial derivatives of u exist almost everywhere on Ω and are locally bounded.

With this regularity theorem it is possible to prove that the upper envelope u of $B(\varphi,g)$ solves (4) with $f = g^n$ in the special case where Ω is the unit ball in $\mathbb{C}^n$. Using the spherical modifications it is possible to obtain generalized solutions for more general domains.

THEOREM 11 (Bedford and Taylor [7]). Let Ω be a strictly pseudoconvex bounded domain in $\mathbb{C}^n$. If $\varphi \in C(\partial\Omega)$, $f \in C(\mathrm{cl}\,\Omega)$, and $f \geq 0$, then there exists a unique plurisubharmonic function $u \subset C(\mathrm{cl}\,\Omega)$ such that

$$(dd^c u)^n = f, \ \Phi(u) = f^{1/n}, \text{ and } u|\partial\Omega = \varphi.$$

Furthermore, if $\partial\Omega$ is smooth, $\varphi \in C^2(\partial\Omega)$, and $f^{1/n} \in \mathrm{Lip}^1(\mathrm{cl}\,\Omega)$, then $u \in \mathrm{Lip}^1(\mathrm{cl}\,\Omega)$.

For more complicated domains it is still very hard to solve the Dirichlet problem (4). We shall give another example of a domain for which this is done.

Suppose that Ω_o is a bounded strictly pseudoconvex domain in $\mathbb{C}^n$. Let further $\Omega_1, \ldots, \Omega_s$ be strictly pseudoconvex and relatively compact subsets of Ω_o, and let the union of their closures be convex in Ω_o. Let $\Omega = \Omega_o \setminus (\mathrm{cl}\,\Omega_1 \cup \ldots \cup \mathrm{cl}\,\Omega_s)$. Then the following theorem holds [9]:

THEOREM 12. <u>There exists a unique function</u> $u \subset \mathrm{Psh}(\Omega) \cap \mathrm{Lip}(\Omega)$, <u>such that</u>

(10) $\quad u = 1$ <u>on</u> $\partial\Omega_o$ <u>and</u> $u = 0$ <u>on</u> $\bigcup_{k=1}^{s} \partial\Omega_k$; $0 < u \leq 1$ <u>and</u> $(dd^c u)^n = 0$ <u>in</u> Ω.

Here it is natural to ask whether the solution of (10) is smooth on $\mathrm{cl}\,\Omega_o \setminus (\Omega_1 \cup \ldots \cup \Omega_s)$. The answer to this question is, in general, negative [5].

5. Capacities on hermitian manifolds and the generalized complex Monge-Ampère equations

Let $\mathbb{M}$ be a complex manifold of complex dimension n endowed with an hermitian metric h and a C^1 tensor field H of type (1,1). In particular we may let H depend on h or take as H an almost complex structure of the tangent bundle $T\mathbb{M}$, for instance the complex structure of $\mathbb{M}$. Let further D be a <u>condenser</u> on $\mathbb{M}$, i.e. a domain whose complement consists of two distinguished disjoint closed sets C_o and C_1 (the <u>condenser plates</u>), $q: \mathbb{M} \to \mathbb{C}$ a continuous mapping (the <u>inhomogeneity function</u>), and p a real number ≥ 1. Consider the class $\mathrm{Adm}\,D$ of all plurisubharmonic C^2-functions u on $\mathrm{cl}\,D$, satisfying the conditions $0 < u(z) < 1$ for $z \in D$, $u|\partial C_o = 0$ and $u|\partial C_1 = 1$. Let [32]:

$$(11) \quad \mathrm{Cap}_p(D,q) = \inf_{\tilde{u} \in \mathrm{Adm}\,D} \left| \int_D q[h(d^c\tilde{u}, d^c\tilde{u})]^{\frac{1}{2}p-1} \det H \, d\tilde{u} \wedge d^c\tilde{u} \wedge (dd^c\tilde{u})^{n-1} \right|,$$

where $h(d^c u, d^c u) = h^{j\bar{k}} u_{:j} u_{:\bar{k}}$, $u_{:j} = (u \circ \mu^{-1})_{|j} \circ \mu$ in any local coordinate system $\mu = (\mu^j)$ on $\mathbb{M}$.

Let further Γ be a homology class of D with real coefficients and $\dim \Gamma = r$. Consider all currents of Γ (more precisely: corresponding to the elements of Γ) in the sense of de Rham, and a locally finite open covering $\mathcal{U} = \{U_j : j \in I\}$ of $\mathbb{M}$. Denote by $\mathrm{adm}(D,\mathcal{U})$ the family of all plurisubharmonic C^2-functions u_j on $U_j \cap D$ defined in each member of the covering which satisfy the following conditions:

(i) the oscillation of u_j in $U_j \cap D$ is less than one,

(ii) $du_j = du_k$ in $U_j \cap U_k \cap D \neq \emptyset$.

Condition (ii) describes a closed real one-form in D. Similarly $d^c u_j$ and $dd^c u_j$ are also well-defined in D. Without ambiguity, we can denote them omitting the indices. Let [31]:

$$(12) \quad \mathrm{cap}_p(D,q,\Gamma,\mathcal{U}) = \sup_{u \in \mathrm{adm}(D,\mathcal{U})} \inf_{T \in \Gamma} |T[q\{h(d^c u, d^c u)\}^{\frac{1}{2}p-1} \det H \, D^r u]|,$$

where

$$D^r u = \begin{cases} d^c u \wedge (dd^c u)^{\frac{1}{2}r-\frac{1}{2}} & \text{for } r \text{ odd}, \\ du \wedge d^c u \wedge (dd^c u)^{\frac{1}{2}r-1} & \text{for } r \text{ even}. \end{cases}$$

For a detailed description of the capacities (11) and (12) we refer to [33], and for an example of their application in [34].

When looking for a complex analogue of the principle of Dirichlet (cf. e.g. [32]) a natural procedure is to take in (12) for Γ the (2n-1) -dimensional homology class of level hypersurfaces $\{z \in \mathrm{cl}\, D: u(z) = \mathrm{const}\}$. (In analogy, for a complex counterpart of the principle of Thomson (cf. e.g. [32]) we had to take in (12) for Γ the orthogonal 1-dimensional homology class.) One should expect that under some reasonable conditions, in particular if we take in (12) for admissible functions only u defined globally with $0 < u(z) < 1$ for $z \in D$ (we write $u \in \mathrm{adm}\, D$), both capacities (12) and (11) will coincide.

The above idea as well as both definitions in the case where $H = J$ (the complex structure of $\mathbb{M}$), $p = 2$, and $q = \mathrm{const}$ is due to Chern, Levine and Nirenberg [16], but the affirmative answer is known only in very special subcases [16], [29], [9]. In the case mentioned the functional minimized attains its minimum for $\tilde{u} = u$ if and only if u fulfils the complex Monge-Ampère equation (7). This equation is a special case of the *generalized complex Monge-Ampère equations*

(13) $\quad dd^c(Fu) \wedge (dd^c u)^{n-1} = 0, \quad F \in C^2(\mathrm{cl}\, D)$

or

(14) $\quad d(G\, d^c u) \wedge (dd^c u)^{n-1} = 0, \quad G \in C^1(\mathrm{cl}\, D)$

which play an analogous role for the general capacity (11) with

(15) $\quad d^c(Fu)' = G\, d^c u, \quad G = q[h(d^c u, d^c u)]^{\frac{1}{2}p-1} \det H.$

In the general case the function u is replaced by a system satisfying the condition (ii). We quote the following two theorems due to Andreotti and Ławrynowicz [2], [3]:

THEOREM 13. *Suppose that*:

(α) D *has a piecewise* C^1-*smooth boundary and compact closure*,

(β) $n \geq 2$, $p = 2$, *and* q *is of the class* C^1,

(γ) u *belongs to* $\mathrm{adm}\, D$ *and satisfies* (14), *where* $G = q \det H$,

(δ) $d(G\, d^c u) = f\, dd^c u$, $f \geq (n-1)^{-1} G$, f *being of the class* C^1.

Then the infimum in (11) *is attained for the* u *in question*.

Remark 1. Theorem 13 remains valid for $n = 1$. In this case the condition (δ) is superfluous.

THEOREM 14. *Suppose that* (α) *and* (β) *hold and that*:

(ε) *the infimum in* (11) *is attained for some* u,

(η) $d(G\, d^c u) = f\, dd^c u$, $G = q \det H$, f *being continuous*.

Then u *satisfies* (14).

R e m a r k 2. Theorem 14 remains valid for $n=1$. In this case the condition (η) is superfluous.

6. Foliation techniques

We move now to the foliation technique, developed in the context of equation (7) by Bedford and Kalka [6]. Using this technique we shall study certain nonlinear partial differential equations involving the complex hessian which can be specified as (13) or (14).

Let $\mathbb{M}$ be a 2n-dimensional C^∞-differentiable manifold and let $T\mathbb{M}$ be its tangent bundle. Let J denote an almost complex structure on $\mathbb{M}$. The spaces $T\mathbb{M}^{1,0}$ and $T\mathbb{M}^{0,1}$ may be defined by the splitting $T\mathbb{M}\otimes_{\mathbb{R}}\mathbb{C}\simeq T\mathbb{M}^{1,0}\oplus T\mathbb{M}^{0,1}$, where $\alpha=\frac{1}{2}(\alpha-iJ\alpha)\oplus\frac{1}{2}(\alpha+iJ\alpha)$. Hence also $T^*\mathbb{M}\otimes_{\mathbb{R}}\mathbb{C}\simeq T^*\mathbb{M}^{1,0}\oplus T^*\mathbb{M}^{0,1}$. Under this splitting, $d=\partial+\bar\partial$. We shall use the notation $\partial_j=\partial/\partial z_j$ and $\bar\partial_j=\partial/\partial\bar z_j$. We denote by

$$\Lambda^{p,q}\mathbb{M}=\Lambda^p(T^*\mathbb{M}^{1,0})\wedge\Lambda^q(T^*\mathbb{M}^{0,1})$$

the spaces of forms of type (p,q) on $\mathbb{M}$, and the spaces of k-forms on $\mathbb{M}$ are given by

$$\Lambda^k\mathbb{M}=\bigoplus_{p+q=k}\Lambda^{p,q}\mathbb{M}.$$

We also extend J^* (the adjoint of J) to

$$\Lambda\mathbb{M}=\bigoplus_k\Lambda^k\mathbb{M}$$

by the rule $J^*f=f$ if f is a 0-form and, in general, by $J^*(\xi\wedge\eta)=J^*\xi\wedge J^*\eta$. If $X\in T\mathbb{M}\otimes_{\mathbb{R}}\mathbb{C}$ is any tangent vector, then $\rfloor X:\Lambda^k\to\Lambda^{k-1}$ is the <u>contraction</u> by X defined by

$$(\omega\rfloor X)(Y_1,\dots,Y_{k-1})=\omega(X,Y_1,\dots,Y_{k-1}),$$

where $\omega\in\Lambda^k\mathbb{M}$ and $\rfloor$ stands for the inner product. If $F\in\Lambda\mathbb{M}$ is an ideal, then

$$\operatorname{Ann}F=\{X\in T\mathbb{M}:\ \omega\rfloor X\in F\ \text{ for all }\ \omega\in F\}.$$

The following lemma can easily be established [6]:

<u>LEMMA 4</u>. <u>Let</u> $F=(\omega_1,\dots,\omega_k)$ <u>be the ideal of</u> $\Lambda\mathbb{M}$ <u>generated by</u> q-<u>forms</u> $\omega_1,\dots,\omega_k$. <u>If</u> $dF\subset F$, <u>then</u> $\operatorname{Ann}F$ <u>is involutive</u>, <u>i.e.</u> <u>for</u> X, $Y\in\operatorname{Ann}F$ <u>we have</u> $[X,Y]\in\operatorname{Ann}F$.

We also have [6]:

<u>LEMMA 5</u>. <u>If</u> $F=(\omega_1,\dots,\omega_k)$ <u>is an ideal of</u> ΛD, $D\subset\mathbb{C}^n$, <u>generated by real</u> (1,1)-<u>forms</u>, <u>then</u> $\operatorname{Ann}F$ <u>is</u> J-<u>invariant</u>.

The following lemma [6] is basic for our considerations:

LEMMA 6. *Let ω_1 and ω_2 be real* (1,1)*-forms with $\omega_2 \geq 0$, and let there exist $a_j, b_j \in T^*D$, $D \subset \mathbb{C}^n$, such that $\omega_1 = \Sigma \pm a_j \wedge Ja_j$, $\omega_2 = -\Sigma b_j \wedge Jb_j$. Then $\dim \operatorname{span}\{a_j, Ja_j, b_j, Jb_j\} = 2p$ if and only if $(\omega_1 + i\omega_2)^p \neq 0$ and $(\omega_1 + i\omega_2)^{p+1} = 0$.*

For the proof we refer to [6].

Remark 3. There are situations in which the condition $\omega_2 \geq 0$ is unnecessary. For instance, if ω_1 and ω_2 can be simultaneously diagonalized, then the conclusion remains valid.

Using Lemma 6 we can obtain a generalization of Theorem 5.1 in [6]:

THEOREM 15. *Suppose that $u: D \longrightarrow \mathbb{C}$, D being a bounded domain in $\mathbb{C}^n$, belongs to $C^3(D)$ and $\operatorname{im} u$ is plurisubharmonic in D. Let further ω be a real* (1,1)*-form of the class $C^3(D)$ such that the rank of ω is equal to k, $1 \leq k \leq n-1$, at every point of D and such that the following conditions hold:*

(16) $\quad \omega^k \wedge (dd^c u)^p = 0, \quad 3 \leq k+p \leq n,$

(17) $\quad \omega^k \wedge (dd^c u)^{p-1} \neq 0,$

(18) $\quad d\omega \in \mathcal{F} = \operatorname{ideal}(\omega, dd^c \operatorname{re} u, dd^c \operatorname{im} u).$

Then there exists a foliation (cf. e.g. [25]) *$\mathcal{L}_{p+k-1}$ of D by complex manifolds of codimension $p+k-1$ with the property that for every leaf $\mathbb{M} \in \mathcal{L}_{p+k-1}$ the functions $\operatorname{im} u|\mathbb{M}$ and $\operatorname{re} u|\mathbb{M}$ are pluriharmonic, but $\partial(\operatorname{im} u)/\partial z_j|\mathbb{M}$ and $\partial(\operatorname{re} u)/\partial z_j|\mathbb{M}$ are holomorphic on $\mathbb{M}$ for each j, $1 \leq j \leq n$.*

Remark 4. If the function u in Theorem 15 is, in addition, continuous on $\operatorname{cl} D$, then the functions $\operatorname{re} u$, $|\partial(\operatorname{re} u)/\partial z_j|$, and $|\partial(\operatorname{im} u)/\partial z_j|$ satisfy the *weak maximum principle* in $\operatorname{cl} D$, i.e. the maximum of $\operatorname{re} u$ on $\operatorname{cl} D$ is equal to the maximum of $\operatorname{re} u$ on ∂D etc.

Proof of Theorem 15. The ideal $\mathcal{F}$ is d-closed and invariant under J. It follows from (17) that $\operatorname{Ann} \mathcal{F}$ has the complex codimension at least $k+p-1$. To show that the codimension of $\operatorname{Ann} \mathcal{F}$ is $k+p-1$, we select forms $c_{k'}, a_{k'}, b_{k'} \in T^*D$ such that

(19) $$\omega = \sum_{j=1}^{k} \pm c_j \wedge Jc_j, \quad dd^c \operatorname{re} u = \sum \pm a_{k'} \wedge Ja_{k'}, \quad dd^c \operatorname{im} u = \sum - b_{k'} \wedge Jb_{k'}.$$

We claim that the real dimension of the span of $\{c_j, Jc_j, a_j, Ja_j, b_j, Jb_j\} = V^*$ is at most $2(p+k-1)$. Let us choose a J-invariant complementary subspace of $V_0^* = \operatorname{span}\{c_j, Jc_j\}$, i.e. $V^* = V_1^* \oplus V_0^*$. Thus we have

$$a_{k'} = a'_{k'} + c'_{k'}, \quad b_{k'} = b'_{k'} + c''_{k'}, \quad \text{where } a'_{k'}, b'_{k'} \in V_1^* \text{ and } c'_{k'}, c''_{k'} \in V_0^*.$$

By the definition of V^*, $\operatorname{span}\{a_{k'}, Ja_{k'}, b_{k'}, Jb_{k'}\} = V^*$. Therefore

$$dd^c \operatorname{re} u = \sum \pm a_{k'} \wedge Ja_{k'} + \sum_{j=1}^{k} (c_j \wedge s_j^{(1)} + Jc_j \wedge s_j^{(2)}),$$

$$dd^c \operatorname{im} u = \sum - b_{k'} \wedge Jb_{k'} + \sum_{j=1}^{k} (c_j \wedge t_j^{(1)} + Jc_j \wedge t_j^{(2)}).$$

Now, if $\dim_{\mathbb{R}} V^* \geq 2(p+k)$, then $\dim_{\mathbb{R}} V_1^* \geq 2p$. Thus $(\Sigma \pm a_{k'} \wedge Ja_{k'} - i\, \Sigma b_{k'} \wedge Jb_{k'})^p \neq 0$ since $\{a_{k'}, Ja_{k'}, b_{k'}, Jb_{k'}\}$ span V_1^*. But this implies that

$$\omega^k \wedge (dd^c u)^p = \Big(\sum_{j=1}^{k} \pm c_j \wedge Jc_j\Big)^k \wedge \Big[\sum \pm a_{k'} \wedge Ja_{k'} - i \sum b_{k'} \wedge Jb_{k'}$$

$$+ \sum (c_j \wedge s_j^{(1)} + Jc_j \wedge s_j^{(2)} + ic_j \wedge t_j^{(1)} + iJc_j \wedge t_j^{(2)})\Big]^p$$

$$= \pm k! \bigwedge_{j=1}^{k} c_j \wedge Jc_j \wedge \Big(\sum \pm a_{k'} \wedge Ja_{k'} - i \sum b_{k'} \wedge Jb_{k'}\Big)^p$$

$$\neq 0,$$

which contradicts (16). Since $\operatorname{Ann} F$ has a constant dimension, it is integrable and this gives the foliation L_{p+k-1}.

If we let $\iota: \mathbb{M} \to D$ denote the inclusion mapping, then $\iota^* u = u|\mathbb{M}$ and, consequently,

$$\partial_{\mathbb{M}} \bar\partial_{\mathbb{M}} (u|\mathbb{M}) = \partial_{\mathbb{M}} \bar\partial_{\mathbb{M}} (\operatorname{re} u|\mathbb{M}) + i \partial_{\mathbb{M}} \bar\partial_{\mathbb{M}} (\operatorname{im} u|\mathbb{M})$$

$$= \iota^* \partial\bar\partial \operatorname{re} u + i(\iota^* \partial\bar\partial \operatorname{im} u) = 0$$

since $T\mathbb{M} \subset \operatorname{Ann}(dd^c \operatorname{re} u, dd^c \operatorname{im} u)$. Finally we have to show that $(\operatorname{im} u)_{|j} |\mathbb{M}$ and $(\operatorname{re} u)_{|j} |\mathbb{M}$ are holomorphic. Let

$$X = \sum_{j=1}^{n} C^j \partial_j \in T\mathbb{M}^{1,0}$$

be any vector field. Since $T\mathbb{M}$, $JT\mathbb{M} \subset \operatorname{Ann}(dd^c \operatorname{re} u, dd^c \operatorname{im} u)$, it follows that

$$(dd^c \operatorname{re} u) \lrcorner X = (dd^c \operatorname{im} u) \lrcorner X = \sum_{j=1}^{n} C^j u_{|j\bar k} = \sum_{j=1}^{n} C^j \operatorname{im} u_{|j\bar k} = 0,$$

and this proves that $(\operatorname{re} u)_{|j} |\mathbb{M}$ and $(\operatorname{im} u)_{|j} |\mathbb{M}$ are holomorphic indeed.

E x a m p l e 1. Suppose that F and u are real-valued C^3-smooth functions on $D \subset \mathbb{C}^n$ such that u is plurisubharmonic on D. Let further $\operatorname{rank}(dd^c(Fu)) = 1$ and the following conditions hold:

$$dd^c(Fu) \wedge (dd^c u)^{n-1} = 0, \quad dd^c(Fu) \wedge (dd^c u)^{n-2} \neq 0.$$

Owing to Theorem 15 we have the foliations of D by complex manifolds of codimension $n-1$.

<u>Example 2</u>. Let u be a real-valued C^3-smooth plurisubharmonic function on D such that the following conditions hold:

$$(dd^c u)^{p+1} = 0, \quad (dd^c u)^p \neq 0.$$

Owing to Theorem 15 we have the foliations of D by complex manifolds of codimension p [6].

<u>Example 3</u>. Let u be as in Theorem 15, satisfying the conditions

$$\partial\bar{u} \wedge \bar{\partial}u \wedge (\partial\bar{\partial}u)^p = 0, \quad \partial\bar{u} \wedge \bar{\partial}u \wedge (\partial\bar{\partial}u)^{p-1} \neq 0.$$

A simple calculation shows that the assumptions of Theorem 15 hold with $\omega = i\,\partial\bar{u} \wedge \bar{\partial}u$ and $k = 1$. This example was investigated in detail in [6].

THEOREM 16. *Let D be a bounded domain in $\mathbb{C}^n$ and u a continuous function on* cl D *satisfying the additional assumptions of Theorem* 15. *Then the functions* re u, $|(\mathrm{re}\, u)_{|j}|$, *and* $|(\mathrm{im}\, u)_{|j}|$ *satisfy the weak maximum principle in* D.

<u>Proof</u>. We are going to prove this theorem indirectly. Assume that, for instance, the function re u does not satisfy the maximum principle, i.e. there exists a point $z_0 \in D$ such that

$$\max_{z \in \mathrm{cl}\, D} \mathrm{re}\, u(z) = \mathrm{re}\, u(z_0), \quad \max_{z \in \partial D} \mathrm{re}\, u(z) < \mathrm{re}\, u(z_0).$$

Further, let $\mathbb{M} \in L_{p+k+1}$ be the leaf of the foliation given by Theorem 15, passing through the point z_0. By Theorem 15 the function $\mathrm{re}\, u|\mathbb{M}$ is pluriharmonic, so it satisfies the maximum principle on $\mathbb{M}$ and, by connectedness of $\mathbb{M}$, the function $\mathrm{re}\, u|\mathbb{M}$ would be identically equal to $\mathrm{re}\, u(z_0)$. Since further $\mathrm{cl}\,\mathbb{M}$ is not a compact subset of D [5], we can find a sequence of $z_n \in \mathbb{M}$ which is convergent to some $\tilde{z} \in \partial D$. By the continuity of $\mathrm{re}\, u|\mathbb{M}$ we would have then $\mathrm{re}\, u(\tilde{z}) = \mathrm{re}\, u(z_0)$ which contradicts our assumption. Similarly we prove the remaining part of our theorem.

The last two theorems are due to J. Kalina.

<u>References</u>

[1] ALEKSANDROV, A. D.: The Dirichlet problem for the equation $\mathrm{Det}\,\|z_{i,j}\| = \psi(z_1,\ldots,z_n, x_1,\ldots,x_n)$, I [in Russian], Vestnik Leningrad. Univ. <u>13</u> (1958), no. 1, 5-24.

[2] ANDREOTTI, A. and J. ŁAWRYNOWICZ: On the generalized complex Monge-Ampère equation on complex manifolds and related questions,

Bull. Acad. Polon. Sci. Sér. Sci. Math. Astronom. Phys. 25 (1977), 943-948.

[3] —— and ——: The generalized complex Monge-Ampère equation and a variational capacity problem, ibid. 25 (1977), 949-955.

[4] BARKER, W.: Kernel functions on domains with hyperelliptic double, Trans. Amer. Math. Soc. 231 (1977), 339-347.

[5] BEDFORD, E. and J. E. FORNAESS: Counterexamples to regularity for the complex Monge-Ampère equation, Invent. Math. 50 (1979), 129-134.

[6] —— and M. KALKA: Foliations and complex Monge-Ampère equations, Comm. Pure Appl. Math. 30 (1977), 543-571.

[7] —— and B. A. TAYLOR: The Dirichlet problem for a complex Monge-Ampère equation, Inventiones Math. 37 (1976), 1-44.

[8] —— and ——: Variational properties of the complex Monge-Ampère equation. I. Dirichlet principle, Duke Math. J. 45 (1978), 375-403.

[9] —— and ——: Variational properties of the complex Monge-Ampère equation. II. Intrinsic norms, Amer. J. Math., to appear.

[10] BELL, S.: Non-vanishing of the Bergman kernel function at boundary points of certain domains in $\mathbb{C}^n$, Ms.

[11] BERGMAN, S.: The kernel function and conformal mapping, 2. ed. (Math. Surveys 5), Amer. Math. Soc., Providence, R.I., 1970.

[12] BOUTET DE MONVEL, L. et S. SJÖSTRAND: Sur la singularité des noyaux de Bergman et de Szegö, Asterisque 34/35 (1976), 123-164.

[13] BREMERMANN, H.: Holomorphic continuation of the kernel function and the Bergman metric in several complex variables, Lectures on functions of a complex variable, Univ. of Michigan Press, Ann Arbor, Mich., 1955, pp. 349-383.

[14] CARAMAN, P.: p-capacity in infinite dimensional spaces, this volume, pp. 69-109.

[15] CHENG, S.-Y. and S.-T. YAU: On the regularity of the Monge-Ampère equation $\det(\partial^2 u/\partial x_i \partial x_j) = F(x,u)$, Comm. Pure Appl. Math. 30 (1977), 41-68.

[16] CHERN, S. S., H. I. LEVINE and L. NIRENBERG: Intrinsic norms on a complex manifold, Global analysis, Papers in honor of K. Kodaira, ed. by D. C. Spencer and S. Iynaga, Univ. of Tokyo Press and Princeton Univ. Press, Tokyo 1969, pp. 119-139; reprinted in: S.-s. CHERN, Selected papers, Springer-Verlag, New York - Heidelberg - Berlin 1978, pp. 371-391.

[17] CHRISTOFFERS, H: Princeton University Thesis, Princeton, N. J., 1976.

[18] DEBIARD, A. et B. GAVEAU: Méthodes de contrôle optimal en analyse complexe. IV. Applications aux algèbres de fonctions analytiques, this volume, pp. 111-142.

[19] DIEDERICH, K.: Über die 1. und 2. Ableitungen der Bergmanschen Kernfunktion und ihr Randverhalten, Math. Ann. 203 (1963), 129-170.

[20] FEFFERMAN, C.: The Bergman kernel and biholomorphic mappings of pseudoconvex domains, Invent. Math. 26 (1974), 1-65.

[21] ——: Monge-Ampère equations, the Bergman kernel, and geometry of pseudoconvex domains, Ann. of Math. 103 (1976), 395-416, and 104 (1976), 393-394.

[22] FOLLAND, G. B. and J. J. KOHN: The Neumann problem for the Cauchy-Riemann complex, Ann. of Math. Studies no. 75, Princeton, N. J., 1972.

[23] GAFFNEY, M.: Hilbert space methods in the theory of harmonic integrals, Trans. Amer. Math. Soc. 78 (1955), 426-444.

[24] GHIŞA, D.: The modulus and the hyperbolic measure, this volume, pp. 169-172.

[25] HOLMANN, H.: On the stability of holomorphic foliations, this volume, pp. 201-211.

[26] HÖRMANDER, L.: L^2-estimates and existence theorems for the $\bar{\partial}$-operator, Acta Math. 113 (1965), 89-152.

[27] HUA Lo-keng: Harmonic analysis of several complex variables in the classical domains [in Chinese], Science Press, Peking 1958; English transl.: Transl. Math. Monographs 6, Amer. Math. Soc., Providence, R. I., 1963.

[28] KALINA, J.: A variational characterization of condenser capacities in $\mathbb{R}^n$ within a class of plurisubharmonic functions, Ann. Polon. Math., to appear.

[29] ——: Biholomorphic invariance of the capacity and the capacity of an annulus, Ann. Polon. Math., to appear.

[30] KERZMAN, N.: The Bergman kernel function. Differentiability at the boundary. Math. Ann. 195 (1972), 149-158.

[31] ŁAWRYNOWICZ, J.: Condenser capacities and an extension of Schwarz's lemma for hermitian manifolds, Bull. Acad. Polon. Sci. Sér. Sci. Math. Astronom. Phys. 23 (1975), 839-844.

[32] ——: Electromagnetic field and the theory of conformal and biholomorphic invariants, Complex analysis and its applications III, International Atomic Energy Agency, Wien 1976, pp. 1-23.

[33] ——: On a class of capacities on complex manifolds endowed with an hermitian structure and their relation to elliptic and hyperbolic quasiconformal mappings, (a) Ann. Polon. Math. 33 (1976), 178 (abstract), (b) Dissertationes Math. 166 (1978), 44 pp. (in extenso)

[34] ——: On biholomorphic continuability of regular quasiconformal mappings, this volume, pp. 341-364.

[35] LIGOCKA, E.: Some remarks on extension of biholomorphic mappings, this volume, pp. 365-378.

[36] ——: How to prove Fefferman's theorem without use of differential geometry, Ann. Polon. Math., to appear.

[37] POGORELOV, A. V.: The Dirichlet problem for the multidimensional analogue of the Monge-Ampère equation [in Russian], Dokl. Akad. Nauk SSSR 201 (1971), 790-793; English transl.: Soviet Math. Dokl. 12 (1971), 1727-1731.

[38] RAMADANOV, I.: Sur une propriété de la fonction de Bergman, C. R. Acad. Bulg. Sci. 20 (1967), 759-762.

[39] SKWARCZYŃSKI, M.: The Bergman function and semiconformal mappings, Bull. Acad. Polon. Sci. Sér. Sci. Math. Astronom. Phys. 22 (1974), 667-673.

[40] ——: A remark on holomorphic isometries with respect to the induced Bergman metrics, this volume, pp. 425-427.

[41] ——: The ideal boundary of a domain in $\mathbb{C}^n$, Ann. Polon. Math., to appear.

[42] WEBSTER, S.: Biholomorphic mappings and the Bergman kernel off the diagonal, Ms.

[43] ZINOV'EV, B. S.: Reproducing kernels for multicircular domains of holomorphy [in Russian], Sibirsk. Mat. Ž. 15 (1974), 35-48 and 236.

J. Kalina, J. Ławrynowicz, E. Ligocka, and M. Skwarczyński

Institute of Mathematics of the
Polish Academy of Sciences, Łódź Branch
Kilińskiego 86, PL-90-012 Łódź, Poland
(J. Kalina and J. Ławrynowicz)

Osiedle Przyjaźń 15 m. 4
PL-00-905 Warszawa Jelonki, Poland
(E. Ligocka)

Institute of Mathematics
University of Warsaw
Pałac Kultury i Nauki, IX p.
PL-00-901 Warszawa, Poland
(M. Skwarczyński)

A VARIATION OF THE MODULUS OF SUBMANIFOLD FAMILIES

Jerzy Kalina and Antoni Pierzchalski (Łódź)

Contents

Introduction

The paper is devoted to the study of extremal properties of the k-moduli of some special hypersurface families on Riemannian manifolds. We examine critical points of the functional

$$\varphi \longmapsto \operatorname{mod}_k M^{\varphi}$$

where φ is a smooth function with nonvanishing differential on a Riemannian manifold and M^{φ} is the family of level hypersurfaces of φ. We show that φ_0 is a critical point if and only if some $(n-1)$-form is closed. The result is obtained by construction of a local variation of φ_0.

1. Preliminaries

Let M,N be oriented Riemannian manifolds of dimension m and n, respectively, and $\varphi : M \longrightarrow N$ a submersion of M onto N. (All manifolds and mappings are assumed to be of the class C^{∞}.) Then for every point $q \in N$ the set $M_q = \varphi^{-1}(q)$ is a closed submanifold of M. The family of all such submanifolds M_q, $q \in N$, is denoted by M^{φ}. It is interesting to investigate moduli of the family and their extremal properties. By the k-modulus of M^{φ} we mean the positive number (or infinity)

$$\text{(0)} \quad \operatorname{mod}_k M^{\varphi} = \inf \int_M \varrho^k dM, \qquad k > 1,$$

where the infimum is taken over all nonnegative Borel functions ϱ on M such that

$$\int_{M_q} \varrho \, dM_q \geq 1 \qquad \text{for} \quad q \in N.$$

This definition is a particular case of the definition of the k-modulus for arbitrary submanifold families (cf. [3]) which is analogous to the Fuglede's definition of the modulus of a family of hypersurfaces in $\mathbb{R}^n$ [1]. In the case of one-dimensional manifolds it reduces to the modulus of a curve family, e.g. in the classical sense of Ahlfors and Beurling. The basic properties of the k-modulus can be found e.g. in [1]. It is shown [3] that

$$\text{(1)} \qquad \operatorname{mod}_k M^{\varphi} = \int_N \Big(\int_{M_q} (I_{\varphi}^{\perp})^{\frac{1}{k-1}} dM_q \Big)^{1-k} dN,$$

where $I_{\varphi}^{\perp}$ is the normal Jacobian of φ, which in the case where N is an open subset of $\mathbb{R}$ reduces to the usual gradient i.e. $I_{\varphi}^{\perp} = |\operatorname{grad} \varphi|$. From now on we assume that $\operatorname{mod}_k M^{\varphi}$ is finite.

2. A variation of the k-modulus

Let $\Phi = \{\Phi_t\}_{-t_o<t<t_o}$ be a family of transformations Φ_t of M such that:

(i) $\Phi_o = \mathrm{id}_M$,

(ii) there exists a compact set $K \subset M$ such that Φ_t is identity outside K,

(iii) the function $(t,p) \longmapsto \Phi_t(p)$ is a smooth function on $(-t_o, t_o) \times M$.

Putting $\varphi_t = \varphi \circ \Phi_t^{-1}$ we obtain a family of submersions of M onto N which we shall call <u>one-parameter deformations</u> of φ. For every t, $-t_o < t < t_o$ the family M^{φ_t} is a family of submanifolds $M_q^t = \varphi_t^{-1}(q)$, $q \in N$ and, consequently, we can assiociate with it its k-modulus.

By Φ-<u>variation</u> of the k-modulus $\operatorname{mod}_k M^{\varphi}$ we mean the derrivative (if it exists)

$$\text{(2)} \qquad \delta_{\Phi} \operatorname{mod}_k M^{\varphi} = \frac{d}{dt}\Big|_{t=0} \operatorname{mod}_k M^{\varphi_t} = \lim_{t \to 0} \frac{1}{t}(\operatorname{mod}_k M^{\varphi_t} - \operatorname{mod}_k M^{\varphi}).$$

We say that φ is a <u>critical point</u> of the functional

$$\varphi \longmapsto \mathrm{mod}_k\, M^{\varphi}$$

if $\delta_{\Phi}\mathrm{mod}_k\, M^{\varphi} = 0$ for every one parameter deformation $\varphi\circ\Phi_t^{-1}$ of φ.

We are going to construct a family of transformations $\Phi = \{\Phi_t\}$, $-t_o < t < t_o$ satisfying (i) - (iii).

Let $p \in M$ and (U,μ) be a coordinate system around p and $f = (f^1,\ldots,f^m)$ be any smooth vector field on $\mathbb{R}^n$ with compact support contained in $\mu(U) \subset \mathbb{R}^m$. Put

$$(3) \qquad \Phi_t = \begin{cases} \mu^{-1} \circ \varrho_t \circ \mu & \text{on } U, \\ \mathrm{id} & \text{on } M \setminus U. \end{cases}$$

where $\varrho_t : \mu(U) \longrightarrow \mu(U)$ is of the form

$$\varrho_t = \mathrm{id} + tf\,, \qquad -t_o < t < t_o .$$

It is easy to see that ϱ_t is a transformation of $\mu(U)$ for sufficiently small t_o, and, consequently Φ_t, $-t_o < t < t_o$ is a transformation of M with prescribed properties.

Now we are going to compute the variation (2) for the one-parameter deformation generated by (3) in the case $N = I$, where I is an open interval:

$$(4) \qquad \delta_{\Phi}\mathrm{mod}_k\, M^{\varphi} = \lim_{t\to 0} \frac{1}{t}\,(\mathrm{mod}_k\, M^{\varphi_t} - \mathrm{mod}_k\, M^{\varphi}) =$$

$$= \lim_{t\to 0} \frac{1}{t}\,\Big(\int_I \Big(\int_{M_q^t} |\mathrm{grad}\,\varphi_t|^{\frac{1}{k-1}}\, dM_q^t\Big)^{1-k} dI$$

$$- \int_I \Big(\int_{M_q} |\mathrm{grad}\,\varphi|^{\frac{1}{k-1}}\Big)^{1-k} dI\Big)$$

$$= (1-k) \lim_{t\to 0} \frac{1}{t} \int_{\varphi(U)} \Big(\frac{R_q(t)}{S_q^k} + o(t)\Big)\, dI,$$

where

$$(5) \qquad R_q(t) = \int_{M_q^t\cap U} |\mathrm{grad}\,\varphi_t|^{\frac{1}{k-1}}\, dM_q^t - \int_{M_q\cap U} |\mathrm{grad}\,\varphi|^{\frac{1}{k-1}}\, dM_q$$

and

$$(6) \qquad S_q = \int_{M_q} |\operatorname{grad}\varphi|^{\frac{1}{k-1}}\, dM_q .$$

It is evident that

$$(7) \qquad \int_{M_q^t \cap U} |\operatorname{grad}\varphi_t|^{\frac{1}{k-1}}\, dM_q^t = \int_{M_q \cap U} |\operatorname{grad}\varphi_t|^{\frac{1}{k-1}} \circ \Phi_t\, \mathfrak{J}_t\, dM_q$$

where $\mathfrak{J}_t$ is the Jacobian of the restriction of Φ_t to the submanifold M_q.

Let (U,μ) be a coordinate system on M. By the rank theorem (see e.g. [2]) we may assume that $\mu(U)$ is a cube $Q = \{x_1,\dots,x_m) \in \mathbb{R}^m : |x_i| < 1,\ i = 1,\dots,m\}$ and $\varphi \circ \mu^{-1} = pr_m$. We may additionally assume without lose of generality that $\partial/\partial x_m$ has the direction of the gradient $\operatorname{grad}\varphi$, ie. $\operatorname{grad}\varphi = h(\partial/\partial x_m)$ for some non-vanishing function h. We also assume that the function

$$p \longmapsto S_{\varphi(p)} = \int_{M_{\varphi(p)}} |\operatorname{grad}\varphi|^{\frac{1}{k-1}}\, dM_{\varphi(p)}, \qquad p \in M$$

is a smooth function on M. This assumption is often fulfilled for example in the case when φ has compact level hypersurfaces or in the case when M is an open subset of $\mathbb{R}^n$ with sufficiently regular boundary. Finally we assume that the Greek indices α, β run over $1,\dots,m$ and Latin indices i,j over $1,\dots,m-1$.

By the above assumption we have

$$(8) \qquad |\operatorname{grad}\varphi_t|^{\frac{1}{k-1}} \circ \Phi_t = |\operatorname{grad}\varphi|^{\frac{1}{k-1}} \Big(1 - \frac{t}{k-1}\frac{\partial f^m}{\partial x_m} + \frac{t}{2(k-1)} f^\alpha \frac{\partial}{\partial x_\alpha} \log g^{mm}\Big) + o(t)$$

and

$$(9) \qquad \mathfrak{J}_t = 1 + t\frac{\partial f^i}{\partial x_j}\delta_i^j + \frac{t}{2} f^\alpha \frac{\partial}{\partial x_\alpha} \log\det[g_{ij}] + o(t)$$

where $g_{\alpha\beta} = g(\frac{\partial}{\partial x_\alpha}, \frac{\partial}{\partial x_\beta})$, $g_{\alpha\beta}\, g^{\beta\gamma} = \delta_\alpha^\gamma$ and the summation convention is used.

Consequently, by (5) and (7) we have

$$R_q(t) = \int_{M_q\cap U} |\operatorname{grad}\varphi|^{\frac{1}{k-1}}\left(1 - \frac{t}{k-1}\frac{\partial f^m}{\partial x_m} + \frac{t}{2(k-1)} f^\alpha \frac{\partial}{\partial x_\alpha}\log g^{mm}\right)$$

$$\times\left(1 + t\frac{\partial f^i}{\partial x_j}\delta^j_i + \frac{t}{2} f^\alpha \frac{\partial}{\partial x_\alpha}\log\det[g_{ij}]\right)dM_q$$

$$- \int_{M_q\cap U} |\operatorname{grad}\varphi|^{\frac{1}{k-1}}\, dM_q + o(t)$$

$$= t\int_{\mu(M_q\cap U)} |\operatorname{grad}\varphi|^{\frac{1}{k-1}}\left(\frac{\partial f^i}{\partial x_j}\delta^j_i - \frac{1}{k-1}\frac{\partial f^m}{\partial x_m}\right.$$

$$\left.+ f^\alpha\frac{\partial}{\partial x_\alpha}\log|\operatorname{grad}\varphi|^{\frac{1}{k-1}}\det[g_{ij}]\right){\mu^{-1}}^*(dM_q) + o(t).$$

Therefore our variation (4) takes the form

$$\delta_\Phi \operatorname{mod}_k M^\varphi = (1-k)\int_{-1}^{1}\left(\int_{\mu(M_q\cap U)} \frac{A^k\circ\mu^{-1}}{|\operatorname{grad}\varphi|}\left(\frac{\partial f^i}{\partial x_j}\delta^j_i - \frac{1}{k-1}\frac{\partial f^m}{\partial x_m}\right.\right.$$

$$\left.\left.+ f^\alpha\frac{\partial}{\partial x_\alpha}\log|\operatorname{grad}\varphi|^{\frac{1}{k-1}}\sqrt{\det[g_{ij}]}\right){\mu^{-1}}^*(dM_q)\right)dq$$

where

$$A(p) = |\operatorname{grad}\varphi|_p^{\frac{1}{k-1}} \Big/ S_{\varphi(p)}, \qquad p\in M,$$

is, by the above assumption, a smooth function while $S_{\varphi(p)}$ is defined by (6).

3. Critical points

By the results of the previous section we have the following explicit form for our variation

$$\delta_\Phi \operatorname{mod}_k M^\varphi = (1-k)\int_Q \left(\frac{\partial f^i}{\partial x_j}\delta^j_i - \frac{1}{k-1}\frac{\partial f^m}{\partial x_m} + \right.$$

$$+ f^{\alpha}\frac{\partial}{\partial x_{\alpha}} \log|\operatorname{grad}\varphi|^{\frac{1}{k-1}} \sqrt{\det[g_{ij}]}\,)A^k \circ \mu^{-1}\sqrt{\det[g_{\alpha\beta}]}\,dx_1\ldots dx_m .$$

By Stokes' theorem we obtain

$$\delta_{\Phi}\operatorname{mod}_k M^{\varphi} = (1-k)\int_Q [-f^i \frac{\partial}{\partial x_i}(A^k \circ \mu^{-1}\sqrt{\det[g_{\alpha\beta}]}\,)$$

$$+ \frac{1}{k-1} f^m \frac{\partial}{\partial x_m}(A^k \circ \mu^{-1}\sqrt{\det[g_{\alpha\beta}]}\,)$$

$$+ A^k \circ \mu^{-1}\sqrt{\det[g_{\alpha\beta}]}\, f^{\alpha}\frac{\partial}{\partial x_{\alpha}} \log|\operatorname{grad}\varphi|^{\frac{1}{k-1}}\sqrt{\det[g_{ij}]}\,]dx_1\ldots dx_m$$

$$= -\int_Q f^m A^k \circ \mu^{-1}\sqrt{\det[g_{\alpha\beta}]}\,(\frac{\partial}{\partial x_m}\log(A^k \circ \mu^{-1}\sqrt{(\det[g_{ij}])^k}\,))dx_1\ldots dx_m .$$

Observe that the variation depends only on f^m and is independent of $f^1,\ldots,f^{m-1}$. It is not surprising because our family of hypersurfaces is invariant under one-parameter deformations generated by vector fields tangent to it. Consequently, $\delta_{\Phi}\operatorname{mod}_k M^{\varphi} = 0$ if and only if

$$\frac{\partial}{\partial x_m}\log(A \circ \mu^{-1}\sqrt{\det[g_{ij}]}\,) = 0$$

or

$$\frac{\partial}{\partial x_m}(A \circ \mu^{-1}\sqrt{\det[g_{ij}]}\,) = 0 . \tag{10}$$

Taking into account a covering of M by a family of coordinate systems with the properties such as those of (U,μ) used above (Section 2) considering in every such coordinate system the form

$$\sqrt{\det[g_{ij}]}\,dx_1 \wedge \ldots \wedge dx_{m-1} \tag{11}$$

we see that every two such forms coincide on the intersection of their domains of definition. Thus (11) defines on M a global $(n-1)$-form ω. Consequently we can write (10) as follows:

$$d(A\omega) = 0. \tag{12}$$

It is easy to notice that the restriction of ω to every level hypersurfaces of φ is the volume element on this hypersurface generated by the Riemannian structure on M.

If now φ_t is an arbitrary one-parameter deformation satisfying (i) - (iii) then one can analogously conclude that the vanishing of its variation is equivalent to (12). We have then the following result:

THEOREM. *A submersion* $\varphi_0 : M \longrightarrow \mathbb{R}$ *is a critical point of the functional*

$$(12) \qquad \varphi \longmapsto \operatorname{mod}_k M^{\varphi}$$

if and only if

$$d(A\omega) = 0$$

where

$$A(p) = \frac{|\operatorname{grad}\varphi_0|^{\frac{1}{k-1}}(p)}{\int_{M(p)} |\operatorname{grad}\varphi_0|^{\frac{1}{k-1}} dM_{\varphi_0(p)}} , \qquad p \in M ,$$

and ω *is the* (n-1)-*form such that its restriction to every level hypersurfaces is its volume form generated by Riemannian structure on* M.

Remark. One can see that the function A realizes the infimum in (0) i.e. it is an extremal function for the family M^{φ_0}.

Further results concerning critical points of the functional (12) will be published in subsequent papers.

References

[1] FUGLEDE, B.: Extremal length and functional completion, Acta Math. 98 (1957), 171-219.

[2] NARASIMHAN, R.: Analysis on real and complex manifolds, North-Holland-Publishing-Company, Amsterdam, 1968.

[3] PIERZCHALSKI, A.: On a module formula on Riemannian manifolds, Ann. Polon. Math (to appear).

Institute of Mathematics of the
Polish Academy of Sciences, Łódź Branch
Kilińskiego 86, PL-90-012 Łódź, Poland

OPERATORS OF FRACTIONAL INTEGRATION

Shyam Lal Kalla (Maracaibo)

Contents

1. INTRODUCTION

Operators of fractional integration play an important role in the theory of integral equations and in problems of mathematical physics. A systematic use of operators of fractional integration enables us to see more clearly the basic structure of any method of solving integral equations and to appreciate more easily the connections which exist among the various methods. Fractional integration is an immediate generalization of repeated integration.

We consider complex-valued functions on the real half line $(0,\infty)$ defined almost everywhere (or everywhere) in $(0,\infty)$. It is often convenient to regard such functions as being defined also in $(-\infty, 0)$, with the value 0 there for formal purposes. Locally integrable functions will mean functions which are locally integrable in $[0,\infty)$; i.e. Lebesgue integrable in $[0,\ell]$, for every ℓ such that $0 < \ell < \infty$.

If $f(x)$ is a locally integrable function, we define the first integral $F_1(x)$ of $f(x)$ by the formula

$$F_1(x) = \int_0^x f(t)\, dt \tag{1}$$

and the subsequent integrals by the recurrence relation

$$F_{n+1}(x) = \int_0^x F_n(t)\, dt \quad ; \quad n = 1, 2, 3, \ldots \tag{2}$$

It is easily proved by induction that for any positive integer n

$$F_{n+1}(x) = \frac{1}{n!} \int_0^x (x-t)^n f(t)\, dt \tag{3}$$

The integral

$$I^\mu f = R_\mu [f(t); x] = \frac{1}{\Gamma(\mu)} \int_0^x (x-t)^{\mu-1} f(t)\, dt \tag{4}$$

is called the Riemann-Liouville fractional integral of order μ. The integral (4) is convergent for a wide class of functions $f(t)$ if $Re(\mu) > 0$. The upper limit x may be real or complex; in the latter case the path of integration is the straight segment $t = xs$, $0 \leqslant s \leqslant 1$. In the case when $\mu = n + 1$, a positive integer, then (4) reduces to (3).

The integral

$$K^\mu f = W_\mu [f(t); x] = \frac{1}{\Gamma(\mu)} \int_x^\infty (t-x)^{\mu-1} f(t)\, dt \tag{5}$$

$$Re(\mu) > 0$$

is called the Weyl fractional integral of order μ. In general x and μ are complex, the path of integration being one of the rays $t = xs$ or $t = x + s$, $s > 0$. We adopt the convention: $I^0 = I$; $K^0 = I$, where I denotes the identity operator.

Existence of $I^\mu f$ (or $K^\mu f$) will mean that $I^\mu f$ exists for almost all x in $(0, \infty)$; so that $I^\mu f$ is a function in the sense defined above. The fundamental existence theorem is the following; it is a corollary of a well known theorem on convolution of integral functions [44].

THEOREM A (Existence) *If* $Re(\mu) > 0$ *and* f *is locally integrable then* $I^{\mu} f$ *exists and is locally integrable.*

THEOREM B (Index law) *If* $Re(a) > 0$, $Re(b) > 0$ *and* f *is locally integrable then* $I^b I^a f = I^{a+b} f$. *That is, the two sides exist and have equal values for almost all positive* x.

Fractional integrals (4) and (5) also occur in the solution of ordinary differential equations [16] where they are called Euler transforms of first and second kind respectively.

Hardy and Littlewood [15] consider the fractional integral

$$f_{\mu}(x) = \int_{-\infty}^{x} f(t)(x-t)^{\mu-1}\, dt \quad , \quad 0 < Re(\mu) < 1 \tag{6}$$

while Love and Young [32] consider the similar fractional integrals in the following form:

$$f_{\mu}^{+}(a,x) = \frac{1}{\Gamma(\mu)} \int_{a}^{x} f(t)\,(x-t)^{\mu-1}\, dt \tag{7}$$

$$f_{\mu}^{-}(x,b) = \frac{1}{\Gamma(\mu)} \int_{x}^{b} f(t)\,(t-x)^{\mu-1}\, dt \quad , \tag{8}$$

$$Re(\mu) > 0 \; , \; a \leqslant x \leqslant b \, .$$

A theory of fractional integrals of functions of several variables has been developed by Riesz [38], which has been used in the solution of partial differential equations [1].

Kober [30] defined fractional integrals of imaginary order by means of Mellin transforms, and discussed various mean convergence properties, an inversion formula and index law. Fractional derivatives of imaginary order are given by Love [31].

The connections between operators of fractional integration (Riemann-Liouville & Weyl) and other integral transforms have been considered by various authors. The relation between fractional integrals and Laplace transform has been considered by Doetsch [5] and Widder [46]. Kober [29] has considered the connections with Mellin transform whereas Erdélyi & Kober [10] have considered the connections with the Hankel transform. Kalla [18,19] has established certain relations between fractional integrals and Varma transform, which generalizes some earlier results [9]. Relations between fractional integrals and Hankel & Meijer transforms are given by Bora & Saxena [2] and Martić [35].

2. ERDÉLYI-KOBER OPERATORS

Operators of fractional integration of a general kind have been considered by Erdélyi [6], Kober [29] and Erdélyi-Kober [10]. Sneddon [42] considers the E-K operators in the following form:

$$I_{\eta,\alpha} f(x) = \frac{2x^{-2\alpha-2\eta}}{\Gamma(\alpha)} \int_0^x (x^2-u^2)^{\alpha-1} u^{2\eta+1} f(u)\, du , \tag{9}$$

$$K_{\eta,\alpha} f(x) = \frac{2x^{2\eta}}{\Gamma(\alpha)} \int_x^\infty (u^2-x^2)^{\alpha-1} u^{-2\alpha-2\eta+1} f(u)\, du , \tag{10}$$

where $Re(\alpha) > 0$, $Re(\eta) > -\frac{1}{2}$.

If we let $\alpha \to 0$ in (9) and (10), we see that

$$I_{\eta,0} = I \quad , \quad K_{\eta,0} = I , \tag{11}$$

where I is the identity operator. We shall mention some basic properties of E-K operators.

(i) $$I_{\eta,\alpha}\, x^{2\beta} f(x) = x^{2\beta} I_{\eta+\beta,\alpha} f(x) \quad ,$$
$$K_{\eta,\alpha}\, x^{2\beta} f(x) = x^{2\beta} K_{\eta-\beta,\alpha} f(x) \quad , \tag{12}$$

(ii) $$I_{\eta,\alpha}\, I_{\eta+\alpha,\beta} = I_{\eta,\alpha+\beta} \quad ,$$
$$K_{\eta,\alpha}\, K_{\eta+\alpha,\beta} = K_{\eta,\alpha+\beta} \quad , \tag{13}$$

The results we have just noted suggest the manner in which we should define the operators $I_{\eta,\alpha}$ and $K_{\eta,\alpha}$ for $\alpha < 0$. From relations (11) and (13), we have formally

$$I_{\eta+\alpha,-\alpha}\, I_{\eta,\alpha} = I \quad , \tag{14}$$

which suggests that, if $\alpha < 0$, we define

$$= I_{\eta,\alpha} f$$

to be the solution of the integral equation

$$I_{\eta+\alpha,-\alpha}\, g = f \ . \tag{15}$$

Similarly, we define

$$h = K_{\eta,\alpha} f \qquad (\alpha < 0) \tag{16}$$

to be the solution of the integral equation

$$K_{\eta+\alpha,-\alpha}\, h = f \ .$$

In 1951, Erdélyi [7] gave another generalization of fractional integrals in the following form:

$$R(\alpha,\beta;m)\; h(x) = \frac{m}{\Gamma(\alpha)} \int_x^\infty t^{-\beta-m\alpha+m-1}\, (t^m - x^m)^{\alpha-1} h(t)\; dt \ , \tag{17}$$

$$S(\alpha,\beta;m)h(x) = \frac{m}{\Gamma(\alpha)} x^{-\beta-m\alpha+m-1} \int_0^x (x^m - t^m)^{\alpha-1} t^\beta h(t)\, dt \quad (18)$$

where $\alpha > 0$, $\beta > -1/2$, $m > 0$ and $h(x) \in L_2(0,\infty)$.

EXAMPLE: We consider the integral equation

$$\int_a^x (x^2 - t^2)^{\lambda/2} P_\nu^{-\lambda}\left(\frac{x}{t}\right) g(t)\, dt = f(x) \quad (19)$$

$$\lambda > -1 \ , \ \nu \geq -1/2 \ , \ 0 < a \leq x \leq b$$

where $P_\nu^{-\lambda}(z)$ is the Legendre function defined as:

$$P_\nu^{-\lambda}(z) = \frac{\{(z-1)/(z+1)\}^{\lambda/2}}{2^\nu\, \Gamma(\nu+1)\Gamma(\lambda-\nu)}$$

$$\int_0^1 x^{\lambda-\nu-1} (1-x)^\nu \{ z + 1 - (z-1)x \}^\nu dx \ ,$$

$$\lambda > \nu > -1 \ , \ z > 1 \ ,$$

and

$$P_\nu^{-\lambda}(z) = \frac{\{(1-z)/(1+z)\}^{\lambda/2}}{2^\nu\, \Gamma(\nu+1)\Gamma(\lambda-\nu)}$$

$$\int_0^1 x^{\lambda-\nu-1} (1-x)^\nu \{z + 1 - (z-1)x\}^\nu dx$$

$$\lambda > \nu > -1 \ , \ -1 < z < 1 \ .$$

$f(x)$ is a known function and $g(t)$ is to be determined.

We define

$$I_{x^n}^{\alpha} f(x) = \frac{1}{\Gamma(\alpha)} \int_a^x (x^n - t^n)^{\alpha-1} f(t)\, n t^{n-1}\, dt \qquad (20)$$

$$\alpha > 0$$

and the extension to $\alpha \leqslant 0$ is made as before. The analogous basic identities remain valid. In particular

$$I_{x^n}^{\alpha} I_{x^n}^{\beta} = I_{x^n}^{\alpha+\beta}$$

remains valid, so that $I_{x^n}^{\alpha}$ and $I_{x^n}^{\beta}$ commute. Note, however, that for $m \neq n$, $I_{x^m}^{\alpha}$ and $I_{x^n}^{\beta}$ do not commute.

Equation (19) can be written as

$$f(x) = I_x^{\lambda-\nu} I_{x^2}^{\nu+1} \left[(2x)^{-\nu-1} g(x)\right], \qquad (21)$$

or

$$I_x^{\nu-\lambda} f(x) = I_{x^2}^{\nu+1} \left[(2x)^{-\nu-1} g(x)\right],$$

$$g(x) = (2x)^{\nu+1} I_x^{-\nu-1} I_x^{\nu-\lambda} f(x).$$

This result is due to Erdélyi [8], where one can find the detailed treatment, including necessary and sufficient conditions for the existence of an integrable solution of (19).

Fox [13] has given an inversion formula for the transform whose kernel is $x^{\nu} K_{\nu}(x)$, where $K_{\nu}(x)$ is the modified Bessel function of the second kind, by the application of fractional integration theory. By using the same technique, Saxena [40] has solved the following integral equation

$$\int_0^{\infty} (xu)^{m-\frac{1}{2}} e^{-\frac{1}{2}xu} W_{k,m}(xu)\, h(u)\, du = V(h,k,m : x) \qquad (22)$$

Operators of fractional integration are employed by several authors to solve a pair of dual integral equations, for example, Erdélyi [7]. Chakrabarti [4] has employed E-K operators to obtain the solution of certain simultaneous pairs of dual integral equations. Operators of fractional integration are used to solve a set of triple integral equations by Jagetya [17], whereas Saxena & Sethi [41] have used them to solve a pair of quadruple integral equations.

The present author [28] has considered an integral equation whose kernel $S_{p,q,r}(x)$ has a Mellin-Barnes type integral representation. The kernel used here is of general character, and consequently various integral equations involving Whittaker functions, Bessel functions, Meijer's G-function etc., as kernels, can be derived as particular cases. By the application of E-K operators, the kernel has been reduced to an exponential function (Laplace transform), which can be inverted by known results.

3. HYPERGEOMETRIC FUNCTION OPERATORS

We define the fractional integration operators by means of the following equations

$$I[f(x)] = I\,[\alpha, \beta, \gamma; m, \mu, \eta, a : f(x)]$$

$$= \frac{\mu\, x^{-\eta-1}}{\Gamma(1-\alpha)} \int_0^x {}_2F_1\left(\alpha, \beta+m; \gamma; \frac{at^\mu}{x^\mu}\right) t^\eta f(t)\, dt, \tag{23}$$

$$R[f(x)] = R\,[\alpha, \beta, \gamma; m, \mu, \delta, a : f(x)]$$

$$= \frac{\mu\, x^{\delta}}{\Gamma(1-\alpha)} \int_x^\infty {}_2F_1\left(\alpha, \beta+m; \gamma; \frac{ax^\mu}{t^\mu}\right) t^{-\delta-1} f(t)\, dt, \tag{24}$$

where ${}_2F_1\,(\alpha,\beta;\gamma;x)$ denote the Gauss hypergeometric function [33] $\alpha,\beta,\gamma,\eta,\delta$ and a are complex parameters.

The operators defined above exist under the following conditions:

(i) $1 \leq p;q < \infty\,,\ p^{-1}+q^{-1} = 1\,,\ \mu > 0\,,\ |\arg(1-a)| < \pi\,,$

(ii) $Re(\alpha) > 0\,,\ Re(\eta) > -q^{-1}\,,\ Re(\delta) > -p^{-1}\,,$

$Re(1+\gamma-\alpha-\beta-m) > 0\,,\ m = 0,1,2,\ldots\,,\ \gamma \neq 0,-1,-2,\ldots$

(iii) $f(x) \in L_p(0,\infty)$.

The last condition ensures that both $I[f(x)]$ and $R[f(x)]$ exist and also that both belong to $L_p(0,\infty)$.

Particular cases. If we set $\gamma = \beta$, $\mu = 1$ and $a = 1$ then (23) and (24) reduce to the operators given by Saxena [39].

On the other hand, if we set $\gamma = \beta$, $m = 0$, $a = 1$, we obtain Erdélyi's operators (17) and (18). Several other operators involving Bessel functions $J_\nu(x)$, confluent hypergeometric function ${}_1F_1$ etc. can be derived from our hypergeometric function operators. The expression for the Mellin transform of these operators, their inversion formulae and some other basic properties of these operators are given in the work of Kalla and Saxena [20,21].

The Mellin transform of $f(x)$ will be denoted by $m\{f(x)\}$. We write $s = p^{-1} + it$, where p and t are real. If $p \geq 1$, $f(x) \in L_p(0,\infty)$ then

$$p = 1\,,\quad m\{f(x)\} = \int_0^\infty x^{s-1} f(x)\,dx\,, \tag{25}$$

$$p > 1 \quad , \quad m\{f(x)\} = \underset{x \to \infty}{\ell.i.m.} \int_{\frac{1}{x}}^{x} x^{s-1} f(x)\, dx \quad , \tag{26}$$

where $\ell.i.m.$ denotes the usual limit in the mean of L_p-spaces.

Recently Virchenko & Makarenko [45] and Makarenko [34] have used hypergeometric function operators to solve some dual and triple integral equations.

4. INTEGRAL OPERATORS INVOLVING FOX'S H-FUNCTION

Now we introduce two operators of fractional integration [24,25] involving Fox's H-function [12] as kernels. The operators introduced here are general and include as particular cases many operators of fractional integration defined by various authors [6,7,10,20] from time to time. We shall study here certain properties of these operators in the form of some theorems.

The H-function due to Fox [12,14] will be defined and represented as follows:

$$H_{p,q}^{m,n}\left[z \,\middle|\, \begin{matrix} ((a_p, A_p)) \\ ((b_q, B_q)) \end{matrix} \right] = H_{p,q}^{m,n}\left[z \,\middle|\, \begin{matrix} (a_1, A_1), \ldots, (a_p, A_p) \\ (b_1, B_1), \ldots, (b_q, B_q) \end{matrix} \right]$$

$$= \frac{1}{2\pi i} \int_L \frac{\prod_{j=1}^{m} \Gamma(b_j - B_j s) \prod_{j=1}^{n} \Gamma(1 - a_j + A_j s)}{\prod_{j=m+1}^{q} \Gamma(1 - b_j + B_j s) \prod_{j=n+1}^{p} \Gamma(a_j - A_j s)} z^s \, ds \tag{27}$$

where an empty product is interpreted as 1, $0 \leqslant m \leqslant q$, $0 \leqslant n \leqslant p$, $A^{'s}$ and $B^{'s}$ are all positive. L is a suitable contour of the Barnes type such that the poles of $\Gamma(b_j - B_j s)$, $j = 1,\ldots,m$, lie on the right-hand side of the contour and those of $\Gamma(1-a_j+A_js)$, $j=1,\dots,n$ lie on the left-hand side of the contour.

Asymptotic expansion and analytic continuation of the H-function have been discussed by Braakshma [3].

We define the fractional integration operators by means of the following relations:

$$R\,[f(x)] = R^{m,n,p,q;a}_{\gamma,r,a_j,A_j;b_j,B_j}\,[f(x)]$$

$$= x^{-\gamma-1}\,\Gamma_R \int_0^x H^{m,n}_{p,q}\left[a\left(\frac{t}{x}\right)^r \,\middle|\, \begin{matrix}((a_p,A_p))\\ ((b_q,B_q))\end{matrix}\right] t^{\gamma} f(t)\,dt \tag{28}$$

in which $r>0$ and

$$\Gamma_R = \left\{r\prod_{j=1}^{q}\Gamma(1-b_j)\right\}\left\{\prod_{j=1}^{m}\Gamma(a_j)\prod_{j=1}^{p}\Gamma(1-a_j)\right\}^{-1} \tag{29}$$

and

$$S\,[f(x)] = S^{t,u,v,w;b}_{\delta,s,c_j,C_j;d_j,D_j}\,[f(x)]$$

$$= x^{\delta}\,\Gamma_S \int_x^{\infty} H^{t,u}_{v,w}\left[b\left(\frac{x}{t}\right)^s \,\middle|\, \begin{matrix}((c_v,C_v))\\ ((d_w,D_w))\end{matrix}\right] t^{-\delta-1} f(t)\,dt \tag{30}$$

in which $s>0$ and

$$\Gamma_S = \left\{s\prod_{j=1}^{w}(1-d_j)\right\}\left\{\prod_{j=1}^{t}\Gamma(c_j)\prod_{j=1}^{v}\Gamma(1-c_j)\right\}^{-1} \tag{31}$$

We shall assume throughout this work that $Re(\gamma)$ and $Re(\delta)$ are both non-negative; hence from [24] for $1 \leq p < \infty$, $f(x) \in L_p(0,\infty)$, we have both $R[f(x)]$ and $S[f(x)]$ in $L_p(0,\infty)$.

Particular cases: As Meijer's G-function, hypergeometric function, Bessel functions etc. are particular cases of the H-function, most of the integral operators given by various authors can be obtained from our operators by giving special values to the parameters. Some of the cases are given below:

(i) Fractional integration operators involving Meijer's G-function [22,23] can be obtained on setting $A_1,\ldots,A_p = B_1,\ldots,B_q = C_1,\ldots,C_v = D_1,\ldots,D_w = 1$.

(ii) If we assign the values $m = t = 1$, $n = u = 2$, $p = v = 2$, $q = w = 2$, $A_1 = A_2 = B_1 = B_2 = C_1 = C_2 = D_1 = D_2 = 1$, $a_1 = c_1 = 1 - \alpha$, $a_2 = c_2 = 1 - \beta - m$, $b_1 = d_1 = 0$, $b_2 = d_2 = 1 - \rho$, then we essentially get the hypergeometric operators (23) and (24). The relation being depicted symbolically as

$$R^{1,2,2,2;a}_{\gamma,r;1-\alpha,1-\beta-m,1,1;0,1-\rho,1,1}[f(x)]$$

$$= \frac{r\, x^{-\gamma-1}}{\Gamma(1-\alpha)} \int_0^x {}_2F_1\left(\alpha, \beta+m; \rho; -a\left(\frac{t}{x}\right)^r\right) t^{\gamma} f(t)\, dt, \tag{32}$$

$$S^{1,2,2,2;a}_{\delta,r,1-\alpha,1-\beta-m,1,1;0,1-\rho,1,1}[f(x)]$$

$$= \frac{r\, x^{\delta}}{\Gamma(1-\alpha)} \int_x^\infty {}_2F_1\left(\alpha, \beta+m; \rho; -a\left(\frac{x}{t}\right)^r\right) t^{-\delta-1} f(t)\, dt. \tag{33}$$

Similarly, the other operators due to Erdélyi, Kober and Sneddon can easily be derived from (28) and (29).

We now establish two theorems which give the expressions for the Mellin transform of the H-function operators.

THEOREM 1 *If* $f(x) \in L_p(0,\infty)$, $1 \leq p \leq 2$ *(or* $f(x) \in M_p(0,\infty)$, *and* $p > 2$*)*, $|\arg a| < \frac{1}{2}\lambda\pi$, $\lambda > 0$, $Re(\gamma) > \max(p^{-1}, q^{-1})$, $p^{-1} + q^{-1} = 1$, $. > 0$ *then*

$$m\left[R\left[f(x)\right]\right] = \frac{\Gamma_R}{r} H^{m+1,n}_{p+1,q+1}\left[a \left| \begin{matrix} ((a_p, A_p)), \left(\frac{r-1-\gamma+s}{r}, 1\right) \\ \left(\frac{s-\gamma-1}{r}, 1\right), ((b_k, B_k)) \end{matrix} \right.\right] m[f(x)]$$

$$= \frac{\Gamma_R}{r} k(s)\, m\left[f(x)\right], \tag{34}$$

where

$$k(s) = H^{m+1,n}_{p+1,q+1}\left[a \left| \begin{matrix} ((a_p, A_p)), \left(\frac{r-1-\gamma+s}{r}, 1\right) \\ \left(\frac{s-\gamma-1}{r}, 1\right), ((b_k, B_k)) \end{matrix} \right.\right] \tag{34 a}$$

and $M_p(0,\infty)$ *denotes the class of all functions* $f(x)$ *of* $L_p(0,\infty)$ *with* $p > 2$, *which are the inverse Mellin transform of functions of* $L_q(-\infty,\infty)$, *and*

$$\lambda = \sum_{j=1}^{n} (A_j) - \sum_{j=n+1}^{p} (A_j) + \sum_{j=1}^{m} (B_j) - \sum_{j=m+1}^{q} (B_j). \tag{35}$$

Proof: We have

$$m\left[R\left[f(x)\right]\right] =$$

$$\int_0^\infty x^{s-1}\left[x^{-\gamma-1}\,\Gamma_R \int_0^x H^{m,n}_{p,q}\left[a\left(\frac{t}{x}\right)^r \left| \begin{matrix} ((a_p, A_p)) \\ ((b_q, B_q)) \end{matrix} \right.\right] t^\gamma f(t)\, dt\right] dx$$

(36)

Changing the order of integration, which is permissible under the conditions stated with the theorem, the theorem then immediately follows on evaluating the x-integral by expressing the H-function as the Mellin-Barnes integral.

Similarly, we can establish the following theorem, which provides us with the Mellin transform of the operator $S[f(x)]$.

THEOREM 2 *Under the conditions of the previous theorem, with* $Re(\delta) > max\ (p^{-1}, q^{-1})$,

$$m\left| S^{t,u,v,w;b}_{\delta,r;c_j,C_j;d_j,D_j} [f(x)] \right|$$

$$= \frac{\Gamma_s}{r} H^{t,u+1}_{v+1,w+1}\left[b \left| \begin{matrix} ((r-s-\delta)/r, 1), ((c_v, C_v)) \\ ((d_w, D_w)), (\frac{s+\delta}{r}, 1) \end{matrix} \right. \right] m\left[f(x)\right]$$

$$= \frac{\Gamma_s}{r} K(s)\, m\, [f(x)]. \tag{37}$$

where $K(s)$ *is defined in* (45 a).

THEOREM 3 *If* $f(x) \in L_p(0,\infty)$, $p^{-1} + q^{-1} = 1$, $g(x) \in L_q(0,\infty)$, $Re(\gamma) > max\ (p^{-1}, q^{-1})$, $r > 0$, $|arg\ a| < \frac{1}{2}\lambda\pi$, $\lambda > 0$, *then*

$$\int_0^\infty g(x)\, R[f(x)]\, dx = \int_0^\infty f(x)\, S[g(x)]\, dx. \tag{38}$$

Proof: (38) immediately follows on interpreting it with the help of (28) and (30).

Let

$$R^{m,n,p,q;a}_{\gamma,r;a_i,A_i;b_i,B_i}\left[f(x)\right] = g(x) \,. \tag{39}$$

If $g(x)$ is a known function and $f(x)$ is to be determined, then this becomes an integral equation. The following two theorems deal with the solution of integral equations of this type.

THEOREM 4 *If*

$$R\left[f(x)\right] = g(x) \,, \tag{40}$$

then

$$f(x) = \frac{1}{r}\,\Gamma_R \int_0^\infty t^{-1}\, g(t)\, h\left(\frac{x}{t}\right) dt \,, \tag{41}$$

where

$$h(x) = \frac{1}{2\pi i} \int_{c-i\infty}^{c+i\infty} \frac{x^{-s}}{k(s)}\, ds \tag{42}$$

for $f(x) \in L_p(0,\infty)$, $1 \le p \le 2$ *(or* $f(x) \in M_p(0,\infty)$, $p > 2$*)*, $|\arg a| < \frac{1}{2}\lambda\pi$, $\mathrm{Re}(\gamma) > \max(p^{-1}, q^{-1})$, $p^{-1} + q^{-1} = 1$, $r > 0$ *and* $\lambda > 0$. $k(s)$ *is as defined in* (34 a).

Proof: Multiplying both sides of (40) by x^{s-1} and then integrating w.r.t. x from 0 to ∞, on using the expression for the Mellin transform of $R[f(x)]$ we obtain the following expression:

$$\frac{r}{\Gamma_R}\, k(s)\; m\left[f(x)\right] = m\left[g(x)\right] .$$

The Mellin inversion theorem leads to the desired result.

Proceeding in the same way as in the previous theorem, we obtain the following result, under the same conditions with $Re(\delta) > max\ (p^{-1}, q^{-1})$.

THEOREM 5 *If*

$$S\,[f(x)] = G(x)\ , \tag{43}$$

then

$$f(x) = \frac{1}{r}\ \Gamma_S \int_0^\infty t^{-1}\ G(t)\ H(\tfrac{x}{t})\ dt\ , \tag{44}$$

where

$$H(x) = \frac{1}{2\pi i} \int_{c-i\infty}^{c+i\infty} \frac{x^{-s}}{K(s)}\ ds \tag{45}$$

and

$$K(s) = H^{m,n+1}_{p+1,q+1}\left[a \,\middle|\, \begin{matrix} \left(\frac{r-s-\delta}{r}, 1\right), & ((a_p, A_p)) \\ ((b_k, B_k)) & , \left(\frac{s+\delta}{r}, 1\right) \end{matrix}\right] \tag{45 a}$$

The product of the operators

In terms of the condensed notations we have

$$R\ |\ S\ |f(x)||\ = x^{-\gamma-1}\ \Gamma_R \int_0^x H^{m,n}_{p,q}\ [a(z/x)^r]\ z^\gamma\ S\ [f(z)]\ dz\ , \tag{46}$$

$$R\ [\ S\ [f(x)]] =$$

$$= x^{-\gamma-1}\ \Gamma_R \int_0^x H^{m,n}_{p,q}\ [a(z/x)^r\ |\ z^\gamma . z^\delta\ \Gamma_S \int_z^\infty H^{t,u}_{v,w}\ |\ b(z/y)^s]$$

$$y^{-\delta-1}\ f(y)\ dy\ dz\ ,$$

$$= \int_0^\infty K(x,y)\, f(y)\, dy \;, \tag{47}$$

where

$$K(x,y) =$$

$$\Gamma_R \Gamma_S x^{-\gamma-1} y^{-\delta-1} \int_0^{\min(x,y)} z^{-1} H^{m,n}_{p,q}\left[a(z/x)^r\right] H^{t,u}_{v,w}\left[b(z/y)^s\right] dz. \tag{48}$$

Various sets of conditions could be imposed in order to justify the interchanges of order of integration . For our purpose, in addition to the assumption that $Re(\gamma)$ and $Re(\delta)$ are non-negative and that r and s are positive , a suitable restriction on the parameters would be the condition that

$$r \min_{1 \leq j \leq m} \left[Re(b_j / B_j)\right] + s \min_{1 \leq j \leq t} \left[Re(d_j / D_j)\right] > -1 \;. \tag{49}$$

To evaluate the integral in (45) we substitute the contour integral representation for the H-functions and invert the order of integration once again. After a little simplification, we obtain

$$K(x,y) = \Gamma_R \Gamma_S \begin{cases} y^{-1} (x/y)^{\delta} H\left[a, b(x/y)^s\right] & \text{if } x < y \,, \\ x^{-1} (y/x)^{\gamma} H\left[a(y/x)^r, b\right] & \text{if } y < x \,, \end{cases} \tag{50}$$

where the parameters of the H-function are alike . $H[x,y]$ stands for the H-function of two variables due to Munot & Kalla [37] and Mittal and Gupta [36].

If we denote by K' the kernel obtained from the composition taken in the other order, i.e. for $S[R[f(x)]]$, it can easily be shown that

$$x\,y\,K'(x,y) = K(y^{-1}, x^{-1}) \;. \tag{51}$$

A detailed treatment of the composition of fractional integration operators can be found in the recent works of Srivastava & Buschman [43] and Kalla [26,27].

We shall now mention some formal properties of the operators $R[f(x)]$ and $S[f(x)]$.

(a) $$x^{\beta}\, R^{m,n,p,q;a}_{\gamma,r;a_j,A_j;b_j,B_j}[f(x)] = R^{-;-}_{\gamma-\beta,r;-;-}[x^{\beta} f(x)]\,. \tag{52}$$

(b) $$x^{\beta}\, S^{t,u,v,w;b}_{\delta,s;c_j,C_j;d_j,D_j}[f(x)] = S^{-;-}_{\delta+\beta,s;-;-}[x^{\beta} f(x)]\,. \tag{53}$$

(c) If $R[f(x)] = g(x)$, then

$$R[f(cx)] = g(cx). \tag{54}$$

(d) If $S[f(x)] = h(x)$, then

$$S[f(cx)] = h(cx), \tag{55}$$

where c is a constant.

The proof of relations (52) to (55) is obvious. The relations (54) and (55) express the homogeneity of the operators. They show that, given a function $f(x)$, it makes no difference whether the operators are applied with respect to x, y or $t = xy$.

We can easily establish the following relation:

$$R^{m,n,p,q;a}_{\gamma,r;a_j,A_j;b_j,B_j}\left[f(x^{-1})\right] = S^{m,n,p,q;a}_{\gamma+1,r;\,-\,;\,-}\left[f(x)\right]. \qquad (56)$$

The above properties of the operators hold whenever their terms are well defined. Otherwise the behaviour of the operators is governed by the following Lemmas:

LEMMA 1 *For a function of* $L_p(0,\infty)$, *both* $R[f(x)]$ *and* $S[f(x)]$ *exist and belong to* $L_p(0,\infty)$.

LEMMA 2 *For a function of* $M_p(0,\infty)$, $p > 0$ *both* $R[f(x)]$ *and* $S[f(x)]$ *belong to* $M_p(0,\infty)$.

LEMMA 3 *The only function* $f(x)$, *satisfying* $R[f(x)] = 0$ (*or* $S[f(x)] = 0$) *for almost all* $x > 0$ *is the null function.*

5. GENERAL OPERATORS OF FRACTIONAL INTEGRATION

Several generalizations of the fractional integration operators have appeared in the literature [6,7,10,11,20,22,24], including those discussed in the previous sections. It is interesting to observe that some of the properties of the operators of fractional integration are similar. Thus we can consider two operators of fractional integration, involving a general function, say $\phi(t/x)$, as the kernel, which is supposed to be a known continuous function. We define the operators as:

$$R[f(x)] = R[f(x),\gamma] = x^{-\gamma-1}\int_0^x t^{\gamma}\,\phi(t/x)\,f(t)\,dt \qquad (57)$$

and

$$S[f(x)] = S[f(x),\delta] = x^{\delta} \int_x^{\infty} t^{-\delta-1} \phi(x/t) f(t)\, dt , \tag{58}$$

where the kernel $\phi(t/x)$ is such that the integrals make sense. The operators (57) and (58) exist under the following conditions:

(i) $1 \leqslant p, q < \infty, p^{-1} + q^{-1} = 1,$

(ii) $Re(\gamma) > -1/q, \quad Re(\delta) > -1/p,$

(iii) $f(x) \varepsilon L_p(0,\infty).$

If we set some specific functions for the kernel $\phi(t/x)$, then we obtain different operators, including those discussed above. These operators are discussed by the author in [26,27].

REFERENCES

[1] BAKER, B.B. and COPSON, E.T. : *The Mathematical Theory of Huygen's Principle*, (Clarendon Press, Oxford 1950).

[2] BORA, S.L. and SAXENA, R.K.: *On fractional integration*, Publ. Inst. Math., Beograd, 11 (25) (1971), 19-22.

[3] BRAAKSMA, B.L.J.: *Asymptotic expansions and analytic continuations for a class of Barnes integrals*, Comp. Mat. 15, (1963) 239-341.

[4] CHAKRABARTI, A.: *On the solution of certain simultaneous pairs of dual integral equations*, ZAMM 54 (1974), 383-387.

[5] DOETSCH, G.: *Theorie und Anwendung der Laplace Transformation* (Springer-Verlag, Berlin 1937).

[6] ERDELYI, A.: *On fractional integration and its applications to the theory of Hankel transforms*, Quart. J. Math. Oxford 11 (1940), 293-303.

[7] ERDELYI, A. : *On some functional transformations* , Univ. e Politec. Torino, Rend. Sem. Mat. 10 (1951), 217-234.

[8] ERDELYI, A. : *An integral equation involving Legendre functions*, SIAM J. Appl. Math. 12 (1964), 15-30.

[9] ERDELYI, A., et al.: *Tables of Integral Transforms , Vols. I and II* (McGraw-Hill, New York 1954).

[10] ERDELYI, A. and KOBER, H.: *Some remarks on Hankel transforms*, Quart. J. Math. Oxford 11 (1940), 212-221.

[11] ERDELYI, A. and SNEDDON, I.N.: *Fractional integration and dual integral equations*, Can. J. Math. 14 (1962), 685-693.

[12] FOX, C.: *The G- and H-functions as symmetrical Fourier kernels* Trans. Amer. Math. Soc. 98 (1961), 395-429.

[13] FOX, C. : *An inversion formula for the kernel* $K_\nu(x)$, Proc. Cambridge Phil. Soc. 61 (1965), 457-467.

[14] GUPTA, K.C.: *On the H-function*, Annal. Soc. Sci. Bruxelles 79 (1965), 98-104.

[15] HARDY, G.H. and LITTLEWOOD, J.E. : *Some properties of fractional integrals*, Math. Z. 27 (1928), 565-606.

[16] INCE, E.L.: *Ordinary Differential Equations* (Dover Publ., New York 1956).

[17] JAGETYA, R.N.: *Solution of dual integral equations by fractional integration*, Math. Edu., 4 (1970), 69-72; *Triple integral equations and fractional integration*, Univ. Nac. Tucumán Rev. Ser. A20 (1970), 41-47.

[18] KALLA, S.L.: *Some theorems of fractional integration* , Proc. Nat. Acad. Sci., India 36A (1966), 1007-1012.

[19] KALLA, S.L.: *Some theorems of fractional integration-II* ,Proc. Nat. Acad. Sci., India 39A (1969), 49-56.

[20] KALLA, S.L. and SAXENA, R.K. : *Integral operators involving hypergeometric functions*, Math. Z. 108 (1969), 231-234.

[21] KALLA, S.L. and SAXENA, R.K. : *Integral operators involving hypergeometric functions-II* , Univ. Nac. Tucumán , Rev. Ser. A24 (1974), 31-36.

[22] KALLA, S.L.: *Fractional integration operators involving hypergeometric functions*, Univ. Nac. Tucumán, Rev. Ser. A20 (1970) 93-100.

[23] KALLA, S.L.: *Fractional integration operators involving hypergeometric functions-II*, Acta Mexicana Cie. Tecn. 3 (1969), 1-5

[24] KALLA, S.L.: *Integral operators involving Fox's H-function*, Acta Mexicana Cie. Tecn. 3 (1969), 117-122.

[25] KALLA, S.L.: *Integral operators involving Fox's H-function-II*, Acta Mexicana Cie. Tecn. (in press).

[26] KALLA, S.L. : *On operators of fractional integration*, Mat. Notae 22 (1970), 89-93.

[27] KALLA, S.L.: *On operators of fractional integration-II* , Mat. Notae 26 (1976)

[28] KALLA, S.L.: *On the solution of an integral equation involving a kernel of Mellin-Barnes type integral*, Kyungpook Math. J. 12 (1972), 93-101.

[29] KOBER, H.: *On fractional integrals and derivatives*, Quart. J. Math. Oxford 11 (1940), 193-211.

[30] KOBER, H.: *On a theorem of Schur and on fractional integrals of purely imaginary order*, Trans. Amer. Math. Soc. 50 (1941), 160-174.

[31] LOVE, E.R. : *Fractional derivatives of imaginary order*, J. London Math. Soc. 3 (1971), 241-259.

[32] LOVE, E.R. and YOUNG, L.C.: *On fractional integration by parts*, Proc. London Math. Soc. 44 (1938), 1-28.

[33] LUKE, Y.L. : *The Special Functions and Their Approximations, Vols. I & II* (Academic Press, New York 1969).

[34] MAKARENKO, L.G. : *Certain triples of integral equations with Watson-type kernels*, Ukrain. Math. J. 27 (1975-76), 564-567.

[35] MARTIC, B.: *A note on fractional integration*, Pub. Inst. Math. Beograd 16 (30), (1973), 111-113.

[36] MITTAL, P.K. and GUPTA, K.C. : *An integral involving generalized functions of two variables*, Proc. Indian Acad. Sci. A75 (1972), 117-123.

[37] MUNOT, P.C. and KALLA, S.L. : *On an extension of generalized functions of two variables*, Univ. Nac. Tucumán, Rev. Ser. A21 (1971), 67-84.

[38] RIESZ, M. : *L'integrales de Riemann-Liouville et le probleme de Cauchy*, Acta Math. 8 (1949), 1-123.

[39] SAXENA, R.K. : *On fractional integration operators*, Math. Z. 96 (1967), 288-291.

[40] SAXENA, R.K. : *An inversion formula for the Varma transform*, Proc. Cambridge Phil. Soc. 62 (1966), 467-471.

[41] SAXENA, R.K. and SETHI, P.L.: *On the formal solution of quadruple integral equations*, Proc. Nat. Acad. Sci. India 42 (1972), 57-61.

[42] SNEDDON, I.N. : *Mixed Boundary Value Problems in Potential Theory* (North Holland Publishing Co., Amsterdam 1966).

[43] SRIVASTAVA, H.M. and BUSCHMAN, R.G.: *Composition of fractional integral operators involving Fox's H-function*, Acta Mexicana Cie. Tecn. 7 (1973), 21-28.

[44] TITCHMARSH, E.C.: *Introduction to the Theory of Fourier Integrals*, 2nd Ed. (Clarendon Press, Oxford 1948).

[45] VIRCHENKO, N.A. and MAKARENKO, L.G. : *Some pairs of integral equations*, Ukrain. Math. J. 27 (1975-76), 648-651.

[46] WIDDER, D.V. : *The Laplace Transform* (Princeton Univ. Press 1941).

División de Postgrado, Facultad de Ingeniería
Universidad del Zulia, Maracaibo, Venezuela

HERMITIAN MANIFOLDS WITH ALMOST PRODUCT STRUCTURES

Shôji Kanemaki (Tokyo)

Contents

Introduction

The product $P^n(C) \times P^m(C)$ of two complex projective spaces and the product $S^{2n+1} \times S^{2m+1}$ of two odd-dimensional spheres are typical examples of Hermitian product manifolds of two manifolds, whose dimensions have the same parity. As our object we consider Hermitian manifolds on each of which a complete system of mutually orthogonal distributions is given. We see how relations among the complex structure J, the metric G and the projectors determining the distributions affect the manifold.

1. Preliminaries

Let M be a 2n-dimensional Hermitian manifold of class C^∞ and denote the Hermitian structure by (J,G), which consists of a linear transformation field J and a Riemannian metric G, and satisfies the conditions

$$J^2 = -I, \quad [J,J] = 0, \quad G(JX,Y) = -G(X,JY)$$

for any vector fields X and Y on M, where I denotes the identity linear transformation field and $[J,J]$ the Nijenhuis tensor field of J, defined by

$$(1.1)\quad [J,J](X,Y) = [JX,JY] - J[JY,Y] - J[X,JY] + J^2[X,Y].$$

We note a familiar result that a necessary and sufficient condition for an almost Hermitian structure (J,G) on a differentiable manifold to be Kählerian is that $\nabla_X J = 0$ for any X, where ∇ denotes the Riemannian connection with respect to G. In fact, on an almost Hermitian manifold the identity equation reduced from (1.1)

$$G([J,J](X,Y),Z) = d\Omega(JX,Y,Z) + d\Omega(X,JY,Z) - 2G((\nabla_Z J)X,JY)$$

is used to prove it, where Ω is the fundamential 2-form of the almost Hermitian structure (J,G), given by $\Omega(X,Y) = G(JX,Y)$.

An almost product structure $\{D(p_a)\}_{a\in(m)}$, $(m) = \{1,2,\ldots,m\}$, on a Riemannian manifold M is a complete system of distributions $D(p_a)$ determined by non-trivial m linear transformation fields p_a, called projectors, satisfying

$$p_a^2 = p_a, \quad p_a p_b = p_b p_a = 0 \ (a \neq b), \quad \sum_{a=1}^{m} p_a = I, \quad G(p_a X,Y) = G(X,p_a Y)$$

for any vector fields X and Y on M. Then these distributions $D(p_a)$ are mutually orthogonal on the manifold. It is fundamental for us to consider an arbitrary and fixed almost product structure consisting of two complementary distributions on a manifold where the almost product structure consisting of m distributions has been given already.

Let $M(\mathfrak{D})$ denote a Hermitian manifold with an almost product structure $\mathfrak{D} = \{D(p),D(q)\}$ whose projectors p and q satisfy the conditions

$$(1.2) \quad p^2=p,\ \ pq=qp=0,\ \ q^2=q,\ \ p+q=I,\ \ G(pX,Y)=G(X,pY),\ \ G(qX,Y)=G(X,qY).$$

Then we have a decomposition law of JX with respect to $\mathfrak{D}$:

$$(1.3) \quad \begin{cases} JpX = f_1 X + qJpX, \\ JqX = f_2 X + pJqX \end{cases}$$

for any vector field X, where we have put $f_1 = pJp$, $f_2 = qJq$. We easily see that f_i $(i = 1,2)$ are skew-symmetric with respect to G and that qJp is related to pJq by $G(qJpX,Y) = -G(X,pJqY)$. It follows from (1.2) and (1.3) that

$$f_1^2X + pX + pJqJpX = 0, \qquad f_2JpX + qJpf_1X = 0,$$

$$f_2^2X + qX + qJpJqX = 0, \qquad f_1JqX + pJqf_2X = 0.$$

Taking a notice of f_1, we shall examine how polynomial of f_1 can be induced by J on a neighbourhood U of a point of M. We choose an integer k $(1 \le k \le n)$ arbitrarily and fix it in mind. We take a field of orthonormal $2k$-farmes $C_1,\ldots,C_{2k}$ and its duals $\gamma^1,\ldots,\gamma^{2k}$ on U, that is, $\gamma^i(X) = G(C_i,X)$, $i \in (k)$. We consider the distributions $D(p)$ and $D(q)$ restricted onto U on which both are given by

$$p = I - q, \quad q = \sum_{i=1}^{k} \gamma^{2i} \otimes C_{2i}, \quad C_{2i-1} = -pJC_{2i}, \quad i = 1,2,\ldots,k.$$

Then it follows from (1.3) that $C_{2i-1} \in D(p)$, $C_{2i} \in D(q)$, and

$$\text{(1.4)} \quad JpX = f_1X + \sum_{i=1}^{k} \gamma^{2i-1}(X)C_{2i}, \quad JC_{2i} = -C_{2i-1} + \sum_{j=1}^{k} \lambda_{ij}C_{2j},$$

where λ_{ij} are scalar fields. The condition $G(JC_{2i},C_{2j}) = -G(C_{2i},JC_{2j})$ implies that $\lambda_{ij} = -\lambda_{ji}$. Accordingly, by (1.4) we obtain

$$f_1^2X = -pX + \sum_{i=1}^{k} \gamma^{2i-1}(X)C_{2i-1},$$

$$\text{(1.5)} \quad f_1C_{2i-1} = -\sum_{j=1}^{k} \lambda_{ij}C_{2j-1}, \quad \gamma^{2i-1}(f_1X) = \sum_{j=1}^{k} \lambda_{ij}\gamma^{2j-1}(X),$$

$$\gamma^{2i-1}(C_{2i-1}) = 1 - \sum_{j=1}^{k} \lambda_{ij}^2, \quad \gamma^{2i-1}(C_{2j-1}) = \sum_{h=1}^{k} \lambda_{ih}\lambda_{hj} \quad (i \ne j).$$

We can see that polynomial of f_1 induced from $J^2 + I = 0$ of J in cases of rank q = 1, 2 and 3 as follows.

i) Rank $q = 1$. We use (1.4) to get $\lambda_{ij} = 0$. Immediately, from (1.5) we have $f_1^2 + p - \gamma^1 \otimes C_1 = 0$, $f_1C_1 = 0$, hence, $f_1^3 + f_1 = 0$ (rank $f_1 = 2n - 2$).

ii) Rank $q = 2$. The relation $\gamma^1(C_1) = \gamma^3(C_3) = 1 - \lambda_{12}^2$ implies that $0 \le \lambda_{12}^2 \le 1$. Thus we have

$$f_1^2 + p - \gamma^1 \otimes C_1 - \gamma^3 \otimes C_3 = 0, \quad f_1C_1 = -\lambda_{12}C_3, \quad f_1C_3 = \lambda_{12}C_1,$$

$$f_1^4 + f_1^2 + \lambda_{12}^2(\gamma^1 \otimes C_1 + \gamma^3 \otimes C_3) = 0;$$

hence,

$f_1^3 + f_1 = 0$ (rank $f_1 = 2n-4$) for $\lambda_{12} = 0$ on U;

$f_1^5 + (\lambda_{12}^2 + 1) f_1^3 + \lambda_{12}^2 f_1 = 0$ (rank $f_1 = 2n-2$) for $0 < \lambda_{12}^2 < 1$ on U;

$f_1^3 + f_1 = 0$ (rank $f_1 = 2n-2$) for $\lambda_{12}^2 = 1$ on U.

iii) R a n k $q = 3$. Consider a vector field $E = \lambda_{23}C_1 + \lambda_{31}C_3 + \lambda_{12}C_5$ and its associated 1-form η. By (1.5) the norm $|E|$ of E has the form $|E| = \sqrt{\lambda_{23}^2 + \lambda_{31}^2 + \lambda_{12}^2}$, say κ, and $f_1E = 0$. In this case, we see that κ must necessarily be $0 \leq \kappa \leq 1$, by applying the property of G being a positive definite metric and Cauchy-Schwarz's formula. Thus we have

$$f_1^2 + p - \gamma^1 \otimes C_1 - \gamma^3 \otimes C_3 - \gamma^5 \otimes C_5 = 0,$$

$$f_1C_1 = -\lambda_{12}C_3 - \lambda_{13}C_5, \quad f_1C_3 = -\lambda_{21}C_1 - \lambda_{23}C_5, \quad f_1C_5 = -\lambda_{31}C_1 - \lambda_{32}C_3,$$

$$f_1^4 + f_1^2 + \kappa^2(\gamma^1 \otimes C_1 + \gamma^3 \otimes C_3 + \gamma^5 \otimes C_5) = 0,$$

hence,

$f_1^3 + f_1 = 0$ (rank $f_1 = 2n-6$) for $\kappa = 0$ on U;

$f_1^5 + (\kappa^2 + 1) f_1^3 + \kappa^2 f_1 = 0$ (rank $f_1 = 2n-4$) for $0 < \kappa < 1$ on U;

$f_1^3 + f_1 = 0$ (rank $f_1 = 2n-4$) for $\kappa = 1$ on U.

We know well that the dimensions of complementary distributions $D(p)$ and $D(q)$ have the same parity. In consideration of this fact we are to deal with the following two cases, J-invariant or J-outspread $\mathfrak{D}$, with a special interest.

<u>CASE I. We assume that an almost product structure $\mathfrak{D}$ on</u> M <u>satisfies a condition</u>: $qJp = 0$. That is, for any vector field X

$$\begin{cases} JpX = f_1X, \\ JqX = f_2X. \end{cases}$$

<u>CASE II. We assume that an almost product structure</u> $(\mathfrak{D}, \mathfrak{F})$ <u>equipped with a global field of orthonormal 2-frames</u> $\mathfrak{F} = \{E_1, E_2\}$,

$JE_1 = E_2$, satisfies a condition: $qJp = \eta^1 \otimes E_2$, where η^1 denotes the associated 1-form with E_1 by G. Then denoting by η^2 the associated 1-form with E_2, we have a decomposition law:

$$\begin{cases} JpX = f_1X + \eta^1(X)E_2, \\ JqX = f_2X - \eta^2(X)E_1. \end{cases}$$

We shall call $\mathfrak{D}$ described in Case I to be J-invariant, and $(\mathfrak{D}, \mathfrak{F})$ described in Case II to be J-outspread, for convenience in this paper.

2. Two theorems

We show in the present section the weakest sufficient condition, in our argument, in order that a Hermitian manifold $M(\mathfrak{D})$ with a J-invariant $\mathfrak{D}$ or $M(\mathfrak{D}, \mathfrak{F})$ with a J-outspread $(\mathfrak{D}, \mathfrak{F})$ be locally a product manifold of two Hermitian manifolds or two normal almost contact metric manifolds. Let us recall now odd-dimensional Riemannian manifolds analogous to almost Hermitian manifolds. An almost contact metric manifold is a $(2n+1)$-dimensional differentiable manifold N on which a set of tensor fields (f,E,η,g) consisting of a $(1,1)$-tensor field f, a vector field E, a 1-form η and a Riemannian metric g satisfies the conditions $f^2 = -I + \eta \otimes E$, $\eta(E) = 1$, $\eta(X) = g(X,E)$, $g(fX,Y) = -g(X,fY)$ for any vector fields X and Y. An almost complex structure J_o can be naturally defined on the product $N \times R$ of N and a real line R by $J_o(X, \lambda d/dt) = (fX + \lambda E, \ -\eta(X)d/dt)$ for a scalar field λ on $N \times R$. An almost contact metric structure is called to be normal if J_o is complex analytic. It is well known that a necessary and sufficient condition for an almost contact metric structure (f,E,η,g) to be normal is that the torsion tensor field $[f,f] + d\eta \otimes E = 0$. Here we continue to state some terminology concerning the above further, which will be treated in the next section. A quasi-Sasakian manifold $N(f,E,\eta,g)$ is a normal almost contact metric manifold with the closed fundametal 2-form F, given by $F(X,Y) = g(fX,Y)$. It has been shown that a necessary and sufficient condition for an almost contact metric manifold $N(f,E,\eta,g)$ to be quasi-Sasakian is that there exists a symmetric linear transformation field ψ on N such that $(\nabla_X f)Y = \eta(Y)\psi X - g(\psi X,Y)E$, $f\psi = \psi f$, $\psi E = E$. A contact metric structure is an almost contact metric structure satisfying a condition $d\eta = 2F$, that is, $\eta \wedge (d\eta)^n \neq 0$. A normal contact metric structure is also called a Sasakian structure. The $(2n+1)$-sphere S^{2n+1} admits a Sasakian structure. A normal almost contact metric structure is cosymplectic if its fundamental 2-form F and the 1-form η are both closed. The $(2n+1)$-dimensional Euclidean space R^{2n+1} and the torus T^{2n+1} $(= S^1 \times \dots \times S^1,\ 2n+1$ factors) are cosymplectic.

Now we state two theorems.

THEOREM 2.1. *Let $M(\mathfrak{D})$ be a Hermitian manifold with a J-invariant $\mathfrak{D}$. If $\nabla_X\phi = 0$ $(\phi = p - q)$ holds for any vector field X on M, then M is locally a product manifold of Hermitian manifolds M_i $(i = 1,2)$ which are the maximal integral manifolds determined by $D(p_i)$, passing through a point, where $p_1 = p$, $p_2 = q$.*

THEOREM 2.2. *Let $M(\mathfrak{D},\mathfrak{F})$ be a Hermitian manifold with a J-outspread $(\mathfrak{D},\mathfrak{F})$. If $\nabla_X\phi = 0$ for any vector field X on M, and*

$$(2.1)\quad \eta^1(X)p(L_{E_2}J)Y - \eta^1(Y)p(L_{E_2}J)X = 0 \qquad \text{for any } X,Y \in D(p),$$

$$(2.2)\quad \eta^2(X)q(L_{E_1}J)Y - \eta^2(Y)q(L_{E_1}J)X = 0 \qquad \text{for any } X,Y \in D(q),$$

then M is locally a product manifold of normal almost contact metric manifolds M_i which are the maximal integral manifolds determined by $D(p_i)$, passing through a point, where L_X denotes the Lie derivation with respect to X.

The proof of Theorem 2.1 is similar to that of Theorem 2.2. We first supply a lemma to prove Theorem 2.2.

LEMMA 2.1. *On a Hermitian manifold $M(\mathfrak{D},\mathfrak{F})$ the identity equations*

$$(2.3)\quad \sum_{i=1}^{2} p[f_i,f_i](X,Y) + d\eta^1(X,Y)E_1 - (\eta^1\wedge\eta^2)(X,Y)p[E_1,E_2]$$
$$- \{d\eta^2((f_1+f_2)X,Y) + d\eta^2(X,(f_1+f_2)Y)\}E_1$$
$$+ \mathfrak{A}_{X,Y}p\{\eta^1(X)(L_{E_2}J)Y - \eta^2(X)(L_{E_1}J)Y + (L_{f_2X}f_1)Y\} = 0,$$

$$(2.4)\quad \sum_{i=1}^{2} q[f_i,f_i](X,Y) + d\eta^2(X,Y)E_2 - (\eta^1\wedge\eta^2)(X,Y)q[E_1,E_2]$$
$$+ \{d\eta^1((f_1+f_2)X,Y) + d\eta^1(X,(f_1+f_2)Y)\}E_2$$
$$+ \mathfrak{A}_{X,Y}q\{\eta^1(X)(L_{E_2}J)Y - \eta^2(X)(L_{E_1}J)Y + (L_{f_1X}f_2)Y\} = 0$$

hold for any vector fields X and Y, where $\mathfrak{A}_{X,Y}$ denotes the alternating sum with respect to X and Y.

P r o o f. Putting $f = f_1 + f_2$ and $T = \eta^1 \otimes E_2 - \eta^2 \otimes E_1$, we make a substitution of $JX = fX + TX$ into (1.1). The Nijenhuis tensor field $[J,J]$ reduces to $[J,J](X,Y) = [JX,JY] - \mathfrak{A}_{X,Y}J[JX,Y] + J^2[X,Y] = [f,f](X,Y) + T^2[X,Y] + [TX,TY] + \mathfrak{A}_{X,Y}\{[TX,fY] - T[fX,Y] - J[TX,Y]\}$. Further, the quantity $[f,f](X,Y)$ reduces to $[f,f](X,Y) = \Sigma_{i=1}^2[f_i,f_i](X,Y) + \mathfrak{A}_{X,Y}\{(L_{f_2X}f_1)Y - f_2(L_Xf_1)Y\}$ by applying the formulae $[X,Y] = L_XY$, $[X,f_iY] = (L_Xf_i)Y + f_i[X,Y]$. Since $[X,\eta^1(Y)E_2] = \{(L_X\eta^1)(Y) + \eta^1([X,Y])\}E_2 + \eta^1(Y)[X,E_2]$, $d\eta^i(X,Y) = X\eta^i(Y) - Y\eta^i(X) - \eta^i([X,Y])$, consequently, we obtain the identity equation

$$[J,J](X,Y) = \sum_{i=1}^{2}([f_i,f_i] + d\eta^i \otimes E_i)(X,Y) - (\eta^1 \wedge \eta^2)(X,Y)[E_1,E_2]$$

$$- \{d\eta^2(fX,Y) + d\eta^2(X,fY)\}E_1 + \{d\eta^1(fX,Y) + d\eta^1(X,fY)\}E_2$$

$$+\mathfrak{A}_{X,Y}\{\eta^1(X)(L_{E_2}J)Y - \eta^2(X)(L_{E_1}J)Y + (L_{f_2X}f_1)Y - f_2(L_Xf_1)Y\}.$$

The required result follows from $p[J,J] = 0$ and $q[J,J] = 0$. Q.E.D.

P r o o f of Theorem 2.2. The condition $\nabla_X\phi = 0$ implies that $[pX,pY] = p[pX,pY]$, $[qX,qY] = q[qX,qY]$; hence, $D(p)$ and $D(q)$ are completely integrable. Therefore, we obtain the maximal integral manifolds M_i corresponding to $D(p_i)$ respectively, passing through a point. Then M_i are almost contact metric manifolds. It suffices us to see the vanishing of the torsion tensor field on each manifold M_i. In fact, it follows from (2.3) described in Lemma 2.1 that $[f_1,f_1](X,Y) + d\eta^1(X,Y)E_1 - \{d\eta^2(f_1X,Y) + d\eta^2(X,f_1Y)\}E_1 + \mathfrak{A}_{X,Y}\eta^1(X)p(L_{E_2}J)Y = 0$ for any $X, Y \in D(p)$. Similarly, from (2.4) we have $[f_2,f_2](X,Y) + d\eta^2(X,Y)E_2 + \{d\eta^1(f_2X,Y) + d\eta^1(X,f_2Y)\}E_2 - \mathfrak{A}_{X,Y}\eta^2(X)q(L_{E_1}J)Y = 0$. Thus we have the conclusion. Q.E.D.

COROLLARY 2.1. *If a Kähler manifold $M(\mathfrak{D})$ with a J-invariant $\mathfrak{D}$ has a property $\nabla_X\phi = 0$ for any vector field X on M, then M is locally a product manifold of Kähler manifolds M_i determined by $D(p_i)$, passing through a point.*

3. Star (*)-property

Consider an almost product structure which imposes a strict condition comparing with any situation deal with those in the preceding section, which we shall represent (*)-property. Namely, suppose that

(*) $\quad \nabla_X \phi = 0, \qquad (\nabla X)\phi = \phi(\nabla X)$

for $\phi = p - q$ and any vector field X on M, and $\mathfrak{D}_*$ means the almost product structure $\mathfrak{D}$ carrying the (*)-property. In this case, we have put an interpretation on ∇X as a linear transformation field which sends Y to $\nabla_Y X$. By the latter condition described in (*) the maximal integral manifolds M_i are totally geodesic since $\nabla_{pX}Y = p\nabla_X Y$, $\nabla_{qX}Y = q\nabla_X Y$. Then $[pX,pY] = p[X,Y]$, $[pX,qY] = 0$, $[qX,qY] = q[X,Y]$ are valid for any X and Y on M. Therefore, M is regarded as the direct product of M_i at each point. It is natural for us to set up (*)-property on M when we require to study Hermitian manifolds according after the theory of Riemannian direct product manifolds.

Immediately, from Theorem 2.2 we infer:

PROPOSITION 3.1. *A Hermitian manifold $M(\mathfrak{D}_*, \mathcal{F})$ with a J-outspread product structure $(\mathfrak{D}_*, \mathcal{F})$ is locally a product of two normal almost contact metric manifolds M_i determined by $D(p_i)$, passing through a point.*

Proof. The equations (2.1) and (2.2) hold identically on the corresponding manifolds M_i. In fact, $(L_{E_2}J)pX = 0$, $(L_{E_1}J)qX = 0$ are true. Q.E.D.

This assertion implies that the complex structure J is Lie-invariant with respect to E_i, $L_{E_i}J = 0$.

Regarding that $G(X,Y) = G(pX,pY) + G(qX,qY)$ holds identically, we are to introduce a notion of purity on tensor fields with respect to an almost product structure $\mathfrak{D}$. We assume, for simpler notation, that any mixed tensor field $T \in \mathcal{T}^r_s(M)$ of contravariant degree r and covariant degree s has the form $T = \Sigma_a(\text{coefficient})_a T'_a \otimes T''_a$, for $T'_a \in \mathcal{T}^0_s(M)$ and $T''_a \in \mathcal{T}^r_0(M)$. Define a homomorphism C_G of $\mathcal{T}^0_0(M)$-modules $\mathcal{T}^r_s(M)$ to $\mathcal{T}^{r-1}_{s+1}(M)$ by

$$C_G(\omega^1 \otimes \dots \otimes \omega^s \otimes X_1 \otimes \dots \otimes X_r) = \omega^1 \otimes \dots \otimes \omega^s \otimes (C_G X_1) \otimes X_2 \otimes \dots \otimes X_r,$$

$C_G X_1$: the associated 1-form with X_1 by G,

i.e., $(C_G X_1)(Y) = G(X_1, Y)$ for any Y,

for any basis $\omega^1 \otimes \dots \otimes \omega^s \otimes X_1 \otimes \dots \otimes X_r \in \mathcal{T}^r_s(M)$. On the contrary, we can consider a homomorphism C^G of $\mathcal{T}^r_s(M)$ $(s \leq 1)$ to $\mathcal{T}^{r+1}_{s-1}(M)$ given by $\omega(X) = G(C^G\omega, X)$ for any 1-form ω, which is the inverse mapping of C_G. A tensor field $T \in \mathcal{T}^r_s(M)$ except for $(r,s)=(0,0)$ is called to be pure with respect to $\mathfrak{D}$ if T satisfies identically

$$(C^r_G T(X_1,\dots,X_s))(X_{s+1},\dots,X_{r+s})$$
$$= \sum_{i=1}^{2} (C^r_G T(p_i X_1,\dots,p_i X_s))(p_i X_{s+1},\dots,p_i X_{r+s})$$

relative to any $X_1,\dots,X_{r+s} \in \mathcal{T}^1_o(M)$, where C^o_G means the identity. In particular, it can be verified that a $(1,s)$-tensor field T is pure if and only if $T(X_1,\dots,X_s) = \Sigma^2_{i=1} p_i T(p_i X_1,\dots,p_i X_s)$. Suppose that a tensor field $T \in \mathcal{T}^o_s(M)$ is pure with respect to $\mathfrak{D}_*$. Since $(\nabla_X T)(X_1,\dots,X_s) = XT(X_1,\dots,X_s) - \Sigma^s_{a=1} T(X_1,\dots,\nabla_X X_a,\dots,X_s)$ $= \Sigma^2_{i=1}(\nabla_{p_i X} T)(p_i X_1,\dots,p_i X_s)$, the covariant derivarive $\nabla T \in \mathcal{T}^o_{s+1}(M)$ is also pure with respect to $\mathfrak{D}_*$. Consequently, we obtain

LEMMA 3.1. *If* $T \in \mathcal{T}^r_s(M)$ *is pure with respect to* $\mathfrak{D}_*$, *then so is* $\nabla T \in \mathcal{T}^r_{s+1}(M)$.

Thus we have the following immediate consequence by definition of the Riemannian curvature tensor field R; $R(X,Y)Z = \nabla_X \nabla_Y Z - \nabla_Y \nabla_X Z - \nabla_{[X,Y]} Z$.

PROPOSITION 3.2. *The curvature tensor field* $R \in \mathcal{T}^1_3(M)$ *on a Riemannian manifold* $M(\mathfrak{D}_*)$ *is written as*

$$R(X,Y)Z = \sum_{i=1}^{2} p_i R(p_i X, p_i Y) p_i Z;$$

hence

$$(\nabla_V R)(X,Y,Z,W) = \sum_{i=1}^{2} (\nabla_{p_i V} R)(p_i X, p_i Y, p_i Z, p_i W),$$

where we have used $R(X,Y,Z,W) = G(R(X,Y)Z,W)$.

We can get its application as follows.

COROLLARY 3.1. *If a Hermitian manifold* $M(\mathfrak{D}_*)$ *with a J-invariant* $\mathfrak{D}_*$ *is symetric (resp. flat), then the factors* M_i *of* M *are symmetric (resp. flat) Hermitian manifolds, and the converse is also true.*

COROLLARY 3.2. *If a Hermitian manifold* $M(\mathfrak{D}_*, \mathfrak{F})$ *with a J-outspread* $(\mathfrak{D}_*, \mathfrak{F})$ *is symmetric (resp. flat), then the factors* M_i *of* M *are symmetric (resp. flat) normal almost contact metric manifolds, and the converse is also true.*

Proposition 3.2 will be available in order to investigate $M(\mathfrak{D}_*)$ and $M(\mathfrak{D}_*, \mathfrak{F})$ in view of Riemannian Geometry, such as the corollaries mentiond above.

THEOREM 3.1. *Let* $M(\mathfrak{D}_*, \mathfrak{F})$ *be a Hermitian manifold with a J-outspread product structure* $(\mathfrak{D}_*, \mathfrak{F})$. *If* $dF_i = 0$, *then there exist symmetric linear transformation fields* ψ_i $(i = 1,2)$ *on* M *such that* $\nabla E_i = f_i\psi_i = \psi_i f_i$, *the factors* M_i *of* M *are quasi-Sasakian manifolds, and the exterior derivative* $d\Omega$ *of the fundamental 2-form* Ω *has the form* $d\Omega = 2(F_1' \wedge \eta^2 - F_2' \wedge \eta^1)$, *where* F_i' *are defined by* $F_i'(X,Y) = G(f_i\psi_i X, Y)$

THEOREM 3.2. *Let* $M(\mathfrak{D}_*, \mathfrak{F})$ *be a Kähler manifold with a J-outspread product structure* $(\mathfrak{D}_*, \mathfrak{F})$. *Then the factors* M_i *of* M *are both cosymplectic manifolds.*

Theorem 3.1 is verified by Proposition 3.1 and $d\eta^i = 2F_i'$. Theorem 3.2 is obtained by the formula: $\nabla_X J = \nabla_X f_1 + \nabla_X f_2 + (\nabla_X\eta^1) \otimes E_2 - (\nabla_X\eta^2) \otimes E_1 + \eta^1 \otimes \nabla_X E_2 - \eta^2 \otimes \nabla_X E_1$ and Theorem 3.1.

Finally, as a trial to make clear an interpretation of the last assertion further we shall show a contrastive proposition.

PROPOSITION 3.3. *Let* $M(\mathfrak{D}, \mathfrak{F})$ *be a Kähler manifold with a J-outspread almost product structure* $(\mathfrak{D}, \mathfrak{F})$. *If*

$$q\nabla_X Y \in D(\eta^2 \otimes E_2), \quad \nabla_X E_2 = (\lambda\eta^1 \otimes E_1 - I)X, \quad \lambda - \text{a scalar field},$$

for any $X, Y \in D(p)$ *on* M, *then the distribution* $D(p)$ *determines the maximal integral manifold* M_1 *passing through a point which is a Sasakian manifold.*

P r o o f. For any $X, Y \in D(p)$ $q\nabla_X Y$ reduces to $q\nabla_X Y = \eta^2(\nabla_X Y)E_2 = -G(\nabla_X E_2, Y)E_2 = \{G(X,Y) - \lambda\eta^1(X)\eta^1(Y)\}E_2$. Then $D(p)$ is completely integerable. It follows from $\nabla J = 0$ that $0 = (\nabla_X J)Y = \nabla_X JY - J\nabla_X Y = \nabla_X(f_1 Y + \eta^1(Y)E_2) - J(p\nabla_X Y + \eta^2(\nabla_X Y)E_2)$ for any $X, Y \in D(p)$. Consequently, $p(\nabla_X f_1)Y = \eta^1(Y)X - G(X,Y)E_1$, $G(\nabla_X E_1 - f_1 X, Y) = 0$ for any $X, Y \in D(p)$. Thus the set (f_1, E_1, η^1, G) of tensor fields restricted onto M_1 defines a Sasakian structure on M_1. Q.E.D.

R e f e r e n c e s

[1] CALABI,E. and B.ECKMANN: A class of compact complex manifolds which are not algebraic, Ann. of Math. 58 (1953), 494-500.

[2] CHEVALLEY,C.: Theory of Lie Groups, Princeton University Press, 1946.

[3] GOLDBERG,S.I. and K.YANO: On normal globally framed f-manifolds, Tôhoku Math.J. 22 (1970), 362-370.

[4] KANEMAKI,S.: Quasi-Sasakian manifolds, Tôhoku Math. J. 29 (1977), 227-233.

[5] ——: On submanifolds of codimension 2 of almost contact metric manifolds I-II, Tensor N.S. 27 (1973), 281-286 and 32 (1978), 323-331.

[6] KOBAYASHI,S. and K.NOMIZU; Foundations of Differential Geometry I-II, Interscience Publishers, 1963-1969.

[7] MORIMOTO,A.: On normal almost contact structures, J. Math. Soc. Japan 15 (1963), 420-436.

[8] SASAKI,S. and Y.HATAKEYAMA: On differentiable manifolds with contact metric structures, J. Math. Soc. Japan 14 (1962), 249-271.

[9] TASHIRO, Y.: On contact structure of hypersurfaces in complex manifolds I, Tôhoku Math. J. 15 (1963), 62-78.

[10] WALKER,A.G.: Connexions for parallel distributions in the large I-II, Quart. J. Math. Oxford 6 (1955), 301-308 and 9 (1958), 221-231.

[11] ———: Almost-product structures, Proc. Symp. Pure Math. III (1961), 94-100.

[12] WILLMORE,T.J.: Parallel distributions on manifolds, Proc. London Math. Soc. 6 (1956), 191-204.

[13] ———: Connexions for systems of parallel distributions, Quart. J.Math. Oxford 7 (1956), 269-276.

[14] YANO,K.: Differential Geometry on Complex and Almost Complex Spaces, Pergamon Press, 1965.

Department of Mathematics
Science University of Tokyo
Wakamiya-cho 26, Shinjuku-ku
Tokyo, Japan

A ONE-PARAMETER FAMILY OF OPERATORS DEFINED ON ANALYTIC FUNCTIONS IN A CIRCLE

Yûsaku Komatu (Tokyo)

Contents

1. Linear integral operator

Let $\mathcal{F}$ denote the class of functions f regular in the unit disk $E = \{|z| < 1\}$ and normalized by $f(0) = f'(0) - 1 = 0$. On the other hand, let σ be a probability measure supported by the interval $I = [0,1]$. Then a linear integral operator $\mathcal{L}$ is defined on $\mathcal{F}$ by means of

$$\mathcal{L} f(z) = \int_0^1 \frac{f(zt)}{t}\, d\sigma(t).$$

Let the Taylor expansion of $f \in \mathcal{F}$ be given by

$$f(z) = \sum_{\nu=1}^{\infty} c_\nu z^\nu \quad \text{with} \quad c_1 = 1.$$

Then we get by direct calculation

$$\mathcal{L} f(z) = \sum_{\nu=1}^{\infty} \alpha_\nu c_\nu z^\nu, \quad \alpha_\nu = \int_0^1 t^{\nu-1} d\sigma(t) \quad (\nu = 1,2,\ldots; \alpha_1 = 1).$$

Since $f \subset \mathcal{F}$ implies $\mathcal{L} f \in \mathcal{F}$, we can apply $\mathcal{L}$ successively for obtaining

$$\mathcal{L}^n f = \mathcal{L}\mathcal{L}^{n-1} f \quad (n=1,2,\dots),$$

where $\mathcal{L}^0 f = f$. It is readily seen that the series form becomes

$$\mathcal{L}^n f(z) = \sum_{\nu=1}^{\infty} \alpha_\nu{}^n c_\nu z^\nu \quad (n=1,2,\dots).$$

2. One-parameter family of operators

We first attempt to interpolate the sequence $\{\mathcal{L}^n\}$ into a family $\{\mathcal{L}^\lambda\}$ depending on a continuous parameter $\lambda \geq 0$. There are, in general, several ways of interpolation. However, we impose here the condition that the additivity $\mathcal{L}^\lambda \mathcal{L}^\mu = \mathcal{L}^{\lambda+\mu}$ should be valid for any pair $\lambda, \mu \geq 0$. In general, if $\{\alpha_\nu\}$ is any sequence satisfying $\alpha_1 = 1$ and $\overline{\lim} \sqrt[\nu]{|\alpha_\nu|} \leq 1$, then the family $\{\mathcal{L}^\lambda\}$ defined for $f(z) = \sum_{\nu=1}^{\infty} c_\nu z^\nu \in \mathcal{F}$ by means of

$$\mathcal{L}^\lambda f(z) = \sum_{\nu=1}^{\infty} \alpha_\nu^\lambda c_\nu z^\nu$$

satisfies certainly the additivity. However, we are concerned here with a particular type of the sequence $\{\alpha_\nu\}$ such that the operator $\mathcal{L}^\lambda$ also admits an integral representation of a similar form as that for $\mathcal{L}$ defined above.

THEOREM 1. *Suppose that $\{\alpha_\nu{}^\lambda\}$ is a non-negative fully monotone (vollmonton) sequence satisfying $\alpha_1 = 1$. Then there exists a probability measure σ_λ, uniquely determined under the normalization $\sigma_\lambda(t) = (\sigma_\lambda(t-0) + \sigma_\lambda(t+0))/2$, such that the operator defined by $\mathcal{L}^\lambda f(z) = \Sigma \alpha_\nu^\lambda c_\lambda z^\nu$ for $f(z) = \Sigma c_\nu z^\nu \in \mathcal{F}$ admits the integral representation*

$$\mathcal{L}^\lambda f(z) = \int_0^1 \frac{f(zt)}{t}\, d\sigma_\lambda(t).$$

Proof. The full monotoneity of $\{\alpha_\nu{}^\lambda\}$ is a sufficient (and also necessary) condition in order that the moment problem of Hausdorff type

$$\int_0^1 t^{\nu-1} d\sigma_\lambda(t) = \alpha_\nu{}^\lambda \quad (\nu = 1,2,\dots)$$

has a solution σ_λ; cf. [2], [5]. The solution is then unique and, in view of $\alpha_1^{\lambda} = 1$, it is a probability measure. By means of this solution we readily have

$$\mathcal{L}^{\lambda} f(z) = \sum_{\nu=1}^{\infty} \alpha_{\nu}^{\lambda} c_{\nu} z^{\nu}$$

$$= \int_0^1 \sum_{\nu=1}^{\infty} c_{\nu} z^{\nu} t^{\nu-1} d\sigma_{\lambda}(t) = \int_0^1 \frac{f(zt)}{t} d\sigma_{\lambda}(t).$$

3. Monotoneity behaviors of the family

We now observe some functionals defined on $\mathcal{F}$ and illustrate some of their behaviors with respect to λ for the family introduced above. For the sake of brevity, we write $f_{\lambda} = \mathcal{L}^{\lambda} f$.

THEOREM 2. *The quantity defined by*

$$M_{\lambda}(r) = \max_{|z|=r} |f_{\lambda}(z)| \qquad (0 \leq r < 1)$$

is decreasing with respect to λ. *Further, it decreases strictly for* $r \in (0,1)$ *provided* $f(z) \not\equiv z$ *and* $\mathcal{L} f \not\equiv f$.

Proof. In view of the Schwarz lemma, we see that $f_{\lambda} \in \mathcal{F}$ implies $|f_{\lambda}(zt)| \leq M_{\lambda}(r) t$ for $|z| = r$ and $z \in I$. Hence we have for any $\lambda \geq 0$ and $\delta > 0$

$$|f_{\lambda+\delta}(z)| = \left| \int_0^1 \frac{f_{\lambda}(zt)}{t} d\sigma_{\delta}(t) \right| \leq M_{\lambda}(r) \qquad (|z| = r),$$

whence follows $M_{\lambda+\delta}(r) \leq M_{\lambda}(r)$. The strictness is readily verified.

THEOREM 3. *The quantities defined by*

$$h_{\lambda}(r) = \min_{|z|=r} \Re \frac{f_{\lambda}(z)}{z} \quad \text{and} \quad H_{\lambda}(r) = \max_{|z|=r} \Re \frac{f_{\lambda}(z)}{z} \qquad (0 \leq r < 1)$$

are increasing and decreasing, respectively, with respect to λ. *More precisely, we have for* $\delta > 0$

$$h_{\lambda+\delta}(r) \geq h_{\lambda}(r) + \Phi(\delta)(1 - h_{\lambda}(r)),$$

$$H_{\lambda+\delta}(r) \leq H_{\lambda}(r) - \Phi(\delta)(H_{\lambda}(r) - 1),$$

where Φ *is given by*

$$\Phi(\delta) = \int_0^1 \frac{1-t}{1+t} d\sigma_{\delta}(t) = 1 - 2 \sum_{\nu=2}^{\infty} (-1)^{\nu} \alpha_{\nu}^{\delta},$$

the α's *being the moments with respect to* σ. *The equality assertion* is the same as in Theorem 2.

P r o o f. For any fixed z with $|z| = r < 1$, $(f_\lambda(zt)/zt - h_\lambda(r))/(1 - h_\lambda(r))$ may be regarded as a function of a complex variable t regular on $\{|t| \leq 1\}$ unless $f(z) = z$. Since it attains the value 1 at the origin and has the non-negative real part, we have in particular

$$\Re \frac{f_\lambda(zt)/zt - h_\lambda(r)}{1 - h_\lambda(r)} \geq \frac{1-t}{1+t} \quad \text{for} \quad t \in I$$

and hence

$$h_\lambda(rt) \geq h_\lambda(r) + \frac{1-t}{1+t}(1 - h_\lambda(r)),$$

which remains to hold trivially for $f(z) = z$. Consequently, in view of

$$\Re \frac{f_{\lambda+\delta}(z)}{z} = \int_0^1 \Re \frac{f_\lambda(zt)}{zt} \, d\sigma_\delta(t) \quad \text{with} \quad \sigma_\delta(1) = 1,$$

we obtain the desired inequality on h. If the equality sign appears for an $r \in (0,1)$, then there exists a z with $|z| = r$ such that we have

$$\Re \frac{f_\lambda(zt)}{zt} = h_\lambda(r) + \frac{1-t}{1+t}(1 - h_\lambda(r)) \quad \text{for} \quad t \in I.$$

Thus, in view of the analyticity and the normalization at 0, the relation

$$\frac{f_\lambda(zt)}{zt} = h_\lambda(r) + \frac{1-t}{1+t}(1 - h_\lambda(r))$$

holds for $t \in E$. Since the left-hand member is bounded there, we must have $h_\lambda(r) = 1$ and hence $f(z) = z$. Next, by applying the result just verified to $2z - f(z)$, we obtain the result on H instead of h.

COROLLARY. The quantities defined in Theorem 3 satisfy for $\delta > 0$

$$h_{\lambda+\delta}(r) \geq h_\lambda(r) + (1 - e^{-\Phi'(0)\delta})(1 - h_\lambda(r)),$$

$$H_{\lambda+\delta}(r) \leq H_\lambda(r) - (1 - e^{-\Phi'(0)\delta})(H_\lambda(r) - 1).$$

P r o o f. Theorem 3 asserts that

$$\frac{h_{\lambda+\delta}(r) - h_\lambda(r)}{\delta} \geq \frac{\Phi(\delta)}{\delta}(1 - h_\lambda(r)),$$

whence follows for $\delta \longrightarrow +0$

$$\frac{\partial}{\partial\lambda} h_\lambda(r) \geq \Phi'(0)(1-h_\lambda(r)).$$

This linear differential inequality can be brought readily into the finite form. In fact, we may write it in the form

$$\frac{\partial}{\partial\lambda}(e^{\Phi'(o)\lambda}h_\lambda(r)) \geq \Phi'(0)e^{\Phi'(o)\lambda}$$

which becomes after integration with respect to λ over the interval $(\lambda, \lambda+\delta)$

$$e^{\Phi'(o)(\lambda+\delta)}h_{\lambda+\delta}(r) - e^{\Phi'(o)\lambda}h_\lambda(r) \geq e^{\Phi'(o)(\lambda+\delta)} - e^{\Phi'(o)\lambda}.$$

This is the desired estimation for h. Similar argument applies also for H.

Now, since the first inequality in Theorem 3 implies that in its corollary, the latter is regarded as weaker than the former. There are two possibilities: either

$$\text{(A)}\quad \Phi(\delta) \geq 1 - e^{-\Phi'(o)\delta} \quad \text{or} \quad \text{(B)} \quad \text{otherwise.}$$

Based on the above-mentioned reason, it will be natural to expect that the case (A) appears provided the class $\{h_\lambda(r)\}(\lambda > 0,\ r \in I,\ f \in \mathcal{F})$ is wide enough. But, in general, (A) is of course invalid without any restriction; for instance, $\Phi(\delta) = 1 - e^{\alpha\delta} - \delta^2$. Accordingly, a question arises for giving a condition necessary and/or sufficient in terms of the class in order that the case (A) holds.

4. A distinguished family

A distinguished family of operators is generated by a special measure $\sigma(t) = t$. The operator $\mathcal{L}$ is then determined by

$$\mathcal{L}f(z) = \int_0^1 \frac{f(zt)}{t}dt$$

and the associated moments are given by

$$\alpha_\nu = \int_0^1 t^{\nu-1}dt = \frac{1}{\nu} \qquad (\nu = 1, 2, \ldots).$$

In the following lines we restrict ourselves exclusively to this distinguished case.

THEOREM 4. *The solution of the moment problem*

$$\int_0^1 t^{\nu-1} d\sigma_\lambda(t) = \frac{1}{\nu^\lambda} \qquad (\nu = 1,\ 2,\ \ldots)$$

is given by the probability measure

$$\sigma_\lambda(t) = \int_0^t \rho_\lambda(\tau)d\tau \quad \text{with} \quad \rho_\lambda(t) = \frac{1}{\Gamma(\lambda)}(\log \frac{1}{t})^{\lambda-1}$$

and hence $\mathfrak{L}^\lambda$ *is represented in the form*

$$\mathfrak{L}^\lambda f(z) = f_\lambda(z) \equiv \frac{1}{\Gamma(\lambda)} \int_0^1 \frac{f(zt)}{t} (\log \frac{1}{t})^{\lambda-1} dt .$$

P r o o f. Basing on the uniqueness of the solution of the moment problem, we have only to verify that σ_λ given in the theorem satisfies the assigned problem, what is an immediate consequence of a familiar formula.

Once the explicit expression of σ_λ having been determined, the factor Φ contained in Theorem 3 is correspondingly specialized.

THEOREM 5. *The factor contained in Theorem 3 is evaluated in the form*

$$\Phi(\delta) = 1 - 2 \sum_{\nu=2}^{\infty} \frac{(-1)^\nu}{\nu^\delta} = 2(1 - 2^{1-\delta})\zeta(\delta) - 1,$$

where ζ *denotes the Riemann zeta function.*

COROLLARY. *The factor contained in the corollary of Theorem 3 is evaluated in the form*

$$1 - e^{-\Phi'(0)\delta} = 1 - (\frac{2}{\pi})^\delta .$$

P r o o f. Direct calculation yields

$$\Phi'(\delta) = 2(1 - 2^{1-\delta})\zeta'(\delta) + 2^{2-\delta}\log 2 \cdot \zeta(\delta),$$

whence follows, in view of $\zeta(0) = -1/2$ and $\zeta'(0) = -(1/2)\log(2\pi)$,

$$\Phi'(0) = \log(2\pi) - 2 \log 2 = \log \frac{\pi}{2} .$$

5. Relation to fractional integration

The operator generated by $\sigma(t) = t$ is closely connected with the fractional integration. In fact, the expression in Theorem 4 can be written in the complex integral form

$$\mathcal{L}^{\lambda}f(z) = \frac{1}{\Gamma(\lambda)} \int_0^z \frac{f(\zeta)}{\zeta} (\log \frac{z}{\zeta})^{\lambda-1} d\zeta,$$

where the integration is taken along the ray from 0 to z and the kernel of the integral denotes the branch which is real positive on the ray. On the other hand, by putting $w = \log z$, we denote by $D^{-\lambda}$ the fractional integration of order λ applicable to any function ψ regular in the half-plane $\{\Re w < 0\}$ and vanishing at the point at infinity. It is defined by

$$D^{-\lambda}\psi(w) = \frac{1}{\Gamma(\lambda)} \int_{\infty}^{w} \psi(\omega)(w-\omega)^{\lambda-1} d\omega,$$

where the integration is taken along the half straight line parallel to the real axis which is contained in the left half-plane. If we put $w = \log z$, $\psi(w) = f(z)$ and compare the above two expressions, we obtain the relation

$$\mathcal{L}^{\lambda}f(z) = D^{-\lambda}f(e^w) \qquad (w = \log z).$$

Since $D^{-\lambda}$ is invertible into the fractional differentiation D^{λ}, the definition region of $\mathcal{L}^{\lambda}$ can be accordingly extended into the negative value of λ. It is readily verified that the series expansion

$$\mathcal{L}^{\lambda}f(z) = \sum_{\nu=1}^{\infty} \frac{c_\nu}{\nu^{\lambda}} z^{\nu},$$

where $f(z) = \sum_{\nu=1}^{\infty} c_\nu z^{\nu} \in \mathfrak{F}$, remains valid for any real λ. In particular, since the sequence $\{1/\nu^{\lambda}\}$ is increasing for $\lambda < 0$, the operator $z^{-1}\mathcal{L}^{\lambda}$ with $\lambda < 0$ may be regarded as a particular Gel'fond-Leont'ev derivative introduced in [1].

6. Some classes of univalent functions

Finally, we consider the subclass $\mathfrak{S}$ of $\mathfrak{F}$ consisting of functions which are univalent in E. Let $\mathfrak{S}^*$ and $\mathfrak{K}$ denote its subclasses which consist of functions starlike with respect to the origin and convex, respectively. On the other hand, let $\mathfrak{F}^+$ be the subclass of $\mathfrak{F}$ consisting of functions f which satisfy $\Re(f(z)/z) > 0$ in E.

<u>THEOREM 6</u>. (i) <u>If</u> $f_\lambda \in \mathfrak{F}^+$, <u>then</u> $f_\mu \in \mathfrak{S}$ <u>for</u> $\mu \geqq \lambda + 1$. (ii) <u>If</u> $zf_{\lambda+1}/f_\lambda \in \mathfrak{F}^+$, <u>then</u> $f_{\lambda+1} \in \mathfrak{S}^*$. (iii) If $f_\lambda \in \mathfrak{S}^*$, <u>then</u> $f_{\lambda+1} \in \mathfrak{K}$.

P r o o f. (i) First, $f_{\lambda+1} = \mathcal{L} f_\lambda$ implies $zf_{\lambda+1}' = f_\lambda \in \mathcal{F}^+$. For $\mu > \lambda+1$, $f_\mu = \mathcal{L}^{\mu-\lambda-1} f_{\lambda+1}$ implies $zf_\mu' = \mathcal{L}^{\mu-\lambda-1}(zf_{\lambda+1}') \in \mathcal{F}^+$. Hence, in view of the Noshiro-Wolff Theorem [3, 6], we have $f_\mu \subset \mathcal{S}$ for $\mu \geq \lambda+1$. (ii) Since $zf_{\lambda+1}' = f_\lambda$, we get $z^2 f_{\lambda+1}'/f_{\lambda+1} = zf_\lambda/f_{\lambda+1} \in \mathcal{F}^+$ which is the condition for $f_{\lambda+1} \in \mathcal{S}^*$. (iii) If $f_\lambda \in \mathcal{S}^*$, then $zf_{\lambda+1}' \in \mathcal{S}^*$ which is the condition for $f_{\lambda+1} \in \mathcal{K}$.

THEOREM 7. <u>Let</u> $f = \Sigma_{\nu=1}^{\infty} c_\nu z^\nu \in \mathcal{F}$ <u>satisfy the condition that there exists a constant</u> N <u>for which</u> $c_\nu = O(\nu^N)$ <u>as</u> $\nu \longrightarrow \infty$. <u>Then we have</u> $\mathcal{L}^\lambda f \subset \mathcal{S}^*$ <u>provided</u> λ <u>is large enough</u>.

P r o o f. We may suppose that there exists a positive integer $M \geq 3$ satisfying $|c_\nu| \leq M\nu^{\lambda-4}$ for $\lambda \geq N+4$. We can then find a value $\Lambda \geq N+4$ such that

$$\sum_{\nu=2}^{M-1} \frac{|c_\nu|}{\nu^{\Lambda-1}} < 2 - \frac{\pi^2}{6} .$$

Thus we obtain for $\lambda \geqq \Lambda$

$$\sum_{\nu=2}^{\infty} \frac{|c_\nu|}{\nu^{\lambda-1}} \leqq \sum_{\nu=2}^{M-1} \frac{|c_\nu|}{\nu^{\Lambda-1}} + \sum_{\nu=M}^{\infty} \frac{M}{\nu^3} < 2 - \frac{\pi^2}{6} + \sum_{\nu=M}^{\infty} \frac{1}{\nu^2} < 1 .$$

This shows that a condition sufficient for $\mathcal{L}^\lambda f \in \mathcal{S}^*$ is satisfied.

COROLLARY. <u>If there exists a</u> κ <u>such that</u> $\inf_{z \in E} \mathfrak{R}(f_\kappa(z)/z) > -\infty$, <u>then</u> $f_\lambda \in \mathcal{S}^*$ <u>provided</u> λ <u>is large enough</u>.

P r o o f. Suppose, by assumption, $\mathfrak{R}(f_\kappa(z)/z) \geqq h_\kappa > -\infty$ in E. Then we see that

$$\frac{f_\kappa(z) - h_\kappa z}{1 - h_\kappa} \equiv z + \sum_{\nu=2}^{\infty} d_\nu z^\nu \in \mathcal{F}^+$$

and hence $|d_\nu| \leq 2$. Consequently, the Taylor coefficients of f_κ are bounded so that, in view of Theorem 7, $f_\lambda = \mathcal{L}^{\lambda-\kappa} f_\kappa \in \mathcal{S}^*$ provided λ is large enough.

THEOREM 8. <u>If</u> $f \in \mathcal{S}$, <u>then</u> $f_\lambda \in \mathcal{S}^*$ <u>at least for</u> $\lambda \geq \lambda_0$ <u>where</u> λ_0 <u>is a certain number less than</u> 4.

P r o o f. Let $f(z) = \sum_{\nu=1}^{\infty} c_\nu z^\nu \in \mathcal{S}$. It is known that its coefficients are estimated by $|c_\nu| \leqq \nu$ $(\nu \leqq 6)$ and, in general, $|c_\nu| \leqq \sqrt{7/6}\,\nu$; cf., for instance, [4]. Thus we get

$$\sum_{\nu=2}^{\infty} \frac{|c_\nu|}{\nu^3} \leqq \sum_{\nu=2}^{6} \frac{1}{\nu^2} + \sqrt{\frac{7}{6}} \sum_{\nu=7}^{\infty} \frac{1}{\nu^2} < \sqrt{\frac{7}{6}} \left(\frac{\pi^2}{6} - 1\right) < 1.$$

Hence we see that a condition $\Sigma_{\nu=2}^{\infty}|c_\nu|/\nu^{\lambda-1} \leqq 1$ sufficient for $f_\lambda \in S^*$ is satisfied for $\lambda \geqq \lambda_0$ with a certain number λ_0 less than 4.

REMARK. The method used here for proving Theorem 8 is not powerful. In fact, even if the Bieberbach's conjecture would have been affirmatively solved, we could only conclude that we can take as λ_0 the unique root of the equation $\zeta(\lambda-2) = 2$ which is contained in the interval (3, 4).

In conclusion, it seems plausible to conjecture that:

(i) If $f \in S$, then $f_\lambda \in S$ at least for $\lambda \geqq 1$;

(ii) If $f \in K$ (or, more generally, $f \in S^*$), then $f_\lambda \in K$ at least for $\lambda \geqq 1$.

References

[1] GELFOND, A.O. and A.F.LEONT'EV: On a generalization of Fourier series, Mat. Sbornik N.S. 29 (71) (1951), 477-500.

[2] HAUSDORFF,F.: Summationsmethoden und Momentfolgen, Math. Z. 9 (1921), 74-109 and 280-299.

[3] NOSHIRO,K.: On the univalency of certain analytic functions, J.Fac. Sci.Hokkaido Imp. Univ. (1) 2 (1934), 89-101.

[4] POMMERENKE,C.: Univalent functions. Studia Math. 25 (Göttingen) (1975), p.24 et seq.

[5] SHOHAT, J.A. and J.D.TAMARKIN: The problem of moments, Math.Surveys 1, Amer. Math. Soc. (New York) (1943), pp. 9 and 11.

[6] WOLFF,J.: L'intégrale d'une fonction holomorphe et à partie réelle positive dans un demi-plan est univalente, C.R. Acad. Paris 198 (1934), 1209-1210.

Department of Mathematics
Chuo University, 1-13-27 Kasuga
Bunkyo-ku, Tokyo 112, Japan

GENERAL VEKUA OPERATORS

Eberhard Lanckau (Karl-Marx-Stadt)

Contents

1. Introduction

In 1948 I.N. Vekua published his "New methods for solving elliptic equations". He wrote a linear partial differential equation of elliptic type and of second order in the plane using complex independent variables z, z^* as

$$\frac{\partial^2}{\partial z \partial z^*}u + A_1(z,z^*)\frac{\partial u}{\partial z} + A_2(z,z^*)\frac{\partial u}{\partial z^*} + A_o(z,z^*)u = 0$$

(notice $4\partial^2/\partial z\partial z^* = \partial^2/\partial x^2 + \partial^2/\partial y^2$, if $z = x + iy$, $z^* = x - iy$), and then he found the (general) solution of this equation by the application of a complex integral transform

$$u(z,z^*) = \int_0^z R(z,z^*,t,t^*)\, f(t)\, dt,$$

$f(z)$ being an arbitrary holomorphic function of z. Here $R = R(z,z^*,t,t^*)$ is the complex Riemann function; Vekua proved its unique existence. We want to show that Vekua's technique is a very powerful tool to solve partial differential equations and that it may be generalized very widely. However, we only give the generalizations of the self adjoint case $A_1 = A_2 = 0$.

2. The Riemann transform

Our concept to solve linear partial differential equations is as follows:

Let $u = u(z,z^*,\tau)$ with $z \in G$, $z^* \in G^*$, $\tau \in T$, G, G^*, T bounded and simply connected, $0 \in G$, $0 \in G^*$, be a solution of

(1) $$L\,u = \frac{\partial^2}{\partial z \partial z^*} u + Su = 0.$$

Here S is a linear operator, and let (1) be self-adjoint with respect to z, z^*. Let F be a set of functions $f(z,\tau)$ holomorphic with respect to z in a certain domain $G_0 \subset G$, $0 \in G_0$, and let a transform Rf exist:

$$R[f(t,\tau)] = h(z,z^*,t,t^*,\tau)$$

with $z \in G_0$, $z^* \in G_0$, $t \in G_0$, $t^* \in G_0$, $\tau \in T_0$ $(0 \in G_0^* \subset G^*)$ for all $f \in F$ with the properties

(I) $(\partial/\partial z)\, R[f(t,\tau)]$, $(\partial^2/\partial z \partial z^*) R[f(t,\tau)]$, $SR[f(t,\tau)]$

exist,

(II) $R[f(t,\tau)]$ integrable with respect to t in G_0,

(III) in $G_0 \times G_0^* \times T_0$:

(2) $LR[f(t,\tau)] = 0$ for all $t \in G_0$, $t^* \in G_0^*$

(3) $(\partial/\partial z^*) R[f(t,\tau)] = 0$ for all $t \in G_0$ and $t = z$.

<u>THEOREM</u>. <u>For all</u> $f \in F$

(4) $$u(z,z^*,\tau) = \int_0^z R[f(t,\tau)]\, dt$$

<u>is a solution of</u> (1) <u>in</u> $G_0 \times G_0^* \times T_0$.

<u>Proof</u> by inserting (4) into (1) using (2),(3).

3. An example

We do not consider generally the existence of the transform R, but we show the high versatility of the above concept by a simple example.

Let S in (1) be an operator, not depending on z, z^*. Some important differential equations of second order in mathematical physics are of this type.

Firstly we define a set of functions F, depending on S.

D e f i n i t i o n. $f(z,\tau) \in F$ iff:

(I) $f(z,\tau)$ is holomorphic with respect to z in G and continuous in $\bar{G}$.

(II) Constants $\alpha > 0$, $c > 0$, $p \geqq 0$ (integer) exist with

(5) $$|S^m f| \leqq \alpha\, c^m (m+p)!^2 \qquad \text{for } m \geqq m_o$$

for all $z \in G$, $\tau \in T_o \subset T$.

Immediately from (5) we have

(6) $S^m f \in F$

We show by examples that F is not empty in the cases of some interesting operators S. In the following examples let condition (I) be satisfied.

E x a m p l e 1 (two-dimensional equation). Let $Su = a\,u$, a constant; thus $S^m u = a^m u$. If $f(z,\tau)$ is bounded in $\bar{T}$, (5) is valid with $T_o = T$.

E x a m p l e 2 (three-dimensional equation).

2.1. Let

(7) $$S = \sum_{k=o}^{2} a_k\, \partial^k/\partial\tau^k, \quad a_o, a_1, a_2 \text{ constant.}$$

$f \in F$ if f is holomorphic with respect to τ in T, continuous in $\bar{T}$, and $T_o \subset T$ is given by

(8) $$|\zeta - \tau| \geqq \delta > 0, \quad \delta < 1, \quad \text{for all } \zeta \in \partial T, \quad \tau \in T_o.$$

P r o o f. Let $k = (k_o, k_1, k_2)$ be a multi-index, $a = (a_o, a_1, a_2)$ a vector. For $m \geqq 0$ with $jk = k_1 + 2k_2$

(9) $$S^m = \sum_{|k|=m} (m!)(k!)\, a^k,\ \partial^{jk}/\partial\tau^{jk}.$$

In $\bar{G} \times \bar{T}$ we have $|f(z,\tau)| \leq M$.

In T_o we may write

(10) $$f(z,\tau) = \frac{1}{2\pi i} \oint_{|\zeta-\tau|=\delta} f(z,\zeta)\frac{d\zeta}{\zeta-\tau},$$

and therefore

$$|\partial^{jk}/\partial\tau^{jk} f| \leq M\,(jk)!\,\delta^{-jk}.$$

Using $b = \max_j |a_j|$ and with $(m!)/(k!) \leqq 3^m$, $jk = 2|k| = 2m$ we find from (9)

$$|S^m f| \leqq M\,(3b)^m \sum_{|k|=m} \delta^{-jk}.$$

The sum has $(m+1)^2$ terms, thus we have with

(11) $(2m)! \leqq 4^m\, m!^2$

$$|S^m f| \leqq M\,(12\, b\, \delta^{-2})^m\, (m+1)!^2,$$

that is (5).

R e m a r k. For $a_2 = 0$ (5) is valid with every $c > 0$. With $a_2 = 0$ we have from (9)

$$|S^m f| \leqq \sum_{k_0+k_1=m} M \cdot \frac{m!}{k_0!\,k_1!} |a_0^{k_0}\, a_1^{k_1}|\, k_1!\; \delta^{-k_1}.$$

Using b as above, we have

$$|S^m f| \leqq M\,(2b)^m\, m! \sum_{k_1=0}^{m} \delta^{-k_1}.$$

The sum has $m+1$ terms, hence

$$|S^m f| \leqq M\,(2b\,\delta^{-1})^m (m+1)!.$$

For every $c > 0$ we get

$$\lim_{m\to\infty} (2b\,\delta^{-1} c^{-1})^m / m! = 0$$

(as terms of a convergent series), and therefore

$$(2b\,\delta^{-1})^m \leqq c^m\, m!$$

for every $c > 0$ and sufficiently large $m > m_0(c)$; thus

$$|S^m f| \leqq M\, c^m\, (m+1)!^2$$

for $m > m_0$.

R e m a r k. The equation (1) with (7) is elliptic for $a_2 > 0$, hyperbolic for $a_2 < 0$, parabolic for $a_1 \neq a_2 = 0$.

2.2. Let

(12) $S = \tau\, \partial^2/\partial\tau^2$.

$f \in F$ if f has the same properties as in Example 2.1 and $T_0 \subset T$ is the same domain.

P r o o f. By induction we find

(13) $$S^m = \sum_{k=o}^{m-1} k!\binom{m}{k}\binom{m-1}{k} \tau^{m-k} \partial^{2m-k}/\partial\tau^{2m-k}.$$

With the Cauchy integral (10) and with the same meaning of M as above, we get with $\tau^* = \max(1, \text{diam } T)$ using $\binom{m}{k} \leqq 2^m$:

$$|S^m f| \leqq M\,(4\,\tau^*)^m \sum_{k=0}^{m-1} k!(2m-k)!\; \delta^{k-2m}.$$

With $k!(2m-k)! = (2m)!$ we find with (11)

$$S^m f \leqq M\,(16\,\tau^*\,\delta^{-2})^m\,(m+1)!^2,$$

that is (5).

R e m a r k. The equation (1) with (12) is of mixed type.

2.3. Let

(14) $$Su = a_o u + a_1 \hat{S}_p u$$

with

$$\hat{S}_p u = \frac{1}{\Gamma(p+1)}\,(\partial/\partial\tau) \int_o^\tau u(z,z^*,\sigma)(\tau-\sigma)^p\, d\sigma,$$

$a_o, a_1, p > 0$ constant.

$f \in F$ if f is bounded in $\bar{T} \times \bar{G}$ and integrable with respect to τ in $T_o = T$, $0 \in T$ (the path of integration in (14) also in T).

P r o o f. We have

$$\hat{S}_p^k = \hat{S}_{kp};$$

from $$\hat{S}_o f = f,$$

$$\hat{S}_{kp} f = \frac{1}{\Gamma(kp)} \int_o^\tau f(z,\sigma)(\tau-\sigma)^{kp-1}\, d\sigma$$

$$(kp > 0)$$

we have with $|f(z,\tau)| \leqq M$:

$$|\hat{S}_p^k f| \leqq \frac{1}{\Gamma(pk)}\, M \cdot \tau^{*pk} \qquad \text{for } k > 0$$

with $\tau^* = \text{diam } T$; and with $g = \max_{x>0} 1/\Gamma(x) > 1$ we have for all $k \geqq 0$:

$$|\hat{S}_p^k f| \leqq g\, M\, \tau^{*pk},$$

and therefore

(15) $$S^m = \sum_{k=o}^{m} \binom{m}{k} a_o^{m-k} a_1^k \hat{S}_p^k ;$$

hence

$$|S^m f| \leqq g\ M\ (2b)^m (m+1)$$

with $b = \max(|a_o|, |a_1|\ \tau^{*p})$. Therefore (5) is valid with every $c > 0$ (the last may be proved as in the remark of Example 2.1.).

<u>Remark</u>. For $p = 1$ equation (1) with (14) may be written as a "pseudoparabolic" equation

$$(\partial^3 / \partial z \partial z^* \partial \tau)\ u + a_o(\partial / \partial \tau) u + a_1 u = 0.$$

<u>Counterexample</u>. Let $S = \partial^3/\partial \tau^3$.
F is not empty, because $e^{\tau} \in F$, but not all holomorphic (with respect to τ in T) functions are in F. For instance with $\tau_o \notin T$

$$f(z, \tau) = 1/(\tau_o - \tau) \notin F,$$

with

$$S^m f = (3m)!\,(\tau_o - \tau)^{-3m-1}$$

(5) is not valid, because for $m \to \infty$

$$(3m)!/(c^m (m+p)!^2) \to \infty$$

for every $c > 0$, $p \geqq 0$ (integer).

<u>2.4</u>. (Equation with transformed argument). Let

$$Su(z,z^*, \tau) = a_o u + a_1 u(z,z^*, \tau + \Delta\tau),\ a_o, a_1, \Delta\tau \text{ constant.}$$

$f \in F$ if f is in τ an entire function of exponential type; here $T_o = T$.

<u>Proof</u>. From

$$S^m f = \sum_{k=o}^{m} \binom{m}{k} a_o^{m-k} a_1^k\ f(z, \tau + k\cdot\Delta\tau)$$

and

$$|f(z, \tau)| \leqq M\ e^{\lambda|\tau|},\quad \lambda > 0,$$

we have

$$|S^m f| \leqq \sum_{k=o}^{m} \binom{m}{k} |a_o^{m-k} a_1^k|\ e^{k|\Delta\tau|\lambda}\ M\ e^{\lambda|\tau|}$$

and with

$$b = 2 \max (|a_0|, |a_1| e^{\lambda \cdot |\Delta \tau|})$$

we find

$$|S^m f| \leqq b^m M e^{\lambda |\tau|}(m + 1)$$
$$\leqq M e^{\lambda \tau^*} b^m (m + 1),$$

with $\tau^* = \text{diam } T$, and therefore (5) may be satisfied with every $c > 0$ (the last again proved as in the remark of Example 2.1).

<u>Example 3</u> (four-dimensional equation). Let

$$\tau = (\tau_1, \tau_2) \quad \text{with} \quad \tau_1 \in T_1, \quad \tau_2 \in T_2, \quad T = T_1 \times T_2.$$

Let

(16) $S = a_1(\partial^p / \partial \tau_1^p) + a_2(\partial^q / \partial \tau_2^q)$, a_1, a_2 constant,

$p = 1, 2; \ q = 1, 2.$

$f(z, \tau_1, \tau_2) \in F$ if f is holomorphic with respect to τ_1 in T_1 and with respect to τ_2 in T_2, and bounded in $\bar{T}$; moreover $T_{o1} \subset T_1$, $T_{o2} \subset T_2$, $T_o = T_{o1} \times T_{o2}$ are given by

$$|\zeta_k - \tau_k| \geqq \delta > 0, \quad \delta < 1, \quad \text{for all} \quad \zeta_k \in \partial T_k, \ \tau_k \in T_{ok},$$

$k = 1, 2.$

<u>Proof</u>. We repeat the calculations of Example 2.1. For $m \geqq 0$:

(17) $$S^m = \sum_{k_1 + k_2 = m} \frac{m!}{k_1! \, k_2!} a_1^{k_1} a_2^{k_2} \partial^{pk_1 + qk_2} / \partial \tau_1^{pk_1} \partial \tau_2^{qk_2} .$$

On the account of the holomorphy of f (and its boundedness) a $M > 0$ exists with

$$|f(z, \tau_1, \tau_2)| \leqq M \qquad \text{in} \quad G \times T.$$

From this follows by Cauchy's inequality:

$$|\partial^{pk_1 + qk_2} f / \partial \tau_1^{pk_1} \partial \tau_2^{qk_2}| \leqq M (pk_1)! \, (qk_2)! \, \delta^{-pk_1 - qk_2};$$

using $m!/(k_1! k_2!) \leqq 2^m$ and

$$(pk_1)! \, (qk_2)! \leqq (pk_1 + qk_2)! \leqq (2m)! ,$$

$$b = 2 \max_{k=1,2} |a_k|.$$

We find from (17)

$$|S^m f| \leq M\, b^m\, (2m)! \sum_{k_1+k_2=m} \delta^{-pk_1-qk_2} .$$

The last sum is $\leq (m+1)\,\delta^{-2m}$, and with (10) we have (5):

$$|S^m f| \leq M\, (4b\,\delta^{-2})^m (m+1)!^2 .$$

Remark 1. Equation (1) with (16) is elliptic for $p = 2$, $a_1 > 0$, $a_2 > 0$; hyperbolic for $p = 2$, $a_1 a_2 < 0$; ultrahyperbolic for $p = 2$, $a_1 < 0$, $a_2 < 0$; parabolic for $p = 1$, $a_1 \neq 0$.

Remark 2. Partial differential operators S of second order with constant coefficients in higher dimensions are to be treated in the same way.

Now we give the transform Rf in the set F.
This construction may be applied for all examples considered above, not depending on the dimension and the type of the equation, and thus we have an unique technique to construct solutions of certain partial differential equations.

THEOREM. *Let* $v = (z - t)(t^* - z^*)$ *with* $z \in G$, $z^* \in G^*$, $t \in G$, $t^* \in G^*$. *The transform* Rf *for the equation* (1) *with an operator* S, *not depending on* z, z^* *in* $D_o \times T_o$, *where* D_o *is a closed polydisc in* $G_z \times G^*_{z^*} \times G_t \times G^*_{t^*}$ $\{(z,z^*,t,t^*): |v| \leq s^2 < 1/c\}$ *is given by*

$$(18) \quad Rf(t,\tau) = \sum_{m=o}^{\infty} \frac{1}{m!^2} v^m S^m f(t,\tau)$$

for all $f \in F$.

Remark. D_o is not empty. Let $d = \frac{1}{2} \min (s, \min_{z\in\partial G} |z|) > 0$ because $0 \in G$), and let $|z|, |z^*|, |t|, |t^*| < d$; then $|v| \leq s^2$ and $D_o = G_{oz} \times G^*_{oz^*} \times G_{ot} \times G^*_{ot^*}$ with $G_o = \{z: |z| \leq d\} \subset G$, $G^*_o = \{z^*: |z^*| \leq d\} \subset G^*$ is not empty.

Proof. 1. Existence: We show the absolute uniform convergence of

$$R_k f(t,\tau) = \sum_{m=o}^{\infty} \frac{1}{m!^2} v^m S^{m+k} f, \qquad k = 0,1,2,\ldots$$

in $D_o \times T_o$, then $Rf = R_o f$. We have

$$|R_k f(t, \tau)| \leqq \sum_{m=o}^{\infty} \frac{1}{m!^2} |v|^m |S^{m+k} f(t, \tau)|$$

and with (5)

$$|R_k f(t, \tau)| \leqq \alpha \sum_{m=o}^{\infty} \frac{(m+p+k)!^2}{m!^2} |cv|^m ,$$

and this series converges uniformly in $|cv| \leq 1 - \varepsilon$, $\varepsilon > 0$.

2. Validity of conditions:

(I) The uniform absolute convergence allows the term-by-term-differentiation (with respect to z, z^* (18) is a power series), and therefore the derivatives

$$(\partial / \partial z^*) \, Rf(t, \tau) = (t - z) \sum_{m=o}^{\infty} \frac{1}{m!(m+1)!} v^m S^{m+1} f(t, \tau),$$

$$(\partial^2 / \partial z \partial z^*) \, Rf(t, \tau) = - R_1 f(t, \tau).$$

exist in $D_o \times T_o$. Further we have

(19) $S^n R_k f = R_{n+k} f$ for $k, n = 0, 1, 1, \ldots$

Proof. $R_{k+n} f$ converges uniformly; therefore with every $\varepsilon > 0$

$$|\sum_{m=N}^{M} \frac{1}{m!^2} v^m S^{n+m+k} f|$$

$$= |\sum_{m=o}^{M} \frac{1}{m!^2} v^m S^{n+k+m} f - \sum_{m=o}^{N-1} \frac{1}{m!^2} v^m S^{n+k+m} f|$$

$$= |\sum_{m=o}^{M} \frac{1}{m!^2} v^m S^{n+k+m} f - S^n \sum_{m=o}^{N-1} \frac{1}{m!^2} v^m S^{k+m} f| < \varepsilon$$

for $N \geqq N(\varepsilon)$ and all $M \geqq N$ in $D_o \times T_o$. Letting $M \to \infty$ we have

$$|R_{k+n} f - S^n \sum_{m=o}^{N-1} \frac{1}{m^2} v^m S^{m+k} f| \leqq \varepsilon;$$

letting $N \to \infty$ we obtain

$$|R_{k+n} f - S^n R_k f| \leqq \varepsilon .$$

Because ε is arbitrary, we have (19), and especially

$$SRf(t, \tau) = R_1 f(t, \tau).$$

(II) With respect to t $Rf(t, \tau)$ is a uniformly convergent series of holomorphic functions $(z - t)^m S^m f(t, \tau)$, and therefore it is holomorphic and may be integrated.

(III) The conditions (2),(3) follow from (I).

(R e m a r k. From

$$(\partial/\partial z)\, Rf(t,\tau) = (t^* - z^*) \sum_{m=0}^{\infty} \frac{1}{m!(m+1)!} v^m S^{m+1} f(t,\tau)$$

follows

$$(\partial/\partial z)\, Rf(t,\tau) = 0 \quad \text{for} \quad t^* = z^* \quad \text{in} \quad D_0 \times T_0 .$$

From (18) we have

$$Rf(t,\tau) = f(t,\tau) \quad \text{for} \quad t = z, \quad t^* = z^* \quad \text{in} \quad G_0 \times T_0 .)$$

3. R e m a r k. $Rf(t,\tau) \in F$ if $f \in F$. Besides the holomorphy with respect to t (see (II)) we have from (19):

$$|S^m Rf| = |R_m f| \leq \sum_{k=0}^{\infty} \frac{1}{k!^2} |v|^k |S^{m+k} f|$$

$$= \alpha\, c^m\, (m+p)!^2 \sum_{k=0}^{\infty} \binom{k+m+p}{k}^2 |cv|^k .$$

From an estimation of

$$(m+p)!^2 \sum_{k=0}^{\infty} \binom{k+m+p}{k}^2 x^k = \left(\frac{d}{dx}\left(x\frac{d}{dx}\right)\right)^{m+p}\left(\frac{x^{m+p}}{1-x}\right)$$

in $x = |cv| = +1 - \varepsilon$, $\varepsilon > 0$, we find again an estimation of the type as (5).

R e m a r k. From the absolute uniform convergence of (18) we have in the domain of convergence for $f \in F$:
Rf is holomorphic with respect to t and τ for the operators (7), (12), (16).
Rf is holomorphic with respect to t and bounded and integrable with respect to τ for the operator (14).

4. Concluding remarks

1. The Riemann transform Rf (18) for equation (1) with (7), (12) may be written (using (9) and (15)) as

$$Rf = \oint_{|\zeta-\tau|=\delta} H(z,z^*,t,t^*,\tau,\zeta)\, f(t,\zeta)\, d\zeta$$

with explicitly known functions H (see [3], [4]) and for equation (1) with (14) (using (15)) as

$$Rf = (\partial/\partial\tau) \int_0^{\tau} H(z,z^*,t,t^*,\tau,\zeta)\, f(t,\zeta)\, d\zeta ,$$

again with a known function H (see [5]).

2. The same concept is to be treated for systems of partial differential equations, written as matrix equations.

References

[1] D.L. COLTON: Partial differential equations in the complex domain, Research Notes in Mathematics 4, Pitman, London - San Francisco - Melbourne 1976.

[2] R.P. GILBERT: Constructive methods for elliptic equations, Lecture Notes in Mathematics 365, Berlin Heidelberg New York 1974.

[3] E. LANCKAU: Bergmansche Integraloperatoren für dreidimensionale Gleichungen und Gleichungssysteme, Beiträge zur Analysis 12 (1978), 99-112.

[4] -: Explizite Lösungen für Rand- und Anfangswertaufgaben gemischten Typs mit drei unabhängigen Veränderlichen, Tagung "Elliptische Differentialgleichungen", Berichte, 143-150, Rostock 1977.

[5] -: Behandlung pseudoparabolischer Differentialgleichungen mit funktionentheoretischen Methoden, to appear in Beiträge zur Analysis.

[6] I.N. VEKUA: New methods for solving elliptic equations, (Moskau 1948) New York 1967.

Sektion-Mathematik
Technische Hochschule Karl-Marx-Stadt
Reichenhainer Straße 41
DDR-9023 Karl-Marx-Stadt, DDR

MESURES PLURIHARMONIQUES ET MESURES ORTHOGONALES DANS LE GROUPE D'HEISENBERG

Guy Laville (Paris)

Table des matières

Résumé

Il est bien connu que le groupe d'Heisenberg H peut être identifié au bord du domaine de Siegel de type II. Le comportement au bord des fonctions holomorphes et pluriharmoniques peut être étudié directement sur H. Une mesure orthogonale est une mesure ν sur H telle que $\int u \, d\nu = 0$ pour toute fonction pluriharmonique et continue u. On étudie les ensembles de mesure nulle pour les mesures orthogonales. Ces ensembles sont comparés à des ensembles où des potentiels associés à des noyaux pluriharmoniques sont infinis. Les ensembles de mesure nulle pour les mesures pluriharmoniques sont aussi étudiés. L'outil principal est ici la solution fondamentale du Laplacien incomplet sur H.

Notations

Le groupe d'Heisenberg, noté H, de dimension réelle 3 est le groupe de Lie $\mathbb{C} \times \mathbb{R}$ avec les coordonnées (z,t) et dont la loi de groupe est:

(1) $(z,t).(z',t') = (z + z', t + t' + 2 \operatorname{Im} z\bar{z}')$.

Définissons les champs de vecteurs

(2) $Z = \dfrac{\partial}{\partial z} + i \bar{z} \dfrac{\partial}{\partial t}$, $\bar{Z} = \dfrac{\partial}{\partial \bar{z}} - i z \dfrac{\partial}{\partial t}$,

(3) $T = \frac{\partial}{\partial t}$.

On a :

$$[Z,\bar{Z}] = -2\,i\,T.$$

Posons

(4) $\Delta_k = \frac{1}{2}(Z\,\bar{Z} + \bar{Z}\,Z)$.

H est identifié avec le bord du domaine de Siegel de type II de l'espace $\mathbb{C}^2$:

$$\{(z_1,z_2) \in \mathbb{C}^2 \,:\, \operatorname{Im} z_2 > |z_1|^2\}.$$

Une fonction (ou une mesure) f définie sur H et à valeurs complexes (resp. réelles) sera dite holomorphe (resp. pluriharmonique) quand elle est la valeur au bord d'une fonction holomorphe (resp. pluriharmonique) de deux variables complexes définie dans le domaine de Siegel. Ceci équivant à dire

(5) $\bar{Z}\,f = 0$ (resp. $Z\,\bar{Z}\,\bar{Z}\,Z\,f = 0$)

(cf.[6]). Nous noterons $\varrho = \varrho(z,t) = |z|^2 - it$, c'est une fonction holomorphe. Comme $\operatorname{Re}\varrho \geq 0$, toutes les expressions telles que $\log\varrho$, $\log(\varrho+1)$ sont bien définis.

σ désignera la mesure de Lebesgue sur $H = \mathbb{C} \times \mathbb{R}$, σ est aussi la mesure de Haar du groupe.

1. Etude des mesures orthogonales

Sur le groupe d'Heisenberg plusieurs noyaux s'introduisent de façon naturelle.

Posons $\varrho = |z|^2 - it$:

$\operatorname{Re}\varrho = |z|^2 > 0.$

considérons les fonctions:

(6) $\operatorname{Re}\,[\log(\frac{\varrho+1}{\varrho})]^n = \operatorname{Re}\,[\log(\frac{|z|^2 + 1 - it}{|z|^2 - it})]^n$

c'est une fonction pluriharmonique puisque c'est la partie réelle d'une fonction holomorphe. De plus, la fonction (6) est toujours minorée

(7) $\operatorname{Re}\frac{1}{\varrho} = \frac{|z|^2}{|z|^4 + t^2}$

est une fonction pluriharmonique et toujours positive

(8) $\frac{1}{|\varrho|} = \frac{1}{(|z|^4 + t)^{\frac{1}{2}}}$

est la solution fondamentale de l'opérateur Δ_K.
Nous verrons que sur le sous-ensemble des fonctions pluriharmoniques, (7) est aussi solution fondamentale de l'opérateur Δ_K.

Les fonctions (6), (7), (8) peuvent s'interpréter comme noyaux grâce à la convolution sur H.

LEMME 1. *Soit ν une mesure orthogonale aux fonctions pluriharmoniques, c'est-à-dire, pour toute fonction pluriharmonique continue h, $\int h\, d\nu = 0$. Soit Γ une fonction holomorphe telle que:*

$$\mathrm{Re}\,\Gamma(0,0) = +\infty,$$

$$\mathrm{Re}\,\Gamma(z,t) \geq 0,$$

$$\Gamma(z,t) \text{ continue hors du point } (0,0).$$

Soit K un compact, tel que pour tout compact $K' \subset K$ il existe une mesure μ telle que $\mu \geq 0$, support de μ inclu dans K' et de plus:

$$\mathrm{Re}\,\Gamma * \mu(z,t) \longrightarrow +\infty \quad \text{quand} \quad (z,t) \longrightarrow (z_0,t_0) \in K',$$

$$(9) \qquad \frac{\mathrm{Im}\,\Gamma * \mu(z,t)}{\mathrm{Re}\,\Gamma * \mu(z,t)} \longrightarrow 0 \quad \text{quand} \quad (z,t) \longrightarrow (z_0,t_0) \in K'.$$

Alors K est de ν-mesure nulle.

Démonstration

Considérons la fonction holomorphe

$$(10) \qquad \varphi = \frac{\Gamma * \mu}{\Gamma * \mu + 1}.$$

φ est bien définie et régulière hors de K', puisque $\mathrm{Re}\,\Gamma * \mu > 0$, le dénominateur ne s'annule jamais.

$$\varphi = \frac{\mathrm{Re}\,\Gamma * \mu + i\ \mathrm{Im}\,\Gamma * \mu}{\mathrm{Re}\,\Gamma * \mu + 1 + i\ \mathrm{Im}\,\Gamma * \mu}.$$

Quand le point (z,t) comverge vers un point de K', φ converge vers 1. φ est donc une fonction continue partout.

Γ est valeur au bord de fonction holomorphe, $\Gamma * \mu$ aussi, donc φ aussi. Donc φ est holomorphe sur H. Enfin:

$$|\Gamma * \mu + 1|^2 \geq |\mathrm{Re}\,\Gamma * \mu|^2 + 1 + |\mathrm{Im}\,\Gamma * \mu|^2$$

d'où $|\varphi| \leq 1$ partout et $|\varphi| < 1$ hors de K'.
ν étant orthogonale et φ^n holomorphe:

$$\int_H \varphi^n\, d\nu = 0,$$

$$\int_H \varphi^n \, d\nu = 0 \longrightarrow \int_{K'} d\nu = \nu(K'),$$

d'oú $\nu(K') = 0$, ceci pour tout $K' \subset K$.

LEMME 2. Sous les mêmes hypothèses que le lemme 1 sauf la condition (9) est remplacé par la condition portant sur Γ : quand $(z,t) \longrightarrow 0$,

$$\frac{\operatorname{Im} \Gamma(z,t)}{\operatorname{Re} \Gamma(z,t)} \longrightarrow 0 .$$

Alors K est de ν-mesure nulle.

Démonstration

Montrons que cette nouvelle condition implique (9). Pour tout $\varepsilon > 0$, il existe une boule anisotrope

$$B = \{(z,t) \in H : |z|^4 + t^2 < a\}$$

telle que $(z,t) \in B$ implique

$$|\operatorname{Im} \Gamma(z,t)| \leq \varepsilon \operatorname{Re} \Gamma(z,t),$$

$$\operatorname{Im} \Gamma * \mu(z,t) = \int_B \operatorname{Im} \Gamma(z',t') \, d\mu(z - z', t - t' - 2 \operatorname{Im} \bar{z}'z)$$

$$+ \int_{H \setminus B} \operatorname{Im} \Gamma(z',t') \, d\mu(z - z', t - t' - 2 \operatorname{Im} \bar{z}'z).$$

La deuxième intégrale du second membre n'a pas de singularité (puisque Γ est continue hors du point $(0,0)$).
Il existe donc une constante M qui majore ce terme:

$$|\operatorname{Im} \Gamma * \mu(z,t)| \leq M + \int_{H \setminus B} |\operatorname{Im} \Gamma(z',t')| \, d\mu(z - z', t - t' - 2 \operatorname{Im} \bar{z}'z)$$

en utilisant l'inégalité entre $\operatorname{Im} \Gamma$ et $\operatorname{Re} \Gamma$ nous avons:

$$|\operatorname{Im} \Gamma * \mu(z,t)| \leq M + \varepsilon \int_H \operatorname{Re} \Gamma(z',t') \, d\mu(z - z', t - t' - 2 \operatorname{Im} \bar{z}'z),$$

$$\left| \frac{\operatorname{Im} \Gamma * \mu(z,t)}{\operatorname{Re} \Gamma * \mu(z,t)} \right| \leq \frac{M}{\operatorname{Re} \Gamma * \mu(z,t)} + \varepsilon .$$

Mais on sait que, quand $(z,t) \longrightarrow (z_o,t_o) \in K'$,

$$\operatorname{Re} \Gamma * \mu(z,t) \longrightarrow +\infty$$

donc si (z,t) est proche de K', on a:

$$\left|\frac{\mathrm{Im}\ \Gamma * \mu(z,t)}{\mathrm{Re}\ \Gamma * \mu(z,t)}\right| \leq 2\varepsilon$$

ce qui montre que la condition (9) est bien remplie, on peut appliquer le lemme 1 et K est de ν-mesure nulle.
Définissons le noyau $\Gamma(z,t)$ de la façon suivante: soit $\zeta \longmapsto \psi(\zeta)$ la transformation conforme du demi-plan $\mathrm{Re}\,\zeta > 0$ dans le domaine:

$$\{w \in \mathbb{C} : \mathrm{Re}\ w > 0 \quad \text{et} \quad \frac{\mathrm{Im}\ w}{\mathrm{Re}\ w} \longrightarrow 0 \quad \text{quand} \quad w \longrightarrow \infty\},$$

domaine que nous supposerons simplement connexe.
ψ existe d'après le théorème de Riemann sur la transformation conforme. Posons:

$$\Gamma(z,t) = \psi(\tfrac{1}{\varrho}) = \psi\left(\frac{1}{|z|^2 - it}\right).$$

<u>THÉORÈME 1</u>. <u>Soit</u> K <u>un compact tel que pour tout</u> $K' \subset K$ <u>il existe une mesure</u> $\mu \geq 0$, <u>support de</u> μ <u>inclu dans</u> K' <u>et</u> $\mathrm{Re}\ \Gamma * \mu(z,t) \longrightarrow +\infty$ <u>quand</u> $(z,t) \longrightarrow (z_o,t_o) \in K'$. <u>Alors, pour toute mesure</u> ν <u>orthogonale aux fonctions pluriharmoniques</u>, K <u>est un ensemble de</u> ν<u>-mesure nulle</u>.

<u>Démonstration</u>
Il suffit d'appliquer le lemme 2 au noyau Γ.

<u>COROLLAIRE 1</u>. <u>Le théorème 1 est vrai avec le noyau:</u>

$$\Gamma = [\log(\frac{1+\varrho}{\varrho})]^n + \left(\frac{\pi^2}{4}\frac{1}{\sin^2\frac{\pi}{2n}}\right)^n.$$

<u>Démonstration</u>
Montrons d'abord $\mathrm{Re}\,\Gamma \geq 0$.
Posons $a + ib = \log(\frac{1+\varrho}{\varrho})$, on a : $a > 0$, $|b| \leq \frac{\pi}{2}$:

$$\mathrm{Re}\,[\log(\tfrac{1+\varrho}{\varrho})]^n = (a^2 + b^2)^n \cos n\theta.$$

La plus petite valeur de cette quantité sera atteinte quand $b = \pm\frac{\pi}{2}$:

$$(a^2 + \tfrac{\pi^2}{4})^n [\cos n \operatorname{Arc} \cos \frac{a}{(a^2 + \frac{\pi^2}{4})^{\frac{1}{2}}}].$$

Le cosinus entre crochets sera positif quand

$$-\frac{\pi}{2} \leq n \operatorname{Arc} \cos \frac{a}{(a^2 + \frac{\pi^2}{4})^{\frac{1}{2}}} \leq \frac{\pi}{2}$$

c'est-à-dire quand

$$\frac{a}{\left(a^2 + \frac{\pi^2}{4}\right)^{\frac{1}{2}}} \geq \cos\frac{\pi}{2n} ,$$

$$a \geq \frac{\pi}{2}\left(\frac{\cos^2\frac{\pi}{2n}}{1 - \cos^2\frac{\pi}{2n}}\right)^{\frac{1}{2}} .$$

On aura la plus petite valeur quand

$$a \leq \frac{\pi}{2}\,\frac{\cos\frac{\pi}{2n}}{\sin\frac{\pi}{2n}}$$

de là $\mathrm{Re}[\log(\frac{1+\varrho}{\varrho})]^n$ se minore par:

$$-\left[\frac{\pi^2}{4}\left(1 + \frac{\cos^2\frac{\pi}{2n}}{\sin^2\frac{\pi}{2n}}\right)\right]^n = -\left(\frac{\pi^2}{4}\,\frac{1}{\sin^2\frac{\pi}{2n}}\right)^n .$$

Montrons maintenant : il existe $C > 0$, tel que , si $|z|^4 + t^2$ est assez petit:

$$\mathrm{Re}[\log(\tfrac{1+\varrho}{\varrho})]^n \geq C\,|\mathrm{Im}[\log(\tfrac{1+\varrho}{\varrho})]^n|^{\frac{n}{n-1}} ,$$

$$(a + ib)^n = \sum_{k=o}^{n} C_n^k\, a^{n-k}\, i^k\, b^k$$

$$= \sum_{p=o}^{[n/2]} C_n^{2p}\, a^{n-2p}\, (-1)^p\, b^{2p}$$

$$+ \sum_{q=o}^{[n/2]} C_n^{2q+1}\, a^{n-2q-1}\, (-1)^q\, b^{2q+1},$$

comme on a $|b| \leq \frac{\pi}{2}$

$$|\mathrm{Im}[\log(\tfrac{1+\varrho}{\varrho})]^n| \leq \begin{cases} ([\frac{n}{2}] + 1)\sup\limits_q C_n^{2q+1}\, (\frac{\pi}{2})^n & \text{si } |a| \leq 1, \\ ([\frac{n}{2}] + 1)\sup\limits_q C_n^{2q+1}\, |a|^{n-1}\, (\frac{\pi}{2})^n & \text{si } |a| \geq 1, \end{cases}$$

$$\leq \begin{cases} C_1 & \text{si } |a| \leq 1, \\ C_1\, |a|^{n-1} & \text{si } |a| \geq 1, \end{cases}$$

$$\frac{1}{a^n}\,\mathrm{Re}[\log(\tfrac{1+\varrho}{\varrho})]^n = \sum_{p=o}^{[n/2]} C_n^{2p}\, a^{-2p}(-1)^p\, b^{2p}.$$

Comme $|b| \leq \frac{\pi}{2}$, quand $(z,t) \longrightarrow 0$, $\log(\frac{1+\varrho}{\varrho}) = a + ib$ est tel que $a \longrightarrow \infty$ donc

$$\frac{1}{a^n}\,\mathrm{Re}[\log(\tfrac{1+\varrho}{\varrho})]^n \longrightarrow 1$$

donc, il existe $A > 0$, tel que si $a \geq A$

$$\frac{1}{2} \leq \frac{1}{a^n}[\mathrm{Re}\,\log(\tfrac{1+\varrho}{\varrho})]^n,$$

$$\frac{a^n}{2} \leq \mathrm{Re}[\log(\tfrac{1+\varrho}{\varrho})]^n$$

On peut choisir $A \geq 1$. Pour $a \geq A$, on a donc,

$$|\mathrm{Im}[\log(\tfrac{1+\varrho}{\varrho})]^n| \leq 2^{\frac{n-1}{n}}\, C_1 (\tfrac{a^n}{2})^{\frac{n-1}{n}}$$

$$\leq 2^{\frac{n-1}{n}}\, C_1 \{\mathrm{Re}[\log(\tfrac{1+\varrho}{\varrho})]^n\}^{\frac{n-1}{n}}.$$

Finalement,

$$\left|\frac{\mathrm{Im}[\log(\frac{1+\varrho}{\varrho})]^n}{\mathrm{Re}[\log(\frac{1+\varrho}{\varrho})]^n}\right| \leq 2^{\frac{n-1}{n}}\, C_1 \frac{1}{(\mathrm{Re}[\log(\frac{1+\varrho}{\varrho})]^n)^{1/n}}.$$

<u>COROLLAIRE 2</u>. <u>Soit</u> K <u>un compact satisfaisant aux hypothèses du théorème</u> 1. <u>Alors il existe une fonction</u> $\tilde{\varphi}$ <u>de l'algèbre</u> A(D) <u>(algèbre des fonctions holomorphes qui sont continues dans l'adhérence du domaine de Siégel</u> D), <u>telle que</u> $|\tilde{\varphi}| \leq 1$ <u>et</u> $\tilde{\varphi}(z,t) = 1$ <u>si</u> $(z,t) \in K$ <u>(fonction pic sur</u> K).

<u>D é m o n s t r a t i o n</u>

La fonction $\varphi : H \longrightarrow \mathbb{C}$ définis par

$$\varphi = \frac{\Gamma * \mu}{\Gamma * \mu + 1}$$

(voir lemmes 1 et 2) est continue dans H. Elle peut se prolonger en une fonction φ à l'aide du noyau de Poisson, dans le domaine D. D'après le théorème de Fatou, les limites admissibles de $\tilde{\varphi}$ sont égales à φ presque partout sur H. Mais φ est continue sur H, donc φ est continue sur $\bar{D}$.

2. Etude des mesures pluriharmoniques

<u>LEMME 3</u>. <u>Soit</u> φ <u>une fonction</u>, $\varphi \in L^1(H) \cap L^\infty(H)$, $\varphi \in \mathcal{C}^\infty(H)$. φ <u>est pluriharmoniques si et seulement si</u>

$$(11)\quad \varphi * \frac{1}{|\varrho|} = \frac{4}{\pi}\varphi * \mathrm{Re}\,\frac{1}{\varrho}.$$

D é m o n s t r a t i o n

Remarquons tout d'abord:

$$\varphi * \frac{1}{|\varrho|}(z,t) = \int \frac{1}{|\varrho|}(z',t')\ \varphi((z,t)(z',t')^{-1})\ d\sigma(z',t')$$

$$\leq \sup_{(z',t')\in B} |\varphi((z,t)(z',t')^{-1})| \int_B \frac{1}{|\varrho|}(z',t')\ d\sigma(z',t')$$

$$+ \sup_{(z',t')\notin B} \frac{1}{|\varrho|}(z',t') \int_H |\varphi(z',t')|\ d\sigma(z',t')$$

où $B = \{(z,t) : |z|^4 + t^2 < 1\}$, donc l'intégrale $\varphi * \frac{1}{|\varrho|}$ converge.

D'autre part, $\mathrm{Re}\,\frac{1}{\varrho} \leq \frac{1}{|\varrho|}$ donc de même l'intégrale $\varphi * \mathrm{Re}\,\frac{1}{\varrho}$ converge.

D'autre part, rappelons que (cf:[2] ou [3])

(12) $\Delta_K \frac{1}{|\varrho|} = -2\pi\delta.$

Pour tout $\varepsilon > 0$,

$$\Delta_K \frac{1}{\varrho+\varepsilon} = \frac{1}{2}(Z\bar{Z} + \bar{Z}Z)\frac{1}{\varrho+\varepsilon}$$

$$= -\frac{1}{2}\bar{Z}\left(\frac{2\bar{z}}{(\varrho+\varepsilon)^2}\right) = -\frac{1}{(\varrho+\varepsilon)^2}.$$

D'où:

(13) $\Delta_K \mathrm{Re}\,\frac{1}{\varrho+\varepsilon} = -\frac{1}{2}\left[\frac{1}{(\varrho+\varepsilon)^2} + \frac{1}{(\bar{\varrho}+\varepsilon)^2}\right].$

$1/\varrho^2$ est le noyau de l'intégrale singulière de Szegö qui envoie L^2 dans L^2. Il est bien connu (cf : [7] ou [8]) que, pour toute fonction pluriharmonique de carré intégrable, donc en particulier pour φ, on a :

(14) $\varphi = \frac{1}{\pi^2}\,\mathrm{VP}\,\varphi * \left(\frac{1}{\varrho^2} + \frac{1}{\bar{\varrho}^2}\right).$

Les convolées étant prises en valeurs principales.

Montrons que:

(15) $\pi^2\varphi = \lim_{\varepsilon\to 0} \varphi * \left(\frac{1}{(\varrho+\varepsilon)^2} + \frac{1}{(\bar{\varrho}+\varepsilon)^2}\right).$

Les valeurs principales des intégrales sont prises au sens suivant:

$$\mathrm{VP}\int_H \psi(z,t)\,dt = \lim_{\substack{R\to\infty\\ \eta\to 0}} \int_{B_R\setminus B_\varepsilon} \psi(z,t)\,dt$$

où $B_R = \{(z,t)\in H : |z|^4 + t^2 < R\}$,

$B_\eta = \{(z,t)\in H : |z|^4 + t^2 < \eta\}$

ce qui correspond aux valeurs principales de la théorie de Korányi-

Vagi (cf. [7]).

D'après la formule (14), il suffit de montrer :

$$\lim_{\varepsilon\to 0} \left(\varphi * \frac{1}{(\varrho+\varepsilon)^2} - \mathrm{V\,P}\,\varphi * \frac{1}{\varrho^2}\right) = 0 \qquad \text{et ceci au point } (0,0).$$

Soit $S = \{(z,t)\in H : |z|^4 + t^2 < 1\}$, montrons

(i) $\int_{H\setminus S} \varphi\left(\frac{1}{(\varrho+\varepsilon)^2} - \frac{1}{\varrho^2}\right) d\sigma \longrightarrow 0$,

(ii) $\mathrm{V\,P} \int_S \varphi(0,0) \left(\frac{1}{(\varrho+\varepsilon)^2} - \frac{1}{\varrho^2}\right) d\sigma \longrightarrow 0$,

(iii) $\mathrm{V\,P} \int_S [\varphi - \varphi(0,0)] \left(\frac{1}{(\varrho+\varepsilon)^2} - \frac{1}{\varrho^2}\right) d\sigma \longrightarrow 0$.

Pour (i):

$$\left|\int_{H\setminus S} \varphi\left(\frac{1}{(\varrho+\varepsilon)^2} - \frac{1}{\varrho^2}\right) d\sigma\right| \le \left\| \frac{1}{(\varrho+\varepsilon)^2} - \frac{1}{\varrho^2} \right\|_{L^\infty(H\setminus S)} \|\varphi\|_{L^1(H\setminus S)}$$

$$\le \varepsilon \left\| \frac{2\varrho+\varepsilon}{\varrho^2(\varrho+\varepsilon)^2} \right\|_{L^\infty(H\setminus S)} \|\varphi\|_{L^1(H\setminus S)} .$$

Pour (ii)

$$\lim_{\varepsilon\to 0} \int_S \frac{1}{(\varrho+\varepsilon)^2} d\sigma = \lim_{\varepsilon\to 0} \int_{|z|^4+t^2<1} \frac{1}{(|z|^2+\varepsilon - it)^2} dz\, d\bar z\, dt$$

$$= \lim_{\varepsilon\to 0} \int_{\substack{(u+\varepsilon)^2 + t^2 < 1 \\ \text{et } u>0}} \frac{1}{(u+\varepsilon - it)^2} du\, dt$$

$$= \lim_{\varepsilon\to 0} \left\{ \int_{\varepsilon^2}^1 \int_{-\frac{\pi}{2}}^{\frac{\pi}{2}} \frac{e^{-2i\theta}}{r^2}\, r\, dr\, d\theta + \int_0^\varepsilon \int_{-1}^1 \frac{1}{(u+\varepsilon-i)^2} dt\, du \right\}$$

$$= \lim_{\varepsilon\to 0} \left\{ \int_{\varepsilon^2}^1 \frac{dr}{r} \int_{-\frac{\pi}{2}}^{\frac{\pi}{2}} e^{-2i\theta}\, d\theta + \int_0^\varepsilon \int_0^1 \frac{(u+\varepsilon)^2 - t^2}{(u+\varepsilon)^2 + t^2} dt\, du \right\}$$

$$= 0 .$$

De façon analogue

$$\int_S \frac{1}{\varrho^2} d\sigma = \lim_{\eta\to 0} \int_{\eta<|z|^4+t^2<1} \frac{1}{(|z|^2 - it)^2} dz\, d\bar z\, dt$$

$$= \pi \lim_{\eta\to 0} \int_\eta^1 \int_{-\frac{\pi}{2}}^{\frac{\pi}{2}} \frac{1}{r} e^{-2i\theta}\, dr\, d\theta$$

$$= 0 .$$

Pour (iii):

$$\lim_{\varepsilon\to 0}\ \lim_{\eta\to 0} \int_{S\setminus B_\eta} \left[\frac{1}{(\rho+\varepsilon)^2} - \frac{1}{\rho^2}\right] [\varphi - \varphi(0,0)]\, d\sigma$$

$$= \lim_{\varepsilon\to 0}\ \lim_{\eta\to 0} \int_{S\setminus B_\eta} \frac{2\varepsilon\rho + \varepsilon^2}{\rho^2(\rho+\varepsilon)^2} \Big[z\frac{\partial\varphi}{\partial z}(0,0) + \bar z\frac{\partial\varphi}{\partial \bar z}(0,0)$$

$$+ t\frac{\partial\varphi}{\partial t}(0,0) + \mathcal{O}(z) + \mathcal{O}(t)\Big]\, d\sigma(z,t).$$

En valeur absolue, on peut majorer par:

$$\lim_{\varepsilon\to 0}\ \lim_{\eta\to 0} \int_{S\setminus B_\eta} \varepsilon^{1/4}\left[\frac{2}{|\rho||\rho+\varepsilon|^{5/4}} + \frac{1}{|\rho|^2|\rho+\varepsilon|^{1/4}}\right]$$

$$\times\Big[|z|\left|\frac{\partial\varphi}{\partial z}(0,0)\right| + |z|\left|\frac{\partial\varphi}{\partial \bar z}(0,0)\right| + |t|\left|\frac{\partial\varphi}{\partial t}(0,0)\right| + \mathcal{O}(z)$$

$$+\mathcal{O}(t)|\Big]\, d\sigma(z,t).$$

Il suffit enfin de remarquer que:

$$\frac{|z|}{|\rho|^{9/4}} \le \frac{1}{|\rho|^{7/4}}, \qquad \frac{|t|}{|\rho|^{9/4}} \le \frac{1}{|\rho|^{5/4}}$$

et $\int_S \frac{1}{(|z|^4 + t^2)^{7/8}}\, dz\, d\bar z\, dt$ est convergente.

D'où finalement:

$$\lim_{\varepsilon\to 0} \int_S \left(\frac{1}{(\rho+\varepsilon)^2} - \frac{1}{\rho^2}\right)[\varphi - \varphi(0,0)]\, d\sigma = 0$$

de là, en réunissant les égalités (i), (ii), (iii) nous obtenons (16) donc (15).

Des formules (15) et (13), on déduit:

$$\varphi(z,t) = -\frac{2}{\pi^2} \lim_{\varepsilon\to 0} \Delta_K \left(\varphi * \operatorname{Re}\frac{1}{\rho+\varepsilon}\right),$$

$$(16) \qquad \varphi(z,t) = -\frac{2}{\pi^2}\, \Delta_K \left(\varphi * \operatorname{Re}\frac{1}{\rho}\right).$$

Considérons maintenant la fonction:

$$\psi = \varphi * \left(\frac{1}{|\rho|} - \frac{4}{\pi}\operatorname{Re}\frac{1}{\rho}\right);$$

cette fonction est bien définie en vertu des convergences démontrées au début de la démonstration de ce lemme. De plus elle est uniformément bornée.

$$\Delta_K\psi = \Delta_K\left(\varphi * \frac{1}{|\rho|}\right) - \frac{4}{\pi}\,\Delta_K\left(\varphi * \operatorname{Re}\frac{1}{\rho}\right)$$

$$= -2\pi\varphi - \frac{4}{\pi}\,\frac{\pi^2}{-2}\,\varphi = 0.$$

D'où, ψ est une fonction constante (cf. [5]).
Etudions le comportement de $\psi(z,t)$ quand $(z,t) \longrightarrow \infty$

(17) $$(\varphi * \frac{1}{|\varsigma|})(z,t) = \int_H \frac{\varphi(z_o,t_o)}{[|z-z_o|^4 + (t-t_o - 2\,\mathrm{Im}\, z\, \bar{z}_o)^2]^{1/2}}\, d\sigma(z_o,t_o).$$

Soit $\varepsilon > 0$. Il existe un compact W de H tel que $W \supset S = \{(z,t) : |z|^4 + t^2 < 1\}$ et

$$\left(\int_{H\setminus W} |\varphi|^{3/2}\, d\sigma\right)^{2/3} \leq \left(\int_{H\setminus S} \frac{1}{|\varsigma|^3}\, d\sigma\right)^{-1/3} \frac{\varepsilon}{2}.$$

W étant ainsi fixé, prenons $|z|$ et $|t|$ assez grand pour que $(z,t) \notin W$ et

$$\sup_{(z_o,t_o)\in W} \frac{1}{[|z-z_o|^4 + (t-t_o - 2\,\mathrm{Im}\, z\, \bar{z}_o)^2]^{1/2}} \leq \left(\int_W |\varphi|\, d\sigma\right)^{-1} \frac{\varepsilon}{2},$$

alors:

$$(\varphi * \frac{1}{|\varsigma|})(z,t) \leq \sup_{(z_o,t_o)\in W} \frac{1}{[|z-z_o|^4 + (t-t_o - 2\,\mathrm{Im}\, z\, \bar{z}_o)^2]^{1/2}} \times$$

$$\times \int_W |\varphi|\, d\sigma + \left(\int_{H\setminus W} |\varphi|^{3/2} d\sigma\right)^{2/3} \left(\int_{H\setminus W} \frac{d\sigma(z_o,t_o)}{[|z-z_o|^4 + (t-t_o - 2\,\mathrm{Im}\, z\, \bar{z}_o)^2]^3}\right)^{1/3},$$

$$|(\varphi * \frac{1}{|\varsigma|})(z,t)| \leq \varepsilon .$$

De là

(18) $$(\varphi * \frac{1}{|\varsigma|})(z,t) \longrightarrow 0 \quad \text{et} \quad (\varphi * \mathrm{Re}\, \frac{1}{\varsigma})(z,t) \longrightarrow 0 .$$

D'où ψ est identiquement nulle et on a bien l'égalité (11).

La réciproque est évidente, il suffit d'appliquer l'opérateur Δ_K aux deux membres de l'égalité (11).

<u>THEOREME 2</u>. <u>Soit</u> λ <u>une mesure pluriharmonique positive et de masse finie; alors</u>

(19) $$\int_H \log\left\{\frac{(1-t) + [|z|^4 + (1-t)^2]^{\frac{1}{2}}}{-t + (|z|^4 + t^2)^{\frac{1}{2}}}\right\} d\lambda(z,t) \leq 2 \int_H d\lambda .$$

<u>Démonstration</u>

Utilisons l'égalité (11) du lemme 3 appliqué à une fonction φ satisfaisant aux hypothèses de ce lemme:

(20) $$\int_H \frac{1}{[|z-z_o|^4 + (t-t_o - 2\,\mathrm{Im}\, z\, \bar{z}_o)^2]^{1/2}}\, \varphi(z,t)\, d\sigma(z,t)$$

$$= \frac{4}{\pi} \int_H \frac{|z-z_o|^2 \quad \varphi(z,t)}{|z-z_o|^4 + (t-t_o - 2 \operatorname{Im} z \bar{z}_o)^2} \, d\sigma(z,t) \ .$$

Faisons $z_o = 0$, et considérons les deux membres comme fonction de t_o; en intégrant entre 0 et 1, nous obtenons:

$$\int_o^1 \frac{dt_o}{\sqrt{|z|^4 + (t-t_o)^2}} = \int_{-t}^{1-t} \frac{dt_o}{\sqrt{|z|^4 + t_o^2}}$$

$$= \log\left\{ \frac{(1-t) + \sqrt{|z|^4 + (1-t)^2}}{- t + \sqrt{|z|^4 + t^2}} \right\}.$$

Remarquons que:

$$1-t \geq - \sqrt{|z|^4 + (1-t)^2},$$

$$1 + 1 - 2t + t^2-t^2 \geq - 2 \sqrt{|z|^4 + (1-t)^2},$$

$$1 + |z|^4 + (1-t)^2 + 2 \sqrt{|z|^4 + (1-t)^2} \geq |z|^4 + t^2 ,$$

$$1 + \sqrt{|z|^4 + (1-t)^2} \geq \sqrt{|z|^4 + t^2} ,$$

$$\frac{1-t + \sqrt{|z|^4 + (1-t)^2}}{-t + \sqrt{|z|^4 + t^2}} \geq 1;$$

le logarithme dans (19) est toujours positif.
Calculons l'autre membre dans la formule (20)

$$\int_o^1 \frac{|z|^2}{|z|^4 + (t-t_o)^2} \, dt_o = \int_{-t}^{1-t} \frac{1}{1 + (t_o^2/|z|^2)^2} \, d(\frac{t_o}{|z|^2})$$

$$= \operatorname{Arc\,tg} \frac{1-t}{|z|^2} + \operatorname{Arc\,tg} \frac{t}{|z|^2}$$

$$= \operatorname{Arc\,tg} \frac{|z|^2}{|z|^4 + t^2 - t}.$$

Finalement l'égalité (20) pour $z_o = 0$ et intégrée en t_o s'écrit:

$$(21) \quad \int \log\left\{ \frac{(1-t) + \sqrt{|z|^4 + (1-t)^2}}{- t + \sqrt{|z|^4 + t^2}} \right\} \varphi(z,t) \, d\sigma(z,t)$$

$$= \frac{4}{\pi} \int \operatorname{Arc\,tg} \left(\frac{|z|^2}{|z|^4 + t^2 - t} \right) \varphi(z,t) \, d\sigma(z,t).$$

De là, pour toute fonction φ pluriharmonique $\varphi \in \mathcal{C}^\infty \cap L^1 \cap L^\infty, \varphi \geq 0$:

$$\int \log \left\{ \frac{(1-t) + \sqrt{|z|^4 + (1-t)^2}}{-t + \sqrt{|z|^4 + t^2}} \right\} \varphi(z,t)\, d\sigma(z,t)$$

$$\leq 2 \int_H \varphi(z,t)\, d\sigma(z,t).$$

Soit maintenant λ une mesure pluriharmonique positive et de masse finie et ψ_n une approximation de l'identité: $\psi_n \in \mathcal{C}^\infty$, à support compact, $\psi_n \geq 0$, $\int \psi_n\, d\sigma = 1$ et $\psi_n \longrightarrow \delta$.

$\lambda * \psi_n$ est pluriharmonique, et appartient à $\mathcal{C}^\infty \cap L^1 \cap L^\infty$

$$\int \lambda * \psi_n\, d\sigma = \int \psi_n\, d\sigma \cdot \int d\lambda = \int d\lambda$$

posons, pour simplifier $h(z,t) = \log \dfrac{(1-t) + \sqrt{|z|^4 + (1-t)^2}}{-t + \sqrt{|z|^4 + t^2}}$.

D'après l'inégalité qui précède, on a:

$$\int h(z,t)(\lambda * \psi_n)(z,t)\, d\sigma(z,t) \leq 2 \int (\lambda * \psi_n)\, d\sigma$$

$$\leq 2 \int d\lambda$$

on peut supposer: $\psi(z,t) = \psi((z,t)^{-1})$,

$$\int (h * \psi_n)(z,t)\, d\lambda(z,t) \leq 2 \int d\lambda$$

d'après le théorème de Fatou:

$$\int \liminf_{n \to \infty} (h * \psi_n)\, d\lambda(z,t)$$

$$\leq \liminf_{n \to \infty} \int (h * \psi_n)\, d\lambda$$

$$\leq 2 \int d\lambda$$

mais $\liminf_{n \to \infty} h * \psi_n = h$ partout d'où $\int h\, d\lambda \leq 2 \int d\lambda$ ce qui est l'inégalité cherchée.

Remarque. On peut écrire l'inégalité (19) sous forme de convolée et on a une majoration uniforme:

$$\sup_{(z,t)} (h * \lambda)(z,t) \leq 2 \int_H d\lambda ;$$

cette inégalité se montrent en effectuant une translation sur la mesure λ.

COROLLAIRE. *Soit λ une mesure pluriharmonique positive de* masse finie. Alors l'ensemble $\{(z,t) \in H : z = a\}$ (i.e. les droites dans la direction t) est de mesure nulle pour λ, pour tout $a \in \mathbb{C}$.

Démonstration

Quand $0 < t < 1$ et $|z| = 0$, on a:

$$- t + (|z|^4 + t^2)^{\frac{1}{2}} = 0,$$

$$1 - t + [|z|^4 + (1 - t^2]^{\frac{1}{2}} \neq 0.$$

Le logarithme dans (19) est donc $+\infty$ pour les points (z,t) tels que $z = 0$ et $t \in]0,1]$. Cet ensemble est donc de mesure nulle pour λ. Donc, l'ensemble $\{(z,t) \in H : z = 0\}$ est aussi de λ-mesure nulle, par translation, ceci est encore vrai pour $a \in \mathbb{C}$.

Références

[1] CHOLLET, A.M.: Ensembles de zéros à la frontière des fonctions analytiques dans la boule de $\mathbb{C}^n$, C. R. Acad. Sci. Paris 277 (1973), 1165-1167.

[2] FOLLAND, G.B.: A fundamental solution for a subelliptic operator, Bull. Am. Math. Soc. 79, 2, March 1973.

[3] —— and E. M. STEIN: Estimates for the $\bar{\partial}_b$ complex and analysis on the Heisenberg group, Comm. Pure Appl. Math. 27 (1974), 429-522.

[4] GREINER, P.C. and E. M. STEIN: Estimates for the $\bar{\partial}$-Neumann problem, Princeton University Press.

[5] LAVILLE, G.: Fonctions pluriharmoniques et solution fondamentale d'un opérateur du 4ème ordre, Bull. Sci. Math. 2ème série 101 (1977), 305-317.

[6] ——: Sur le calcul de la fonction conjugée à plusieurs variables complexes, Bull. Sci. Math. 2ème série 102 (1978), 257-272.

[7] KORÁNYI, A. and S. VAGI: Singular integrals in homogeneous spaces and some problems of classical analysis, Ann. Scuola Norm. Sup. Pisa 25 (1971).

[8] ——, ——, and G. WELLAND: Remarks on the Cauchy integral and the conjugate function in generalized half planes, J. Math. Mech. 19 (1969), 1069-1081.

[9] FORELLI, F.: Measures whose Poisson integrals are pluriharmonic I, Ill. Math. J. 18 (1974), 373-388.

[10] ——: Measures whose Poisson integrals are pluriharmonic II, Ill. Math. J. 19 (1975), 584-592.

[11] HENKIN, G.M. and E. M. CHIRKA: Boundary properties of holomorphic functions of several complex variables, Journal of Soviet Math. 5, 5, May 1976.

Mathématiques
Université Pierre et Marie Curie
4, Place Jussieu
F-75230 Paris Cedex 05, France

ON BIHOLOMORPHIC CONTINUABILITY OF REGULAR QUASICONFORMAL MAPPINGS

Julian Ławrynowicz (Łódź)

Contents

Summary

Almost nothing is known about biholomorphic and even holomorphic continuability of quasiconformal mappings into complex manifolds whose real dimension is twice the dimension of the original manifold. The same applies to biholomorphic continuability of plane quasiconformal mappings into $\mathbb{C}^2$. The problem is very important, e.g. in various approaches to the physics of elementary particles involving the complex geometry of the natural world.

Quasiconformal mappings may be treated as homeomorphic solutions of certain systems of homogenous elliptic partial differential equations which in the two-dimensional case reduce to the Beltrami differential equation, whereas in higher dimensions are nonlinear. Therefore we prefer to utilize here the geometrical variational techniques. The modern geometrical variational method of extremal length is due to Ahlfors and Beurling. On complex manifolds some analogues have been developed by Kobayashi and by Chern, Levine and Nirenberg. For hermitian manifolds it has been introduced by the present author in the Bull. Acad. Polon. Sci. Sér.Sci. Math. Astronom. Phys. 23 (1975), 839-844, and then developed in later papers. It has proved to be quite useful for the problem in question.

The inverse problem of quasiconformality of projections of

biholomorphic mappings was studied in the same Bulletin 23 (1975), 845-851. Here, we firstly extend some results of this paper and make a suitable restriction for further purposes: we distinguish the class of mappings transforming infinitesimal spheres onto (infinitesimal) circular ellipsoids. The extensions obtained still depend on one order of quasiconformality, denoted here by Q. We show, however, that in order to formulate the problem of biholomorphic continuability we need a dependence on two orders of quasiconformality, denoted by Q and Q′. Finally it appears that there are three types of continuability theorems.

Introduction

Almost nothing is known about biholomorphic and even holomorphic continuability of quasiconformal mappings into complex manifolds whose real dimension is twice the dimension of the original manifold [12]. The same applies to biholomorphic continuability of plane quasiconformal mappings into $\mathbb{C}^2$.

The problem is very important, e.g. in various modern approaches to the physics of elementary particles [23], [18], [19], [9], [10], [4], [21], [26] involving the complex geometry of the natural world. In fact, in the case where an external particle is introduced, the light cones of the space-time change in a bounded way according to a finite velocity of this particle and this suggests the use of quasiconformal mappings.

These mappings may be treated as homeomorphic solutions of certain systems of homogenous elliptic partial differential equations [6] which in the two-dimensional case reduce to the Beltrami differential equation [5], whereas in higher dimensions the systems are overdetermined and the equations are nonlinear. Therefore we prefer to utilize here the geometrical variational techniques.

The modern geometrical variational method of extremal length is due to Ahlfors and Beurling [1]. On complex manifolds some analogues have been developed by Kobayashi [13] and by Chern, Levine and Nirenberg [8]. For hermitian manifolds it has been introduced in [14] and then developed in [16], [2], and [3]. It has proved to be quite useful for the problem in question.

Let us remark that on Hilbert spaces analogues mappings have been studied in [22], and a counterpart of extremal length — in [25].

The inverse problem of quasiconformality of projections of biholomorphic mappings was studied in [15] and [16]. In Section 1 we extend some results of those papers and make a suitable restriction for further purposes: we distinguish the class of mappings transforming infinitesimal spheres onto (infinitesimal) circular ellipsoids. The extensions obtained still depend on one order of quasiconformality, denoted there by Q. We show, however, that in order to formulate properly the problem of biholomorphic continuability we need a dependence on two orders of quasiconformality, denoted by Q and Q′: the corresponding extension is established in Section 2. Finally it appears that there are three types of continuability theorems which are established in Section 3 — the main section of this paper.

1. Spacial quasiconformal mappings transforming infinitesimal spheres onto circular ellipsoids

Assumption 1.1. Suppose that $\mathbb{L} = \mathbb{M}_1 \times \mathbb{M}_2$ is the product complex manifold endowed with the hermitian structure

(1.1) $h(v,w) = \hat{g}(v,w) + \hat{g}(Jv,Jw); \quad v,w \in T_z\mathbb{L}$,

(1.2) $\hat{g}(v,w) = \hat{g}_1(pr_z^1 v, pr_z^1 w) + g_2(pr_z^2 v, pr_z^2 w)$,

where $\mathbb{M}_1$ and $\mathbb{M}_2$ are m-dimensional Riemannian manifolds of metrics g_1 and g_2, respectively, $m>1$, and J is the complex structure of $\mathbb{L}$. Manifolds with boundary are also considered.

Here $T_z\mathbb{L}$ denotes the tangent space of $\mathbb{L}$ at a point $z \in \mathbb{L}$, and pr_z^1 denotes the mapping induced by the projection $z \longmapsto x \in \mathbb{M}_1$ between the corresponding tangent spaces, pr_z^2 being defined analogously:

$$pr_z^1 : T_z\mathbb{L} \longrightarrow T_x\mathbb{M}_1, \quad pr_z^2 : T_z\mathbb{L} \longrightarrow T_y\mathbb{M}_2.$$

In a physical context $\mathbb{M}_1$ and $\mathbb{M}_2$ should be rather denoted by $\mathbb{M}_{\underline{e}}$ and $\mathbb{M}_{\underline{n}}$, respectively. The subscripts $\underline{e}$ in $\mathbb{M}_{\underline{e}}$ and $\underline{n}$ in $\mathbb{M}_{\underline{n}}$ originate in a description of elementary particles [18], [19]: $\mathbb{M}_{\underline{e}}$ and $\mathbb{M}_{\underline{n}}$ are connected with the external electromagnetic and nuclear fields, respectively.

Remark 1.1. The choice (1.1)-(1.2) of the function $(g_1,g_2) \longmapsto h$ is not essential for further consideration at all.

We recall that a sense-preserving homeomorphism $F : \mathbb{M}_1 \longrightarrow \mathbb{M}_2$ is said to be Q-quasiconformal, $1 \leqslant Q < +\infty$, if

$$(1/Q)\operatorname{cap} \mathfrak{D} \leqslant \operatorname{cap} F[\mathfrak{D}] \leqslant Q \operatorname{cap} \mathfrak{D},$$

for each condenser $\mathfrak{D}$ whose closure is compact on $\mathbb{M}_1$, and

$$\operatorname{cap} \mathfrak{D} = \inf_{u \in \operatorname{adm} \mathfrak{D}} \int_{\mathfrak{D}} |g_1(\operatorname{grad}_{\mathbb{M}_1} u, \operatorname{grad}_{\mathbb{M}_1} u)|^{\frac{1}{2}m} d\tau(\mathbb{M}_1)$$

A condenser on $\mathbb{M}_1$ is a domain on $\mathbb{M}_1$ whose complement consists of two distinguished disjoint closed sets (the condenser plates). A proof of existence of the unique Lebesgue measure $\tau(\mathbb{M}_1)$ can be found in [24]. The class adm $\mathfrak{D}$ of admissible functions consists of functions $u : \mathbb{M}_1 \longrightarrow \mathbb{R}$ such that:

1° u is continuous, $u \circ \gamma_o$ is absolutely continuous, and $\underline{D}(u \circ \gamma_o) = \underline{D}u \circ \underline{D}\gamma_o$ almost everywhere for the parametrization by arc length γ_o of all rectifiable curves on $\mathfrak{D}$, where $\underline{D}\gamma$ denotes the derivatives of γ,

2° $\| \mathrm{grad}_{\mathbb{M}_1} u \|^m(x) = |g_1((\mathrm{grad}_{\mathbb{M}_1} u)(x), (\mathrm{grad}_{\mathbb{M}_1} u)(x))|^{\frac{1}{2}m}$ is $\tau(\mathbb{M}_1)$-integrable over any compact subset of $\mathbb{M}_1$,

3° u equals 0 on one condenser plate C_o and 1 on the other C_1, say.

We shall often need the following lemma (cf.[7],p.209, and [20]):

LEMMA 1.1. *If $F:\mathbb{M}_1 \longrightarrow \mathbb{M}_2$ is a sense-preserving homeomorphism and either $\mathrm{cap}\, F[\mathfrak{D}] \leqslant Q\mathrm{cap}\,\mathfrak{D}$ for each condenser $\mathfrak{D}$ whose closure is compact on $\mathbb{M}_1$ or $\mathrm{cap}\, F[\mathfrak{D}] \geqslant (1/Q)\mathrm{cap}\,\mathfrak{D}$ for each condenser $\mathfrak{D}$ whose closure is compact on $\mathbb{M}_1$, then F is Q^{m-1}-quasiconformal. The estimate obtained for the order of quasiconformality is precise.*

Assumption 1.2. Suppose that $\mathbb{L}_o$ is an m-dimensional hermitian manifold of metric h and complex structure J, endowed with an arbitrary C^1 tensor field H of type (1.1). Let D be a condenser on $\mathbb{L}_o$, $q:\mathbb{L}_o \longrightarrow \mathbb{C}$ - a continuous mapping (the inhomogeneity function), and p - a real number of $(1;+\infty)$. Finally, let $\mathcal{U} = \{U_j : j \in I\}$ be a locally finite open covering of $\mathbb{L}_o$ and Γ a homology class of D with real coefficients, $\dim \Gamma = r$; the elements of Γ being viewed as currents in the sense of de Rham.

Denote by $\mathrm{adm}(D,\mathcal{U})$ the class of all plurisubharmonic C^2-functions u_j defined in $U_j \cap D$ which satisfy the following conditions:

(i) the oscillation of u_j in $U_j \cap D$ is less than one,

(ii) $du_j = du_k$ in $U_j \cap U_k \cap D = \emptyset$,

where $d = \partial + \bar\partial$. Condition (ii) describes a closed real one-form in D. Similarly $d^c u_j$ and $dd^c u_j$, where $d^c = i(\bar\partial - \partial)$, give rise to global forms, well defined in D. Without ambiguity, we can denote them omitting the indices. Let

$$(1.3)\quad \mathrm{cap}_p(D,q,\Gamma,\mathcal{U}) = \sup_{u \in \mathrm{adm}(D,\mathcal{U})} \inf_{T \in \Gamma} |T[q[h(d^c u, d^c u)]^{\frac{1}{2}p-1} \det H\, \underline{D}^r u]|,$$

where

$$\underset{\sqcup}{D}^r u = \begin{cases} d^c u \wedge (dd^c u)^{\frac{1}{2}r-\frac{1}{2}} & \text{for } r \text{ odd}, \\ du \wedge d^c u \wedge (dd^c u)^{\frac{1}{2}r-1} & \text{for } r \text{ even}, \end{cases}$$

and the branch of $h^{\frac{1}{2}p}$ is chosen so that $1^{\frac{1}{2}p} = 1$. If $\mathrm{adm}(D,\mathcal{U}) = \emptyset$, we put $\mathrm{cap}_p(D,q,\Gamma,\mathcal{U}) = 0$. It can be proved [14], [16] what follows:

LEMMA 1.2. *Let $\mathbb{L}_o$ be an m-dimensional hermitian manifold of metric h which is not supposed to be positive definite and let Assumption 1.2 hold. If q is of the class C^1 and for any u defined globally on $\mathrm{cl}\, D$ and satisfying $0 < u(z) < 1$ for $z \in D$, $u|\partial C_o = 0$ and $u|\partial C_1 = 1$, there exists a C^1-solution Ψ: $\mathrm{cl}\, D \longrightarrow \mathbb{C}$ of the system of differential equations $d^c(\Psi u) = \Phi\, d^c u$, where $\Phi = q[h(d^c u, d^c u)]^{\frac{1}{2}p-1}\det H$, then $\mathrm{cap}_p(D,q,\Gamma,\mathcal{U}) < +\infty$.*

LEMMA 1.3. *Let $\mathbb{L}_o$ be an m-dimensional hermitian manifold of metric h which is not supposed to be positive definite and let Assumption 1.2 hold. Then (for an arbitrary q) the capacity (1.3) is a biholomorphic invariant. More exactly, let $f:\mathbb{L}_o \longrightarrow \mathbb{L}'_o$ be a biholomorphic mapping and f_* the induced homomorphism on homology classes, where $\mathbb{L}'_o = f[\mathbb{L}_o]$ is an hermitian manifold endowed with:*

(i) the hermitian metric h' given, in local coordinates $\mu = (\mu^j)$, by the relations

(1.4) $h'^{l\bar{n}} = (h^{j\bar{k}} \circ f^{-1})(f^l_{:j} \circ f^{-1})(\bar{f}^n_{:\bar{k}} \circ f^{-1})$,

where $u_{:j} = (u \circ \mu^{-1})_{|j} \circ \mu$ and $_{|j}$ denotes differentiation with respect to the j-th coordinate, or, equivalently, by

(1.5) $h'_{l\bar{n}} = (h_{j\bar{k}} \circ f^{-1})\, f^{-1}{}^j_{:l}\, \bar{f}^{-1}{}^k_{:n}$,

(ii) the C^1 tensor field H' of type (1,1) given accordingly, in local coordinates, by the relations

(1.6) $H'^l_n = (H^j_k \circ f^{-1})(f^l_{:j} \circ f^{-1})\, f^{-1}{}^k_{:n}$.

Then $\mathrm{cap}_p(f[D], q \circ f^{-1}, f_*\Gamma, f[\mathcal{U}]) = \mathrm{cap}_p(D,q,\Gamma,\mathcal{U})$.

In particular, we may let H depend on h or to take as H an almost complex structure of the tangent bundle $T\mathbb{L}_o$, for instance the complex structure J. The role of the additional structure H can be seen from the following result [15], [16].

Assumption 1.3 Suppose that $\mathbb{L}_o$ is a one-dimensional hermitian manifold of metric h. Denote by $\mathbb{M}_o$ the corresponding real manifold (in the sense of the theorem of Korn and Lichtenstein) endowed with the riemannian metric g given, in local coordinates, by the formulae $g_{11} = g_{22} = h_{1\bar{1}}$, $g_{12} = g_{21} = 0$. In addition Assumption 1.2 holds with $p \in [1;3]$ and $r = 1$. Let further D be a condenser on $\mathbb{L}_0$ and $\mathfrak{D}$ the same (i.e. embedded) condenser on $\mathbb{M}_0$. We also set $\Phi(\varphi) = q[h(\varphi,\varphi)]^{\frac{1}{2}p-1}H_1^1$ with $h(\varphi,\psi) = h^{1\bar{1}}\varphi_1\psi_{\bar{1}}$ and $h^{1\bar{1}}h_{1\bar{1}} = 1$, for any C^1-differentiable differential one-forms φ and ψ on $\mathbb{L}_0$.

LEMMA 1.4. *Suppose that Assumption 1.3 is satisfied. For any function $f:\mathbb{L}_0 \longrightarrow \mathbb{C}$ of the class C^1 let ψ denote a one-form of the class C^1 on D which, for every $\gamma \in \Gamma$, has the reproducing property*

$$\iint_D f\Phi(\varphi)\varphi \wedge \bar{\psi}^c = \int_\gamma f\Phi(\varphi)\varphi,$$

where φ is an arbitrary one-form of the class C^1 on D and, as usually, $\psi^c = -(\psi_1 dz - \psi_{\bar{1}} d\bar{z})$, $\bar{\psi} = \overline{\psi_{\bar{1}}}dz + \overline{\psi_1}d\bar{z}$. Suppose also that

$$[1/f\Phi(\psi)]\{h(d[f\Phi(\psi)],\psi) + h(\psi, d[f\Phi(\psi)])\} \leqslant 0.$$

Then, for some H, there are positive constants C and $C(a)$, $0 < a \leqslant 1$, independent of D (including the condenser plates) and depending only on $\mathbb{L}_0$ (including h), p, and q, such that

$$[1/C(a)][\lambda_p(\Gamma,|q|)]^a \leqslant \operatorname{cap}_p(D,q,\Gamma,\mathcal{U}) \leqslant C[\lambda_{\tilde{p}}(\Gamma,\tilde{q})]^{\frac{1}{2}},$$

where $\tilde{p} = 2/(p-1)$, $\tilde{q} = |q|^{2/(p-1)}$, and $\lambda_{\tilde{p}}(\Gamma,\tilde{q})$ denotes the (p,q)-extremal length of Γ.

Remark 1.2. In our context instead of recalling the definition of $\lambda_{\tilde{p}}(\Gamma,\tilde{q})$ we only notice that it reduces to $\operatorname{cap}\mathfrak{D}$ if $p = 2$, $q = 1$, and Γ is the homology class of $\mathfrak{D}$ with real coefficients, represented by a Jordan curve separating in $\mathfrak{D}$ the condenser plates.

Suppose now that Assumption 1.2 holds. With every condenser $\mathfrak{D}$ whose closure is compact on $\mathbb{M}_1$ associate a condenser $\mathfrak{D} \times F[\mathfrak{D}]$ (provided that it exists) whose plates depend on the plates C_0 and C_1 of $\mathfrak{D}$ as well as on the plates $F[C_0]$ and $F[C_1]$ of $F[\mathfrak{D}]$ in an arbitrary way, and denote, for short, by $\operatorname{cap}(\mathfrak{D} \times F[\mathfrak{D}])$ the corresponding capacity (1.3) with $D = \mathfrak{D} \times F[\mathfrak{D}]$; $F:\mathbb{M}_1 \longrightarrow \mathbb{M}_2$ being a homeomorphism.

Assumption 1.4. Suppose that Assumptions 1.1 and 1.2 hold, $\mathbb{L}' = \mathbb{M}_1' \times \mathbb{M}_2'$ is an hermitian manifold analogous to $\mathbb{L}$, there exists a biholomorphic mapping $f:\mathbb{L}_0 \longrightarrow \mathbb{L}_0' = f[\mathbb{L}_0]$, $\mathbb{L}_0 \subset \mathbb{L}$, $\mathbb{L}_0' \subset \mathbb{L}'$, and the metrics h and $h'|\mathbb{L}_0'$ are related, in local coordinates $\mu = (\mu^j)$, by the formulae (1.4), where $u_{:j} = (u \circ \mu^{-1})_{|j} \circ \mu$, or, equivalently, by (1.5), while the tensor fields H and $H'|\mathbb{L}'_0$ are related by the formulae (1.6).

Assumption 1.5. Suppose next that for some points O_1 of $\mathbb{M}_1$ and O_2 of $\mathbb{M}_2$ we have a relation of the form

(1.7) $G \circ U(x,O_2) = V(O_1,F(x)), \quad x \in \mathbb{M}_1,$

where U and V are the projections of f:

(1.8) $f(x,y) = (U(x,y),\ V(x,y)); \quad U(x,y) \in \mathbb{M}'_1, \quad V(x,y) \in \mathbb{M}'_2,$

for every $x \in \mathbb{M}_1$ and $y \in \mathbb{M}_2$; F being a homeomorphism.

Remark 1.3. In other words, we suppose that we have the commutative diagram

$$\begin{array}{ccc} \mathbb{M}_1 & \xrightarrow{F} & \mathbb{M}_2 \\ U(\ ,O_2)\downarrow & & \downarrow V(O_1,\ ). \\ \mathbb{M}'_1 & \xrightarrow[G]{} & \mathbb{M}'_2 \end{array} \tag{1.9}$$

Assumption 1.6. Finally we suppose that there are positive constants c_1, c_2, and a such that for every condenser $\mathfrak{D}$ whose closure is compact on $\mathbb{M}_1$ there exist: a capacity $\operatorname{cap}(\mathfrak{D} \times \mathfrak{F})$ and an analogous capacity $\operatorname{cap}(\mathfrak{U} \times \mathfrak{V})$, where

(1.10) $\mathfrak{F} = F[\mathfrak{D}], \quad \mathfrak{U} = U[\mathfrak{D} \times \{O_2\}], \quad \mathfrak{V} = V[\{O_1\}\ \ F[\ \ ]],$

which fulfils the estimates

(1.11) $c_1^{-1}\operatorname{cap}^a\mathfrak{D} + c_2^{-1}\operatorname{cap}^a\mathfrak{F} \leq \operatorname{cap}(\mathfrak{D} \times \mathfrak{F}) \leq c_1\operatorname{cap}^a\mathfrak{D} + c_2\operatorname{cap}^a\mathfrak{F}$

and

(1.12) $c_1^{-1}\operatorname{cap}^a\mathfrak{U} + c_2^{-1}\operatorname{cap}^a\mathfrak{V} \leq \operatorname{cap}(\mathfrak{U} \times \mathfrak{V}) \leq c_1\operatorname{cap}^a\mathfrak{U} + c_2\operatorname{cap}^a\mathfrak{V},$

where $\operatorname{cap}^a\mathfrak{D} = (\operatorname{cap}\mathfrak{D})^a$ etc.

Relations (1.11) and (1.12) originate in extending the situation described in Lemma 1.4. We have[16]:

PROPOSITION 1.1. If Assumptions 1.4-1.6 are fulfilled and the mappings F and G are Q-quasiconformal, then the mappings $U(\ ,O_2)$ and $V(O_1,\ )$ are K_1 and K_2-quasiconformal, respectively, where

(1.13) $K_1 \leq \left(\dfrac{c_1 + c_2Q^a}{c_1^{-1} + c_2^{-1}Q^{-a}}\right)^{1/a}, \quad K_2 \ \ \left(\dfrac{c_1Q^a + c_2}{c_1^{-1}Q^{-a} + c_2^{-1}}\right)^{1/a}.$

<u>Both estimates in</u> (1.13) <u>are precise</u>.

We are going to prove the following:

<u>PROPOSITION 1.2</u>. <u>If Assumptions</u> 1.4-1.6 <u>are fulfilled</u>, <u>then the mappings</u> F <u>and</u> G <u>are</u> <u>Q-quasiconformal</u>, <u>where</u>

$$(1.14)\quad Q \leqslant \left(\frac{c_2-c_2^{-1}}{c_1^{-1}-c_1}\right)^{\frac{m-1}{a}} \quad \underline{\text{or}} \quad Q \leqslant \left(\frac{c_1-c_1^{-1}}{c_2^{-1}-c_2}\right)^{\frac{m-1}{a}}$$

<u>provided that</u>

$$(1.15)\quad c_1 < 1, \quad c_2 > 1$$

<u>or</u>

$$(1.16)\quad c_1 > 1, \quad c_2 < 1,$$

<u>respectively</u>. <u>Both estimates in</u> (1.14) <u>are precise</u>.

<u>COROLLARY 1.1</u>. <u>Under the hypotheses of Proposition</u> 1.2 <u>the mappings</u> $U(\ ,O_2)$ <u>and</u> $V(O_1,\ )$ <u>are</u> K_1 <u>and</u> K_2<u>-quasiconformal</u>, <u>respectively</u>, <u>where</u>

$$(1.17)\quad K_1 \leqslant \left[\frac{c_1(c_1^{-1}-c_1)^{m-1}+c_2(c_2-c_2^{-1})^{m-1}}{c_1^{-1}(c_1^{-1}-c_1)^{-m+1}+c_2^{-1}(c_2-c_2^{-1})^{-m+1}}\right]^{\frac{1}{a}} (c_1^{-1}-c_1)^{-2\frac{m-1}{a}}$$

$$(1.18)\quad K_2 \leqslant \left[\frac{c_1(c_1^{-1}-c_1)^{-m+1}+c_2(c_2-c_2^{-1})^{-m+1}}{c_1^{-1}(c_1^{-1}-c_1)^{m-1}+c_2^{-1}(c_2-c_2^{-1})^{m-1}}\right]^{\frac{1}{a}} (c_2-c_2^{-1})^{2\frac{m-1}{a}}$$

<u>in the case of</u> (1.15), <u>whereas</u>

$$(1.19)\quad K_1 \leqslant \left[\frac{c_1(c_1-c_1^{-1})^{-m+1}+c_2(c_2^{-1}-c_2)^{-m+1}}{c_1^{-1}(c_1-c_1^{-1})^{m-1}+c_2^{-1}(c_2^{-1}-c_2)^{m-1}}\right]^{\frac{1}{a}} (c_1-c_1^{-1})^{2\frac{m-1}{a}},$$

$$(1.20)\quad K_2 \leqslant \left[\frac{c_1(c_1-c_1^{-1})^{m-1}+c_2(c_2^{-1}-c_2)^{m-1}}{c_1^{-1}(c_1-c_1^{-1})^{m+1}+c_2^{-1}(c_2^{-1}-c_2)^{-m+1}}\right]^{\frac{1}{a}} (c_2^{-1}-c_2)^{-2\frac{m-1}{a}}$$

<u>in the case of</u> (1.16). <u>The estimates</u> (1.17)-(1.20) <u>are precise</u>.

The above corollary follows directly from Proposition 1.1 and 1.2.

<u>P r o o f o f P r o p o s i t i o n 1.2</u>. In the case of (1.15) from (1.11) and (1.12) we get, for every condenser $\mathscr{D}$ whose closure is compact on $\mathbb{M}_1$, the precise estimates

$$(1.21)\qquad \operatorname{cap}\mathfrak{F} \geqslant \Big(\frac{c_1^{-1}-c_1}{c_2-c_2^{-1}}\Big)^{\frac{1}{a}}\operatorname{cap}\mathfrak{D} \quad\text{and}\quad \operatorname{cap}\mathfrak{V} \geqslant \Big(\frac{c_1^{-1}-c_1}{c_2-c_2^{-1}}\Big)^{\frac{1}{a}}\operatorname{cap}\mathfrak{U},$$

respectively. Hence, by Lemma 1.1 we obtain the first estimate in (1.14) which is precise. Similarly, in the case of (1.16) we obtain the precise estimates

$$(1.22)\qquad \operatorname{cap}\mathfrak{F} \leqslant \Big(\frac{c_1-c_1^{-1}}{c_2^{-1}-c_2}\Big)^{\frac{1}{a}}\operatorname{cap}\mathfrak{D} \quad\text{and}\quad \operatorname{cap}\mathfrak{V} \leqslant \Big(\frac{c_1-c_1^{-1}}{c_2^{-1}-c_2}\Big)^{\frac{1}{a}}\operatorname{cap}\mathfrak{U},$$

and by Lemma 1.1 again, the second estimate in (1.14) which is precise as well, as desired.

R e m a r k 1.4. In [16] Proposition 1.1 is proved under a weaker assumption that $\mathbb{M}_1$, $\mathbb{M}_2$, $\mathbb{M}_1'$, and $\mathbb{M}_2'$ are pseudo-riemannian, but of the same index. Also the symmetry of g_1, g_2, g_1', and g_2' is not needed, for it is enough to replace (1.1) e.g. by

$$(1.23)\qquad h(v,w) = g(v,w) - i\,g(Jv,w) + i\,g(v,Jw) + g(Jv,Jw);\ v,w \in T_z\mathbb{L}.$$

Therefore the same generalization applies, in principle, to Proposition 1.2 and Corollary 1.1. It seems, however, that for essentially pseudo-riemannian manifolds [17] the estimates (1.11) and (1.12) with (1.15) or (1.16) may never be satisfied. Such a conclusion would suit very well to the physical picture given in [18] and [19].

Let us return to our main topic: biholomorphic continuability of quasiconformal mappings. We should like to prove theorems of the following kind: If, under some additional hypotheses (if necessary), for a certain point O_2 of $\mathbb{M}_2$ we have

$$(1.24)\qquad \begin{array}{ccc} \mathbb{M}_1 & (1.11) & \mathbb{M}_2 \\ {\scriptstyle U(\ ,O_2)}\downarrow & & \\ \mathbb{M}_1' & (1.12') & \mathbb{M}_2' \end{array} \quad\text{or}\quad \begin{array}{ccc} \mathbb{M}_1 & (1.11) & \mathbb{M}_2 \\ {\scriptstyle U(\ ,O_2)}\downarrow{\scriptstyle K_1\text{-qc}} & & \\ \mathbb{M}_1' & \xrightarrow[Q'\text{-qc}]{G} & \mathbb{M}_2' \end{array}$$

$$\text{or}\quad \begin{array}{ccc} \mathbb{M}_1 & \xrightarrow[Q\text{-qc}]{F} & \mathbb{M}_2 \\ {\scriptstyle U(\ ,O_2)}\downarrow{\scriptstyle K_1\text{-qc}} & & \\ \mathbb{M}_1' & (1.12') & \mathbb{M}_2', \end{array}$$

where <u>qc</u> stands for <u>quasiconformal</u> and (1.12′) = (1.25) for

$$(1.25)\qquad c_1'^{-1}\operatorname{cap}^a\mathfrak{U} + c_2'^{-1}\operatorname{cap}^a\mathfrak{V} \leqslant \operatorname{cap}(\mathfrak{U}\times\mathfrak{V}) \leqslant c_1'\operatorname{cap}^a\mathfrak{U} + c_2'\operatorname{cap}^a\mathfrak{V},$$

then $U(\ ,O_2)$ is biholomorphically continuable to a unique

$$f = U + iV : \mathbb{L}_o(\mathbb{M}_1) \longrightarrow \mathbb{L}_o'(\mathbb{M}_1') = f[\mathbb{L}_o(\mathbb{M}_1)]$$

so that, for $O_1 = F^{-1}(O_2)$, we have

$$\begin{cases} F \text{ and } G \text{ are } Q\text{-qc}, \\ U(\ ,O_2) \text{ is } K_1\text{-qc}, \\ V(O_1,\ ) \text{ is } K_2\text{-qc}; \end{cases} \quad \text{or} \quad \begin{cases} F \text{ is } Q\text{-qc}, \\ V(O_1,\ ) \text{ is } K_2\text{-qc}, \\ (1.25) \text{ holds}; \end{cases}$$

$$\text{or} \quad \begin{cases} G \text{ is } Q'\text{-qc}, \\ V(O_1,\ ) \text{ is } K_2\text{-qc}, \\ (1.11) \text{ holds}; \end{cases}$$

respectively, with $Q, Q', K_1, K_2, c_1, c_1', c_2$, and c_2' estimated explicitly and precisely.

It appears that a sufficient condition to fill these schemes is to add the hypotheses that $\mathbb{M}_1$ is even-dimensional and, after complexification, contained in $\mathbb{C}^{\frac{1}{2}m}$, whereas $U(\ ,O_2)$ is biholomorphic.

The first two hypotheses suit well to the physical desire of including at the first instance to considerations the four-dimensional (real) spaces of a particle corresponding to the Minkowski space of observations (possibly with some holes).

As regards the third hypotheses, since $U(\ ,O_2)$ is already K_1-quasiconformal (in two latter cases by definition; in the first case this fact can be deduced from (1.11) and (1.25) - cf. Theorem 3.1 below), the Hartogs theorem yields that it is ensured by a more convenient assumption that the complexified $U(\ ,O_2)$ is holomorphic in each variable separately. But $U(\ ,O_2)$ is K_1-quasiconformal, so it transforms infinitesimal spheres onto infinitesimal ellipsoids (cf. e.g.[7], p.128) whose projections to the complex planes in question are infinitesimal circles. These planes at a point $U(x_o, O_2), x_o = (x_o^1, \dots x_o^m) \in \mathbb{M}_1$, are tangent to the images under U of the planes

$$(1.26) \quad \{x \in \mathbb{M}_1 : x^j + i\, x^{\frac{1}{2}m+j} = x_o^j + x_o^{\frac{1}{2}m+j}, \quad j = 1, \dots, k-1, \\ k+1, \dots, \tfrac{1}{2}m\},\ k = 1, \dots, \tfrac{1}{2}m,$$

respectively. We say that the above described infinitesimal ellipsoids are <u>circular</u> or, more precisely, <u>circular</u> with respect to the complex planes in question.

Thus we are led to the following

<u>Definition</u>. A quasiconformal mapping U of $\mathbb{M}_1 \subset \mathbb{R}^m$, m even, onto $\mathbb{M}_2 \subset \mathbb{R}^m$ is called <u>circular</u> [or: is said to transform

infinitesimal spheres onto (infinitesimal) <u>circular</u> ellipsoids] if for every $x_0 \in \mathbb{M}_1$, there exist $\frac{1}{2}m$ mutually perpendicular planes $\mathbb{C}_1,\dots,\mathbb{C}_{\frac{1}{2}m}(\mathbb{C}_1 \times \dots \times \mathbb{C}_{\frac{1}{2}m} = C^{\frac{1}{2}m})$ such that $U|\mathbb{M}_1 \cap \mathbb{C}_k$, $k = 1,\dots,\frac{1}{2}m$, are holomorphic [or, equivalently, such that U transforms infinitesimal spheres onto (infinitesimal) circular ellipsoids].

The planes $\mathbb{C}_1,\dots,\mathbb{C}_{\frac{1}{2}m}$ are given by the formulae (1.26). Thus we are looking for a suitable Cartesian coordinate system in $\mathbb{C}^{\frac{1}{2}m}$ assuring the required property.

2. Generalization of the propositions on quasiconformal projections of biholomorphic mappings

If we look at the schemes (1.24) it appears that the choice of the condition (1.12′) = (1.25) instead of (1.12) in the first case of (1.24) is essential for the sake of completeness since in the second and third case we really obtain, in general, $c_1' \neq c_1$ and $c_2' \neq c_2$, so the choice of their maxima instead of them looses precision. Thus before we pass in the next section to our main topic we give the corresponding generalizations of Propositions 1.1 and 1.2 as well as of Corollary 1.1. To this end we modify Assumption 1.6 as follows:

A s s u m p t i o n 2.1. Suppose that there are positive constants c_1, c_1', c_2, c_2', and a such that for every condenser $\mathfrak{D}$ whose closure is compact on $\mathbb{M}_1$ there exist: a capacity $\operatorname{cap}(\mathfrak{D} \times \mathfrak{F})$ and an analogous capacity $\operatorname{cap}(\mathfrak{U} \times \mathfrak{V})$, where $\mathfrak{F}$, $\mathfrak{U}$, and $\mathfrak{V}$ are defined by (1.10), which fulfil the estimates (1.11) and (1.25), where $\operatorname{cap}^a \mathfrak{D} = (\operatorname{cap}\mathfrak{D})^a$ etc.

PROPOSITION 2.1. <u>If Assumptions</u> 1.4, 1.5, <u>and</u> 2.1 <u>are fulfilled and the mappings</u> F <u>and</u> G <u>are</u> Q <u>and</u> Q′-<u>quasiconformal, respectively, then the mappings</u> $U(\ ,O_2)$ <u>and</u> $V(O_1,\ )$ <u>are</u> K_1 <u>and</u> K_2-<u>quasiconformal, respectively, where</u>

$$(2.1)\quad K_1 \leq \max{}^{1/a}\left(\frac{c_1 + c_2 Q^a}{c_1'^{-1} + c_2'^{-1} Q'^{-a}},\ \frac{c_1' + c_2' Q'^a}{c_1^{-1} + c_2^{-1} Q^{-a}}\right),$$

$$(2.2)\quad K_2 \leq \max{}^{1/a}\left(\frac{c_1 Q^a + c_2}{c_1'^{-1} Q'^{-a} + c_2'^{-1}},\ \frac{c_1' Q'^a + c_2'}{c_1^{-1} Q^{-a} + c_2^{-1}}\right),$$

<u>and</u> $\max^{1/a}(\alpha,\beta) = [\max(\alpha,\beta)]^{1/a}$. <u>Both estimates</u> (2.1) <u>and</u> (2.2) <u>are precise</u>.

P r o o f. By Lemma 1.3, owing to the relations (1.8) and (1.10), we have

$$\operatorname{cap}(\mathfrak{U} \times \mathfrak{V}) = \operatorname{cap}(\mathfrak{D} \times \mathfrak{F}).$$

Hence (1.11) and (1.22) yield

(2.3) $c_1^{-1}\operatorname{cap}^a\mathfrak{D} + c_2^{-1}\operatorname{cap}^a\mathfrak{F} \leqslant c_1'\operatorname{cap}^a\mathfrak{U} + c_2'\operatorname{cap}^a\mathfrak{V}$

and

(2.4) $c_1'^{-1}\operatorname{cap}^a\mathfrak{U} + c_2'^{-1}\operatorname{cap}^a\mathfrak{F} \leqslant c_1\operatorname{cap}^a\mathfrak{D} + c_2\operatorname{cap}\mathfrak{V}$

On the other hand, the estimates

(2.5) $(1/Q)\operatorname{cap}\mathfrak{D} \leqslant \operatorname{cap}\mathfrak{F} \leqslant Q\operatorname{cap}\mathfrak{D}, \quad (1/Q')\operatorname{cap}\mathfrak{U} \leqslant \operatorname{cap}\mathfrak{V} \leqslant Q'\operatorname{cap}\mathfrak{U},$

being a consequence of the Q-quasiconformality of F and Q'-quasiconformality of G together with the relations (1.8) and (1.10), imply

(2.6) $\operatorname{cap}\mathfrak{F} \geqslant Q^{-1}\operatorname{cap}\mathfrak{D}, \quad \operatorname{cap}\mathfrak{V} \leqslant Q'\operatorname{cap}\mathfrak{U},$

(2.7) $\operatorname{cap}\mathfrak{F} \leqslant Q\operatorname{cap}\mathfrak{D}, \quad \operatorname{cap}\mathfrak{V} \geqslant Q'^{-1}\operatorname{cap}\mathfrak{U}$

and, eqivalently,

(2.8) $\operatorname{cap}\mathfrak{D} \geqslant Q^{-1}\operatorname{cap}\mathfrak{F}, \quad \operatorname{cap}\mathfrak{U} \leqslant Q'\operatorname{cap}\mathfrak{V},$

(2.9) $\operatorname{cap}\mathfrak{D} \leqslant Q\operatorname{cap}\mathfrak{F}, \quad \operatorname{cap}\mathfrak{U} \geqslant Q'^{-1}\operatorname{cap}\mathfrak{V}.$

Therefore from (2.3) and (2.6) we get

(2.10) $(c_1^{-1} + c_2^{-1}Q^{-a})\operatorname{cap}^a\mathfrak{D} \leqslant (c_1' + c_2'Q'^{a})\operatorname{cap}^a\mathfrak{U}.$

Similarly, from (2.4) and (2.7) we obtain

(2.11) $(c_1'^{-1} + c_2'^{-1}Q^{-a})\operatorname{cap}^a\mathfrak{U} \leqslant (c_1 + c_2Q^a)\operatorname{cap}^a\mathfrak{D}.$

Finally, inequalities (2.10) and (2.11) yield

(2.12) $$\left(\frac{c_1^{-1} + c_2^{-1}Q^{-a}}{c_1' + c_2'Q'^{a}}\right)^{1/a}\operatorname{cap}\mathfrak{D} \leqslant \operatorname{cap}\mathfrak{U} \leqslant \left(\frac{c_1 + c_2Q^a}{c_1'^{-1} + c_2'^{-1}Q'^{a}}\right)^{1/a}\operatorname{cap}\mathfrak{D},$$

and this, since $U(\ ,O_2)$ is a sense-preserving homeomorphism, gives the desired K_1-quasiconformality of it with K_1 satisfying (2.1). The above consideration shows also that this estimate is precise.

Similarly, from (2.3), (2.8) and (2.4), (2.9) we obtain

(2.13) $$\left(\frac{c_1^{-1}Q^{-a} + c_2^{-1}}{c_1'Q'^{a} + c_2'}\right)^{1/a}\operatorname{cap}\mathfrak{F} \leqslant \operatorname{cap}\mathfrak{V} \leqslant \left(\frac{c_1Q^a + c_2}{c_1'^{-1}Q'^{-a} + c_2'^{-1}}\right)^{1/a}\operatorname{cap}\mathfrak{F},$$

and hence we analogously infer the K_2-quasiconformality of $V(O_1,\ )$ with K_2 satisfying (2.2) which is also precise, thus completing the proof.

PROPOSITION 2.2. *If Assumptions* 1.4, 1.5, *and* 2.1 *are fulfilled then the mappings* F *and* G *are* Q *and* Q'-*quasiconformal, respectively, with* Q *estimated by* (1.14) *provided that we have* (1.15) *or* (1.16), *respectively, and* Q' *estimated by*

$$(2.14)\quad Q' \leqslant \left(\frac{c_2' - c_2'^{-1}}{c_1'^{-1} - c_1'}\right)^{\frac{m-1}{a}} \quad \text{or} \quad Q' \leqslant \left(\frac{c_1' - c_1'^{-1}}{c_2'^{-1} - c_2'}\right)^{\frac{m-1}{a}}$$

provided that

$$(2.15)\quad c_1' < 1, \quad c_2' > 1$$

or

$$(2.16)\quad c_1' > 1, \quad c_2' < 1,$$

respectively. Both estimates in (1.14) *as well as in* (2.14) *are precise.*

Proof. The proof of the Q-quasiconformallity of F with Q estimated precisely by (1.14) is identical with that of Proposition 1.2. In order to prove the Q'-quasiconformality of G with Q' estimated precisely by (2.14) we have to replace in the proof mentioned the estimates of $\operatorname{cap} \mathcal{V}$ appearing in (1.21) and (1.22) by

$$\operatorname{cap} \mathcal{V} \geqslant \left(\frac{c_1'^{-1} - c_1'}{c_2' - c_2'^{-1}}\right)^{\frac{1}{a}} \operatorname{cap} \mathcal{U} \quad \text{and} \quad \operatorname{cap} \mathcal{V} \leqslant \left(\frac{c_1' - c_1'^{-1}}{c_2'^{-1} - c_2'}\right)^{\frac{1}{a}} \operatorname{cap} \mathcal{U}$$

according to the cases (2.15) and (2.16), respectively.

COROLLARY 2.1. *Under the hypotheses of Proposition* 2.2 *the mappings* $U(\;, O_2)$ *and* $V(O_1, \;)$ *are* K_1 *and* K_2-*quasiconformal, respectively, where*

$$(2.17)\quad K_1 \leqslant (c_1^{-1} - c_1)^{-\frac{m-1}{a}} (c_1'^{-1} - c_1')^{-\frac{m-1}{a}} \times$$

$$\times \max^{\frac{1}{a}}\left[\frac{c_1(c_1^{-1} - c_1)^{m-1} + c_2(c_2 - c_2^{-1})^{m-1}}{c_1'^{-1}(c_1'^{-1} - c_1')^{-m+1} + c_2'^{-1}(c_2' - c_2'^{-1})^{-m+1}}, \frac{c_1'(c_1'^{-1} - c_1')^{m-1} + c_2'(c_2' - c_2'^{-1})^{m-1}}{c_1^{-1}(c_1^{-1} - c_1)^{-m+1} + c_2^{-1}(c_2 - c_2^{-1})^{-m+1}}\right],$$

$$(2.18)\quad K_2 \leqslant (c_2 - c_2^{-1})^{\frac{m-1}{a}} (c_2' - c_2'^{-1})^{\frac{m-1}{a}} \times$$

$$\times \max^{\frac{1}{a}}\left[\frac{c_1(c_1^{-1} - c_1)^{-m+1} + c_2(c_2 - c_2^{-1})^{-m+1}}{c_1'^{-1}(c_1'^{-1} - c_1')^{m-1} + c_2'^{-1}(c_2' - c_2'^{-1})^{m-1}}, \frac{c_1'(c_1'^{-1} - c_1')^{-m+1} + c_2'(c_2' - c_2'^{-1})^{-m+1}}{c_1^{-1}(c_1^{-1} - c_1)^{m-1} + c_2^{-1}(c_2 - c_2^{-1})^{m-1}}\right]$$

in the case of (1.15) *with* (2.15);

$$(2.19)\quad K_1 \leqslant (\bar c_1^{-1} - c_1)^{-\frac{m-1}{a}} (c_1' - c_1'^{-1})^{\frac{m-1}{a}} \times$$

$$\times \max^{\frac{1}{a}}\left[\frac{c_1(\bar c_1^{-1} - c_1)^{m-1} + c_2(c_2 - \bar c_2^{-1})^{m-1}}{c_1'^{-1}(c_1' - c_1'^{-1})^{m-1} + c_2'^{-1}(c_2'^{-1} - c_2')^{m-1}}, \frac{c_1'(c_1' - c_1'^{-1})^{-m+1} + c_2'(c_2'^{-1} - c_2')^{-m+1}}{\bar c_1^{-1}(\bar c_1^{-1} - c_1)^{-m+1} + \bar c_2^{-1}(c_2 - \bar c_2^{-1})^{-m+1}}\right],$$

$$(2.20)\quad K_2 \leqslant (c_2 - \bar c_2^{-1})^{\frac{m-1}{a}} (c_2'^{-1} - c_2')^{-\frac{m-1}{a}} \times$$

$$\times \max^{\frac{1}{a}}\left[\frac{c_1(\bar c_1^{-1} - c_1)^{-m+1} + c_2(c_2 - \bar c_2^{-1})^{-m+1}}{c_1'^{-1}(c_1' - c_1'^{-1})^{-m+1} + c_2'^{-1}(c_2'^{-1} - c_2')^{-m+1}}, \frac{c_1'(c_1' - c_1'^{-1})^{m-1} + c_2'(c_2' - c_2'^{-1})^{m-1}}{\bar c_1^{-1}(\bar c_1^{-1} - c_1)^{m-1} + \bar c_2^{-1}(\bar c_2^{-1} - c_2)^{m-1}}\right]$$

in the case of (1.15) *with* (2.16);

$$(2.21)\quad K_1 \leqslant (c_1 - \bar c_1^{-1})^{\frac{m-1}{a}} (c_1'^{-1} - c_1')^{\frac{m-1}{a}} \times$$

$$\times \max^{\frac{1}{a}}\left[\frac{c_1(c_1 - \bar c_1^{-1})^{-m+1} + c_2(\bar c_2^{-1} - c_2)^{-m+1}}{c_1'^{-1}(c_1'^{-1} - c_1')^{-m+1} + c_2'^{-1}(c_2' - c_2'^{-1})^{-m+1}}, \frac{c_1'(c_1'^{-1} - c_1')^{m-1} + c_2'(c_2' - c_2'^{-1})^{m-1}}{\bar c_1^{-1}(c_1 - \bar c_1^{-1})^{m-1} + \bar c_2^{-1}(\bar c_2^{-1} - c_2)^{m-1}}\right],$$

$$(2.22)\quad K_2 \leqslant (\bar c_2^{-1} - c_2)^{-\frac{m-1}{a}} (c_2' - c_2'^{-1})^{\frac{m-1}{a}} \times$$

$$\times \max^{\frac{1}{a}}\left[\frac{c_1(c_1 - \bar c_1^{-1})^{m-1} + c_2(\bar c_2^{-1} - c_2)^{m-1}}{c_1'^{-1}(c_1'^{-1} - c_1')^{m-1} + c_2'^{-1}(c_2' - c_2'^{-1})^{m-1}}, \frac{c_1'(c_1'^{-1} - c_1')^{-m+1} + c_2'(c_2' - c_2'^{-1})^{-m+1}}{\bar c_1^{-1}(c_1 - \bar c_1^{-1})^{-m+1} + \bar c_2^{-1}(\bar c_2^{-1} - c_2)^{-m+1}}\right]$$

in the case of (1.16) *with* (2.15);

$$(2.23)\quad K_1 \leqslant (c_1 - \bar c_1^{-1})^{\frac{m-1}{a}} (c_1' - c_1'^{-1})^{\frac{m-1}{a}} \times$$

$$\times \max^{\frac{1}{a}}\left[\frac{c_1(c_1 - \bar c_1^{-1})^{-m+1} + c_2(\bar c_2^{-1} - c_2)^{-m+1}}{c_1'^{-1}(c_1' - c_1'^{-1})^{m-1} + c_2'^{-1}(c_2'^{-1} - c_2')^{m-1}}, \frac{c_1'(c_1' - c_1'^{-1})^{-m+1} + c_2'(c_2'^{-1} - c_2')^{-m+1}}{\bar c_1^{-1}(c_1 - \bar c_1^{-1})^{m-1} + \bar c_2^{-1}(\bar c_2^{-1} - c_2)^{m-1}}\right],$$

$$(2.24)\quad K_2 \leqslant (\bar c_2^{-1} - c_2)^{-\frac{m-1}{a}} (c_2'^{-1} - c_2)^{-\frac{m-1}{a}} \times$$

$$\times \max^{\frac{1}{a}}\left[\frac{c_1(c_1 - \bar c_1^{-1})^{m-1} + c_2(\bar c_2^{-1} - c_2)^{m-1}}{c_1'^{-1}(c_1' - c_1'^{-1})^{-m+1} + c_2'^{-1}(c_2'^{-1} - c_2')^{-m+1}}, \frac{c_1'(c_1' - c_1'^{-1})^{m-1} + c_2'(c_2'^{-1} - c_2')^{m-1}}{\bar c_1^{-1}(c_1 - \bar c_1^{-1})^{-m+1} + \bar c_2^{-1}(\bar c_2^{-1} - c_2)^{-m+1}}\right]$$

in the case of (1.16) with (2.16), where $\max^{1/a}(\alpha,\beta) = \max(\alpha,\beta)]^{1/a}$. The estimates (2.17)-(2.24) are precise.

The above corollary follows directly from Propositions 2.1 and 2.2.

We conclude this section with rewriting the estimates appearing in Proposition 1.2, Corollary 1.1, Proposition 2.2, and Corollary 2.1 in a more concise form using the notation

$$/s/ = 1/(s^{-1} - s) \quad \text{for} \quad 0<s<1, \quad /s/ = s-s^{-1} \quad \text{for} \quad s>1.$$

Thus we have:

Remark 2.1. The estimates (1.14) are equivalent to

(2.25) $Q \leqslant (/c_1///c_2/)^{(m-1)/a}$.

Remark 2.2. The estimates (1.17) and (1.19) are equivalent to

$$(2.26) \quad K_1 \leqslant \left[\frac{c_1/c_1/^{-m+1} + c_2/c_2/^{m-1}}{c_1^{-1}/c_1/^{m-1} + c_2^{-1}/c_2/^{-m+1}}\right]^{\frac{1}{a}} /c_1/^{2\frac{m-1}{a}}.$$

B. The estimates (1.18) and (1.20) are equivalent to

$$(2.27) \quad K_2 \leqslant \left[\frac{c_1/c_1/^{m-1} + c_2/c_2/^{-m+1}}{c_1^{-1}/c_1/^{-m+1} + c_2^{-1}/c_2/^{m+1}}\right]^{\frac{1}{a}} /c_2/^{2\frac{m-1}{a}}.$$

Remark 2.3. The estimates (2.14) are equivalent to

(2.28) $Q' \leqslant (/c_1'///c_2'/)^{(m-1)/a}$.

Remark 2.4. A. The estimates (2.17), (2.19), (2.21), and (2.23) are equivalent to

(2.29)

$$K_1 \leqslant (/c_1//c_1'/)^{\frac{m-1}{a}} \max^{\frac{1}{a}}\left[\frac{c_1/c_1/^{-m+1} + c_2/c_2/^{m-1}}{c_1'^{-1}/c_1'/^{m-1} + c_2'^{-1}/c_2'/^{-m+1}}, \frac{c_1'/c_1'/^{-m+1} + c_2'/c_2'/^{m-1}}{c_1^{-1}/c_1/^{m-1} + c_2^{-1}/c_2/^{-m+1}}\right].$$

B. The estimates (2.18), (2.20), (2.22), nad (2.24) are equivalent to

$$(2.30)\quad K_2 \leqslant (|c_2|/|c_2'|)^{\frac{m-1}{a}} \max^{\frac{1}{a}} \left[\frac{c_1|c_1|^{m-1} + c_2|c_2|^{-m+1}}{c_1'^{-1}|c_1'|^{-m+1} + c_2'^{-1}|c_2'|^{m-1}}, \right.$$

$$\left. \frac{c_1'|c_1'|^{m-1} + c_2'|c_2'|^{-m+1}}{c_1^{-1}|c_1|^{-m+1} + c_2^{-1}|c_2|^{m-1}} \right].$$

3. Biholomorphic continuability of spacial quasiconformal mappings transforming infinitesimal spheres onto circular ellipsoids

We are going to formulate in terms of circular quasiconformal mappings and prove three theorems on biholomorphic continuability which make precise the three cases in (1.24), respectively.

Assumption 3.1. Suppose that Assumption 1.1 holds with m even, $\mathbb{M}_1$ contained, after complexification $\mathbb{M}_1 \longmapsto \mathbb{M}_1^c$, in $\mathbb{C}^{\frac{1}{2}m}$, and u being a mapping of $\mathbb{M}_1$ onto an analogous Riemannian manifold $\mathbb{M}_1'$, which, after complexification $u(x) \longmapsto u_c(x_c)$, is biholomorphic. Here $\mathbb{M}_1^c$ consists of all $x_c \in \mathbb{C}^{\frac{1}{2}m}$ such that $x \in \mathbb{M}_1$ and

$$x_c = (x^1 + ix^{\frac{1}{2}m+1}, \ldots, x^{\frac{1}{2}m} + ix^m), \quad u_c = (u^1 + iu^{\frac{1}{2}m+1}, \ldots, u^{\frac{1}{2}m} + iu^m).$$

Let further $\mathbb{M}_2$ consist of those $y \in \mathbb{R}^m$ for which there is an $x \in \mathbb{M}_1$ such that

$$(3.1)\quad z_c \equiv (z^1 + iz^{\frac{1}{2}m+1}, \ldots, z^{\frac{1}{2}m} + iz^m) \in \mathbb{M}_1^c \quad \text{and}$$
$$\text{and} \quad \bar{z}_c \equiv (\bar{z}^1, \ldots, \bar{z}^m)_c \in \mathbb{M}_1^c; \quad z = x + iy.$$

Finally, let $\mathbb{L}_0$ consist of all $(x,y) \in \mathbb{L}$, satisfying (3.1).

Consider next a situation where both Assumptions 3.1 and 1.2 hold and let f be a biholomorphic mapping of $\mathbb{L}_0$ onto $\mathbb{L}_0' \subset \mathbb{L}'$, $\mathbb{L}'$ being an hermitian manifold analogous to $\mathbb{L}$.

Assumption 3.2. Suppose that the metrics h and $h'|\mathbb{L}_0'$ are related by the formulae (1.4) or, equivalently, by (1.5), while the tensor fields H and $H'|\mathbb{L}_0'$ are related by the formulae (1.6).

Remark 3.1. - Clearly, now $:j = |j$.

THEOREM 3.1. A. - *If Assumption 3.1 is fulfilled, u admits the unique biholomorphic continuation (1.8) onto $\mathbb{L}_0$, where*

$$(3.2)\quad U_c(x,y) = \tfrac{1}{2}[u_c(z_{c\bar{c}}) + \overline{u_c(\bar{z}_{c\bar{c}})}], \quad V_c(x,y) = -\tfrac{1}{2}i[u_c(z_{c\bar{c}}) - \overline{u_c(\bar{z}_{c\bar{c}})}],$$

and

(3.3) $s_{\bar{c}} = (\mathrm{re}\, s^1, \ldots, \mathrm{re}\, s^{\frac{1}{2}m}, \ldots, \mathrm{im}\, s^1, \ldots, \mathrm{im}\, s^{\frac{1}{2}m})$ *for* $s \in \mathbb{C}^{\frac{1}{2}m}$.

Moreover, if $\mathbb{M}_2'$ *consists of those* $\eta \in \mathbb{R}^m$ *for which there is a* $\xi \in \mathbb{M}_1'$ *such that*

(3.4) $\zeta_c \in \mathbb{M}_1'^{c}$ *and* $\bar{\zeta}_c \in \mathbb{M}_1'^{c}$, *where* $\zeta = \xi + i\eta$,

then $f[\mathbb{L}_0] = \mathbb{L}_0'$, *where* $\mathbb{L}_0'$ *consists of all* $(\xi, \eta) \in \mathbb{L}' = \mathbb{M}_1' \times \mathbb{M}_2'$, *satisfying the conditions* (3.4).

B. If, in addition, a relation of the form (1.7) *is fulfilled together with Assumptions* 3.2 *and* 2.1, *then the mappings* F *and* G *are* Q *and* Q′*-quasiconformal, respectively, with* Q *estimated by* (1.14) *provided that we have* (1.15) *or* (1.16), *respectively, and* Q′ *estimated by* (2.14) *provided that we have* (2.15) *or* (2.16), *respectively. Furthermore, the mappings* $U(\ ,O_2)$ *and* $V(O_1,\ )$ *are* K_1 *and* K_2*-quasiconformal, respectively, where* K_1 *and* K_2 *are estimated by*

(2.17) *and* (2.18) *in the case of* (1.15) *with* (2.15),
(2.19) *and* (2.20) *in the case of* (1.15) *with* (2.16),
(2.21) *and* (2.22) *in the case of* (1.16) *with* (2.15),
(2.23) *and* (2.24) *in the case of* (1.16) *with* (2.16),

where $\max^{1/a}(\alpha, \beta) = [\max(\alpha, \beta)]^{1/a}$. *The estimates* (1.14), (2.14), *and* (2.17)-(2.24) *are precise.*

Proof. *A.* The proof of Theorem 3.1.A is essentially contained in [11]. Namely, it is clear that the relations (3.2) together with (3.1) and (3.3)′ determine the unique holomorphic continuation (1.8) of u. With the notation

$$(U + iV)_{\bar{c}} \equiv (U_c^1, \ldots, U_c^{\frac{1}{2}m}, V_c^1, \ldots, V_c^{\frac{1}{2}m}),$$

by (3.1) and (3.3), for $k = 1, \ldots, \frac{1}{2}m$ we have

$$(U + iV)_{\bar{c}c}^k = (U + iV)_{\bar{c}}^k + i(U + iV)_{\bar{c}}^{\frac{1}{2}m+k} = U_c^k + iV_c^k.$$

This, together with (3.2), yields $(U + iV)_{c\bar{c}}^k(x,y) = u_c^k(z_{c\bar{c}})$. Similarly,

$$[\overline{(U + iV)_{\bar{c}}}]_c^k(x,y) = \overline{U_c^k(x,y)} + i\overline{V_c^k(x,y)} = u_c^k(\bar{z}_{cc}).$$

Hence, if $(x,y) \in \mathbb{L}_0$, we have not only (3.1), but also similar relations

$$[(U + iV)_{\bar{c}}]_c(x,y) \in \mathbb{M}_1'^{c} \quad \text{and} \quad [\overline{(U + iV)_{\bar{c}}}]_c(x,y) \in \mathbb{M}_1'^{c},$$

so $f(z) \in \mathbb{L}_0'$, i.e. $f[\mathbb{L}_0] \subset \mathbb{L}_0'$.

On the other hand, if $(\xi, \eta) \in \mathbb{L}_o'$, there exist points λ and μ in $\mathbb{M}_1$ such that

$$\zeta_c = u_c(\lambda) \quad \text{and} \quad \bar{\zeta}_c = u_c(\mu), \quad \text{where} \quad \zeta = \xi + i\eta .$$

Hence

$$\xi = U_c((\lambda + i\mu)^{c\bar{c}}), \quad \eta = V_c((\lambda + i\mu)^{c\bar{c}}),$$

where $(\lambda, \mu) \longmapsto (\lambda + i\mu)^{c\bar{c}}$ is the inverse mapping of $(x,y) \longmapsto$ $\longmapsto (x,iy)_{c\bar{c}}$, namely,

$$(\lambda + i\mu)^{c\bar{c}} = (\tfrac{1}{2}(\mu + \lambda)_{\bar{c}}, \tfrac{1}{2}(i\mu - i\lambda)_{\bar{c}}).$$

Therefore $f^{-1}(\zeta) \in \mathbb{L}_o$, i.e. $f[\mathbb{L}_o] \supset \mathbb{L}_o'$.

Finally we construct for the mapping u^{-1} the corresponding holomorphic continuation $\varphi: \mathbb{L}_o' \longrightarrow \mathbb{L}_o$ and observe that both $f \circ \varphi | \mathbb{M}_1'$ and $\varphi \circ f | \mathbb{M}_1$ are identity mappings, so $\varphi = f^{-1}$, as desired.

B. By Theorem 3.1.A we state that Assumptions 1.4 and 1.5 are fulfilled, so we can apply Proposition 2.2 and Corollary 2.1 which immediately give the desired result.

Remark 3.2. Let $\mathcal{O}_k$ denote the sheaf of germs of holomorphic functions of $\mathbb{C}^k$ and π_k the natural projection $\mathcal{O}_k \longrightarrow \mathbb{C}^k$. If, under the hypotheses of Theorem 3.1.A, $\mathbb{M}_1^c$ is a domain of holomorphy (in $\mathbb{C}^{\frac{1}{2}m}$), then $\mathbb{L}_o$ has the following properties:

(a) $\mathbb{L}_o$ is a connected domain of holomorphy (in $\mathbb{C}^m$) containing $\mathbb{M}_1^c$;

(b) for every real-analytic pluriharmonic function $\tilde{u}$ on $\mathbb{M}_1$ there is an analytic function $\tilde{f}$, arbitrarily continuable over $\mathbb{L}_o$ (i.e. $\pi_m[\tilde{f}] = \mathbb{L}_o$) such that for any x of $\mathbb{M}_1$ the germ $\tilde{u}_x$ belongs to $\tilde{f}$;

(c) there is a real-analytic pluriharmonic function u^o on $\mathbb{M}_1$ whose continuation f^o in the sense of (b) is such that for any (x,y) of $\mathbb{L}_o$, any $f^o{}_{(x,y)}$ of $\pi_m^{-1}(x,y) \cap f^o$, and any φ of $f^o{}_{(x,y)}$, φ cannot be holomorphically extended on any polydisc $\mathbb{D}$ if $\mathbb{D} \setminus \mathbb{L}_o \neq \emptyset$.

Conversely, if $\mathbb{L}_o$ satisfies (a) then $\mathbb{M}_1$ is a domain of holomorphy.

For proofs of both facts we refer to [11].

Assumption 3.3. Suupose that Assumption 1.1 holds with m even, $\mathbb{M}_1$ contained in $\mathbb{R}^m$, and u being a circular K_1-quasiconformal mapping of $\mathbb{M}_1$ onto an analogous Riemannian manifold $\mathbb{M}_1'$. Let further $\mathbb{M}_2$ consist of those $y \in \mathbb{R}^n$ for which there is an

$x \in \mathbb{M}_1$ such that the condition (3.1) is satisfied. Next, let $\mathbb{L}_0$ consist of all $(x,y) \in \mathbb{L}$, satisfying (3.1). Finally, let $f:\mathbb{L}_0 \longrightarrow \mathbb{L}_0' = f[\mathbb{L}_0]$ be the unique biholomorphic continuation of u, described by the formulae (1.8), (3.2), and (3.3), where $\mathbb{L}_0'$ consists of all $(\xi, \eta) \in \mathbb{L}' = \mathbb{M}_1' \times \mathbb{M}_2'$, satisfying the conditions (3.4), whereas $\mathbb{M}_2'$ consists of those $\eta \in \mathbb{R}^m$ for which there is a $\xi \in \mathbb{M}_1'$ such that (3.4) holds.

THEOREM 3.2. *Suppose that Assumptions 3.3 and 3.2 as well as a relation of the form (1.7) are fulfilled, where G is Q'-quasiconformal. Moreover, suppose that there are positive constants c_1, c_2, and a such that for every condenser $\mathfrak{D}$ whose closure is compact on $\mathbb{M}_1$ there exists a capacity $\operatorname{cap}(\mathfrak{D} \times \mathfrak{F})$, where $\mathfrak{F} = F[\mathfrak{D}]$, which fulfills the estimates (1.11), where $\operatorname{cap}^a \mathfrak{D} = (\operatorname{cap} \mathfrak{D})^a$ etc. Then the mappings F and $V(0_1, \ )$ are Q and K_2-quasiconformal, respectively, where Q and K_2 are estimated by*

$$(3.5)\quad Q \leqslant \left(\frac{c_2 - c_2^{-1}}{c_1^{-1} - c_1}\right)^{\frac{m-1}{a}}, \quad K_2 \leqslant \left[\frac{c_2 - c_2^{-1}}{c_1^{-1}(Q'K_1)^{-a} - c_1(Q'K_1)^a}\right]^{\frac{m-1}{a}}$$

in the case of (1.15), whereas by

$$(3.6)\quad Q \leqslant \left(\frac{c_1 - c_1^{-1}}{c_2^{-1} - c_2}\right)^{\frac{m-1}{a}}, \quad K_2 \leqslant \left[\frac{c_1(Q'K_1)^a - c_1^{-1}(Q'K_1)^{-a}}{c_2^{-1} - c_2}\right]^{\frac{m-1}{a}}$$

in the case of (1.16). The estimates (3.5) and (3.6) are precise. Furthermore, the image capacity $\operatorname{cap}(\mathfrak{U} \times \mathfrak{V})$, corresponding to $\operatorname{cap}(\mathfrak{D} \times \mathfrak{F})$ in the sense of Lemma 1.3, where $\mathfrak{U}$ and $\mathfrak{V}$ are defined in (1.10), is related to $\operatorname{cap} \mathfrak{U}$ and $\operatorname{cap} \mathfrak{V}$ by (1.25), where c_1' and c_2' satisfy the precise estimates

$$(3.7)\quad c_1' \leqslant c_1 K_1^a, \quad c_2' \leqslant c_2(c_2 - c_2^{-1})^{m-1}[c_1^{-1}(Q'K_1)^{-a} - c_1(Q'K_1)^a]^{-m+1}$$

and

$$(3.8)\quad c_1' \leqslant c_1 K_1^a, \quad c_2' \leqslant c_2(c_2^{-1} - c_2)^{-m+1}[c_1(Q'K_1)^a - c_1^{-1}(Q'K_1)^{-a}]^{m-1}$$

in the cases of (1.15) and (1.16), respectively.

Proof. By (3.2) and (3.1) the mapping $U(\ , 0_2)$ is K_1-quasiconformal and, on the other hand, G is Q'-quasiconformal, so, by (1.7), we have the precise estimates

$$(1/K_1)\operatorname{cap}\mathfrak{D} \leqslant \operatorname{cap}\mathfrak{U} \leqslant K_1 \operatorname{cap}\mathfrak{D},$$

$$(1/Q')\operatorname{cap}\mathfrak{U} \le \operatorname{cap}\mathfrak{V} \le Q'\operatorname{cap}\mathfrak{U}$$

for an arbitrary condenser $\mathfrak{D}$ whose closure is compact on $\mathbb{M}_1$, where $\mathfrak{U}$ and $\mathfrak{V}$ are defined in (1.10). Therefore from (1.11) we get

$$(3.9)\quad \operatorname{cap}(\mathfrak{D}\times\mathfrak{F}) - c_2^{-1}\operatorname{cap}^a\mathfrak{F} \ge c_1^{-1}K_1^{-a}\operatorname{cap}^a\mathfrak{U} \ge c_1^{-1}(Q'K_1)^{-a}\operatorname{cap}^a\mathfrak{V},$$

$$(3.10)\quad \operatorname{cap}(\mathfrak{D}\times\mathfrak{F}) - c_2\operatorname{cap}^a\mathfrak{F} \le c_1K_1^a\operatorname{cap}^a\mathfrak{U} \le c_1(Q'K_1)^a\operatorname{cap}^a\mathfrak{V},$$

where $\mathfrak{F}$ is defined in (1.10), and hence

$$[c_1^{-1}(Q'K_1)^{-a} - c_1(Q'K_1)^a]\operatorname{cap}^a\mathfrak{V} \le (c_2 - c_2^{-1})\operatorname{cap}^a\mathfrak{F}.$$

In the case of (1.15), by Lemma 1.1, from the above inequality we derive the second estimate in (3.5) which is precise. Besides, in this case we get, by Lemma 1.1 again, the first estimate in (3.5) which is precise as well. Furthermore, by Lemma 1.3, owing to the relations (1.8) and (1.10), we have

$$\operatorname{cap}(\mathfrak{U}\times\mathfrak{V}) = \operatorname{cap}(\mathfrak{D}\times\mathfrak{F}).$$

Hence (3.9) and (3.10) together with the estimates for $\operatorname{cap}\mathfrak{F}$ in terms of $\operatorname{cap}\mathfrak{V}$ yield

$$(3.11)\quad \operatorname{cap}(\mathfrak{U}\times\mathfrak{F}) \ge c_1^{-1}K_1^{-a}\operatorname{cap}^a\mathfrak{U} + \frac{c_2^{-1}}{c_2 - c_2^{-1}}[c_1^{-1}(Q'K_1)^{-a} - c_1(Q'K_1)^a]^{-m+1}\operatorname{cap}^a\mathfrak{V},$$

i.e. we arrive at (1.25), where c_1' and c_2' satisfy the precise estimates (3.7).

In the case of (1.16) the proof is analogous. The relations (3.11) and (3.12) should be replaced by

$$(3.13)\quad \operatorname{cap}(\mathfrak{U}\times\mathfrak{V}) \geqslant c_1^{-1}K_1^{-a}\operatorname{cap}^a\mathfrak{U} + \frac{c_2^{-1}}{(c_2^{-1} - c_2)^{-m+1}}[c_1(Q'K_1)^a - c_1^{-1}(Q'K_1)^{-a}]^{-m+1}\operatorname{cap}^a\mathfrak{V},$$

$$(3.14)\quad \operatorname{cap}(\mathfrak{U}\times\mathfrak{V}) \leqslant c_1K_1^a\operatorname{cap}^a\mathfrak{U} + \frac{c_2}{c_2^{-1} - c_2}[c_1(Q'K_1)^a - c_1^{-1}(Q'K_1)^{-a}]\operatorname{cap}^a\mathfrak{V}.$$

THEOREM 3.3. *Suppose that Assumptions 3.3 and 3.2 as well as a relation of the form (1.7) are fulfilled, where F is Q-quasiconformal. Moreover, suppose that there are positive constants c_1', c_2', and a such that for every condenser $\mathfrak{D}$ whose closure is compact on $\mathbb{M}_1$ there exists a capacity $\operatorname{cap}(\mathfrak{U}\times\mathfrak{V})$, where $\mathfrak{U}$ and $\mathfrak{V}$ are defined in (1.10), which fulfils the estimates (1.25),* *where* $\operatorname{cap}^a \mathfrak{U} = (\operatorname{cap} \mathfrak{U})^a$ *etc. Then the mappings* G *and* $V(O_1,\ )$ *are* Q' *and* K_2*-quasiconformal, respectively, where* Q' *and* K_2 *are estimated by*

$$(3.15)\quad Q' \leq \left(\frac{c_2' - c_2'^{-1}}{c_1'^{-1} - c_1}\right)^{\frac{m-1}{a}}, \quad K_2 \leq \left[\frac{c_2' - c_2'^{-1}}{c_1'^{-1}(Q K_1)^{-a} - c_1'(Q K_1)^a}\right]^{\frac{m-1}{a}}$$

in the case of (2.15), *whereas by*

$$(3.16)\quad Q' \leq \left(\frac{c_1' - c_1'^{-1}}{c_2'^{-1} - c_2'}\right)^{\frac{m-1}{a}}, \quad K_2 \leq \left[\frac{c_1'(Q K_1)^a - c_1'^{-1}(Q K_1)^{-a}}{c_2'^{-1} - c_2'}\right]^{\frac{m-1}{a}}$$

in the case of (2.16). *The estimates* (3.15) *and* (3.16) *are precise. Furthermore, the preimage capacity* $\operatorname{cap}(\mathfrak{D} \times \mathfrak{F})$, *corresponding to* $\operatorname{cap}(\mathfrak{U} \times \mathfrak{V})$ *in the sense of Lemma* 1.3, *where* $\mathfrak{F} = F[\mathfrak{D}]$, *is related to* $\operatorname{cap} \mathfrak{D}$ *and* $\operatorname{cap} \mathfrak{F}$ *by* (1.11), *where* c_1 *and* c_2 *satisfy the precise estimates*

$$(3.17)\quad c_1 \leq c_1' K_1^a, \quad c_2 \leq c_2'(c_2' - c_2'^{-1})^{m-1}[c_1'^{-1}(Q K_1)^{-a} - c_1'(Q K_1)^a]^{-m+1}$$

and

$$(3.18)\quad c_1 \leq c_1' K_1^a, \quad c_2 \leq c_2'(c_2'^{-1} - c_2')^{-m+1}[c_1'(Q K_1)^a - c_1'^{-1}(Q K_1)^{-a}]^{m-1}$$

in the cases of (2.15) *and* (2.16), *respectively.*

First proof. By (3.2) and (3.1) the mapping $U(\ , O_2)$ is K_1-quasiconformal and, on the other hand, F is Q-quasiconformal, so, by (1.7), we have the precise estimates

$$(1/K_1)\operatorname{cap} \mathfrak{D} \leq \operatorname{cap} \mathfrak{U} \leq K_1 \operatorname{cap} \mathfrak{D},$$

$$(1/Q)\operatorname{cap} \mathfrak{D} \leq \operatorname{cap} \mathfrak{F} \leq Q \operatorname{cap} \mathfrak{D}$$

for an arbitrary condenser $\mathfrak{D}$ whose closure is compact on $\mathbb{M}_1$, where $\mathfrak{U}$ and $\mathfrak{F}$ are defined in (1.10). Therefore from (1.25) we get

$$(3.19)\quad \operatorname{cap}(\mathfrak{U} \times \mathfrak{V}) - c_2'^{-1} \operatorname{cap}^a \mathfrak{V} \geq c_1'^{-1} K_1^{-a} \operatorname{cap}^a \mathfrak{D} \geq c_1'^{-1}(Q K_1)^{-a} \operatorname{cap}^a \mathfrak{F},$$

$$(3.20)\quad \operatorname{cap}(\mathfrak{U} \times \mathfrak{V}) - c_2' \operatorname{cap}^a \mathfrak{V} \leq c_1' K_1^a \operatorname{cap}^a \mathfrak{D} \leq c_1'(Q K_1)^a \operatorname{cap}^a \mathfrak{F},$$

where $\mathfrak{V}$ is defined in (1.10), and hence

$$(c_2' - c_2'^{-1}) \operatorname{cap}^a \mathfrak{V} \geq [c_1'^{-1}(Q K_1)^{-a} - c_1'(Q K_1)^a] \operatorname{cap}^a \mathfrak{F}.$$

In the case of (2.15), by Lemma 1.1, from the above inequality we derive the second estimate in (3.15) which is precise. Besides, in this case we get, by Lemma 1.1 again, the first estimate in (3.15) which is precise as well. Furthermore, by Lemma 1.3, owing to the relations (1.8) and (1.10), we have

$$\mathrm{cap}(\mathfrak{D} \times \mathfrak{F}) = \mathrm{cap}(\mathfrak{U} \times \mathfrak{V}).$$

Hence (3.19) and (3.20) together with the estimates for $\mathrm{cap}\,\mathfrak{V}$ in terms of $\mathrm{cap}\,\mathfrak{F}$ yield

$$(3.21)\quad \mathrm{cap}(\mathfrak{D} \times \mathfrak{F}) \geq c_1'^{-1} K_1^{-a} \mathrm{cap}^a \mathfrak{D} + \frac{c_2'^{-1}}{c_2' - c_2'^{-1}} [c_1'^{-1}(Q K_1)^{-a} - c_1'(Q K_1)^a] \mathrm{cap}^a \mathfrak{F},$$

$$(3.22)\quad \mathrm{cap}(\mathfrak{D} \times \mathfrak{F}) \leq c_1' K_1^a \mathrm{cap}^a \mathfrak{D} + \frac{c_2'}{(c_2' - c_2'^{-1})^{-m+1}} [c_1'^{-1}(Q K_1)^{-a} - c_1'(Q K_1)^a]^{-m+1} \mathrm{cap}^a \mathfrak{F},$$

i.e. we arrive at (1.11), where c_1 and c_2 satisfy the precise estimates (3.17).

In the case of (2.16) the proof is analogous. The relations (3.21) and (3.22) should be replaced by

$$(3.23)\quad \mathrm{cap}(\mathfrak{D} \times \mathfrak{F}) \geq c_1'^{-1} K_1^{-a} \mathrm{cap}^a \mathfrak{D} + \frac{c_2'^{-1}}{(c_2'^{-1} - c_2')^{-m+1}} [c_1'(Q K_1)^a - c_1'^{-1}(Q K_1)^{-a}]^{-m+1} \mathrm{cap}^a \mathfrak{F},$$

$$(3.24)\quad \mathrm{cap}(\mathfrak{D} \times \mathfrak{F}) \leq c_1' K_1^a \mathrm{cap}^a \mathfrak{D} + \frac{c_2'}{c_1'^{-1} - c_2'} [c_1'(Q K_1)^a - c_1'^{-1}(Q K_1)^{-a}] \mathrm{cap}^a \mathfrak{F}.$$

S e c o n d p r o o f. It is sufficient to prove that the inverse of a circular K-quasiconformal mapping is again a circular K-quasiconformal mapping and then apply Theorem 3.2 to be inverse mapping u^{-1} instead of u.

R e f e r e n c e s

[1] L.V. AHLFORS and A. BEURLING: Conformal invariants and function-theoretic null-sets, Acta Math. 83 (1950), pp. 101-129.

[2] A. ANDREOTTI and J. ŁAWRYNOWICZ: On the generalized complex Monge-Ampère equation on complex manifolds and related questions, Bull. Acad. Polon. Sci. Sér. Sci. Math. Astronom. Phys. 25 (1977), pp. 943-948.

[3] A. ANDREOTTI and J. ŁAWRYNOWICZ: The generalized complex Monge-Ampère equation and a variational capacity problem, ibid. 25

(1977), pp. 949-955.

[4] M.F. ATIYAH: Geometry of Yang-Mills fields, Lezioni Fermiane, Ann. Scuola Norm. Sup. Pisa Cl. Sci., to appear.

[5] E. BELTRAMI: Delle variabili complesse sopra una superficie qualunque, Ann. Mat. Pura Appl. (2) 1 (1867/8), pp. 329-336.

[6] B. BOJARSKI and T. IWANIEC: Topics in quasiconformal theory in several variables, Proc. of the First Finnish-Polish Summer School in Complex Analysis at Podlesice. Part II, ed. by J. Ławrynowicz and O. Lehto, Uniwersytet Łódzki, Łódź 1978, pp. 21-44.

[7] P. CARAMAN: n-dimensional quasiconformal (qcf) mappings , Editura Academiei-Abacus Press, Bucureşti-Tunbridge Wells, Kent 1974.

[8] S.S. CHERN, H.I. LEVINE and L.NIRENBERG: Intrinsic norms on a complex manifold, Global analysis, Papers in honor of K. Kodaira, ed. by D.C. Spencer and S. Iynaga, Univ. of Tokyo Press and Princeton Univ. Press, Tokyo 1969, pp. 141-148.

[9] H. GRAUERT: Statistical geometry and space-time, Comm. Math. Phys. 49 (1976), pp. 155-160.

[10] H. GRAUERT: Statistische Geometrie. Ein Versuch zur geometrischen Denkung physikalischer Felder, Nachrichten Akad. Wiss. Göttingen 1976, pp. 13-32.

[11] M. JARNICKI: Analytic continuation of pluharmonic functions, Zeszyty Nauk. Uniw. Jagielloń. 441 Prace Mat. No. 18 (1977), 45-51.

[12] C.O. KISELMAN: Prolongement des solutions d'une équation aux dérivées partielles à coefficients constants, Bull. Soc. Math. France 97 (1969), pp. 329-356.

[13] S. KOBAYASHI: Hyperbolic manifolds and holomorphic mappings, Marcel Dekker, Inc., New York 1970.

[14] J. ŁAWRYNOWICZ: Condenser capacities and an extension of Schwarz's lemma for hermitian manifolds, Bull. Acad. Polon. Sci. Sér. Sci. Math. Astronom. Phys. 23 (1975),pp. 839-844.

[15] J. ŁAWRYNOWICZ: On quasiconformality of projections of biholomorphic mappings, ibid. 23 (1975), pp. 845-851.

[16] J. ŁAWRYNOWICZ: On a class of capacities on complex manifolds endowed with an hermitian structure and their relation to elliptic and hyperbolic quasiconformal mappings, Dissertationes Math. to appear.

[17] J. ŁAWRYNOWICZ and W. WALISZEWSKI: Conformality and pseudo-riemannian manifolds, Math. Scand. 28 (1971), pp. 45-69.

[18] J. ŁAWRYNOWICZ and L. WOJTCZAK: A concept of explaining the properties of elementary particles in terms of manifolds, Z. Naturforsch. 29a (1974), pp. 1407-1417.

[19] J. ŁAWRYNOWICZ and L. WOJTCZAK: On an almost complex manifold approach to elementary particles, ibid. 32a (1977), pp. 1215-1221.

[20] J. LELONG-FERRAND: Etude d'une classe d'applications liées à des homomorphismes d'algèbres de fonctions, et généralisant les quasi conformes, Duke Math. J. 40 (1973), pp. 163-186.

[21] R. PENROSE: The complex geometry of the natural world, Proc. of the Internat. Congress of Mathematicians Helsinki 1978, to appear.

[22] G. PORRU: Mappe con distorsione triangolare limitata in spazi normati, Boll. Un. Mat. Ital. (5) 14-A (1977), pp. 599-607.

[23] A.D. SAKHAROV: The topological structure of elementary charges and the CPT-symmetry [in Russian], A Memorial Volume to I.E. Tamm, Nauka, Moscow 1972, pp. 242-247.

[24] K. SUOMINEN: Quasiconformal maps in manifolds, Ann. Acad. Sci. Fenn. Ser. A I 393 (1966), 39pp.

[25] E. VESENTINI: Invariant metrics on convex cones, Ann. Scuola Norm. Sup. Pisa Cl. Sci. (4) 3 (1976), pp. 671-696.

[26] R.O. WELLS, Jr.: Complex manifolds and mathematical physics, Bull. Amer. Math. Soc.

Institute of Mathematics of the
Polish Academy of Sciences, Łódź Branch
Kilińskiego 86, PL-90-012 Łódź, Poland

SOME REMARKS ON EXTENSION OF BIHOLOMORPHIC MAPPINGS

Ewa Ligocka (Warszawa)

In the papers [8] and [13] there were given conditions on Bergman kernel functions $K_D(z,t)$ and $K_G(w,s)$ of bounded domains D and G in $\mathbb{C}^n$, which guarantee that if $h: D \longrightarrow G$ is a biholomorphic mappings from D onto G then it can be extended to a smooth diffeomorphism between the closures $\bar{D}$ and $\bar{G}$. In [8] and [13] it was proved that these conditions are valid for strictly pseudoconvex domains with C^∞-boundary (by using the deep results of [2]). In [8] it was also checked that these conditions hold also for complete, circular, strictly starlike domains and are invariant under cartesian product. Now we shall state these conditions in a more general form. It permits us to study analytic extensions of biholomorphic mappings and to find connections between our conditions and properties of orthogonal projection onto the space of square integrable holomorphic functions and solution of $\bar{\partial}$ - equation.

We begin with the following definitions.

Definition 1. Let D be a bounded domain and let $D \subset E \subset \bar{D}$. We say that a function f defined on D is of class C^k on E iff all derivatives of f up to order k can be extended to E as continuous functions.

Definition 2. We shall say that boundary ∂D of a bounded domain D satisfies the minimal regularity conditions if it is locally a graph of some Lipschitz function from $\mathbb{R}^{2n-1}$ into $\mathbb{R}$. That means that for every $z \in \partial D$ there exists an open neighborhood U of z, the system of coordinates $x_1, \dots x_{2n}$ in $\mathbb{C}^n = \mathbb{R}^{2n}$ and the Lipschitz function $\varphi: \mathbb{R}^{2n-1} \longrightarrow \mathbb{R}$ such that

$$U \cap D = U \cap \{x \in R^{2n} : x_{2n} > \varphi\ (x_1, \dots x_{2n-1})\}.$$

We need in fact the following two properties of the domains whose boundary satisfies the minimal regularity conditions:

1) Every differentiable function on D, with bounded derivatives can be extended to a function continuous on $\bar{D}$;
2) Every function of class C^k on $\bar{D}$ can be extended to the a function of class C^k on $\mathbb{C}^n$.

These properties are proved in [12] (VI, Theorem 5 and the proof of Theorem 5'). Property 2 can be also obtained from the general Whitney extension theorem for k-jets (see [9]) by proving the a function of class C^k on $\bar{D}$ determines a k-jet on D which satisfies Whitney condition.

Now we shall formulate our conditions concerning Bergman kernel function $K_D(z,t)$ of a domain D:

A_k) There exists an open dense set $U \subset D$ such that $K_D(z,t)$ is a function of class C^k on $\bar{D} \times U$, $1 \leq k < \infty$;

B) For every $z_o \in \partial D$ there exists $n+1$ points $a_o, \ldots a_n \in U$ such that

$$\det \begin{bmatrix} K_D(z,a_o) & \ldots\ldots & K_D(z,a_n) \\ \dfrac{\partial K_D(z,a_o)}{\partial z_1} & \ldots\ldots & \dfrac{\partial K_D(z,a_n)}{\partial z_1} \\ \vdots & & \vdots \\ \dfrac{\partial K_D(z,a_o)}{\partial z_n} & \ldots\ldots & \dfrac{\partial K_D(z,a_n)}{\partial z_n} \end{bmatrix} (z_o) \neq 0 .$$

We shall say that $K_D(z,t)$ satisfies condition A_∞ iff it satisfies condition A_k for every natural k.

It can be checked by simple computations that the condition B can be stated in the following equivalent form: For every $z_o \in \partial D$ there exist a point $a_o \in U$ and points $a_1, \ldots a_n \in U$ such that $K_D(z,a_o) \neq 0$ and

$$\det \left[\frac{\partial}{\partial z_j} \frac{K_D(z,a_i)}{K_D(z,a_o)} \right] (z_o) \neq 0$$

Hence, we obtain the following

PROPOSITION 1. *If ∂D satisfies the minimal regularity condition and $K_D(z,t)$ satisfies the condition A_k, then the condition B is equivalent to the following: For every $z_o \in \partial D$ there exist $n+1$ points $a_o, \ldots, a_n \in U$ such that $K_D(z_o,a_o) \neq 0$ and the functions*

$K_D(z,a_i)/K_D(z,a_o)$ <u>can be extended to a system of local coordinates of the class</u> C^k <u>in some neighborhood of</u> z_o. <u>In other words, it means that the mapping</u> $z \longrightarrow K_D(z,a_o),\dots K_D(z,a_n)$ <u>can be extended to a</u> C^k<u>-diffeomorphic imbedding of a neigborhood of</u> z_o <u>into the projective space</u> $\mathbb{CP}^n$.

Now we shall prove that the conditions A_k and B are weaker than the analogous conditions of [8] and [13].

<u>PROPOSITION 2</u>. <u>If</u> $K_D(z,t)$ <u>satisfies the condition</u> A_K, $k \geq 2$, <u>and for every</u> $z_o \in \partial D$ <u>there exists</u> $a \in U$ <u>such that</u>

$$K_D(z_o,a) \neq 0 \quad \text{and} \quad \det\left[\frac{\partial^2 \ln K_D}{\partial z_j \partial \bar{t}_i}(z_o,a)\right] \neq 0,$$

<u>then the condition</u> B) <u>holds</u>.

<u>P r o o f</u>. There exists a neighborhood W of z_o in D and a neighborhood V of a in U such that

$$K_D(z,t) \neq 0 \quad \text{and} \quad \det\left[\frac{\partial^2 \ln K_D}{\partial z_i \partial \bar{t}_j}(z,t)\right] \neq 0 \quad \text{on} \quad W \times V.$$

We have:

$$\frac{\partial}{\partial z_i}\frac{K_D(z,t)}{K_D(z,a)} = \frac{K_D(z,t)}{K_D(z,a)}\frac{\partial}{\partial z_i}[\ln K_D(z,t) - \frac{\partial}{\partial z_i}\ln K_D(z,a)]$$

$$= \frac{K_D(z,t)}{K_D(z,a)}\cdot g_i(z,t)$$

Now, as in the proof of Theorem 1,

$$\det\left[\frac{\partial}{\partial \bar{t}_j} g_i(z,t)\right] = \det \frac{\partial^2}{\partial z_i \partial \bar{t}_j}[\ln K_D(z,t)] \neq 0 \quad \text{in} \quad W \times V,$$

so the functions $g_j(z_o t)$ form local coordinates on some neighborhood of a. Then there exist points $a_1,\dots a_n \in V$ such that $\det[g_i(z_o,a_j)] \neq 0$

$$\text{Hence} \quad \det\left[\frac{\partial}{\partial z_i}\frac{K_D(z,a_j)}{K_D(z,a)}\right](z_o) = \frac{K_D(z_o,a_1)\dots K_D(z_o a_n)}{K_D(z_o,a)^n}\det[g_i(z_o,a_j)] \neq 0$$

and the condition B) holds.

Now we shall state

THEOREM 1.

Let D and G be bounded domains in $\mathbb{C}^n$, whose boundaries ∂D and ∂G satisfy the minimal regularity conditions. Assume that the Bergman kernel functions $K_D(z,t)$ and $K_G(w,s)$ satisfy the conditions A_k and B. Then every biholomorphic mapping between D and G extends to a C^k-diffeomorphism between some open neighborhoods of $\bar{D}$ and $\bar{G}$.

Proof. Suppose that h maps D biholomorphically onto G. Let $U \subset D$ and $W \subset G$ be dense open sets in the condition A_k. Take $V = U \cap h^{-1}(W)$. Since V is dense in U and $h(V)$ is dense in W, the conditions A_k and B remain valid if we replace U by V and W by $h(V)$.

Now we can proceed in the similar way as Webster in [13]. At first we prove that there exists $m > 0$ and $M > 0$ such that $m < |Jh(z)| < M$ for $z \in D$, where $Jh(z)$ denotes a Jacobi determinant of h at z. Suppose that there exists a sequence of points $z_n \in D$ such that $Jh(z_n) \longrightarrow 0$. We can assume that $z_n \longrightarrow z_o \in \partial D$. By the condition B there exists a point $a_o \in V$ such that $K_D(a_o, z_o) \neq 0$. By the transformation rule for the Bergman function we have

$$K_D(z_n, a_o) = K_G(h(z_n), h(a_o)) . \quad Jh(z_n) \cdot \overline{Jh(a_o)}$$

Since, by the condition A_k, $K_G(S, h(a_o))$ is bounded on G, then $\lim K_D(z_n, a_o) = K_D(z_o, a_o) = 0$ and we get a contradiction.
Hence $|Jh(z)|$ must be bounded from below on D. The same consideration applied to the mapping h^{-1} shows that $Jh(z)$ is also bounded from above on D.

Now we shall prove that the first derivatives of h are bounded functions on D. Suppose that it is not true. This means that there exists a sequence $z_n \in D$ such that

$$\sup_{i,j} \left| \frac{\partial h_i}{\partial z_j} (z_n) \right| \longrightarrow \infty \quad \text{as} \quad n \longrightarrow \infty.$$

We can also assume that $z_n \longrightarrow z_o \in \partial D$ and $h(z_n) = S_n \longrightarrow S_o \in \partial G$. Now, let $b_o, \ldots b_n$ be points from $h(V)$ such that $K_G(S_o, b_o) \neq 0$ and

$$\det \left[\frac{\partial}{\partial S_j} \frac{K_G(S, b_i)}{K_G(S, b_o)} \right]_{\substack{i=1,\ldots,n \\ j=1,\ldots,n}} (S_o) \neq 0.$$

Put $a_i = h^{-1}(b_i) \in V$, $i = 0, \ldots n$.

From the transformation rule for the Bergman function we obtain the following matrix equality:

$$\left[\frac{\partial}{\partial z_j} \frac{K_D(z,a_i)}{K_D(z,a_o)} (z_n)\right] = \left[\frac{\overline{Jh(a_i)}}{Jh(a_o)} \frac{\partial}{\partial S_j} \frac{K_G(S,b_i)}{K_G(S,b_o)}(S_n)\right] \circ \left[\frac{\partial h_i}{\partial z_j}(z_n)\right].$$

There exists an integer N such that the terms of matrix on the left side are uniformly bounded on the set $\{z_n\}_{n>N}$, the first matrix on the right side is invertible and the terms of its inverse are uniformly bounded on the set $\{S_n\}_{n>N}$. It implies that the terms of the matrix $[(\partial/\partial z_j)h_i(z_n)]$ are uniformly bounded on the set $\{z_n\}_{n>N}$ and we get a contradition.

Hence the derivatives of h are bounded on D. Since ∂D satisfies the minimal regularity conditions, then h can be continued to the continuous function on $\bar{D}$. We can prove in the same way that h^{-1} can be extended to continuous function on $\bar{G}$ so h extends to a homeomorphism between $\bar{D}$ and $\bar{G}$. Now we can make use of the local coordinates introduced in Proposition 1. Let $z_o \in \partial D$ and let $a_o, \ldots a_n \in V$ be such that $K_D(z_o,a_o) \neq 0$ and

$$\det\left[\frac{\partial}{\partial z_j} \frac{K_D(z,a_i)}{K_D(z,a_o)}(z_o)\right] \neq 0.$$

It follows from the first part of the proof that $K_G(h(z_o),h(a_o)) \neq 0$ and

$$\det\left[\frac{\partial}{\partial S_j} \frac{K_G(S,h(a_o))}{K_G(S,h(a_o))}(h(z_o))\right] \neq 0.$$

Then there exist neighborhoods V of z_o and W of $h(z_o)$ such that $h(D \cap V) = (G \cap W)$ and the functions

$$u_i = \frac{K_D(z,a_i)}{K_D(z,a_o)} \quad \text{and} \quad v_i = \frac{K_G(S,h(a_i))}{K_G(S,h(a_o))}, \; i = 1,\ldots,n$$

can be extended to local C^k-coordinates on V and W, respectively. By the transformation rule for the Bergman function the mapping $h|_{D \cap V}$ can be expressed in these coordinates as linear mapping with diagonal matrix $u_i = v_i \overline{Jh(a_i)}/Jh(a_o)$. It implies that h can be extended to a C^k-mapping between V and W. This ends the proof of Theorem 1.

R e m a r k 1. The assumptions of Theorem 1 are invariant with respect to Cartesian product. This means that if the conditions A_k, B and the minimal regularity condition hold for D_1 and D_2, then they are satisfied also by $D_1 \times D_2$. The invariance of the conditions A_k and B can be proved in exactly the same way as in Remark 5 in [8]. Now suppose that the minimal regularity condition holds for ∂D_1 and ∂D_2. We must prove it for $\partial(D_1 \times D_2) = \partial D_1 \times \bar{D}_2 \cup \bar{D}_1 \times \partial D_2$. It is clear that it is sufficient to check this condition for points $z = (z_1, z_2) \in \partial D_1 \times \partial D_2$. Let $U_1, U_2, \varphi_1, \varphi_2$ be such that $U_1 \cap D_1 = U_1 \cap \{x \in R^{2m} : x_{2m} > \varphi_1(x_1, \dots, x_{2m-1})\}$ and $U_2 \cap D_2 = U_2 \cap \{y \in R^{2n} : y_{2n} > \varphi_2(y_1, \dots, y_{2n-1})\}$, $z_1 \in U_1$, $z_2 \in U_2$ and φ_1, φ_2 are Lipschitz functions. Then we have

$$(U_1 \times U_2) \cap (D_1 \times D_2) = (U_1 \times U_2) \cap (\{x \in \mathbb{R}^{2m} : x_{2m} > \varphi_1(x_1, \dots x_{2m-1})\}$$
$$\times \{y \in \mathbb{R}^{2n} : y_{2n} > \varphi_2(y_1, \dots y_{2n-1})\}).$$

This last set can be expressed as follows:

$$\{(x,y) \in \mathbb{R}^{2m} \times \mathbb{R}^{2n} : x_{2m} + y_{2n} - \varphi_1(x_1, \dots x_{2m-1}) - \varphi_2(y_1, \dots y_{2n-1})$$
$$> |x_{2m} - y_{2n} - \varphi_1(x_1, \dots x_{2m-1}) + \varphi_2(y_1, \dots y_{2n-1})|\}$$

so by taking the coordinates

$$w = x_{2m} + y_{2n},\ V = x_{2m} - y_{2n},\ x_1, \dots x_{2m-1},\ y_1, \dots y_{2n-1}$$

and

$$\varphi(v, x_1, \dots x_{2m-1}, y_1, \dots y_{2n-1}) = \varphi_1(x_1, \dots x_{2m-1}) + \varphi_2(y_1, \dots y_{2n-1})$$
$$+ |v - \varphi_1(x_1, \dots x_{2m-1}) + \varphi_2(y_1, \dots y_{2n-1})|$$

we obtain

$$(U_1 \times U_2) \cap (D_1 \times D_2) = (U_1 \times U_2) \cap \{p \in \mathbb{R}^{2m} \times \mathbb{R}^{2n} : w > \varphi(v, x_1, \dots x_{2m-1},$$
$$y_1, \dots y_{2m-1})\}.$$

Thus the minimal regularity conditions are satisfied.

R e m a r k 2. Theorem 1 is a generalization of Proposition 2 of [13] and Theorems 1, 1' and Remark 3 of [8]. The main difference between the proof of Theorem 1 and the proof of this fact is in using the local coordinates introduced in Proposition 1. These coordinates permit us to avoid the inductive proof of smoothness of h at the boundary and as will be shown in the sequel, are useful in studying analytic extension of biholomorphic mappings. It should be mentioned that the use of such quotiens of Bergman kernel functions or equivalently mappings into projective space defined by Bergman function was sugested to me by the papers of Skwarczyński [10] [11], who used similar objects to construct an invariant compactification of a domain D.

THEOREM 2. *Suppose that the domain D and G satisfy the assumptions of Theorem 1. Let U and W denote the open dense sets from of the condition A_k for D and G, respectively. If for every $t \in U$ and $w \in W$ the expression $K_D(z,t)$ extends analytically to some neighborhood of $\bar{D}$ (depending on t) and $K_G(S,w)$ extends analytically to some neighborhood of $\bar{G}$ (depending on w), then every biholomorphic mapping from D onto G can be extended to biholomorphic mapping between neighborhoods of $\bar{D}$ and $\bar{G}$.*

Proof. By Theorem 1 a biholomorphic mapping h can be extended to a C^∞-diffeomorphism between neighborhoods of D and G. Let $z_o \in \partial D$. As in the proof of Theorem 1 we can find $a_o, \dots, a_n$ so that the functions $u_i = K_D(z,a_i)/K_D(z,a_o)$ are local holomorphic coordinates in a neighborhoo W of z_o, and the functions $v_i = K_G(S,h(a_i))/K_G(S,h(a_o))$ are local holomorphic coordinates in a neighborhood V of $h(z_o)$ and $h(W \cap D) = V \cap$ The extension of h in these coordinates has the form $v_i = \overline{J\ h(a_o)}/J\ h(a_i) \cdot u_i$. It implies that h can be extended analytically through every point of ∂D. The same considerations imply that h^{-1} can be extended analytically through every point of ∂G. Hence h can be extended to a biholomorphic mapping on some open neighborhood of $\bar{D}$.

From Remark 2 of [8] and the above Theorem 2 there follows

COROLLARY 1. *If D and G are the bounded circular complete and stictly starlike domains in $\mathbb{C}^n$ and h maps biholomorphically D onto G, then h can be extended to the biholomorphic mapping between neighborhoods of $\bar{D}$ and $\bar{G}$.*

Remark 2 of [8] shows that no regularity conditions on ∂D and ∂G are nedeed in this case.

Remark 3. At present there are known the following classes of domains in $\mathbb{C}^n$ for which the assumptions of Theorem 2 are fulfilled:

1) plane domains with the real analytic boundaries,
2) strictly pseudoconvex domains in $\mathbb{C}^n$ with real analytic boundaries,

3) complete circular strictly starlike domains,
4) Cartesian products of domains from previous three classes,
It is suprising that these classes of domains have almost nothing in common (exept a "good boundary behavior" of Bergman functions). In particular the regularity of the boundary is neither necessary nor sufficient condition for the properties A_k and B. In [13] an example of plane domain with C^1 boundary was given, for which these conditions are not valid. On the other hand these circular domains can have very bad boundaries.

We shall study now the connections between the property A_k, orthogonal projection on the space of a holomorphic square integrable function and the $\bar{\partial}$-problem.

THEOREM 3. The condition A_k, $1 \leq k < \infty$ is equivalent to the following statement: There exists an open dense subset $U_o \subset D$ such that for every function V of a class C^∞ with compact support contained in U_o the orthogonal projection KV on the space $L^2H(D)$ of square integrable holomorphic functions D is the function of class C^k on $\bar{D}$.

P r o o f. Suppose that condition A_k is satisfied. Let U be an open dense subset of D such that $K_D(z,t)$ is of class C^k on $U \times \bar{D}$. Since $K_D(z,t) = \overline{K_D(t,z)}$, $K_D(z,t)$ is also of class C^k on $\bar{D} \times U$. Let V be a C^∞-function with supp $V \subset\subset U$.
We have

$$KV(z) - \int_D K_D(z,t)V(t)dt = \int_{\text{supp } V} K_D(z,t)V(t)dt.$$

Observe that $K_D(z,t)$ and all its derivatives up to order k are bounded on $D \times (\text{supp } V)$. Thus by the theorem on differentation of integrals and the Lebesgue theorem $KV(z)$ is of class C^k on $\bar{D}$. Now let us assume that there exists an open set U D such that for every C^∞-function V, supp $V \subset\subset U$, the function KV is of class C^k on D.

Take some $z_o \in U$ and let G be a domain with C^∞ buondary such that $z_o \in G \subset \bar{G} \subset U$. Now let φ be a function of class C on D such that supp $\varphi = G$, $\varphi > 0$ on G. By the Martinelli- Bochner formula we have for every $f \in H(G) \cap C(\bar{G})$ and every $z \in G$

$$f(z) = \frac{(n-1)!}{(2\pi i)^n} \int_{\delta G} f(t) \sum_{k=1}^{n} (-1)^{k-1} \frac{\bar{t}_k - \bar{z}_k}{|t-z|^{2n}} d\bar{t}_1 \ldots \widehat{d\bar{t}_k} \ldots d\bar{t}_n \wedge dt_1 \ldots dt_n$$

$$= \frac{(n-1)!}{(2\pi i)^n} \int_{\partial G} f(t) . \sum_{k=1}^{n} (-1)^{k-1} \frac{\bar{t}_k - \bar{z}_k}{(|t-z|^2+\varphi(t))^n} d\bar{t}_1 \ldots \widehat{d\bar{t}_k} \ldots d\bar{t}_n \wedge dt_1 \ldots dt_n$$

Since $t > 0$ on G we can use the Stokes formula and obtain after simple computations

$$f(z) = \frac{n!}{(2\pi i)^n} \int_G f(t) \cdot \frac{\varphi(t) - \sum_{k=1}^{n} \frac{\delta\varphi}{\delta t_k}(\bar{t}_k - \bar{z}_k)}{(|t-z|^2 + \varphi(t))^{n+1}} dt =$$

$$= \frac{n!}{(2\pi i)^n} \int_D f(t) \frac{\varphi(t) - \sum_{k=1}^{n} \frac{\delta\varphi}{\delta t_k}(\bar{t}_k - \bar{z}_k)}{(|t-z|^2 + \varphi(t))^{n+1}} dt = \int_D f(t) \cdot H_\varphi(z,t) dt$$

Note, that for every $z \in G$, $\operatorname{supp} H_\varphi(z,t) \subset \bar{G}$. Then for $f \in L^2H(D)$ we have

$$\int_D (K_D(z_o,t) - H\varphi(z_o,t)) \cdot f(t) dt = f(z_o) - f(z_o) = 0$$

so the function $\overline{K_D(z_o t)} - \overline{H_\varphi(z_o t)}$ is orthogonal to $L^2H(D)$. It implies that $\overline{K_D(z_o t)}$ is an orthogonal projection of $\overline{H_\varphi(z_o, t)}$ on $L^2H(D)$ and therefore by the assumptions the function $K_D(z_o\ t)$ is of class C^k on $\bar{D}$.

Observe now that for every $z_o \in G$ and $f \in L^2H(D)$ we have

$$\frac{\delta^{|\alpha|} f}{\delta_z^{\alpha}}(z_o) = \int_D H_\varphi(z,t) \frac{\delta^{|\alpha|} f}{\delta_t^{\alpha}}(t) dt = (-1)^{|\alpha|} \int_D \frac{\delta^{|\alpha|}}{\delta_t^{\alpha}} H_\varphi(z,t) f(t) dt$$

and by the reproducing property of the Bergman kernel function

$$\frac{\delta^{|\alpha|} f}{\delta_z^{\alpha}}(z_o) = \int_D \frac{\delta^{|\alpha|} K_D}{\delta_z^{\alpha}}(z_o t) f(t) dt.$$

Thus the function $(\partial^{|\alpha|}/\partial z^\alpha) K_D(z_o,t)$ is an orthogonal projection of $(-1)^{|\alpha|}(\partial^{|\alpha|}/\partial t^\alpha) H(z_o,t)$ and therefore the function $(\partial^{|\alpha|}/\partial z^\alpha) K_D(z_o,t)$ is class C^k on $\bar{D}$.

Thus we proved that for every $z_o \in U$ and α, β such that $|\alpha| + |\beta| \leq k$ the functions $(\partial^{|\alpha|+|\beta|}/\partial \bar{t}^\beta \partial z^\alpha) K_D(z_o,t)$ can be extended to continuous functions on D. Now we shall prove the following fact: If $f(z,t)$ is a function on $U \times \bar{D}$ holomorphic on $U \times D$ and such that for every $z_o \in U$ $f(z_o,t)$ is continuous on $\bar{D}$, then there exists an open dense set $W \subset U$ such that $f(z,t)$ is continuous on $W \times D$. Denote by F_k the set

$$\{z \in U : \sup_{t \in D} |f(z,t)| \leq k\}$$

The sets F_k are closed in U and $\cup_{k=1}^{\infty} F_k = U$. It follows from Baire's theorem, that for every open set $V \subset U$ there exists a number k for which int $(F_k \cap V) \neq \phi$. Thus, the set W consisting of all $z \in U$ which has a neighbohood V_z such that $f(z,t)$ is bounded on $V_z \times \bar{D}$ is open and dense in U.

Now, let $(z_n, t_n) \longrightarrow (z_o t_o)$ in $W \times \bar{D}$. Suppose that $f(z_n, t_n) \not\longrightarrow f(z_o, t_o)$. Then, taking a suitable subsequence, we can assume that there exists an $\varepsilon > 0$ such that $|f(z_n, t_n) - f(z_o, t_o)| > \varepsilon$ for all $n \in N$. Since $f(z,t)$ is bounded on $V_{z_o} \times \bar{D}$, we can find subsequence t_{n_k} such that the sequence of functions $f(z, t_{n_k})$ converges almost uniformly on V_{z_o} to a function $h(z)$. We have $h(z) = f(z, t_o)$ since for every z, $f(z,t)$ is continuous on $\bar{D}$, so $f(z, t_{n_k})$ converges to $f(z, t_o)$ uniformly on the set $(\{z_n\} \cup \{z_o\}) \cap V_z$. Consequently,

$$\lim_{k \to \infty} f(z_{n_k}, t_{n_k}) = f(z_o, t_o).$$

We obtain a contradiction which shows that $f(z,t)$ is continuouos on $W \times \bar{D}$. Now we can take $(\partial^{|\alpha|+|\beta|} / \partial \bar{t}^{\beta} \partial z^{\alpha}) K_D(z,t)$ as a function $f(z,t)$ and the above fact we can find an open dense set $W_{\alpha,\beta}$ in U such that $(\beta^{|\alpha|+|\beta|} / \partial \bar{t}^{\beta} \partial z^{\alpha}) K_D(z,t)$ is continuous on $W_{\alpha,\beta} \times \bar{D}$. Now if we take

$$U_o = \bigcap_{\substack{\alpha,\beta \\ |\alpha|+|\beta| \leq k}} W,$$

then $K_D(z,t)$ is a function of class C^k on $U_o \times \bar{D}$, so the condition A_k is satisfied indeed.

COROLLARY 2. *If there exists an open dense subset* $U \subset D$ *such that every* C^∞*-function* V *on* D, supp $V \subset\subset U$, *the projection* KV *is of class* C^∞ *on* $\bar{D}$, *then the condition* A_∞ *is satisfied.*

COROLLARY 3. *If there exists an open dense subset* $U \subset D$ *such that for every* $\bar{\partial}$*-closed* $(0,1)$ *form* u *of class* C^∞ *on* D *such that* supp $u \subset\subset U$ *the solution of the equation* $\bar{\partial} v = u$ *which is orthogonal to the functions of* $L^2H(D)$ *is a function of class* C^k *on* $\bar{D}$, *then the condition* A_k *is satisfied.*

Remark 3. The last corollary indicates that the condition A_k is connected with the $\bar{\partial}$-problem in a very peculiar way. Usually in the applications of the $\bar{\partial}$-problem, the main role is played by the existence of a good solution of the $\bar{\partial}$-prolblem e.g. which is smoth on $\bar{D}$, or the existence of an operator inverse to $\bar{\partial}$ which is continuous with respect to the Sobolev norms, or the Hölder norm, or the L^p norm. In our case when u is a $\bar{\partial}$-closed $(0,1)$ form with compact support in D, such a solution and operator always exist and are given by

$$v(z) = \frac{(n-1)!}{(2\pi i)^n} \int_D u(t) \sum_k (-1)^k \frac{\bar{t}_i - \bar{z}_i}{|t-z|^{2n}} d\bar{t}_1 \ldots d\bar{t}_k \ldots d\bar{t}_n \; dt_1 \ldots dt_n$$

as it follows easily from the Koppelman formula. However this good solution is not orthogonal to $L^2H(D)$, so our problem consists of regularity of the peculiar solution of the equation $\bar{\partial}v = u$ (which always exists), namely this solution which is orthogonal to $L^2H(D)$. Now we shall ask the following two questions:

Problem 1. Suppose that $f(z,t)$ is a function defined on $U \times \bar{D}$, where U is an open set and D is a domain. Let $f(z,t)$ be a holomorphic function on $U \times D$ such that for every $z \in U$, $f(z,t)$ is continuous on $\bar{D}$. Is it true then that f is continuous on $U \times \bar{D}$? Does it hold in particular when $U \subset D$ and $f(z,t) = K_D(z,t)$ or $f(z,t) = (\partial^{|\alpha|+|\beta|}/\partial\bar{t}^{\beta}\partial z^{\alpha}) K_D(z,t)$?

Problem 2. Let $U \subset D$. Suppose that $K_D(z,t)$ is continuous on $U \times \bar{D}$. Then, by symmetry, $K_D(z,t)$ is continuous on $U \times \bar{D} \cup \bar{D} \times U$. Is then $K_D(z,t)$ continuous on $D \times \bar{D}$?

The positive answer of these question would permit us to remove all considerations about open dense sets from Theorem 3 and the previous part of the paper, and the whole situation would become simpler and more elegant. Observe that the property A_k is basic in our approach to the problem of extensions of biholomorphic mapping.
It should be mentioned that if the Bergman kernel function of a domain D satisfies the condition A_k then the set of points $z_o \in \partial D$ for which

$$\det \begin{bmatrix} K_D(z_o, a_i) \\ \frac{\partial}{\partial z} K_D(z_o, a_i) \end{bmatrix} \equiv 0$$

cannot be too big. As it has been shown by Webster [13] the condition A_k alone can give information on extensions of biholomorphic mappings across a part of boundary.

Remark on Bell's result. Recently Bell [1] (his paper was presented at the Conference on Several Complex Variables, Princeton, April 1978) proved the following very interesting fact:

BELL'S "DENSITY LEMMA". Let D be a bounded domain in $\mathbb{C}^n$ with C^{∞}-boundary. Denote by $W^s(D)$ the usual Sobolev space and by $H^s(D)$ its subspace consisting of holomorphic functions. Let K be an orthogon projector from $L^2(D)$ onto the space $L^2H(D)$ of square integrable holomorphic functions. Assume μ that there exist $S > 2n$ and $M \geq 0$ such that K is a bounded operator from $W^{S+M}(D)$ into $H^S(D)$. Then

$K(W^{S+M}(D)) \subset K(C_o^\infty(D)) \subset H^S(D)$. <u>The closure is taken in the space</u> $H^S(D)$.

Bell uses his lemma to prve that if K is a bounded operator from $W^{S+M}(D)$ into $H^S(D)$, $S \geq 2n$, then the Bergman function $K_D(z,t)$ is continuous on $\bar{D} \times D$ and for every $z \in \bar{D}$ there exists $a \in D$ such that $K_D(z,a) \neq 0$.

However, Bell's lemma has <u>far more essential consequences</u>.

It permits to prove, in the same manner as in Bell's paper, the following

<u>THEOREM 4</u>. <u>Let</u> D <u>be a bounded domain with</u> C^∞<u>-boundary</u>. <u>Suppose that the assumptions of Bell's lemma are satisfied for</u> $S > 2n+1$. <u>Then</u>:

1. <u>The Bergman function</u> $K_D(z,t)$ <u>is of class</u> C^k <u>on</u> $\bar{D} \times D$ <u>for</u> $k = S - 2n - 1$, <u>so the condition</u> A_k <u>holds for</u> D;
2. <u>The condition</u> B <u>is satisfied</u>.

<u>P r o o f</u>. It follows the Sobolev lemma that $H^S(D) \subset C^k(\bar{D})$, and there exists $c > 0$ such that

$$\sup_{z \in D} \left(|f(z)| + \sum_{|\alpha| \leq k} \left|\frac{\partial^{|\alpha|} f}{\partial z^\alpha}(z)\right|\right) \leq c\|f\|_S \quad \text{for every } f \in H^S(D).$$

In particular if $z_o \in \partial D$ then $(\partial^{|\alpha|}/\partial z^\alpha) f(z_o)$ is a continuous linear functional $H^S(D)$ (for $|\alpha| \leq k$). For every compact $K \subset D$ we can take a domain G with C^∞-boundary such that $K \subset G \subset\subset D$, a function φ with zeros of infinite order on ∂G, and construct the function $H_\varphi(z,t)$ as in the proof of Theorem 3. We have $K_D(z,t) = K(\overline{H_\varphi(t,z)})$ for $t \in G$, so by the boundedness of K, $W^{S+M} \longrightarrow H^S$, $\sup_{t \in k} \|K_D(\cdot,t)\|_S < \infty$. Hence $K_D(z,t)$ is of class C^k on $\bar{D} \times D$ and the condition A^k holds. Now we can proceed in the same manner as in the proof of Corollary in Bell's paper to prove that the functions $K_D(\cdot,t)$, $t \in D$, are linearly dense in the space $K(C_o^\infty(D)) \subset H^S(D)$.

If $f \in \overline{K(C_o^\infty(D))}$ and is orthogonal to the family of functions $\{K_D(\cdot,t)\}_{t \in D}$, then for every $u \in C_o^\infty(D)$ we have

$$0 = \sum_{|\alpha| \leq S} \int_D u(t) \left(\int_D \frac{\partial^{|\alpha|} f}{\partial z^\alpha}(z) \cdot \frac{\partial^{|\alpha|}}{\partial z^\alpha} K_D(z,t)\,dz\right) dt$$

$$= \sum_{|\alpha| \leq S} \int_D \frac{\partial^{|\alpha|} f}{\partial z^\alpha}(z) \left(\int_D u(t) \cdot \frac{\partial^{|\alpha|}}{\partial z^\alpha} K_D(z,t)\,dt\right) dz = \langle K(U), f \rangle_S$$

so $f = 0$. The domain D is bounded in $\mathbb{C}^n$, so the functions $1, z_1, \ldots z_n$ belong to $W^{S+M}(D)$ and hence to $K(W^{S+M}(D))$. It follows from Bell's lemma that $1, z_1, \ldots, z_n$ belong to $\overline{K(C_o^\infty(D))}$. Suppose now that the condition B is not satisfied. Then there exists a point $z_o \in \partial D$ for which

$$F(t_o,\dots,t_n) = \det \begin{bmatrix} K_D(z_o,t_o), & \dots & K_D(z_o,t_n) \\ \frac{\partial}{\partial z_1} K_D(z_o,t_o) & \dots & \frac{\partial}{\partial z_1} K_D(z_o,t_n) \\ \vdots & & \vdots \\ \frac{\partial}{\partial z_n} K_D(z_o,t_o) & \dots & \frac{\partial}{\partial z_n} K_D(z_o,t_n) \end{bmatrix} \equiv 0 \quad \text{on } D^{n+1}$$

Since the determinant is a n-linear and continuous function of its columns, so it implies that for every $(n+1)$-tuple $g_o(z),\dots,g_n(z)$ of functions from $\overline{K(C_o^\infty(D))}$ we have

$$\det \begin{bmatrix} g_o(z_o) & \dots & g_n(z_o) \\ \frac{g_o}{z_i}(z_o) & \dots & \frac{g_n}{z_i}(z_o) \end{bmatrix} = 0, \quad i = 1,\dots,n.$$

By putting $g_o = 1,\ g_1 = z_1,\dots,g_n = z_n$ we obtain a contradiction.

The fact proved above together with Theorem 1 of this paper shows the great importance of Bell's lemma in study of boundary behavior of biholomorphic mapping. This lemma replaces the difficult study of asymtotic expansion for Bergman function used in the original proof of Fefferman's theorem [4] (see also [2]). It should be mentioned that the proof of Bell's lemma is short, elegant and rather elementary.

It follows from the Kohn's results [7] that if D is a bounded domain with C^∞-boundary and the subelliptic estimates for the $\bar\partial$-Neumann operator holds for D, then the assumptions of Bell's lemma are satisfied for every s, so the conditions A_∞ and B hold for D. So now the proof of Fefferman's theorem consists of the subelliptic estimates for the $\bar\partial$-Neumann problem on strictly pseudoconvex domains which were proved by Kohn (see [5]), the Bell's lemma [1], the theorem proved in this Remark, and Theorem 1 of this paper.

It occurs that Fefferman's theorem can be obtained as a relatively elementary consequences of Kohn's results [5]. Moreover, these subelliptic estimates were proved by Diederich and Fornaess [3] for bounded pseudoconvex domains with real-analytic boundary. Thus we obtain

THEOREM 5. Let D be a bounded pseudoconvex domain with real-analytic boundary. Then the conditions A_∞ and B hold for D.

COROLLARY 4. Let D and G be bounded pseudoconvex domains with real-analytic boundaries. Every biholomorphic mapping from D onto G extends to a C^∞-diffeomorphism between some neighborhood of $\bar D$ and $\bar G$.

This corollary improves the result of Bedford and Fornaess (Duke

Math. J. 1978) who proved that such a mapping can be extended to a homeomorphism between $\bar{D}$ and $\bar{G}$ if D and G are in $\mathbb{C}^2$. (Recently Diederich and Fornaess [14] proved that if D is a pseudoconvex domain in $\mathbb{C}^n$ with C^2-boundary and G is a pseudoconvex domain in $\mathbb{C}^n$ with real analytic boundary then every proper holomorphic mapping from D onto G extends continuously on $\bar{D}$). It remains an open problem, whether such a mapping can be extended to a biholomorphic mapping between some neighborhoods of $\bar{D}$ and $\bar{G}$.

The complete and detailed proof of Fefferman's theorem will appear in a joint paper S.Bell, E.Ligocka: A Simplification and Extension of Fefferman's Theorem on Biholomorphic Mappings (in preparation).

References

[1] BELL, S.R.: Non-vanishing of the Bergman kernel function at boundary points of certain domains in $\mathbb{C}^n$, Math.Ann. (to appear).

[2] BOUTET DE MONVEL, L. et S.SJÖSTRAND: Sur la singularité des noyaux de Bergman et de Szegö, Asterique 34, 35 (1976), 123-164.

[3] DIEDERICH, K. and J.E.FORNAESS: Pseudoconvex domains with real-analytic boundaries, Ann. of Math. 107 (1978), 371-384.

[4] FEFFERMAN, C.: The Bergman kernel and bihomolomorphic mappings of pseudoconvex domains, Invent Math. 26 (1974), 1-65.

[5] FOLLAND, G. and J.J.KOHN: The Neumann problem for the Cauchy-Riemann complex, Ann of Math. Studies no 75, Princeton Univ. Press 1972.

[6] HENKIN, G.M. and E.M.ČIRKA: Boundary properties of holomorphic function of several complex variables, Ms.

[7] KOHN, J.J.: Sufficient conditions for subelliplicity on weakly pseudoconvex domains, Proc.Math.Acad.Sci. USA 74, n. 6 (1977), 2214-2216.

[8] LIGOCKA, E.: How to prove Feffermann's theorem without use of differential geometry, Ann.Pol.Math. (to appear).

[9] MALGRANGE, B.: Ideals of differentiable functions, Oxford 1966.

[10] SKWARCZYŃSKI, M.: Biholomorphic invariants related to the Bergman function, Dissertations Math. (to appear).

[11] ——: The ideal boundary of a domain in $\mathbb{C}^n$, Ann.Pol.Math. (to appear).

[12] STEIN, E.M.: Singular integrals and differentiability properties of functions, Princeton 1970.

[13] WEBSTER, S.: Biholomorphic mappings and the Bergman kernel off the diagonal, Inventiones Math. 51 (1979), 155-169.

[14] DIEDERICH, K. and J.E.FORNAESS: Proper holomorphic mappings onto pseudoconvex domains with real analytic boundary (to appear).

Osiedle Przyjaźń 15 m.4
PL-00-905 Warszawa Jelonki
Poland

THE ISOPERIMETRIC INEQUALITY AND SOME EXTREMAL PROBLEMS IN H^1

Miodrag Mateljević (Beograd)

In this note we continue the study of an application of the isoperimetric inequality to some extremal problems in Hardy space H^1, initiated in [3]. We shall use the usual notation of Hardy spaces; cf. [1].

An estimate of Hardy-Littlewood, concerning the identity mapping from the space $H^{1/2}$ into $B_{1/2}$, is improved. We show that the norm of the mapping equals π. Some generalisations and consequences of this result are obtained. Applying the duality relation in some cases we get better estimates from those which were obtained by the isoperimetric inequality. These results are the best possible because in all cases the extremal functions are obtained.

We shall use the following result which can easily be derived from the Theorem 2 of [3].

<u>THEOREM A</u>. <u>If</u> $f \in H^1$, <u>then</u>

$$\pi^{-1} \iint_{|z| \le 1} |f(z)|^2 dxdy \le \|f\|_1^2$$

<u>with equality if and only if</u> f <u>has the form</u>

$$f(z) = \alpha(1-\beta z)^{-2} \qquad (\alpha \in C,\ |\beta| < 1).$$

<u>THEOREM 1</u>. <u>If</u> $f \in H^P$ $(0 < p < \infty)$, <u>then</u>

$$(1) \qquad \pi^{-1} \iint_{|z|<1} |f(z)|^{2P} dxdy \le \{\frac{1}{2\pi} \int_0^{2\pi} |f(e^{i\theta})|^P d\theta\}^2$$

<u>with equality if and only if</u> f <u>has the form</u>

$$f(z) = \alpha(1-\ z)^{-2/P} \qquad (\alpha \in C,\ |\beta| < 1).$$

<u>P r o o f</u>. If $f \in H^P$ $(0 < p < \infty)$ and has no zeros in $|z| < 1$,

then $f^p \in H^1$, and (1) follows from Theorem A. If f has zeros, we factor out the Blaschke product $B(z)$ to obtain a nonvanishing function $g(z) = f(z)/B(z)$ which is again of class H^p. Thus

$$\pi^{-1} \iint_{|z|<1} |f(z)|^{2p} dxdy \le \pi^{-1} \iint_{|z|<1} |g(z)|^{2p} dxdy \le \{\frac{1}{2\pi} \int_0^{2\pi} |g(e^{i\theta})|^p d\theta\}^2$$
$$= \{\frac{1}{2\pi} \int_0^{2\pi} |f(e^{i\theta})|^p d\theta\}^2.$$

The equality in (1) holds if and only if it holds in all previous inequalities. Hence $|B(z)| = 1$ for $|z| < 1$. Combining this with Theorem A we obtain that f has the form given in Theorem 1.

The space B_p $(0 < p < \infty)$ consists of all functions f analytic in $|z| < 1$ for which

$$\| f\|_{B_p} = \iint_{|z|<1} |f(z)|(1-|z|)^{1/p-2} dxdy < \infty.$$

The next corollary is a particular case of Theorem 1 for $p = 1/2$.

COROLLARY 1. *If* $f \in H^{1/2}$, *then* $\| f\|_{B_{1/2}} \le \pi \| f\|_{1/2}$ *with equality if and only if* f *has the form*

$$f(z) = \alpha(1-\beta z)^{-4} \quad (\alpha \in C, \ |\beta| < 1).$$

R e m a r k 1. This corollary improves an estimate of Hardy-Littlewood ([2], Theorem 31). They showed that $H^{1/2} \subset B_{1/2}$ and the identity map from $H^{1/2}$ into $B_{1/2}$ is continuous, i.e. that $\| f\|_{B_{1/2}} \le C\| f\|_{1/2}$ for some constant C independent of $f \in H^{1/2}$. A short proof of this result was given in [4]; pp.117-118, where it was also proved that the constant C may be specified as $3\pi^2$.

The Bergman space B_2 consists of all functions f analytic in $|z| < 1$ for which

$$\| f\|^2_{B_2} \equiv \pi^{-1} \iint_{|z|<1} |f|^2 dxdy < +\infty.$$

It is easy to see that B_2 is a Hilbert space, where the scalar product of two functions $f,h \in B_2$ is defined by the relation

$$(f,h) = \pi^{-1} \iint_{|z|<1} f\bar{h} dxdy.$$

THEOREM 2. *If* $f,h \in H^1$ *then*

$$|\pi^{-1} \iint_{|z|<1} f\bar{h} dxdy| \le \| f\|_1 \| h\|_1 \tag{2}$$

<u>with equality if and only if</u> f <u>and</u> h <u>have the form</u>

$$f = \alpha(1-\beta z)^{-2}, \quad h = \lambda f \quad (\alpha,\lambda \in C,\ |\beta| < 1).$$

<u>P r o o f</u>. Suppose $f,h \in H^1$. By Schwarz's inequality and Theorem A,

$$|(f,h)| \le \|f\|_{B_2}\|h\|_{B_2} \le \|f\|_1 \|h\|_1 .$$

This proves the inequality (2). The equality in (2) holds if and only if it holds in all previous inequalities. The other part of Theorem 2 now follows from Theorem A and the fact that in Schwarz's inequality equality holds if and only if $h = \lambda f$ $(\lambda \in C)$.

Applying Theorem 2 to $h(z) = (1-\bar{\beta}z)^{-2}$ $(|\beta| < 1)$, we obtain

<u>COROLLARY 2</u>. <u>If</u> $f \in H^1$ <u>and</u> $|\beta| < 1$, <u>then</u>

$$|f(\beta)| \le \frac{\|f\|_1}{1-|\beta|^2}$$

<u>with equality if and only if</u> f <u>has the form</u> $f(z) = \alpha(1-\beta z)^{-2}$ $(\alpha \in C)$.

Hence we have (cf. [1], Exercises, p.144):

<u>COROLLARY 3</u>. <u>If</u> $f \in H^p$ $(0 < p < \infty)$ <u>and</u> $|\beta| < 1$, <u>then</u>

$$|f(\beta)| \le (1-|\beta|^2)^{-1/p}\|f\|_p$$

<u>with equality if and only if</u> f <u>has the form</u>

$$f(z) = \alpha(1-\beta z)^{-2/p} \quad (\alpha \in C).$$

<u>LEMMA 1</u>. <u>If</u> $f(z) = \Sigma_{n=0}^{\infty} a_n z^n \in H^1$, $h(z) = \Sigma_{n=0}^{\infty} c_n z^n \in H^1$ <u>and</u> $\kappa(z) = \int_0^z h(\xi)d\xi$ $(|z| < 1)$, <u>then</u>

$$(3) \qquad \pi^{-1} \iint_{|z|<1} f\bar{h}\,dxdy = \frac{1}{2\pi i}\int_{|z|=1} f\bar{\kappa}\,dz$$

<u>P r o o f</u>. Since $f \in H^1$ and $\kappa(z)$ is absolutely continuous on $|z| = 1$, applying Parseval's relation ([5], pp.89-91), we find

$$(4) \qquad \frac{1}{2\pi}\int_0^{2\pi} f(e^{i\theta})\,\overline{\kappa(e^{i\theta})e^{-i\theta}}\,d\theta = \Sigma_{n=0}^{\infty} \frac{a_n\bar{c}_n}{n+1} .$$

On the other hand, we have

$$\pi^{-1}\iint_{|z|<1} f\bar{h}\,dxdy = \frac{1}{\pi}\int_0^1 r\,dr\int_0^{2\pi}(\Sigma a_n r^n e^{in\theta})\cdot(\Sigma\bar{c}_n r^n e^{-in\theta})\,d\theta$$

(5)

$$= \Sigma_{n=0}^{\infty}\frac{a_n\bar{c}_n}{n+1}\ .$$

The equality (3) follows from (4) and (5).

The next corollary is an immediate consequence of Theorem 2 and Lemma 1.

COROLLARY 4. *If* $f,\kappa'\in H^1$, *then*

$$\left|\frac{1}{2\pi}\int_{|z|=1} f\bar{\kappa}\,dz\right|\le\|f\|_1\|\kappa'\|_1$$

with equality if and only if f *and* κ' *have the form*

$$\kappa'(z) = \alpha(1-\beta z)^{-2},\quad f = \gamma(1-\beta z)^{-2}\quad(\alpha,\gamma\in C,\ |\beta|<1).$$

Combining Lemma 1 and the duality relation ([1], p.130) with the inequality (2), we obtain

THEOREM 3. *If* $h\in H^1$ *and* $\kappa(z) = \int_0^z h(\xi)d\xi\quad(|z|<1)$, *then*

(6) $$\sup_{f\in H^1,\|f\|_1\le 1}\left|\pi^{-1}\iint_{|z|<1} f\bar{h}\,dxdy\right| = \min_{g\in H^\infty}\|\bar{\kappa}-g\|_\infty\le\|h\|_1 .$$

It is clear that the relation (6) gives a better estimate than (2).

COROLLARY 5. *If* $f,h\in H^1$ *and* $\kappa(z) = \int_0^z h(\xi)d\xi\quad(|z|<1)$, *then*

$$\left|\pi^{-1}\iint_{|z|<1} f\bar{h}\,dxdy\right|\le\|\kappa\|_\infty\|f\|_1 .$$

Proof. Since $\min_{g\in H^\infty}\|\bar{\kappa}-g\|\le\|\bar{\kappa}\|_\infty = \|\kappa\|_\infty$, the proof follows from the preceding theorem.

THEOREM 4. *If* $f'\in H^1$, *then*

(7) $$\pi^{-1}\iint_{|z|<1}|f'(z)|^2dxdy\le\|f'\|_1\|f\|_\infty$$

with equality if and only if f *has the form*

$$f(z) = B\prod_{i=1}^{n}(z-\alpha_i)(1-\bar{\alpha}_i z)^{-1}\quad(n\in N,\ |\alpha_i|<1,\ B\in C).$$

Proof. In view of Lemma 1, we have

$$\pi^{-1}\iint_{|z|<1}|f'(z)|^2dxdy = \frac{1}{2\pi}\int_0^{2\pi} f'(e^{i\theta})\overline{f(e^{i\theta})}e^{i\theta}d\theta$$

$$\leq \frac{1}{2\pi}\int_0^{2\pi} |f'(e^{i\theta})f(e^{i\theta})e^{i\theta}|d\theta \leq \|f'\|_1 \|f\|_\infty .$$

Hence, to hold equality in (7), it is necessary and sufficient that

$$(8) \qquad f'(e^{i\theta})\overline{f(e^{i\theta})}\, e^{i\theta} \geq 0 \qquad \text{a.e.}$$

and

$$|f(e^{i\theta})| = \|f\|_\infty \qquad \text{a.e.}$$

Since $f' \in H^1$, f is continuous on $|z| = 1$. Thus $|f(z)| = \|f\|_\infty$ on $|z| = 1$. Hence, either f has finitely many zeros or f vanishes identically on $|z| \leq 1$. We may suppose that f has finitely many zeros since otherwise the theorem is trivial. Now if we factor out the Blaschke product $B(z)$, we shall obtain a nonvanishing function $g(z) = f(z)\ B(z)$ which is continuous on $|z| \leq 1$ and satisfies $|g(z)| = \|g\|_\infty$ on $|z| = 1$. Applying the maximum principle to g and $1/g$, we conclude that $g(z) \equiv B$ in $|z| \leq 1$, where B is a complex constant. Thus, if in (7) holds equality, f has the form given by Theorem 4.

Conversely, let $f(z) = B \prod_{i=1}^{n} B_i(z)$ $(n \in N)$, where $B_i(z) = (z-\alpha_i)(1-\bar{\alpha}_i z)^{-1}$ $(|\alpha_i| < 1)$. Then

$$f'(e^{i\theta})\,\overline{f(e^{i\theta})}e^{i\theta} = |B|^2 \frac{d}{d\theta} \arg f(e^{i\theta})$$

and

$$\frac{d}{d\theta} \arg B_i(e^{i\theta}) = (1-|\alpha_i|^2)|1-\bar{\alpha}_i e^{i\theta}|^{-2} \quad (i = 1,2,\dots,n)$$

Hence it follows that f satisfies (8) and this completes the proof.

References

[1] DUREN,P.L.: Theory of H^p Spaces, Academic Press, New York 1970.

[2] HARDY,G.H. and J.E.LITTLEWOOD: Some properties of fractional integrals II. Math. Z. 34 (1932), 403-439.

[3] MATELJEVIĆ,M.: The isoperimetric inequality in the Hardy class H^1, Mat.Vesnik 3(16)(31) (1979).

[4] SHAPIRO,J.H.: Remarks on F-spaces of analytic functions (Lecture Notes in Mathematics), SpringerVerlag, Berlin-Heidelberg-New York 1977.

[5] ZYGMUND,A.: Trigonometric Series, Cambridge Univ. Press, London-New York, 1959.

Miodrag Mateljević

Institut za matematiku
Prirodno-matematički fakultet
Studentski trg 16
YU-11000 Beograd, Jugoslavija

SUR LES BASES POLYNOMIALES SEMI-SIMPLES DE L'ESPACE H(K)

Nguyen Thanh Van (Toulouse)

Table des matières

INTRODUCTION.

Soit K un compact de $\mathbb{C}^n$. On désigne par H(K) l'espace des germes de fonctions analytiques au voisinage de K, muni de la topologie usuelle $H(K) = \lim\limits_{\overrightarrow{\Omega}} \operatorname{ind} H(\Omega)$, où les Ω sont les ouverts de $\mathbb{C}^n$ qui contiennent K.

Définition.

a) Une suite $(B_k)_{k\in\mathbb{N}}$ de polynômes de n variables complexes est dite semi-simple lorsqu'il existe une bijection α de $\mathbb{N}$ sur $\mathbb{N}^n$ telle que pour tout $k\in\mathbb{N}$ on ait :

- $|\alpha(k)| \leq |\alpha(k+1)|$ (pour $\alpha = (\alpha_1,\ldots,\alpha_n)\in\mathbb{N}^n$ on pose $|\alpha|=\alpha_1+\ldots+\alpha_n$)

- $B_k(z) = \sum\limits_{j=0}^{k} \lambda_{k,j}\, z^{\alpha(j)}$ avec $\lambda_{k,k}\neq 0$ ($z^{\alpha(j)} = z_1^{\alpha_1(j)}\ldots z_n^{\alpha_n(j)}$)

b) Une base $(B_k)_{k\in\mathbb{N}}$ de H(K), au sens de Schauder, est dite polynomiale semi-simple lorsque (B_k) est une suite de polynômes semi-simple.

Pour $n = 1$ (et $\lambda_{k,k} = 1$, ce qui n'est pas essentiel) les bases polynomiales semi-simples sont les bases simples de 1ère espèce suivant la terminologie de [3], dont le chapitre 2 contient une étude de ces bases et celles dites "de Pincherle". L'existence d'une base simple de H(K), de la forme $B_k(z) = (z - a_1)\ldots(z - a_k)$ ($B_o \equiv 1$) où (a_k) est une suite de points de K, a été établie dans les cas suivants :

- $\Omega = \overline{\mathbb{C}} \setminus K$ est connexe et la fonction de Green de Ω avec pôle au point infini, prolongée par 0 sur K, est continue (Leja [1] (1957) et Walsh [9] (1935)).

- Cap K = 0 (Zaharjuta [10] (1970)).

Dans sa communication à la Conférence sur les fonctions analytiques tenue à Cracovie en 1974, Zaharjuta a prouvé l'existence d'une base polynomiale semi-simple pour H(K) dans le cas où K est un compact polynomialement convexe de $\mathbb{C}^n$ dont la fonction extrémale $\phi_K(z)$ de Siciak est continue. Nous donnerons ici une autre démonstration de ce résultat. Nous étudierons aussi des cas où l'on peut construire une base polynomiale semi-simple (B_k) comme une suite orthonormale dans $L^2(K,\mathcal{M})$, où $\mathcal{M}$ est une certaine mesure de Radon positive sur K.

Zaharjuta s'est posée la question suivante [10] : si K est un compact de $\mathbb{C}$ de capacité > 0, à complémentaire connexe et non régulier (i.e. K est effilé en un point au moins de sa frontière), H(K) possède-t-il une base (B_k) de la forme $B_k(z) = (z - a_1) \ldots (z - a_k)$?

Nous montrerons que la réponse est non, de manière plus générale : si $K_1, K_2 \ldots K_n$ sont des compacts de $\mathbb{C}$ de capacité positive et si $H(K_1 \times K_2 \times \ldots \times K_n)$ possède une base polynomiale semi-simple, alors chaque K_j est régulier.

<u>Problème ouvert</u> : Soit K un compact polynomialement convexe et non polaire dans $\mathbb{C}^n$ tel que H(K) possède une base polynomiale semi-simple. La fonction extrémale ϕ_K est-elle continue ?

Sur les bases polynomiales semi-simples de l'espace H(K)

I - EXISTENCE ET PROPRIETES DES BASES POLYNOMIALES SEMI-SIMPLES.

Dans cette section K désigne toujours un compact polynomialement convexe de $\mathbb{C}^n$ régulier au sens de la continuité sur $\mathbb{C}^n$ de la fonction extrémale ϕ_K de Siciak :

$$\phi_K(z) = \text{Sup}\{|P(z)|^{1/d^\circ P} \;\; P \text{ polynôme de degré } \geqslant 1, \; \|P\|_K \leqslant 1\}.$$

Nous renvoyons à Siciak [7] et [8] pour les propriétés de ϕ_K. Notons toutefois que ϕ_K est continue sur $\mathbb{C}^n$ si elle est continue en tout point de K.

Pour tout $r > 1$ on pose $\Omega_r = \{z \in \mathbb{C}^n : \phi_K(z) < r\}$,

c'est un ouvert de $\mathbb{C}^n$ contenant K et $\bigcap_{r>1} \Omega_r = K$.

1. Il est bien connu que $H(K) \hookrightarrow C(K)$ où $\hookrightarrow$ désigne l'injection continue définie par l'opération de restriction. Puisque H(K) est nucléaire, on sait d'après Mityagin ([2], Prop. 3a) qu'il existe un espace hilbertien $\mathcal{H}$ tel que

$$H(K) \hookrightarrow \mathcal{H} \hookrightarrow C(K).$$

Soit $\mathcal{H}_{\mathcal{P}}$ le sous espace fermé de $\mathcal{H}$ engendré par les polynômes, puisque les polynômes sont denses dans H(K) (K étant polynomialement convexe) on a $H(K) \hookrightarrow \mathcal{H}_{\mathcal{P}}$.

Théorème 1. (Zaharjuta [12]) Soit α une bijection de $\mathbb{N}$ sur $\mathbb{N}^n$ telle que $|\alpha(k)| \leqslant |\alpha(k+1)|$ $\forall$ k. Soit (B_k) la base orthonormale de $\mathcal{H}_{\mathcal{P}}$ obtenue à partir de la suite $\{e_{\alpha(k)}\}$ $(e_{\alpha(k)}(z) = z^{\alpha(k)})$ par le procédé d'orthonormalisation de Hilbert-Schmidt. Alors (B_k) est une base polynomiale semi-simple de H(K) et de $H(\Omega_r)$ pour tout $r > 1$. De plus pour tout $r > 1$ et tout $\varepsilon \in]0, r[$. On a :

$$C(r, \varepsilon).(r - \varepsilon)^{|\alpha(k)|} \leqslant \|B_k\|_{\overline{\Omega}_r} \leqslant C.r^{|\alpha(k)|} \;\; \forall k$$

où C et $C(r, \varepsilon)$ sont des constantes finies > 0 indépendantes de k.

Démonstration.

a) Zaharjuta a prouvé ce théorème par la technique d'échelles hilbertiennes. Nous donnons ici une autre démonstration selon une idée de Walsh ([9], p.128-131), utilisant le théorème de Bernstein-Walsh dans C^n que nous rappelons maintenant.

Pour $f \in C(K)$ et $q \in \mathbb{N}$ on pose

$$d_K(f, \mathcal{P}_q) = \text{Inf } \{\|f - P\|_K : P \in \mathcal{P}_q\}$$

où $\mathcal{P}_q$ désigne l'ensemble des polynômes de degré $\leq q$.

Théorème. (Siciak 1962, voir [7] ou [8]). Pour tout $r > 1$, Une fonction $f \in C(K)$ est la restriction à K d'une fonction holomorphe sur Ω_r si et seulement si

$$\limsup_{q \to \infty} \left[d_K(f, \mathcal{P}_q)\right]^{1/q} \leq \frac{1}{r}$$

b) On sait d'après le procédé de Hilbert-Schmidt que $B_k = \sum_{0}^{k} \lambda_{k,j}\, e_{\alpha(j)}$ avec $\lambda_{k,k} \neq 0$. Le fait que $\mathcal{H} \hookrightarrow C(K)$ donne : $\|B_k\|_K \leq C\|B_k\|_{\mathcal{H}} = C \quad (C = c^{te})$.

D'après l'Inégalité de Bernstein-Walsh (conséquence immédiate de la définition de ϕ_K) :

$$(0) \qquad \|B_k\|_{\overline{\Omega}_\rho} \leq C. \rho^{|\alpha(k)|} \quad \forall \rho > 1.$$

Montrons que (B_k) est une base de $H(\Omega_r)$, pour tout $r > 1$. Soit $f \in H(\Omega_r)$, alors $f \in \mathcal{H}_{\mathcal{B}}$ et par conséquent f s'écrit de manière unique en série $\Sigma\, C_k\, B_k$ convergente dans $\mathcal{H}_{\mathcal{B}}$. Il suffit de prouver que cette série converge dans $H(\Omega_r)$. Il est bien connu que pour tout $q \in \mathbb{N}$ il existe un polynôme P_q de degré $\leq q$ tel que $\|f - P_q\|_K = d_K(f, \mathcal{P}_q)$. Soit $\tau \in]1, r[$, puisque $f \in H(\Omega_r)$ il résulte du Théorème et de l'Inégalité de Bernstein-Walsh que :

$$(1) \quad \limsup (\|f - P_q\|_{\overline{\Omega}_\tau})^{1/q} \leq \frac{\tau}{r} \quad ;$$

pour tout $k \geqslant 1$ on a : $C_k = \langle f, B_k \rangle_{\mathcal{H}} = \langle f - P_{|\alpha(k)|-1}, B_k \rangle$ car B_k est orthogonal à tout polynôme de degré $< |\alpha(k)|$, donc

$|C_k| = |\langle f - P_{|\alpha(k)|-1}, B_k \rangle| \leqslant \|B_k\|_{\mathcal{H}} \|f - P_{(\alpha(k)|-1}\|_{\mathcal{H}}$;

Or $\|B_k\|_{\mathcal{H}} = 1$ et $H(\Omega_\tau) \hookrightarrow H(K) \hookrightarrow \mathcal{H}$, donc :

(2) $|C_k| \leqslant C' \|f - P_{|\alpha(k)|-1}\|_{\overline{\Omega}_\tau}$ $(C' = C^{te})$

Pour tout $\varepsilon > 0$ on a d'après (1) et (2) :

$$|C_k| \leqslant C(r, \tau, \varepsilon) \left(\frac{\tau}{r}\right)^{|\alpha(k)|} (1 + \varepsilon)^{|\alpha(k)|} \quad \forall k$$

Puisque τ est choisi arbitrairement dans $]1, r[$ on a

(3) $|C_k| \leqslant C'(r, \varepsilon) \left(\frac{1+\varepsilon}{r}\right)^{|\alpha(k)|}$ $\forall k$.

(0) et (3) prouvent que $\Sigma\, C_k B_k$ converge normalement sur tout compact de Ω_r.

(B_k) est donc une base polynomiale semi-simple de $H(\Omega_r)$, $\forall r > 1$. Il en résulte immédiatement qu'elle est une base (p.s-s) de $H(K)$. Il reste à prouver l'inégalité $C(r, \varepsilon)(r - \varepsilon)^{|\alpha(k)|} \leqslant \|B_k\|_{\overline{\Omega}_r}$. On remarque d'abord que d'après le raisonnement précédent, étant donnée une suite (C_k) de nombres complexes, la série $\Sigma\, C_k B_k$ converge dans $H(\Omega_r)$ si et seulement si $\limsup_{k\to\infty} |C_k|^{1/|\alpha(k)} \leqslant \frac{1}{r}$. Le contraire de l'inégalité à prouver (pour tout $r > 1$) implique ceci : il existe $r_0 > 1$ tel que $\liminf (\|B_k\|_{\overline{\Omega}_{r_0}})^{1/|\alpha(k)|} < r_0$, ce qui entraine (raisonnement classique) qu'il existe une suite (C_k) telle que

- $\limsup |C_k|^{1/|\alpha(k)|} = \frac{1}{r_0}$

- $\Sigma\, C_k B_k$ converge dans $H(\Omega_{r_1})$ pour un certain $r_1 > r_0$. Donc contradiction.

2. Nous considérons maintenant des cas où (B_k) peut être construit comme une suite orthonormale dans $L^2(K,\mu)$, où μ est une certaine mesure de Radon positive sur K.

<u>Définition de la Condition (L^*)</u> Soit μ une mesure de Radon positive sur K, on dit que le couple (K,μ) vérifie la condition (L^*) lorsque pour toute famille $(P_i)_{i \in I}$ de polynômes vérifiant :

$$\begin{cases} \sup_i |P_i(z)| < +\infty \text{ pour } z \in K \text{ sauf peut-être sur un ensemble} \\ \text{de } \mu\text{-mesure nulle} \end{cases}$$

et tout $\varepsilon > 0$, il existe un ouvert $\Omega \supset K$ et une constante M tels que

$$\|P_i\|_\Omega \leq M(1+\varepsilon)^{d^\circ P_i} \quad \forall i$$

Lorsque $K \subset \mathbb{C}$ (K est toujours supposé régulier) et μ = mesure harmonique sur K, on sait [4] que (K,μ) vérifie (L^*).

<u>Problème ouvert</u> Etant donné un compact $K \subset \mathbb{C}^n$ polynomialement convexe et régulier (au sens de la continuité de ϕ_K), existe-t-il une mesure de Radon positive μ sur K telle que (K,μ) vérifié (L^*) ?

Pleśniak vient d'obtenir un intéressant résultat sur ce sujet.

<u>Théorème</u>. (Pleśniak [5]). Soit K un compact de $\mathbb{C}^n$ (resp. $\mathbb{R}^n$). Supposons que pour tout point $a \in K$ il existe une courbe analytique réelle Γ telle que $a \in \Gamma$ et $\Gamma \cap (K \setminus \{a\}) \subset \operatorname{Int} K$. Alors (K, λ_{2n}) (resp. (K, λ_n)) vérifie la condition (L^*), où λ_{2n} (resp. λ_n) désigne la mesure de Lebesgue 2n-dimensionnelle (resp. n-dimensionnelle).

<u>Théorème 1 bis</u>. On suppose que (K,μ) vérifie (L^*) (K un compact polynomialement convexe régulier de $\mathbb{C}^n$, μ une mesure de Radon positive sur K). Soit α une bijection de $\mathbb{N}$ sur $\mathbb{N}^n$ telle que $|\alpha(k)| \leq |\alpha(k+1)|$ $\forall k$. Soit (B_k) la suite orthonormale dans

$L^2(K,\mu)$ construite à partir de la suite $\{e_{\alpha(k)}\}$ par le procédé de Hilbert-Schmidt. Alors (B_k) est une base polynomiale semi-simple de H(K) et de $H(\Omega_r)$ pour tout $r > 1$. De plus

$$\lim_{k\to\infty} (\|B_k\|_{\overline{\Omega}_\rho})^{1/|\alpha(k)|} = \rho \quad \forall \rho > 1.$$

<u>Démonstration abrégée</u>. On note d'abord que la construction de (B_k) par le procédé de Hilbert-Schmidt est possible car lorsque (K,μ) vérifie (L^*) la suite $\{e_{\alpha(k)}\}$ considérée <u>dans $L^2(K,\mu)$</u> est un système linéairement indépendant dans cet espace. On montre par un raisonnement élémentaire ([4], p.86-89) que

$$(*) \quad \limsup (\|B_k\|_K)^{1/|\alpha(k)|} \leq 1.$$

A une légère modification prés, la démonstration se fait exactement comme au paragraphe 1 (dém. du Th. 1).

<u>Remarque</u>. Lorsque (B_k) existe (i.e. $\{e_{\alpha(k)}\}$ est linéairement indépendante dans $L^2(K,\mu)$) et vérifie $(*)$, on peut montrer que (K,μ) vérifie (L^*) (voir [4], p.86-89). Dans ([4], p. 87) il faut aussi supposer a priori que la suite (z^k) soit linéairement indépendante dans $L^2(K,\mu)$.

3. <u>Théorème 2 (de prolongement)</u>. Soit K un compact de $\mathbb{C}^n$ polynomialement convexe et régulier. Si (Q_k) est une base polynomiale semi-simple de H(K), alors elle est une base de $H(\Omega_r)$ pour tout $r > 1$.

<u>Démonstration</u>.

D'après l'hypothèse, il existe une bijection α de $\mathbb{N}$ sur $\mathbb{N}^n$ telle que $|\alpha(k)| \leq |\alpha(k+1)|$ et

$$Q_k = \sum_{0}^{k} a_{k,j}\, e_{\alpha(j)} \text{ avec } a_{k,k} \neq 0.$$

a) Cas où $K = \overline{U}_\rho$, polydisque fermé de centre 0 et de rayon $\rho \geqslant 0$.

On peut supposer $a_{k,k} = 1$, quitte à remplacer Q_k par $\frac{1}{a_{k,k}} Q_k$. Si (Q_k) est une base de $H(U_r)$ (resp. $H(\overline{U}_r)$) alors on sait ([3], p.188) qu'elle est équivalente à la base $\{e_{\alpha(k)}\}$, i.e. l'application $f = \Sigma\, C_k\, Q_k \longrightarrow g = \Sigma\, C_k\, e_{\alpha(k)}$ est un automorphisme vectoriel topologique de $H(U_r)$ (resp. $H(\overline{U}_r)$).

Soit $(b_{k,j})$ la matrice (infinie) inverse de $(a_{k,j})$, on a $b_{k,k} = 1$ et $b_{k,j} = 0$ si $k < j$.

Lemme. i) Soit $0 < R < \infty$. (Q_k) est une base de $H(U_R)$ si et seulement si pour tout $r < R$ il existe $\tau < R$ et $M \geqslant 0$ tels que

$$\sum_{0}^{k} (|a_{k,j}| + |b_{k,j}|) r^{|\alpha(j)|} \leqslant M.\ \tau^{|\alpha(k)|} \quad \forall\, k.$$

ii) Soit $0 \leqslant R < \infty$. (Q_k) est une base de $H(\overline{U}_R)$ si et seulement si pour tout $R' > R$ il existe $\tau > R$ et $M \geqslant 0$ tels que

$$\sum_{0}^{k} (|a_{k,j}| + |b_{k,j}|)\ \tau^{|\alpha(j)|} \leqslant M.(R')^{|\alpha(k)|} \quad \forall\, k.$$

La démonstration de ce lemme est identique à celle de ([3], p.198, Th.2). Ce lemme et la propriété $|\alpha(k)| \leqslant |\alpha(k+1)|$ donnent immédiatement, par un calcul simple : si (Q_k) est une base de $H(\overline{U}_\rho)$, alors elle est une base de $H(U_r)$ pour tout $r > \rho$.

b) Cas général.

Soit (B_k) la base polynomiale semi-simple du théorème 1 correspondante à la même bijection α de $\mathbb{N}$ sur $\mathbb{N}^n$. Il résulte du Théorème 1 et du Théorème de Banach-Steinhaus que l'application T qui à $f = \Sigma\, C_k\, B_k$ fait correspondre $g = \Sigma\, C_k\, e_{\alpha(k)}$ est un isomorphisme simultané des espaces $H(K)$ et $H(\Omega_r)$ sur les espaces $H(\overline{U}_1)$ et $H(U_r)$ ($\forall\, r > 1$). Posons $\tilde{Q}_k = T(Q_k)$. $(\tilde{Q}_k)$ est une base polynomiale semi-simple de $H(\overline{U}_1)$, donc d'après a) elle est une base de $H(U_r)$ ($\forall r > 1$) ; par conséquent $\{Q_k\} = \{T^{-1}(\tilde{Q}_k)\}$ est une base de $H(\Omega_r)$.

II - REGULARITÉ DES K_j QUAND $H(K_1 \times K_2 \times \ldots \times K_n)$ POSSEDE UNE BASE POLYNOMIALE SEMI-SIMPLE.

1) Théorème 3. Soit $K = K_1 \times K_2 \times \ldots \times K_n$ où les K_j sont des compacts de $\mathbb{C}$ de capacité positive. Si $H(K)$ possède une base polynomiale semi-simple, alors $H(K)$ est isomorphe à $H(\overline{U})$ ($\overline{U}$ = le polydisque unité fermé).

Démonstration.

Soit (Q_k) une base polynomiale semi-simple de $H(K)$. On sait que $Q_k = \sum_0^k a_{k,j}\, e_{\alpha(j)}$ avec $a_{k,k} \neq 0$, où α est une bijection de N sur N^n telle que $|\alpha(k)| \leq |\alpha(k+1)|$. On pose :

$$\tilde{Q}_k = \frac{1}{a_{k,k}} Q_k = e_{\alpha(k)} + \sum_0^{k-1} \sigma_{k,j}\, e_{\alpha(j)}$$

$(\tilde{Q}_k)$ est une base de $H(K)$ et on a

Lemme 1. Soit $R_j = \text{Cap}\, K_j$. Pour toute suite (C_k) de nombres complexes, les propriétés suivantes sont équivalentes

(i) $\Sigma\, C_k\, \tilde{Q}_k$ converge dans $H(K)$

(ii) $|C_k| \leq C(\ell)\, \dfrac{\ell^{|\alpha(k)|}}{R_1^{\alpha_1(k)} \ldots R_n^{\alpha_n(k)}}$ $\forall k$, pour un certain $\ell \in]0, 1[$ ($C(\ell) = C^{te}$ dépendant de ℓ).

La démonstration du Lemme 1 s'appuie sur le lemme suivant.

Lemme 2. (Retherford et Mc Arthur [6]). Soit E un espace tonnelé, soit (B_k) une base de Schauder de E et soit Γ une famille de semi-normes continues engendrant la topologie de E. Alors pour tout $\mu \in \Gamma$ il existe $\nu \in \Gamma$ et une constante $C(\mu, \nu)$ tels que si $x \in E$ et $x = \Sigma x_k B_k$, alors

$$\sup_{k,\ell} \mu\left(\sum_{k}^{\ell} x_j B_j\right) \leq C(\mu, \nu)\, \nu(x)$$

(i) $\Longrightarrow$ (ii). Puisque les polynômes sont denses dans H(K), le complémentaire de K_j est connexe $(j = 1, 2, \ldots, n)$ et K est polynomialement convexe. On choisit $E_1, E_2, \ldots, E_n$ des compacts réguliers et à complémentaire connexe, tels que $\text{Cap } E_j > \text{Cap } K_j$ et $E_1 \times E_2 \ldots \times E_n \subset \mathcal{U}$, où $\mathcal{U}$ est un ouvert de $\mathbb{C}^n$ contenant K sur lequel la série $\Sigma C_k Q_k$ converge uniformément. Pour tout $j = 1, 2 \ldots n$, on désigne par ϕ_j la fonction extrémale de E_j et par $(L_k^{(j)})$ la suite des polynômes de Leja associée à E_j. On pose pour tout $\alpha = (\alpha_1, \ldots, \alpha_n) \in \mathbb{N}^n$:

$$L_\alpha(z) = L_{\alpha_1}^{(1)}(z_1) \ldots L_{\alpha_n}^{(n)}(z_n) \quad (z = (z_1, \ldots, z_n) \in \mathbb{C}^n)$$

et pour tout $r > 1$:

$$\Delta_r = \Delta_r^{(1)} \times \ldots \times \Delta_r^{(n)} \text{ où } \Delta_r^{(j)} = \{w \in \mathbb{C} : \phi_j(w) < r\}.$$

On sait que pour chaque j, $(L_k^{(j)})$ est une base polynomiale simple de $H(\Delta_r^{(j)})$. On en déduit par un raisonnement classique que $(L_\alpha)_{\alpha \in \mathbb{N}^n}$ est une base de $H(\Delta_r)$. On note que pour tout $k \geq 1$:

$$\tilde{Q}_k = L_{\alpha(k)} + \sum_{0}^{k-1} \lambda_{k,j} L_{\alpha(j)}.$$

Choisissons maintenant r assez petit pour que $\bar{\Delta}_r \subset \mathcal{U}$. On a d'après le lemme 2 : pour tout $\rho \in]1, r[$, il existe $\tau \in]1, r[$ tel que

$$\|L_{\alpha(k)}\|_{\Delta_\rho} \leq C(\rho, \tau) \|\tilde{Q}_k\|_{\Delta_\tau} \quad \forall k$$

On a donc pour tout k

$$|C_k| \, \|L_{\alpha(k)}\|_{\Delta_\rho} \leq C(\rho, \tau) \, |C_k| \, \|\tilde{Q}_k\|_{\Delta_\tau} \leq M(\rho, \tau) < \infty.$$

On sait d'autre part : $\lim\limits_{k\to\infty} \left(\|L_k^{(j)}\|_{\Delta_\rho^{(j)}}\right)^{1/k} = \rho \text{ Cap } E_j > \text{Cap } K_j.$

Par conséquent il existe $\ell \in]0, 1[$ tel que

$$|C_k| \leq C(\ell) \, \frac{\ell^{|\alpha(k)|}}{R^{\alpha(k)}} \quad \forall k$$

où $R^{\alpha(k)} = R_1^{\alpha_1(k)} \ldots R_n^{\alpha_n(k)}$.

<u>(ii) ⟹ (i)</u>. Pour chaque j = 1, 2,... n on choisit un compact E_j tel que $\mathring{E}_j \supset K_j$ et Cap $E_j < \frac{1}{\ell}$ Cap K_j. Soit $(T_k^{(j)})$ la suite des polynômes de Tchebicheff associée à E_j. On pose $T_\alpha(z) = T_{\alpha_1}^{(1)}(z_1) \ldots T_{\alpha_n}^{(n)}(z_n)$ pour $\alpha = (\alpha_1,\ldots,\alpha_n) \in \mathbb{N}^n$ et $z = (z_1,\ldots, z_n) \in \mathbb{C}^n$, on a pour $k \geq 1$

$$T_{\alpha(k)} = \tilde{Q}_k + \sum_{0}^{k-1} B_{k,j} \tilde{Q}_j.$$

Soit μ une seminorme continue sur H(K) soient ν et $C(\mu , \nu)$ comme dans le lemme 2, on a

$$\mu(\tilde{Q}_k) \leq C(\mu , \nu) \, \nu(T_{\alpha(k)}) \quad \forall k$$

Puisque $H(\mathring{E}) \hookrightarrow H(K)$ $(E = E_1 \times \ldots \times E_n)$ on a

$$\nu(T_{\alpha(k)}) \leq C' \|T_{\alpha(k)}\|_E \quad \forall k \ (C' = C^{te}).$$

Or $\lim(\|T_k^{(j)}\|_{E_j})^{1/k} = \text{Cap } E_j < \frac{1}{\ell} \text{ Cap } K_j$, donc

$$\Sigma|C_k| \, \mu(\tilde{Q}_k) \leq C'.C(\mu , \nu) \, \Sigma|C_k| \, \|T_{\alpha(k)}\|_E < \infty.$$

La série $\Sigma \, C_k \tilde{Q}_k$ converge donc absolument dans H(K). (Fin de la démonstration du lemme 1).

Il résulte immédiatement du lemme 1 et du Théorème de

Banach-Steinhaus que l'application $\mathcal{J}$ qui à $f = \Sigma\, C_k\, \tilde{Q}_k$ fait correspondre $\mathcal{J}(f) = \Sigma\, C_k\, R^{\alpha(k)}\, e_{\alpha(k)}$ est un isomorphisme vectoriel topologique de $H(K)$ sur $H(\bar{U})$.

<u>Conséquence</u>. Dans les hypothèses du Théorème 3, les K_j sont réguliers.

En effet puisque $H(K)$ est isomorphe à $H(\bar{U})$ on sait d'après Zaharjuta [11] que K est $\mathbb{C}^n$-régulier, d'autre part K est polynomialement convexe donc pour tout D ouvert borné contenant K, la fonction $h^*_{K,D}$ est nulle sur K, $h^*_{K,D}$ = Reg. sup. $h_{K,D}$ où

$h_{K,D}(z) = \mathrm{Sup}\{\varphi(z) : \varphi \in PSH(D),\ \varphi \leq 1 \text{ sur } D \text{ et } \varphi \leq 0 \text{ sur } K\}$.

Or cette propriété équivaut à la régularité de K(continuité de ϕ_K, voir [8], p. 135, prop. 6.1), et puisque $K = K_1 \times \ldots \times K_n$ la régularité de K équivaut à la régularité de tous les K_j.

2) <u>Remarques</u>

a) Par la même méthode on peut montrer : si $D_1 \ldots D_n$ sont des ouverts bornés de $\mathbb{C}$ tels que $H(D_1 \times \ldots \times D_n)$ possède une base polynomiale semi-simple, alors $H(D_1 \times \ldots \times D_n)$ est isomorphe à $H(U)$ et chaque D_j est à complémentaire connexe et à frontière régulière pour le Problème de Dirichlet.
Pour $n = 1$ l'isomorphisme de $H(D)$ et $H(U)$ a été démontré dans ([3], p.185, Prop. 3).

b) On obtient une caractérisation des compacts K de $\mathbb{C}$ de capacité > 0 tels que $H(K)$ possède une base polynomiale simple : $\bar{\mathbb{C}} \setminus K$ est un domaine à frontière régulière pour le Problème de Dirichlet.

c) Par dualité ([3], chap.2) on a l'énoncé suivant : soit Ω un domaine borné de $\mathbb{C}$, $H(\Omega)$ possède une base de Pincherle (ϕ_p) ("de Pincherle" veut dire qu'il existe $a \in \Omega$ tel que $\phi_p^{(p)}(a) \neq 0$ et $\phi_p^{(j)} = 0$ pour $j < p$) si et seulement si la frontière de Ω est régulière pour le Problème de Dirichlet.

d) Problème ouvert. Soit D un ouvert de $\mathbb{C}$ à complémentaire connexe et à frontière régulière pour le Problème de Dirichlet. H(D) possède-t-il une base polynomiale simple ?

Note Nous profitons de l'occasion pour signaler une lacune de notre article [3]. Le Théorème 1 du Chapitre 1 de cet article (page 175) est démontré seulement pour E compact ou ouvert (de $\mathbb{C}^m$). Par une faute d'inattention cette hypothèse n'a pas figuré dans l'énoncé. Le Théorème serait probablement vrai sans cette hypothèse.

BIBLIOGRAPHIE

[1] LEJA F. : Sur certaines suites liées aux ensembles plans et leur application à la représentation conforme, Ann. Soc. Pol. Math, 4 (1957).

[2] MITYAGIN B. : Approximative dimension and bases in nuclear spaces, Russian Math. Surveys, 16(1961).

[3] NGUYEN THANH VAN : Bases de Schauder dans certains espaces de fonctions holomorphes, Ann. Inst. Fourier, 22 (1972).

[4] NGUYEN THANH VAN : Familles de polynômes ponctuellement bornées, Ann. Pol. Math., 31 (1975).

[5] PLESNIAK W. : On some polynomial condition of the type of Leja in $\mathbb{C}^N$, Proceedings of the 7th. Conference on Analytic Functions (Kozubnik 1979).

[6] RETHERFORD J.M. et Mc ARTHUR C.W. : Some remarks on bases in linear topological spaces, Math.Annalen, 164 (1966).

[7] SICIAK J. : On some extremal functions and their application in the Theory of analytic functions of several complex variables, Trans. Amer. Math. Soc. 105 (2) (1962).

[8] SICIAK J. : Extremal Plurisubharmonic Functions in C^n, Proceedings of the 1st Finnish-Polish Summer School in Complex Analysis, Univ. of Lodz 1977.

[9] WALSH J.L. : Interpolation and Approximation, Amer. Math. Soc. Colloq. Publ., 3rd edition (1960).

[10] ZAHARJUTA V.P. : Spaces of functions of one variable analytic in open sets and on compacta, Math. USSR Sbornik, 11 (1970).

[11] ZAHARJUTA V.P. : Fonctions plurisousharmoniques extrémales, échelles hilbertiennes et isomorphismes d'espaces de fonctions analytiques de plusieurs variables complexes (en russe), Théorie des fonctions, Analyse fonctionnelle et leurs applications, 19 et 21 (1974).

[12] ZAHARJUTA V.P. : Fonctions plurisousharmoniques extrémales, polynômes orthogonaux et Théorème de Bernstein - Walsh pour les fonctions analytiques de plusieurs variables complexes (en russe), Ann. Pol. Math., 33 (1976).

NGUYEN THANH VAN
UER MIG
Université Paul Sabatier
118, route de Narbonne
31077 TOULOUSE CEDEX
FRANCE

ON SOME POLYNOMIAL CONDITIONS OF THE TYPE OF LEJA IN $\mathbb{C}^N$

Wiesław Pleśniak (Kraków)

Contents

Abstract

We show that the property of a compact set E in $\mathbb{C}^N$ to satisfy conditions of the type of Leja's famous polynomial condition is invariant under a large class of holomorphic mappings from a neighborhood of the polynomial envelope of E, $\hat{E}$, to $\mathbb{C}^M$ $(M \leq N)$ containing, in particular, all open holomorphic mappings. This yields new examples of sets E in $\mathbb{C}^N$ satisfying the conditions under consideration.

1. Introduction

Let E be a Borel subset of the space $\mathbb{C}^N$ of N complex variables and let μ be a nonnegative set function defined on a family of subset of E, containing all Borel subsets, such that $\mu(\phi) = 0$. In the sequel such a function μ will be simply called the measure. A condition (C) is said to be satisfied μ-almost everywhere on E (μ.a.e.) if $\mu(\{x \in E : \text{(C) is not satisfied at } x\}) = 0$. The pair (E,μ) is said to satisfy condition (L^*) (resp. (L_0^*)), if for every subfamily $\mathcal{F}$ of the family $\mathcal{P}$ of all polynomials from $\mathbb{C}^N$ to $\mathbb{C}^1$, with $\sup\{|f(z)| : f \in \mathcal{F}\} < \infty$ μ-a.e. on E, and for every $b > 1$ there exists an open neighborhood U of the closure of E, $\bar{E}$, and a constant $C > 0$ such that

$$\|f\|_U \leq Cb^{\deg f} \quad \text{for all } f \text{ in } \mathcal{F}$$

(resp. if for every family $\mathcal{F} \subset \mathcal{P}$ with $\sup\{|f(z)| : f \in \mathcal{F}\} < \infty$ μ-a.e. on E and for every $b > 1$ there exists a constant $C > 0$ such that

$$\|f\|_E \leq Cb^{\deg f} \quad \text{for all } f \text{ in } \mathcal{F}),$$

$\|f\|_F$ denoting the supremum norm of f on a set F.

If μ is the counting measure ($\mu(A)$ = number of elements of A) then (L^*) becomes simply the classical polynomial condition of Leja.

It is seen that a pair (E,μ) satisfies (L^*) if and only if it satisfies (L_o^*) and E satisfies the following condition:

(B) (Bernstein's inequality). For each $b > 1$ there exists a neighborhood U of $\bar{E}$ such that for each $p \in \mathcal{P}$,

$$\|p\|_U \leq b^{\deg p} \max\{1, \|p\|_E\} .$$

If E is compact, the above condition is equivalent to the L-regularity of E (see [9], [11]), i.e. to the continuity in $\mathbb{C}^N$ of Siciak's extremal function associated with E,

$$\Phi_E(z) = \sup\{|p(z)|^{1/\deg p} : p \in \mathcal{P}, \|p\|_E \leq 1, \deg p \geq 1\}$$

for $z \in \mathbb{C}^N$.

With respect to various applications in complex analysis (cf. [8], [9], [10], [3] and [4]) it is an important matter to know for what pairs (E,μ) (resp. for what compact sets E in $\mathbb{C}^N$) the conditions (L^*) and (L_o^*) (resp. (B)) are satisfied. If $N = 1$, this problem is rather well explored. For instance in this case the function $\log \Phi_E$ is known to be equal to the Green function with pole at ∞ of the unbounded connected component $D_\infty(E)$ of the set $\mathbb{C}^1 \setminus E$, whence the problem of the L-regularity of a plane compact set E is equivalent to the problem of the regularity of E with respect to the Green function of $D_\infty(E)$ (see e.g. [8]). As regards the condition (L^*) it is worth while to mention the classical example due to Leja [2]:

1.1. Example. If E is a rectifiable Jordan arc in $\mathbb{C}^1$ and μ is the length measure on E, then the pair (E,μ) satisfies (L^*).

For other examples of pairs satisfying (L^*) see [3] and [10].

If $N > 1$, the problem of characterizing pairs (E,μ) (resp. compact sets E) that satisfy conditions (L^*) and (L_o^*) (resp. condition (B)) is much more difficult (for some examples in this case see [9], [10] and [6]). In this note we shall present two theorems yielding a new information about the above question. We shall also prove a criterion of the L^*-regularity (with respect to the Lebesgue measure) which essentially enlarges the list of known examples of pairs satisfying (L^*).

2. Main results

2.1. THEOREM [5]. *Let* E *be a compact set in* $\mathbb{C}^N$ *and* h *an open holomorphic mapping defined in an open neighborhood* U *of the polynomial envelope of* E, $\hat{E}$, *with values in* $\mathbb{C}^M$ $(M \leq N)$. *Then, if* E *is* L-*regular, so is the set* $h(E)$.

In order to state the second theorem we shall need the notion of the L-capacity. Let $\mathcal{L}$ denote the class of all pluarisubharmonic functions u on $\mathbb{C}^N$ such that

$$u(z) \leq m + \log(1 + |z|), \quad z \in \mathbb{C}^N,$$

with the real constant m depending on u, where $|z| = \max\{|z_j| : 1 \leq j \leq N\}$. For a subset E of $\mathbb{C}^N$, we define the *extremal function* V_E associated with E by

$$V_E(z) = \sup\{u(z) : u \in \mathcal{L}, \ u \leq 0 \ \text{on} \ E\}$$

for $z \in \mathbb{C}^N$ (see [11], [9]). If E is compact, the function V_E is known to be equal to the function $\log \Phi_E$ (see [11], [9]). Let V_E^* denote the upper regularization of V_E,

$$V_E^*(z) = \limsup_{w \to z} V_E(w), \quad z \in \mathbb{C}^N,$$

and set

$$c(E) = \limsup_{|z| \to \infty} |z| \exp[-V_E^*(z)].$$

The number $c(E)$ is called the L-*capacity* of E. If $N = 1$, $c(E)$ is equal to the logarithmic capacity of E (see [9]).

2.2. THEOREM. *Let* E *be a compact set in* $\mathbb{C}^N$ *and* h *an open holomorphic mapping from an open neighborhood* U *of* E *to* $\mathbb{C}^N$ $(M \leq N)$. *Let* μ *and* ν *be measures on* E *and* $h(E)$, *respectively, satisfying the following hypotheses:*

(H_1) *If* $c(E) > 0$, *then for every Borel subset* F *of* E *with* $\mu(E \setminus F) = 0$ *we have* $c(F) > 0$,

(H_2) <u>for each Borel subset</u> I <u>of</u> $h(E)$, $\nu(I)=0$ <u>implies</u> $\mu(h^{-1}(I)\cap E)=0$.
<u>Then if the pair</u> (E,μ) <u>satisfies condition</u> (L^*) <u>(resp.</u> (L_o^*)<u>), so does the pair</u> $(h(E),\nu)$.

A crucial role while proving both theorems is played by the following two lemmas. The first one is a version for families of holomorphic functions of a theorem due to Siciak (see [8]) being a generalization to the N-dimensional case of the known Bernstein-Walsh theorem.

2.3. LEMMA ([5], Lemma 2.1). <u>Let</u> $A(U)$ <u>denote the Banach space of all bounded holomorphic functions defined in an open set</u> U <u>in</u> $\mathbb{C}^N$, <u>equipped with the supremum norm on</u> U. <u>For each polynomially convex, compact set</u> $E\subset U$ <u>there exist constants</u> $C>0$ <u>and</u> $a\in(0,1)$, <u>both</u> C <u>and</u> a <u>independent of</u> $f\in A(U)$ <u>and</u> n, <u>such that</u>

$$\mathrm{dist}_E(f,\mathcal{P}_n) := \inf\{\|f-p\|_E : p\in\mathcal{P}_n\} \le C\|f\|_U\, a^n$$

<u>for all</u> f <u>in</u> $A(U)$ <u>and</u> $n=1,2,\ldots$, $\mathcal{P}_n$ <u>denoting the set of all polynomials from</u> $\mathbb{C}^N$ <u>to</u> $\mathbb{C}^1$ <u>of degree at most</u> n.

The second lemma easily follows from the recent result by Josefson [1] proving Lelong's conjecture according to which every locally polar subset of $\mathbb{C}^N$ is globally polar (with respect to the plurisubharmonic functions in $\mathbb{C}^N$).

2.4. LEMMA ([5], Lemma 2.5). <u>Let</u> h <u>be an open holomorphic mapping in an open set</u> $U\subset\mathbb{C}^N$ <u>with values in</u> $\mathbb{C}^M$ $(M\le N)$. <u>If</u> E <u>is a subset of</u> U <u>with a positive</u> L-<u>capacity</u> $c(E)$, <u>then</u> $c(h(E))>0$.

The detailed proof of Theorem 2.1 can be found in [5]. In this note we shall only outline the proof of Theorem 2.2.

<u>S k e t c h o f t h e p r o o f</u>. If suffices to show that, for a given sequence $\{f_n\}$ of polynomials from $\mathbb{C}^M$ to $\mathbb{C}^1$ with $\deg f_n\le n$ such that

$$(+)\qquad \sup_n\{|f_n(w)|\}<\infty \qquad \nu\text{-a.e. on } h(E),$$

to every $b>1$ there corresponds an open neighborhood G of $h(E)$ and a constant $A>0$ such that

$$\|f_n\|_G\le Ab^n,\qquad n=1,2,\ldots$$

It follows from (+) and (H_2) that

$$M(z) := \sup_n\{|f_n(h(z))|\} < \infty \qquad \mu\text{-a.e. on } E.$$

Hence by (H_1), since the countable union S of sets S_i with $c(S_i) = 0$ has the L-capacity $c(S) = 0$ (see [9], Theorem 3.6 and Corollary 3.9), there exists a positive integer k such that $c(E_k) > 0$, where

$$E_k := \{z \in E : M(z) \leq k\}.$$

Then by Lemma 2.4, $c(h(E_k)) > 0$ whence the extremal function $\Phi_{h(E_k)}$ is locally bounded in $\mathbb{C}^M$. We may also assume that h is bounded in U. Then

$$\| f_n \circ h\|_U = \| f_n\|_{h(U)} \leq kA_1^n$$

for $n \geq 1$, where $A_1 = \sup_{h(U)} \Phi_{h(E_k)}(z)$. Consequently, by Lemma 2.3, there exist: a constant $a \in (0,1)$, a constant $A_2 > 0$, a positive integer s and polynomials p_n of degree at most sn such that

$$(++) \qquad \| f_n \circ h - p_n\|_E \leq A_2 a^n$$

for $n = 1,2,\ldots,$ whence

$$\sup_n\{|p_n(z)|\} < \infty \qquad \mu\text{-a.e. on } E.$$

Since $(E,\mu) \in (L^*)$, for every $b > 1$ there exists an open neighborhood V of E and a constant $A_3 > 0$ such that

$$\| p_n\|_E \leq \| p_n\|_V \leq A_3 b^{n/2}, \qquad n \geq 1.$$

Hence, by (++),

$$\| f_n\|_{h(E)} = \| f_n \circ h\|_E \leq A_4 b^{n/2},$$

where $A_4 = 2\max\{A_2,A_3\}$. By Theorem 2.1, since the set E is L-regular, so is the set $h(E)$, whence there exists an open neighborhood W of $h(E)$ and a constant $A_5 > 0$ such that

$$\| f_n\|_W \leq A_5 b^{n/2} \| f_n\|_{h(E)} \leq A_6 b^n$$

for $n \geq 1$, where $A_6 = A_4A_5$. The theorem is proved.

2.5. Remark. In both Theorems 2.1 and 2.2 the assumption that h is open can be essentially weakened. For instance it suffices to assume either

(H_3) (version for (L^*)) For each bounded open subset V of $\mathbb{C}^N$ such that $E \subset V \subset \bar{V} \subset U$, the extremal function $\Phi_{h(\bar{V})}$ is continuous at every point of $h(E)$,

or

(H_4) (version for (L_o^*)) for each connected component V of U with $c(E \cap V) > 0$, we have $c(h(V)) > 0$.

3. The case of Lebesgue measure

In the sequel λ_k will denote the k-dimensional Lebesgue measure. By the local summability of plurisubharmonic functions, if E is a subset of $\mathbb{C}^N$ (resp. $\mathbb{R}^N$, where the space $\mathbb{R}^N$ is treated as a subset of $\mathbb{C}^N$ consisting of those points $z = (z_1,\dots,z_N)$ that $\operatorname{Im} z_j = 0$ for $j = 1,\dots,N$) with $\lambda_{2N}(E) > 0$ (resp. $\lambda_N(E) > 0$), then $c(E) > 0$. Hence if $\mu = \lambda_{2N}$ (resp. $\mu = \lambda_N$) the assumption (H_1) is fulfilled. It can also be proved (see [7], Proposition 4.3 and Remark 4.4) that the assumption (H_2) is satisfied for $\mu = \lambda_{2N}$, $\nu = \lambda_{2M}$ and a holomorphic mapping h such that

$$(+++) \quad \operatorname{int}\{z \in U : \operatorname{rank}_z h < M\} = \emptyset$$

(resp. for $\mu = \lambda_N$, $\nu = \lambda_M$ nad h satisfying (+++), if $E \subset \mathbb{R}^N$ and $h(E) \subset \mathbb{R}^M$). It is also known ([7], Lemma 5.5) that for U connected and h such that $c(h(U)) > 0$, we have $c(\{z \in U : \operatorname{rank}_z h < M\}) = 0$. Hnece, if U is connected, assumption (H_3) implies assumption (+++). Consequently, by Remark 2.5, from Theorem 2.2 we derive

3.1. COROLLARY. Let E be a polynomially convex, compact set in $\mathbb{C}^N$ (resp. any compact set in $\mathbb{R}^N$). Let h be a holomorphic mapping in a connected open neighborhood U of E with values in $\mathbb{C}^M$ $(M \leq N)$, satisfying (H_3) (and additionally, if $E \subset \mathbb{R}^N$, such that $h(E) \subset \mathbb{R}^M$). Then if the pair (E,λ_{2N}) (resp. (E,λ_N)) satisfies (L^*), so does the pair $(h(E),\lambda_{2N})$ (resp. $(h(E),\lambda_M)$).

In order to obtain a similar corollary for condition (L_o^*) it suffices to replace in the above statement condition (H_3) by (H_4) and (L^*) by (L_o^*).

4. A criterion of the L^*-regularity

Suppose E is a compact set in $\mathbb{C}^N$ (resp. $\mathbb{R}^N$) with the following property.

(P) For every point $a \in E$ there exists a 2N-dimensional (resp. N-dimensional) parallelepiped P such that $a \in P \subset E$. Then, by Fubini's theorem, it follows from Example 1.1 that the pair (E, λ_{2N}) (resp. (E, λ_N)) satisfies (L^*). It can be shown (see [7], Attainment Criterion 6.3) that the L^*-regularity of E (with respect to the Lebesgue measure) still holds when the above condition (P) is replaced by the following weaker assumption:

(I) For every point $a \in E$ there exists a line segment I such that $a \in I$ and $I \setminus \{a\} \subset \operatorname{int} E$.

We shall now prove an essentially stronger version of the above criterion.

4.1. CRITERION (of analytic accesibility). *Let E be a compact set in $\mathbb{C}^N$ (resp. $\mathbb{R}^N$). Suppose that for every point $a \in E$ there exists a real analytic manifold M of dimension 1 such that $a \in M$ and $M \cap (E \setminus \{a\}) \subset \operatorname{int} E$. Then the pair (E, λ_{2N}) (resp. (E, λ_N)) satisfies (L^*), and consequently E is L-regular.*

P r o o f. Fix a point $a \in E$. By the assumptions there exists: a neighborhood V of a, a neighborhood W of $0 \in \mathbb{C}^N$ and a biholomorphism h of V onto W such that $h(a) = 0$ and $h(V \cap M) = W \cap \{(z_1, \dots, z_N) \in \mathbb{C}^N : y_1 = z_2 = \dots = z_N = 0\}$ where $z_j = x_j + y_j$ for $j = 1, \dots, N$. Then we have $h(V \cap M \cap (E \setminus \{a\})) \subset \operatorname{int} h(V \cap E)$. Thus, for each $r > 0$, the origin is attainable by a line-segment from the interior of the set $F_r := B(0,r) \cap h(V \cap E)$, where $B(z,r)$ denotes the closed ball with centre z and radius r, whence by [7], Attainment Criterion 6.3, each set F_r is L^*-regular at 0 with respect to the Lebesgue measure λ_{2N} (resp. λ_N, if $E \subset \mathbb{R}^N$). This means that for every family $\mathcal{F} \subset \mathcal{P}$ with $\sup\{|f(z)| : f \in \mathcal{F}\} < \infty$ a.e. on F_r, and for every $b > 1$ there exists a neighborhood U of 0 and a constant $C > 0$ (U and C may depend on r) such that

$$\|f\|_U \leq C b^{\deg f}, \quad f \in \mathcal{F}.$$

In particular, from this it follows that each set F_r (which is compact for sufficiently small r) is L-regular at 0 (i.e. the extremal function Φ_{F_r} is continuous at 0), and by [5], Theorem 3.12 (local version of Theorem 2.1 above), so is the set $B(a,s) \cap E$ for all $s > 0$ sufficiently small. We have to prove that each set $B(a,s) \cap E$ is also

L^*-regular at a with respect to the measure λ_{2N} (resp. λ_N). This, however, can be easily done by a similar argument to that of the proof of Theorem 2.2. (Observe that for sufficiently small r, the polynomial envelope of F_r, $\hat{F}_r$, is contained in W, and therefore we can apply Lemma 2.3). Thus the set E satisfies (L^*) at a (with respect to the Lebesgue measure) and an application of the Borel-Lebesgue theorem completes the proof of the criterion.

References

[1] JOSEFSON, B.: On the equivalence between locally polar and globally polar sets for plurisubharmonic functions on $\mathbb{C}^n$, Arkiv för Matematik 16 (1978), 109-115.

[2] LEJA, F.: Sur les suites de polynômes bornées presque partout sur la frontière d'un domaine, Math. Ann. 108 (1933), 517-524.

[3] NGUYEN THANH VAN: Familles de polynômes ponctuellement bornées, Ann. Polon. Math. 31 (1975), 83-90.

[4] PLEŚNIAK, W.: Quasianalytic functions in the sense of Bernstein, Dissertationes Math. 147, (1977), pp. 1-70.

[5] ——: Invariance of the L-regularity of compact sets in $\mathbb{C}^N$ under holomorphic mappings, Trans. Amer. Math. Soc. 246 (1978), 373-383.

[6] ——: A criterion of the L-regularity of compact sets in $\mathbb{C}^N$, Zeszyty Nauk. Uniw. Jagiello. 21 (1979), 97-103.

[7] ——: Invariance of some polynomial conditions for compact subsets of $\mathbb{C}^N$ under holomorphic mappings, Zeszyty Nauk. Uniw. Jagiello. 22 (to appear).

[8] SICIAK, J.: On some extremal functions and their applications in the theory of analytic functions of several complex variables, Trans. Amer. Math. Soc. 105 (1962), 322-357.

[9] ——: Extremal plurisubharmonic functions in $\mathbb{C}^N$, Procedings of the First Finnish-Polish Summer School in Complex Analysis at Podlesice, Vol. I, University of Łódź, Łódź 1977, pp.115-152.

[10] ——: On some inequalities for polynomials, Zeszyty Nauk. Uniw. Jagiello. 21 (1979), 7-10.

[11] ZAHARJUTA, V.P.: Extremal plurisubharmonic functions, orthogonal polynomials and Bernstein-Walsh theorem for analytic functions of several complex variables, Ann. Polon. Math. 33 (1976), 137-148 (Russian).

Institute of Mathematics
Jagiellonian University
Reymonta 4, PL-30-059 Kraków, Poland

THE BOUNDARY CORRESPONDENCE UNDER MAPPINGS WITH BOUNDED TRIANGULAR DILATATION IN REAL NORMED SPACES

Giovanni Porru* (Cagliari)

Contents

Introduction

Let H be a real normed space with norm $\|\cdot\|$, and let S be a subset of H . Following H. Renggli [4] we give the

Definition. The mapping $T:S \longrightarrow H$ is said to have a bounded triangular dilatation (b.t.d.) if and only if there is a real number C , $C \geq 1$, such that for every triple $X, Y, Z \in S$, the condition $\|X-Y\| \leq \|Y-Z\|$ implies $\|TX-TY\| \leq C\|TY-TZ\|$.

If $H = R^n$, then there is a strict connection between mappings with b.t.d. and quasiconformal mappings. In fact, Renggli [5] established the following result. If Ω, Ω' are domains of R^n , the homeomorphism $f:\Omega \longrightarrow \Omega'$ has locally b.t.d. if and only if it is locally quasiconformal or antiquasiconformal.

If H is a general real normed space, we can define quasiconformal mappings between domains of H , for example, by using the Gehring metric definition [1] . It is easy to show that a homeomorphism $T:\Omega \to \Omega'$ (Ω and Ω' being domains of H) with b.t.d. is quasiconformal according to the metric definition. We do not know if a quasiconformal homeomorphism $T:\Omega \longrightarrow \Omega'$ has locally b.t.d.

If Ω , Ω' are domains of R^n, the homeomorphism $f:\Omega \longrightarrow \Omega'$ is

---

* The A. is a member of GNFA of the CNR.

quasiconformal if and only if it is a Θ-mapping [2]. We can define the Θ-mappings also if Ω and Ω' are domains of H [4]. Later, we will show that if the homeomorphism $T : \Omega \longrightarrow \Omega'$ (Ω, Ω' domains of H) is a Θ-mapping, then T and T^{-1} have locally b.t.d..

Now we consider the mapping $T : S \longrightarrow H$ $(S \subset H)$ with b.t.d.. Following H. Renggli [5] we could show that if $H = R^n$, and if S is unbounded, then either T is constant on S, or

$$\lim_{\substack{X \in S \\ \|X\| \to \infty}} \|TX\| = \infty.$$

If H has infinite dimension, this result is not true, as we may see in Example 2 of [3]. Further, if $H = R^n$, then T is continuous on S. Also this latter result is not true if H has not finite dimension, as we may see in Example 1 of [3]. By this example it is easy to note also that if $T : S \longrightarrow S'$ (S and S' being subsets of an infinite dimensional real normed space H) has b.t.d., then, the reverse mapping $T^{-1} : S' \longrightarrow S$ has not, in general, b.t.d.. Therefore, if we wish to have a class of mappings defined on a subset of H, with properties similar to these we have when $H = R^n$, it is necessary to restrict the previous definition. That is we consider mappings $T : S \longrightarrow S'$ such that T and T^{-1} have b.t.d..

1. The boundary correspondence

We begin by establishing some preliminary lemmas.

LEMMA 1.1. *Let* $T : S \longrightarrow H$ *have* b.t.d., *and let* $X_i, X_o \in S$ $(i = 1,2,\dots)$ *with* $\lim_i X_i = X_o$. *If* $\{TX_i\}$ *is relatively compact in* H, *then*

$$\lim_i TX_i = TX_o.$$

Proof. Let $\{TX_{i_k}\}$ be a Cauchy sequence. For $\varepsilon > 0$, let $n = n(\varepsilon)$ be such that, if $k > h = n(\varepsilon)$:

$$\| TX_{i_k} - TX_{i_h} \| < \varepsilon / C^2.$$

Further, let $m(h) = \bar{n}(\varepsilon) \geq n(\varepsilon)$ be such that, for $k > m(h)$

$$\| X_o - X_{i_k} \| \leq \| X_{i_k} - X_{i_h} \| .$$

From the latter it follows, for $k > \bar{n}(\varepsilon)$

$$(1.1) \quad \| TX_o - TX_{i_k} \| \leq C \| TX_{i_k} - TX_{i_h} \| \leq \varepsilon/C.$$

Now fix i_k such that (1.1) holds and take $\nu = \nu(i_k)$ such that, for $i > \nu$:

$$\| X_i - X_o \| \leq \| X_o - X_{i_k} \| .$$

Hence, recalling (1.1), we obtain

$$\| TX_i - TX_o \| \leq C \| TX_o - TX_{i_k} \| \leq \varepsilon,$$

and the lemma is proved.

LEMMA 1.2. *Let* $T : S \longrightarrow H$ *have* b.t.d., *and let* $X_i \in S$ $(i = 1,2,\ldots)$. *If* $\{TX_i\}$ *is a non constant Cauchy sequence, then* $\{X_i\}$ *is bounded.*

P r o o f. Suppose $\{X_i\}$ is unbounded. Then there exists a subsequence $\{X_{i_k}\}$ such that

$$\| X_1 - X_{i_k} \| \leq \| X_{i_k} - X_{i_{k+1}} \| ,$$

from which it follows that

$$(1.2) \quad \| TX_1 - TX_{i_k} \| \leq C \| TX_{i_k} - TX_{i_{k+1}} \| .$$

Since $\{TX_{i_k}\}$ is a Cauchy sequence, (1.2) implies

$$\lim_k TX_{i_k} = TX_1.$$

Let X_{i_r} be such that $TX_{i_r} \neq TX_1$. Arguing as before, we prove that

$$\lim_k TX_{i_k} = TX_r.$$

But, as $TX_1 \neq TX_r$, we obtain a contradiction, and the lemma follows.

LEMMA 1.3. *Let* $T : S \longrightarrow H$ *have* b.t.d., *where* S *is a convex set of* H. *If* $X,Y,Z \in S$ *are such that*

$$\| X - Y \| \leq m \| Y - Z \| \qquad (m \text{ - a positive integer}),$$

then we have

(1.3) $\qquad \| TX - TY\| \le C \frac{C^m-1}{C-1} \| TY - TZ\|$,

where we let $\frac{C^m-1}{C-1} = m$ *if* $C = 1$.

P r o o f. The inequality (1.3) is true (by definition) if $m=1$. Suppose that it holds for $m-1$. If $\| X - Y\| \le \| Y - Z\|$, the lemma is trivial. Assume $\| X - Y\| > \| Y - Z\|$ and consider the point

$$X_o = Y + \frac{\| Y - Z\|}{\| X - Y\|} (X - Y) .$$

We have

$$m\| X_o - Y\| = m\| Y - Z\| \ge \| X - Y\| = \| X - X_o\| + \| X_o - Y\| ,$$

that is

$$\| X - X_o\| \le (m-1)\| X_o - Y\| .$$

Hence, by the hypothesis it follows that

(1.4) $\qquad \| TX - TX_o\| \le C \frac{C^{m-1}-1}{C-1} \| TX_o - TY\|$.

Since $\| X_o - Y\| = \| Y - Z\|$, we have also

$$\| TX_o - TY\| \le C\| TY - TZ\| .$$

Using the latter inequality and (1.4) we obtain

$$\| TX - TY\| \le \| TX - TX_o\| + \| TX_o - TY\| \le (C^2 \frac{C^{m-1}-1}{C-1} + C) \| TY - TZ\| ,$$

and (1.3) follows.

LEMMA 1.4. *Let* $T : S \longrightarrow H$ *have b.t.d. with* S *a convex set, and let* $X_i \in S$ $(i = 1,2,\ldots)$. *If* $\{TX_i\}$ *is a non constant Cauchy sequence, then* $\{X_i\}$ *is also a Cauchy sequence.*

P r o o f. Suppose that $\{X_i\}$ is not a Cauchy sequence. Let $r > 0$ be such that

(1.5) $\qquad \| X_{i_h} - X_{i_k}\| \ge r, \quad h,k = 1,2,\ldots$

As $\{TX_{i_h}\}$ is a non constant Cauchy sequence, by Lemma 1.2 there exists $L > 0$ such that

(1.6) $\| X_1 - X_{i_h} \| \leq L, \qquad h = 1,2,\dots$

The inequalities (1.5) and (1.6) imply:

(1.7) $\| X_1 - X_{i_h} \| \leq \frac{L}{r} \| X_{i_h} - X_{i_k} \| .$

If m is an integer such that $m \geq L/r$, from (1.7) and Lemma 1.3 we obtain

$$\| TX_1 - TX_{i_h} \| \leq C \frac{C^m - 1}{C-1} \| TX_{i_h} - TX_{i_k} \| .$$

Hence

(1.8) $\lim_h TX_{i_h} = TX_1 .$

Taking X_2 in (1.6) instead of X_1, and arguing as above, we obtain

$$\lim_h TX_{i_h} = TX_2 .$$

This result and (1.8) are in contradiction because a non constant mapping with b.t.d. on a connected set is injective [3]. The lemma is therefore proved.

Now we may prove the following

THEOREM 1.1. *Let S, S' be two convex sets of H, and let $T : S \longrightarrow S'$ be a bijective mapping such that T and T^{-1} have b.t.d. Then T is a homeomorphism of S onto S'.*

P r o o f. Let X_i, $X_o \in S$ $(i = 1,2,\dots)$, and let $\lim X_i = X_o$. By Lemma 1.4 applied to T^{-1}, $\{TX_i\}$ is a Cauchy sequence. By Lemma 1.1 we have $\lim TX_i = TX_o$; hence $T : S \longrightarrow S'$ is continuous. In the same way, it can be proved that also $T^{-1} : S' \longrightarrow S$ is continuous.

In order to prove the continuability on the boundary, we need the following

LEMMA 1.5. *Let S, S' be two convex sets of H, and let $T : S \longrightarrow S'$ be a bijective mapping such that T and T^{-1} have b.t.d. If $X_i, Y_i \in S$ $(i = 1,2,\dots)$, if $\{X_i\}$ and $\{Y_i\}$ are Cauchy sequences and if $\lim \| X_i - Y_i \| = 0$, then, $\{TX_i\}$ and $\{TY_i\}$ are Cauchy sequences and $\lim \| TX_i - TY_i \| = 0$.*

P r o o f. The first part of this lemma is a consequence of Lemma 1.4. Now, for $i = 1,2,\dots$ we choose $i_k > i$ and such that (we may suppose $\lim Y_{i_k} \neq Y_i$, $i = 1,2,\dots$):

$$\| X_{i_k} - Y_{i_k} \| \leq \| Y_{i_k} - Y_i \|$$

from which it follows that

$$\| TX_{i_k} - TY_{i_k} \| \leq C \| TY_{i_k} - TY_i \| .$$

If we take $\varepsilon > 0$ and choose $n = n(\varepsilon)$ such that, for $i > n$,

$$\| TY_{i_k} - TY_i \| < \varepsilon / C,$$

we have $\| TX_{i_k} - TY_{i_k} \| < \varepsilon$, and the lemma follows.

Finally we may prove the main result of this section.

THEOREM 1.2. *Suppose that H is a real Banach space, and suppose that S,S' are two convex domains of H. If $T : S \longrightarrow S'$ is a bijective mapping such that T and T^{-1} have b.t.d., then we can define a homeomorphism $\bar{T} : \bar{S} \longrightarrow \bar{S}'$ (we denote by $\bar{S}$ the colosure of S) such that $\bar{T}$ and $\bar{T}^{-1}$ have b.t.d. and such that the restriction of $\bar{T}$ to S equals T.*

Proof. Let $P \in \partial S$ (= the boundary of S), and let $X_i \in S$ $(i = 1,2,\ldots)$ with $\lim X_i = P$. By Lemma 1.4, $\{TX_i\}$ is a Cauchy sequence, and, because H is complete, there exists a $P' \in \partial S'$ such that $\lim TX_i = P'$. We define $\bar{T}P = P'$. By Lemma 1.5 such a definition does not depend on the sequence $\{X_i\}$; in fact, if $Y_i \in S$ and if $\lim Y_i = P$, then $\lim TY_i = P'$. Using Lemma 1.5 we see also that if $P, Q \in \partial S$ and $P \neq Q$, then $\bar{T}P \neq \bar{T}Q$. Hence we can define a bijective mapping $\bar{T}$ of $\bar{S}$ on $\bar{S}'$. Such a mapping has b.t.d. In fact, if $X,Y,Z \in \bar{S}$, and $\| X - Y \| \leq \| Y - Z \|$, we may find three sequences $X_i, Y_i Z_i \in S$ such that $\lim X_i = X$, $\lim Y_i = Y$, $\lim Z_i = Z$ and such that $\| X_i - Y_i \| \leq \| Y_i - Z_i \|$. From this we have

$$\| TX_i - TY_i \| \leq C \| TY_i - TZ_i \| .$$

Hence, by Lemma 1.4, we obtain:

$$\| \bar{T}X - \bar{T}Y \| \leq C \| \bar{T}Y - \bar{T}Z \| .$$

In the same way it can be proved that $\bar{T}^{-1}$ has b.t.d. The conclusion is now a consequence of Theorem 1.1.

Remark. The hypothesis of convexity of S used in Lemma 1.3, and consequently in Theorems 1.1 and 1.2, can be weakened. In fact, for the proof of Lemma 1.3 it is sufficient to suppose that, if $X,Y,Z \in S$, there exist $(h-1)$ points $Y_1, Y_2, \ldots, Y_{h-1} \in S$, with h being an increasing function of the ratio $\|X - Y\| / \|Y - Z\|$, such that

$$\|X - Y_1\| = \|Y_1 - Y_2\| = \ldots = \|Y_{h-1} - Y\| \leq \|Y - Z\| .$$

Then, if $T : S \longrightarrow H$ has b.t.d. and if $\|X - Y\| \leq m\|Y - Z\|$, we have

$$\|TX - TY_1\| \leq C\|TY_1 - TY_2\| \leq \ldots \leq C^{h(m)}\|TY - TZ\| .$$

From the latter we obtain

$$\|TX - TY\| \leq \|TX - TY_1\| + \|TY_1 - TY_2\| + \ldots + \|TY_{h-1} - TY\|$$

$$\leq (C^{h(m)} + C^{h(m)-1} + \ldots + C)\|TY - TZ\| .$$

Hence

$$\|TX - TY\| \leq C\,\frac{C^{h(m)}-1}{C-1}\|TY - TZ\| \quad \left(\text{where } \frac{C^{h(m)}-1}{C-1} = h(m) \text{ if } C=1\right).$$

We observe that if S is a convex set, then we may take $h(m) = m$. The previous condition, with h suitable, is satisfied if S is a connected set formed by the union of a finite number of convex sets.

2. Connection between mappings with bounded triangular dilatation and Θ-mappings

The Θ-mappings were introduced for $H = R^n$ by Gehring [2]. For general H, the definition of Θ-mappings has been given in [4]. We recall:

Definition. A homeomorphism T of S onto S' (S and S' being domains of H) is said to be a Θ-mapping if there exists a function $\Theta(t)$ which is continuous and increasing in $0 \leq t < 1$ with $\Theta(0) = 0$, such that the following conditions hold:

i) If $X_o \in S$ and $\|X - X_o\| < \inf_{Y \in \partial S} \|X_o - Y\| \equiv \rho(X_o, \partial S)$, then

$$\frac{\|TX - TX_o\|}{\rho(TX_o, \partial S')} \leq \Theta\left(\frac{\|X - X_o\|}{\rho(X_o, \partial S)}\right);$$

ii) the restriction of T to any subdomain $\Delta \subset S$ satisfies i).

Now we prove the following

<u>THEOREM 2.1</u>. <u>Let</u> $T : S \longrightarrow S'$ (S <u>and</u> S' <u>being domains of</u> H) <u>be a</u> Θ-<u>mapping. Then, for every</u> $X_o \in S$, <u>there exists a neighbourhood</u> $B(X_o,r_o)$ <u>such that</u> T <u>has b.t.d. on</u> $B(X_o,r_o)$.

<u>Proof</u>. Let $X_o \in S$ and $0 < r < \frac{1}{2}\rho(X_o,\partial S)$. Let $X,Y,Z \in B(X_o,r/2)$ be such that $\|X - Y\| \leq \|Y - Z\|$. Define

$$M = \frac{X+Y}{2}, \qquad N = \frac{3Y-X}{2}.$$

Obviously M, $N \in B(X_o,r)$. Let $\Delta = S - \{N\}$, $\Delta' = S' - \{TN\}$. We have

$$\rho(M,\partial\Delta) = \|M-N\|, \quad \|X-M\| = \|X-Y\|/2 = \|M-N\|/2 = \rho(M,\partial\Delta)/2,$$

$$\rho(TM,\partial\Delta') \leq \|TM - TN\|.$$

Hence

$$\frac{\|TX-TM\|}{\rho(TM,\partial\Delta')} \leq \Theta\left(\frac{\|X-M\|}{\rho(M,\partial\Delta)}\right),$$

what implies

$$(2.1) \qquad \|TX - TM\| \leq \Theta(\tfrac{1}{2})\, \|TM - TN\|.$$

In the same way we obtain

$$(2.2) \qquad \|TY - TM\| \leq \Theta(\tfrac{1}{2})\, \|TM - TN\|.$$

From (2.1) and (2.2) we derive

$$(2.3) \qquad \|TX - TY\| < 2\Theta(\tfrac{1}{2})\, \|TM - TN\|.$$

Now let $\Delta = S - \{Z\}$, $\Delta' = S' - \{TZ\}$. We have

$$\rho(Y,\partial\Delta) = \|Y-Z\| \geq \|X-Y\|, \quad \|M-Y\| = \|X-Y\|/2 \leq \rho(Y,\partial\Delta)/2,$$

$$\rho(TY,\partial\Delta') \leq \|TY - TZ\|.$$

Hence

$$\frac{\|TM-TY\|}{\rho(TY,\partial\Delta')} \leq \Theta\left(\frac{\|M-Y\|}{\rho(Y,\partial\Delta)}\right)$$

implies

(2.4) $\quad \| TM - TY\| < \Theta(\frac{1}{2}) \| TY - TZ\|$.

Analogously we obtain

(2.5) $\quad \| TN - TY\| < \Theta(\frac{1}{2}) \| TY - TZ\|$.

From (2.4) and (2.5) we have

$$\| TM - TN\| < 2\Theta(\tfrac{1}{2}) \| TY - TZ\|.$$

The latter inequality and (2.3) imply

$$\| TX - TY\| < 4\Theta^2(\tfrac{1}{2}) \| TY - TZ\| ,$$

and the theorem is proved.

R e m a r k. With the method used in the proof of Theorem 2.1 we can prove that if $T : H \longrightarrow H$ is a Θ-mapping, then T has b.t.d. on H.

We complete this section with the following

THEOREM 2.2. *Let* $T^{-1} : S' \longrightarrow S$ *(*S *and* S' *being domains of* H*) be a* Θ*-mapping. Then, for every* $X_o \in S$*, there exists a neighbourhood* $B(X_o,r_o)$ *such that* T *has b.t.d. on* $B(X_o,r_o)$.

P r o o f. Let $X_o \in S$ and $0 < r < \rho(X_o,\partial S)/3$ be such that, for $X,Y \in B(X_or)$, $\| TX - TY\| \le \rho(TY,\partial S')$. Let $X,Y,Z \subset B(X_o,r)$ under the condition $\| X - Y\| \le \| Y - Z\|$. Let $\Delta = S - \{X\}$. We have $\rho(Y,\partial\Delta) = \| Y - X\|$, $\rho(TY,\partial\Delta') = \| TY - TX\|$. We may write

$$1 \le \frac{\| Z - Y\|}{\| Y - X\|} = \frac{\| T^{-1}(TZ)-T^{-1}(TY)\|}{\rho(T^{-1}(TY),\partial\Delta)} \le \Theta(\frac{\| TZ-TY\|}{\rho(TY,\partial\Delta')}) = \Theta(\frac{\| TZ-TY\|}{\| TY-TX\|}) .$$

As $\Theta(t)$ is increasing, we have

$$\| TX - TY\| \le (\Theta^{-1}(1))^{-1}\| TY - TZ\| ,$$

and the theorem is proved.

COROLLARY. *Let* $T : S \longrightarrow S'$ *be a* Θ*-mapping. Then,* T *and* T^{-1} *have locally b.t.d.*

3. Composition of mappings with bounded triangular dilatation

If $T : S \longrightarrow S'$ is a bijective mapping such that T and its inverse have b.t.d., then $T^{-1} : S' \longrightarrow S$ *is a bijective mapping such* that T^{-1} and its inverse have b.t.d. (by definition). The identity mapping $TX = X$ has b.t.d. Further, we have the following

THEOREM 3.1. *Let* S, S', S'' *be three convex sets of* H. *If* $T_1 : S \longrightarrow S'$ *and* $T_2 : S' \longrightarrow S''$ *are mappings with b.t.d., then* $T_2\ T_1 : S \longrightarrow S''$ *is a mapping with b.t.d.*

Proof. Let $X, Y, Z \in S$ be such that $\|X - Y\| \leq \|Y - Z\|$. We have

$$(3.1)\qquad \|T_1X - T_1Y\| \leq C_1 \|T_1Y - T_1Z\| .$$

Let m be an integer with $m \geq C_1$. Lemma 1.3 and the estimate (3.1) imply

$$\|T_2T_1X - T_2T_1Y\| \leq C_2 \frac{C_2^m - 1}{C_2 - 1} \|T_2T_1Y - T_2T_1Z\| .$$

Hence the theorem follows.

Remark. The mappings T defined in convex sets, such that T and T^{-1} have b.t.d., form a group. Obviously, we may consider more general sets as we observe in the Remark of the first section.

References

[1] CARAMAN, P.: Quasiconformal mappings in real normed space, Rev. Roum. Math. Pure Appl.24 (1979), 33-78.

[2] GEHRING, F.W.: The Carathéodory convergence theorem for quasiconformal mappings in space, Ann. Acad. Sci. Fenn. A.I. 336/11 (1963), 21 pp.

[3] PORRU, G.: Mappe con distorsione triangolare limitata in spazi normati, Boll. Un. Mat. It. 14A (1977), 599-607.

[4] ——: O-mappings and quasiconformal mappings in normed spaces, Rend. Sem. Mat. Univ. Padova 57 (1977), 173-182.

[5] RENGGLI, H.: Deppelverhältnisse und quasikonforme Abbildungen, Comment. Math. Helv. 43 (1968), 161-175.

[6] ——: On triangular dilatation, Proc. Romanian-Finnish Seminar on Teichmüller Spaces and Quasiconformal Mappings, Braşov, Romania (1969).

Istituto di Matematica
Università di Cagliari
I-09100 Cagliari, Italia

P-REGULARITY OF SETS IN $\mathbb{C}^n$

Azimbaĭ Sadullaev (Taškent)

Summary. In this paper the problem of P-regularity of compacts in $\mathbb{C}^n$ is considered. With the help of the notion of P-regularity a sufficient condition for P-regularity of compacts is given. An example of a nonregular Jordan domain in the real plane $\mathbb{R}^2 = \{(z,w) \in \mathbb{C}^2 : \operatorname{Im} z = \operatorname{Im} w = 0\}$ is constructed.

In analogy to the notion of regularity of compacts in the Euclidean space $\mathbb{R}^n$ [1], the so-called P-regularity *) of a compact set K situated in the complex space $\mathbb{C}^n$ is defined with help of the extremal function of Siciak [2]:

(1) $$\Phi(z,K) = \sup\{|P(z)|^{1/\deg P}\},$$

where the supremum is taken over all polynomials $P(z)$ satisfying the condition $|P(z)| \leq 1$ on K. The equality

(2) $$\ln \Phi(z,K) = V(z,K) \overset{\text{def}}{=} \sup_{u \in L(K)} u(z)$$

has been proved by Zaharjuta [5], [6] in the case of compact sets K such that the function $\Phi(z,K)$ is continuous in $\mathbb{C}^n$ and by Siciak [7] in the general case.

Let us denote by $L(K)$ the class of such plurisubharmonic functions $u(z)$ in $\mathbb{C}^n$ that $u(z) \leq 0$ on K and $\overline{\lim_{z\to\infty}}[u(z)/\ln|z|] \leq 1$, where $|z| = (|z_1|^2 + \dots + |z_n|^2)^{\frac{1}{2}}$ is the Euclidean norm.

0.1. Definition. (cf. [7], [8], [9]). A compact set $K \subset \mathbb{C}^n$ is called P-regular **) at the point $z^0 \in K$ if the function

*) In the papers of Siciak and Zaharjuta the terms "L-regularity" and "C^n-regularity" are used. Here we utilize the term "P-regularity" in connection with the terms "P-measure" and "P-capacity", introduced by the present author in [3] and [4].

**) In the case $n = 1$ the above definition of regularity slightly differs from the classical one. However, for a polynomially convex compact set both definitions are equivalent.

(3) $V^*(z,K) = \overline{\lim_{w \to z}} V(w,K)$

equals zero at this point.

If K is regular at its every point, then $V^*(z,K)$ is continuous in $\mathbb{C}^n$ and the identity $V^*(z,K) \equiv V(z,K)$ holds (cf. [6], [7]).

0.2. Remark. Analogously, the notion of P-regularity may be introduced for an arbitrary set E and a point z^0 of the closure $\bar{E}$.

The regularity of a compact set plays an important role in the potential theory and in problems of approximation of functions by polynomials (cf. [10], [2], and [7]), also in the problems concerned with separate analytic functions (cf. [11] and [12]) etc.

There is a series of tests allowing to state the regularity of plane sets (cf. [1]).

Notice the following sufficient conditions of P-regularity of compact sets in $\mathbb{C}^n$.

Siciak [7] proved that if $K = K_1 \times K_2$ and $z^{(j)} \in K_j$ is a regular point of the compact set K_j, $j = 1, 2$, then the point $(z^{(1)}, z^{(2)})$ is a regular point of the compact set K.

For compact sets $K \subset \mathbb{R}^n = \{\operatorname{Im} z_j = 0,\ j = 1, 2, \ldots, n\}$ Gončar observed that if an open segment (a, z^0) is contained in the interior of K, i.e. $(a, z^0) \subset \operatorname{int} K$, then z^0 is a regular point of K.

In the papers [13] and [14] of Pleśniak some properties of P-regular points of compact sets in C^n are proved. In particular, the invariance of P-regularity by open biholomorphic mappings is proved and certain conditions for P-regularity are given.

A number of sufficient conditions for P-regularity of compact sets in $\mathbb{C}^n$ are to be found in the papers [15] and [16].

In the present paper with the help of the notion of P-regularity a condition for P-regularity of an arbitrary set $E \subset \mathbb{C}^n$ is given (Section 1). This condition is close to the Gončar and Pleśniak conditions. In Section 2 a Jordan domain $D \subset \mathbb{R}^2 = \{(z,w) \in \mathbb{C}^n : \operatorname{Im} z = \operatorname{Im} w = 0\}$ is constructed so that $\bar{D}$ is not a regular compact set. This problem was verbally proposed by Siciak.

1. 1.1. Definition. A set $E \subset \mathbb{C}^n$ is called P-separated at the point $z^0 \in \bar{E}$ if either z^0 is an isolated point of E or z^0 is not isolated, but there exist a neighbourhood $U \ni z^0$ and a plurisubharmonic function $u(z)$ in U such that $\overline{\lim_{z \to z^0,\, z \in E}}\, u(z) < u(z^0)$.

It is not difficult to see that the property for a set E to be P-separated at a point z_0 is a local property of this set. Besides, in Definition 1.1 we may suppose that $u(z) \in L$, i.e. $\overline{\lim}_{z \to \infty} [u(z)/\ln|z|] \leq 1$.

If K is a polynomially convex compact set in $\mathbb{C}$, then it is separated at $z^o \in K$ if and only if z^o is not a regular point of K (cf. [1]). In particular, an arbitrary continuum K is not separated at any point $z^o \in K$.

As the example of Favorov [17] shows, in the space $\mathbb{C}^n$ the situation is quite different. Moreover, as we will see below, the notions of P-separability and P-regularity in $\mathbb{C}^n$, $n > 1$, are essentially different.

1.2. Definition. The set $\Gamma = \{z = z(t), \alpha < t < \beta\}$, where $\alpha < \beta$ and $z(t) = (z_1(t), \ldots, z_n(t))$ is an analytic vector-function of the real parameter t, is called a real analytic curve in $\mathbb{C}^n$.

Hereafter, the trivial case $z(t) \equiv \text{const}$ will be beyond any consideration.

1.3. PROPOSITION. A real analytic curve Γ is not P-separated at any point $z^o \in \Gamma$.

Proof. Let $u(z) \in L$ be an arbitrary plurisubharmonic function. Since $z(t)$ is a real-analytic function, then there exists a neighbourhood $G \supset (\alpha, \beta)$, $G \subset \mathbb{C}$, and a holomorphic vector-function $\tilde{z}(w)$ such that $\tilde{z}(w)|_{w=t} \equiv z(t)$. The function $u \circ \tilde{z}(w)$ is subharmonic in G. Therefore for an arbitrary point $t^o \in (\alpha, \beta)$ we have

$$\overline{\lim_{t \to t^o}} u \circ \tilde{z}(t) = u \circ \tilde{z}(t^o)$$

since, as we observed before, every continuum on the plane is not separated. Consequently, $\overline{\lim}_{z \to z^o, z \in \Gamma} u(z) = u(z^o)$, q.e.d.

Notice that an arbitrary real analytic curve in $\mathbb{C}^n$, $n > 1$, is a P-polar set; therefore it cannot be P-regular.

The proposition proved allows us to obtain the following condition of regularity.

1.4. THEOREM. Let $E \subset \mathbb{C}^n$ be an arbitrary set. If there exists a real analytic curve $z = z(t)$, $t \in (\alpha, \beta)$, such that for a certain $t^o \in (\alpha, \beta)$ the point $z(t)$ is a regular point of E for every $t < t^o$, then the point $z^o = z(t^o)$ is also a regular point of E.

2. In this section we are going to construct an example of a nonregular compact set $K \subset \mathbb{C}^2$, bounded by a Jordan curve lying in the real plane $\mathbb{R}^2 \subset \mathbb{C}^2$. We denote by $(z, w) = (x + iy, u + iv)$ the variables in $\mathbb{C}^2$ and consider the real plane $\mathbb{R}^2 = \{y = v = 0\}$ in $\mathbb{C}^2$.

2.1. LEMMA. Let

$$f(z)=\sum_{k=0}^{\infty}a_{n_k}z^{n_k}$$

be a gap Ostrowski series (i.e. $n_k/n_{k+1}\to 0$ as $k\to\infty$) with the radius of convergence $R=1$. Then the graph $A=\{w-f(z)=0\}\subset \mathbb{C}^2$ is P-separated at every boundary point $(z_o,w_o)\in\bar{A}$, where $|z_o|=1$.

Proof. Let

$$P_{n_m}(z)=\sum_{k=0}^{m}a_{n_k}z^{n_k}. \tag{4}$$

If $M=\max_{|z|=\frac{2}{3}}|f(z)|$, then for $|z|\le\frac{1}{3}$ the inequalities

$$|f(z)-P_{n_m}(z)|\le\sum_{k=m+1}^{\infty}|a_{n_k}|\cdot\frac{1}{3^{n_k}}\le M\sum_{k=m+1}^{\infty}\frac{1}{2^{n_k}}\le\frac{2M}{2^{n_{m+1}}} \tag{5}$$

hold. Clearly, in the bidisc $U(0,\frac{1}{3})=\{|z|<\frac{1}{3},\ |w|<\frac{1}{3}\}$ we have

$$|w-P_{n_m}(z)|\le|w|+|P_{n_m}(z)|\le\frac{1}{3}+M+\frac{2M}{2^{n_{m+1}}}\le C \tag{6}$$

where C is a constant independent of m. Consequently, by the Bernstein-Walsh inequality [7], the estimate

$$|w-P_{n_m}(z)|^{1/n_m}\le 3c\tau \tag{7}$$

holds in any bidisc $U(0,\tau)$, $\tau\ge\frac{1}{3}$.

Consider the sequence

$$u_m(z,w)=\frac{1}{n_{m+1}}\ln|w-P_{n_m}(z)|. \tag{8}$$

It has the following properties:

(a) If $(z,w)\in A$, $|z|\le\frac{1}{3}$, then, beginning from some m_o the inequality $u_m(z)\le\ln\frac{2}{3}$ holds. Indeed, by (5), we have

$$|w-P_{n_m}(z)|^{1/n_{m+1}}=|f(z)-P_{n_m}(z)|^{1/n_{m+1}}\le\frac{1}{2}(2M)^{1/n_{m+1}}\le\frac{2}{3}.$$

(b) For every $\tau>0$ and every $\varepsilon>0$ there exists a number m_o such that for $m>m_o$ and $(z,w)\in U(0,\tau)$ the inequality $u_m(z,w)<\varepsilon$ holds. The proof follows from the relation (7).

(c) If $|z_o|=1$, then for an arbitrary w_o we have

$$\overline{\lim_{m\to\infty}}\,u_m(z_o,w_o)=0.$$

In fact, suppose that $\overline{\lim\limits_{m\to\infty}}\,\frac{1}{n_{m+1}}\ln|w_o-P_{n_m}(z_o)|<\alpha<0$. Then, beginning from some m_o we have

$$|w_o - P_{n_m}(z_o)| < e^{\alpha n_{m+1}}.$$

Hence it follows that

$$P_{n_m}(z_o) \to w_o \quad \text{and} \quad |\sum_{k=m+1}^{\infty} a_{n_k} z_o^{n_k}| < e^{\alpha n_{m+1}}.$$

Therefore

$$(9) \qquad |a_{n_{m+1}}| - |\sum_{k=m+2}^{\infty} a_{n_k} z_o^{n_k}| \le |\sum_{k=m+1}^{\infty} a_{n_k} z_o^{n_k}| < e^{\alpha n_{m+1}}$$

and, consequently,

$$|a_{n_{m+1}}| \le e^{\alpha n_{m+1}} + e^{\alpha n_{m+2}} \le 2e^{\alpha n_{m+1}}$$

what contradicts $R = 1$.

Fix now a point $(z_o, w_o) \in \bar{A}$, $|z_o| = 1$, and consider a subsequence u_{m_j} satisfying the conditions

$$u_{m_j}(z_o, w_o) > -\frac{1}{2^{j+1}} \quad \text{and} \quad u_{m_j}(z, w) < \frac{1}{2^{j+1}} \quad \text{in the bidisc } U(0,j).$$

Let

$$(10) \qquad F(z,w) = \sum_{j=1}^{\infty} (u_{m_j}(z,w) - \frac{1}{2^{j+1}}).$$

The sum $F(z,w)$ expresses a plurisubharmonic function since for every τ there exists an index $j_o(\tau)$ such that for $j > j_o(\tau)$ terms

$$u_{m_j}(z,w) - \frac{1}{2^{j+1}}$$

of the series (10) are negative and

$$F(z_o, w_o) > -\sum_{j=1}^{\infty} \frac{1}{2^j} = -1, \quad \text{i.e. } F \not\equiv -\infty.$$

If $(z,w) \in A$, $|z| < \frac{1}{3}$, then, by (a),

$$F(z,w) = \sum_{m_j \le m_o} (u_{m_j}(z,w) - \frac{1}{2^{j+1}}) + \sum_{m_j > m_o} (u_{m_j}(z,w) - \frac{1}{2^{j+1}})$$

$$\le \sum_{m_j \le m_o} (u_{m_j}(z,w) - \frac{1}{2^{j+1}}) + \sum_{m_j > m_o} (\ln\frac{2}{3} - \frac{1}{2^{j+1}}) = -\infty.$$

Clearly, $F(z,w) = -\infty$ everywhere on A. Thus

(11) $$\overline{\lim_{\substack{(z,w)\to(z_o,w_o)\\ (z,w)\in A}}} F(z,w) < F(z_o,w_o).$$

This completes the proof.

2.2. Remark. By a slightly different procedure in obtaining Lemma 2.1 it is possible to construct a function $F(z,w)\in L$, satisfying the condition (11).

2.3. Remark. From Lemma 2.1 it is easy to obtain the following familiar result: If a gap series satisfies the condition $n_k/n_{k+1} \to 0$ as $k\to\infty$, then the region of convergence of the sum $f(z)$ is the unit disc $\{|z|<1\}$.

2.4. Example. Let

(12) $$f(z) = \sum_{k=0} \frac{1}{(k!)^{\ln k!}} z^{k!}.$$

The function $f(z)$ is holomorphic in the unit disc and smooth on its closure. By Lemma 2.1 the graph A is P-separated at every point $(z_o, f(z_o))$, $|z_o| = 1$.

We are going to construct a function $F(z,w)\in L$, such that $F(z,w) = -\infty$ on A, but $F(1, f(1)) \geq -1$. Let $G = \{(z,w)\in \mathbb{C}^2\colon F(z,w) < -2\}$. Then G is a neighbourhood of A and the intersection $G\cap \mathbb{R}^2$, where $\mathbb{R}^2 = \mathbb{R}^2(x,u) = \{(z,w)\in\mathbb{C}^2\colon y=v=0\}$, is a neighbourhood (in $\mathbb{R}^2$) of the curve $u=f(x)$, $0\leq x<1$. Since $F(1, f(1))\geq -1$ then $(1, f(1))\notin G\cap\mathbb{R}^2$. As $f(x)\in C^\infty$ on the segment $[0,1]$, then there exists a domain $D\subset G\cap\mathbb{R}^2$ such that D is also a neighbourhood of the curve $u=f(x)$, $0\leq x<1$, and the boundary D is smooth everywhere except perhaps for a single point $(1, f(1))$. Besides, the domain D may be chosen so that $\bar{D}\setminus(1, f(1))\subset G$. Let $K=\bar{D}$. Then we have (cf. [7], Proposition 3.11):

$$V^*[(z,w);K] = V^*[(z,w);K\setminus(1, f(1))].$$

Since $F(z,w)+2\in L(K\setminus(1, f(1)))$, then $V^*[(z,w);K]\geq F(z,w)+2$. At the point $(1, f(1))$ we have $V^*[(1, f(1));K]\geq F(1, f(1))+2\geq 1$, i.e. $(1, f(1))$ is an irregular point of the compact set K. Notice that all other points of K are regular.

Институт математики
им. В. И. Романовского
Академии наук УзССР
Астрономический пул. 11
SU-700 052 Ташкент 52, СССР

References

[1] N. S. LANDKOF: Fundamentals of modern potential theory [in Russian], Izd. "Nauka", Moscow 1966.

[2] J. SICIAK: On some extremal functions and their applications in the theory of analytic functions of several complex variables, Trans. Amer. Math. Soc. 105 (1962), no. 2, 322-357.

[3] A. SADULLAEV: A boundary uniqueness theorem in $\mathbb{C}^n$ [in Russian], Mat. Sb. (N. S.) 101 (143) (1976), no. 4, 568-583.

[4] A. SADULLAEV: Defect divisors in the sense of Valiron [in Russian], Mat. Sb. (N. S.) 108 (150) (1979), no. 4, 567-580.

[5] V. P. ZAHARJUTA: Spaces of analytic and harmonic functions of several variables [in Russian], Tezisy Dokladov na Vsesojuznoĭ Konferencii po Teorii Funkciĭ, Harkov 1971, pp. 74-78.

[6] V. P. ZAHARJUTA: Extremal plurisubharmonic functions, orthogonal polynomials, and the Bernšteĭn-Walsh theorem for functions of several complex variables [in Russian], Proceedings of the Sixth Conference on Analytic Functions (Kraków, 1974), Ann. Polon. Math. 33 (1976/77), no. 1-2, 137-148.

[7] J. SICIAK: Extremal plurisubharmonic functions in $\mathbb{C}^n$, Proceedings of the First Finnish-Polish Summer School in Complex Analysis at Podlesice, University of Łódź, Łódź 1977, pp. 115-152.

[8] J. SICIAK: Separately analytic functions and envelopes of holomorphy of some lower dimensional subsets of $\mathbb{C}^n$, Ann. Polon. Math. 22 (1969), 145-171.

[9] V. P. ZAHARJUTA: Extremal plurisubharmonic functions, Hilbert scales, and the isomorphism of spaces of analytic functions of several variables I [in Russian], Teor. Funkciĭ Funkcional. Anal. i Priložen. Vyp. 19 (1974), 133-157 and 161.

[10] J. L. WALSH: Interpolation and approximation by rational functions in the complex domain, 2nd ed. (American Math. Soc. Colloq. Publ. 20), American Math. Soc., Boston 1960; Russian translation: Izdatelstvo Inostrannoĭ Literatury, Moscow 1961.

[11] J. SICIAK: Analyticity and separate analyticity of functions defined on lower dimensional subsets of $\mathbb{C}^n$, Zeszyty Naukowe U. J. 13 (1969), 53-70.

[12] V. P. ZAHARJUTA: Separately analytic functions, generalizations of the Hartogs theorem, and envelopes of holomorphy [in Russian], Mat. Sb. (N. S.) 101 (143) (1976), no. 1, 57-76 and 159.

[13] W. PLEŚNIAK: Invariance of the L-regularity of compact sets in $\mathbb{C}^n$ under holomorphic mappings, Trans. Amer. Math. Soc. 246 (1978), 373-383.

[14] W. PLEŚNIAK: A criterion of the L-regularity of compact sets in $\mathbb{C}^n$, Zeszyty Naukowe U. J. 21 (1979), 97-103.

[15] R. M. DUDLEY and B. RANDOL: Implications of pointwise bounds on polynomials, Duke Math. J. 29 (1962), no. 3, 455-458.

[16] M. S. BAOUEDI and C. GAULAOUIC: Approximation of analytic functions on compact sets and Bernstein's inequality, Trans. Amer. Soc. 189 (1974), 251-261.

[17] S. Ju. FAVOROV: On a problem of V. S. Vladimirov [in Russian], Funkcional. Anal. i Priložen. Tom 12 Vyp. 3 (1978), 90 pp.

A REMARK ON HOLOMORPHIC ISOMETRIES WITH RESPECT TO THE INDUCED BERGMAN METRICS

Maciej Skwarczyński (Warszawa)

Let D be a bounded domain in $\mathbb{C}^n$, and denote by $K_D(z,t)$ the Bergman function of D. It is well known that the form

$$\omega = i\partial\bar{\partial}\log K_D(z,z)$$

is invariant under the operator J of complex structure. In other words $\omega(JX, JY) = \omega(X,Y)$ for all tangent vectors X and Y. Also ω is a real form since $\partial\bar{\partial} = -\bar{\partial}\partial$. The associated symmetric tensor given by $g(X,Y) = \omega(X, JY)$ is positive definite, and therefore defines a Kähler metric in D. This Bergman tensor can be written as

$$g = \frac{1}{2}\sum_{i,j=1}^{n}(T_{i\bar{j}}\,dz_i\otimes d\bar{z}_j + T_{\bar{i}j}\,d\bar{z}_i\otimes dz_j),$$

where

$$T_{i\bar{j}} = \frac{\partial^2}{\partial z_i\partial\bar{z}_j}\log K_D(z,z) \quad\text{and}\quad T_{\bar{i}j} = \overline{T_{i\bar{j}}}.$$

For all points $z, t\in D$ we define the distance $d_D(z,t)$ as the infimum of the length of all paths which join z and t, and are contained in D. For details cf. [1]. We may also consider another distance function in D, defined by the explicit formula (cf. [2]):

$$\varrho_D(z,t) = \left[1 - \left(\frac{K_D(z,t)\,K_D(t,z)}{K_D(z,z)\,K_D(t,t)}\right)^{\frac{1}{2}}\right]^{\frac{1}{2}}.$$

For a subdomain $U\subset D$ we shall consider the restriction of each of the above distance functions to U. Let us consider now two domains D_i, $i = 1, 2$, and subdomains $U_i\subset D_i$, $i = 1, 2$. Assume that $f\colon U_1\to U_2$ is a real analytic mapping. It was shown in [3] that if

$$\varrho_{D_2}(f(z), f(t)) = \varrho_{D_1}(z,t)$$

for all $z, t \in U_1$, then

$$d_{D_2}(f(z), f(t)) = d_{D_1}(z,t)$$

for all $z, t \in U_1$.

The purpose of this note is to state a result in the opposite direction:

THEOREM. *Assume that* $f: U_1 \to U_2$ *is a holomorphic mapping such that*

$$d_{D_2}(f(z), f(t)) = d_{D_1}(z,t)$$

for all $z, t \in U_1$. *Then*

$$\varrho_{D_2}(f(z), f(t)) = \varrho_{D_1}(z,t)$$

for all $z, t \in U_1$.

Proof. By assumption, $\omega_{D_1} = f^* \omega_{D_2}$. Hence

$$\partial\bar\partial \log K_{D_1} = f^*(\partial\bar\partial \log K_{D_2}).$$

Obviously

$$\log(K_{D_2} \circ f) = f^*(\log K_{D_2}).$$

Note that f^* commutes with ∂ and $\bar\partial$ since f is holomorphic. Therefore by applying $\partial\bar\partial$ to both sides we obtain

$$\partial\bar\partial \log(K_{D_2} \circ f) = f^*(\partial\bar\partial \log K_{D_2}) = \partial\bar\partial \log K_{D_1}.$$

It follows that the function $\log(K_{D_2} \circ f) - \log K_{D_1}$ is pluriharmonic and can be written locally as $h + \bar h$ for some holomorphic function h. Therefore in a neighbourhood of some point $p \in U_1$

$$K_{D_2}(f(z), f(z)) = K_{D_1}(z,z)\, e^{h(z)}\, e^{\overline{h(z)}}.$$

Using for the second time the fact that f is holomorphic we see that in a neighbourhood of the point $(p,p) \in U_1 \times U_1$

$$K_{D_2}(f(z), f(t)) = K_{D_1}(z,t)\, e^{h(z)}\, e^{\overline{h(t)}}.$$

It follows that

$$\varrho_{D_2}(f(z), f(t)) = \varrho_{D_1}(z,t)$$

for z, t in some neighbourhood of $p \in U_1$. Hence f is an isometry there. The real-analytic extension of an isometry with respect to ϱ_{D_1} and ϱ_{D_2} is an isometry with respect to ϱ_{D_1} and ϱ_{D_2}; cf. [3]. This proves that

$$\varrho_{D_2}(f(z), f(t)) = \varrho_{D_1}(z,t)$$

for all $z, t \in U_1$. The proof is completed.

I do not know whether the above theorem is true without the assumption that f is holomorphic.

References

[1] BERGMAN, S.: The kernel function and conformal mapping, Math. Surveys No. 5, Amer. Math. Soc. 1970.

[2] SKWARCZYNSKI, M.: The invariant distance in the theory of pseudoconformal transformations and the Lu Qi-keng conjecture, Proc. Amer. Math. Soc. 22 (1969), 305-310.

[3] ———: The Bergman function and semiconformal mappings, Bull. Acad. Polon. Sci. Sér. Sci. Math. Astronom. Phys. 22 (1974), 667-673.

Institute of Mathematics
University of Warsaw
Pałac Kultury i Nauki, IX p.
PL-00-901 Warszawa, Poland

THEOREMS ON HOLOMORPHIC BISECTIONAL CURVATURE AND PSEUDOCONVEXITY ON KÄHLER MANIFOLDS

Osamu Suzuki* (Tokyo)

Contents

Introduction

In the theory of several complex variables pseudoconvexity is well known and in hermitian geometry holomorphic bisectional curvature is introduced by Goldberg and Kobayashi ([3]). In this paper we shall show three theorems concerning pseudoconvexity and holomorphic bisectional curvature. At first we shall be concerned with the complement of compact analytic sets and q-convex domains on kähler manifolds and prove Theorems I,II (§ 3). Secondly we shall consider the local behavior of an energy function and give a differential geometric interpretation of holomorphic bisectional curvature (Theorem III in § 4). In order to prove the theorems, we shall prepare a lemma, which gives us an explicit formula of laplacians of lengths and energies for a special family of curves (§ 2).

1. Preliminaries

In this section we prepare some propositions concerning normal coordinates for real analytic kähler metrics and give the definition of q-convex functions for continuous functions. We use the notations

* The author is supported by Humboldt Foundation and Taniguchi Foundation during the preparation of this paper.

in [5].

Let M be a kähler manifold with a real analytic kähler metric g. Then we can show the following Lemma, which is due to Takeuchi ([9])(see [7], [8]).

LEMMA (1.1). (i) *For any point* $p_0 \in M$ *and for any positive integer* α_0, *there exist positive constants* δ, M_0 *and a neighborhood* V *of* p_0 *such that the following holds: For any point* $p \in V$ *and for any geodesic* σ *through* p, *there exists a neighbourhood* U *of* p *and a system of local coordinates* $z^1, z^2, \ldots, z^n$ *at* p *in* U *satisying the following* $(1)\sim(4)$:

(1) σ *is expressed as* $\operatorname{Im} z^1 = 0,\ z^2 = 0, \ldots, z^n = 0$,

(2) $\{|\operatorname{Re} z^1| < \delta,\ \operatorname{Im} z^1 = z^2 = \ldots = 0\} \subset U$,

(3) *The metric tensor with respect to* $z^1, \ldots, z^n$ *can be expressed as*

$$2g_{1,\bar{1}} = 1 - K_{1\bar{1}1\bar{1}}(0)(z^1)^2 - K_{1\bar{1}1\bar{1}}(0)(\bar{z}^1)^2 + 2\sum K_{1\bar{1}\gamma\bar{\delta}}(0)z_\gamma \bar{z}_\delta + \mathcal{O}(|z|^3),$$

$$2g_{\alpha,\bar{\beta}} = \delta_{\alpha\beta} + 2\sum K_{\alpha\bar{\beta}\gamma\bar{\delta}}(0)z_\gamma \bar{z}_\delta + \mathcal{O}(|z|^3)$$

$$(\alpha \neq 1 \ \text{and} \ \beta \neq 1),$$

(4) *Let* $\Phi : I_\delta \longrightarrow U$ *be defined by* $z^1 = t,\ z^2 = 0, \ldots, z^n = 0$, *where* $I_\delta = \{t \in \mathbb{R} : |t| < \delta\}$. *Then* $2g_{1,\bar{1}} \circ \Phi(t) - 1$, $\| D_t^\varepsilon (g_{\alpha,\bar{\beta}} \circ \Phi(t)\| \leq M_0$ *and* $\| D_t^\varepsilon (g^{\alpha,\bar{\beta}} \circ \Phi(t))\| \leq M_0$ *for* $\varepsilon \leq \alpha_0$, *where* $(g^{\alpha,\bar{\beta}})$ *is the inverse of* $(g_{\alpha\bar{\beta}})$. *Moreover*, $\mathcal{O}(|z|^3)$ *can be bounded with a positive constant* M' *which depends on* δ *and* M_0, *i.e.*, $\mathcal{O}(|z|^3) \leq M'|z|^3$.

(ii) *For any compact set* K *there exist positive numbers* δ *and* M_0 *satisfying* $(1)\sim(4)$ *for any point* p *in* K.

We use this lemma in real form.We write $z^i = x^i + \sqrt{-1}\, y^i$ and $X_i = \partial/\partial x^i$ and $Y_i = \partial/\partial y^i$ $(i = 1,2,\ldots,n)$. We also denote the connection coefficients with respect to x^i and y^i by Γ^i_{jk} (see [7], p. 198). We summarize basic properties:

PROPOSITION (1.2). *We obtain the following equalities:*

(I) $g(X_i, X_j) \circ \Phi = \delta_{i,j} + 2\operatorname{Re} K_{j\bar{i}1\bar{1}}(0)t^2 + \mathcal{O}(t^3)$,

$g(X_i, Y_j) \circ \Phi = 2\operatorname{Im} K_{i\bar{j}1\bar{1}}(0)t^2 + \mathcal{O}(t^3)$,

(II) $\Gamma^{k}_{i\,j}\circ\Phi = 2\mathrm{Re}\, K_{\bar{k}i\bar{1}j}(0)t + \mathcal{O}(t^2)$
$\Gamma^{n+k}_{i\,j}\circ\Phi = 2\mathrm{Im}\, K_{\bar{k}i\bar{1}j}(0)t + \mathcal{O}(t^2)$ $\quad (k,i,j) \neq (1,1,1)$,

where $|\mathcal{O}(t^3)| \leqq M't^3$ and $|\mathcal{O}(t^2)| \leqq M't^2$ hold with some positive constant M' which depends only on M_0 and δ.

From (1) in (1.1),

(III) $\mathrm{Re}\, K_{j\bar{1}1\bar{1}}(0) = 0$ and $\mathrm{Im}\, K_{j\bar{1}1\bar{1}}(0) = 0 \quad (j = 2,3,\ldots,n)$.

Since $\Gamma^{i}_{jk}\circ\Phi = \Gamma^{i}_{kj}\circ\Phi$ hold,

(IV) $2\mathrm{Re}\, K_{\bar{1}k\bar{1}1}(0) = 2\mathrm{Re}\, K_{\bar{1}1\bar{1}k}(0)$,

$2\mathrm{Im}\, K_{\bar{1}k\bar{1}1}(0) = 2\mathrm{Im}\, K_{\bar{1}1\bar{1}k}(0)$.

(V) $g(R(X_i,X_1)X_1,X_k) = 2\mathrm{Re}\, K_{\bar{k}1\bar{i}1} + 2\mathrm{Re}\, K_{\bar{k}1i\bar{1}}$,

$g(R(Y_i,X_1)X_1,Y_k) = 2\mathrm{Re}\, K_{\bar{k}1i\bar{1}} - 2\mathrm{Re}\, K_{k\bar{1}i\bar{1}}$,

$g(R(X_i,X_1)X_1,Y_k) = 2\mathrm{Im}\, K_{\bar{k}1\bar{i}1} + 2\mathrm{Im}\, K_{\bar{k}1i\bar{1}}$,

$g(R(Y_i,X_1)X_1,X_k) = 2\mathrm{Im}\, K_{\bar{k}1\bar{i}1} - 2\mathrm{Im}\, K_{\bar{k}1i\bar{1}}$.

The proofs are easy and may be omitted.

Here we are concerned with q-convex functions. Extending the definition of pseudoconvex functions in [9], we define q-convex functions for continuous functions on a complex manifold M. We take a point p in M and choose local coordinates $z^1, z^2, \ldots, z^n$ at p. Let $\tau : \mathbb{D}_\varepsilon \longrightarrow M$ be an imbedding of $\mathbb{D}_\varepsilon = \{\lambda \in \mathbb{C}: |\lambda| < \varepsilon\}$ in M with $\tau(C) = p$. τ can be written as $\tau = (f_1, f_2, \ldots, f_n)$. We write

(1.3) $(\partial f_j/\partial\lambda)(0) = a_j \quad (j = 1,2,\ldots,n)$.

In this section, we assume that $|a_1|^2 + |a_2|^2 + \ldots + |a_n|^2 = 1$. For an imbedding τ we consider the complex line L_τ which is defined by

(1.4) $z^1 = a_1\lambda, \ldots, z^n = a_n\lambda$,

which we call the complex line L_τ associated to τ. We write $\lambda = re^{i\theta}$. For a continuous function f which is defined on a domain D in M and for an imbedding τ at $p \in D$, we set

$$\mu^{\tau}_{r}(f)(p) = \frac{1}{2\pi}\int_0^{2\pi} f\circ\tau(re^{i\theta})\, d\theta$$

and

$$\Delta_\tau f(p) = \lim_{r\to 0} \frac{\mu_r^\tau f(p) - f(p)}{r^2} .$$

Remark. If $f \in C^2$-class, then $\Delta_\tau f(p) = 1/4 \frac{\partial^2 f\circ\tau}{\partial\lambda\partial\bar\lambda}(0)$. We take a complex subspace L of $T_p(M)$ and set

(1.5) $H_L f(p) = \inf_{\tau,\, \tau_0 \in L} \Delta_\tau(f)(p),$

where $\tau_0 = \tau_* (\partial/\partial\lambda)_p$. If we choose $T_p(M)$ as L, we call $H_L f(p)$ the hessian of f, i.e.,

$$Hf(p) = H_{T_p(M)} f(p).$$

Now we define q-convex functions:

Definition (1.6). A continuous function f on D is called a q-convex function at $p \in D$, if there exists a $(n-q+1)$-dimensional complex subspace L in $T_p(M)$ such that

$$H_L\, f(p) \geqq 0.$$

Especially, if the inequality holds, we call it a strongly q-convex function.

By using q-convex functions, we define q-convex domains:

Definition (1.7). (1) A relatively compact domain D is called locally q-convex, if for every point $p \in \partial D$, there exist a neighbourhood of p and a q-convex function φ such that $D \cap U = \{\varphi < 0\}$. If φ is of C^∞-class, then D is called a C^∞-q-convex domain.

(2) A C^∞-q-convex domain is called smooth if $d\varphi(p) \neq 0$.

2. A lemma

In this section we shall give an explicit formula for Laplacian of length functions or energy functions for a certain complex family of curves.

Let M be a kähler manifold with a real analytic kähler metric and let A be a q-dimensional non-singular analytic set in M. We choose a neighbourhood U of A such that there exists a geodesic G which attains the distance between p and A for every point

$p \in U$. The other end point of σ is denoted by r. By using Lemma (1.1), we can choose local coordinates $z^1, z^2, \ldots, z^n$ at r for σ. We may assume that A can be expressed as follows:

$$z^i = \beta^i(z^{n-q+1}, \ldots, z^n) \qquad (i = 1,2,\ldots,n-q).$$

From (3) in (1.1), we may assume that

$$(2.1) \quad \frac{\partial \beta^i}{\partial z^j}(0) = 0 \qquad (i = 1,2,\ldots,n-q \text{ and } j = n-q+1,\ldots,n).$$

By l we denote the length of σ. We choose a complex line L through p:

$$(2.2) \quad L: z^1 = 1 + a_1\lambda, \quad z^2 = a_2\lambda, \ldots, \quad z^n = a_n\lambda,$$

We write $\lambda = \xi + \sqrt{-1}\,\eta$. We consider the following family of lines:

$$(2.3) \quad \Psi: \begin{cases} z^1 = \frac{t}{l}(1+a_1\lambda) + (1-\frac{t}{l})\,\beta^1(a_{n-q+1}\lambda, \ldots, a_n\lambda), \\ z^i = \frac{t}{l}\,a_i\lambda + (1-\frac{t}{l})\,\beta^i(a_{n-q+1}\lambda, \ldots, a_n\lambda) \\ \qquad\qquad (i = 2,3,\ldots,n-q), \\ z^{n-q+k} = a_{n-q+k}\lambda \qquad (k = 1,2,\ldots,q). \end{cases}$$

We get a line C_λ from L to A for every fixed λ. We denote the length function and the energy function by $d^*(\xi, \eta)$ and $e^*(\xi, \eta)$, respectively, where the energy function is defined by

$$(2.4) \quad e^*(\xi, \eta) = \frac{1}{2}\int_0^l \|X\|^2\,dt,$$

where $X = \Psi_*(\partial/\partial t)$ [6]. For X and $Y \in T_p(M)$, we write

$$H'(X,Y) = K(X,Y,X,Y) + K(X,JY,X,JY),$$

where J denotes the complex structure of M. If X and Y are unit vectors, we write $H(X,Y)$ for $H'(X,Y)$, which is nothing but the holomorphic bisectional curvature of X and Y. We write

$$U^0 = \frac{t}{l}(\mathrm{Re}\,a_1\,X_1 + \mathrm{Im}\,a_1\,Y_1),$$

$$U^1 = \sum_{i=2}^{n-q}\frac{t}{l}(\mathrm{Re}\,a_i\,X_i + \mathrm{Im}\,a_i\,Y_i),$$

$$U^2 = \sum_{k=1}^{q}(\mathrm{Re}\,a_{n-q+k}\,X_{n-q+k} + \mathrm{Im}\,a_{n-q+k}\,Y_{n-q+k}),$$

$$A = \sum_{i=1}^{n-q} \{(\mathrm{Re}\ a_i)^2 + (\mathrm{Im}\ a_i)^2\},$$

and

$$A' = A - \{(\mathrm{Re}\ a_1)^2 + (\mathrm{Im}\ a_1)^2\}.$$

Then we can get the following lemma:

LEMMA (2.5).

(i) $\frac{\partial d^*}{\partial \xi}(0,0) = \frac{\partial e^*}{\partial \xi}(0,0) = -\mathrm{Re}\ a_1 + \mathcal{O}(l^3)$,

$\frac{\partial d^*}{\partial \eta}(0,0) = \frac{\partial e^*}{\partial \eta}(0,0) = \mathrm{Im}\ a_i + \mathcal{O}(l^3)$.

(ii)
$$\Delta d^*(0,0) = \frac{A}{l} + \frac{A'}{l} + \int_0^1 H'(X,U^0)\circ \Psi(0,0,t)\, dt - \int_0^1 \{H'(X,Y) + H'(X,U^1)\}\circ \Psi(0,0,t)\, dt + \int_0^1 H'(X,U^2)\circ \Psi(0,0,t)\, dt + \mathcal{O}(l^2),$$

$$\Delta e^*(0,0) = 2\frac{A}{l} - 2\int_0^1 \{H'(X,Y) + H'(X,U^1)\}\circ \Psi(0,0,t)\, dt + \int_0^1 H'(X,U^2)\circ \Psi(0,0,t)\, dt + \mathcal{O}(l^2),$$

where $\mathcal{O}(l^3)$ and $\mathcal{O}(l^2)$ can be bounded with a positive constant M' which depends only on M_0 and δ, i.e., $|\mathcal{O}(l^3)| \leqq M'l^3$ and $|\mathcal{O}(l^2)| \leqq M'l^2$.

It is easily seen that the terms $\mathcal{O}(l^2)$, $\mathcal{O}(l^3)$ etc. in the following of this paper satisfy the above condition. Therefore we will not mention it in the following.

We set

$$D_1 = \frac{\partial}{\partial t}, \qquad D_2 = \frac{\partial}{\partial \xi}, \qquad \text{and} \qquad D_3 = \frac{\partial}{\partial \eta}.$$

Respectively we write

$$X = \Psi_*(D_1), \qquad Y = \Psi_*(D_2) \qquad \text{and} \qquad Z = \Psi_*(D_3).$$

Since Ψ is a holomorphic mapping of λ for fixed t, we see that

(2.6) $Z = J\,Y.$

In order to prove Lemma 2.4 we prepare the following

PROPOSITION (2.7). We obtain

(i) $$\frac{\partial d^*}{\partial \xi}(0,0) = \frac{\partial e^*}{\partial \xi}(0,0) = g(Y,X)\Big|_0^1 ,$$

$$\frac{\partial d^*}{\partial \eta}(0,0) = \frac{\partial e^*}{\partial \eta}(0,0) = g(Z,X)\Big|_0^1 .$$

(ii) $$\Delta d^*(0,0) = \int_0^1 \{g(\nabla_{D_1}\tilde{Y}, \nabla_{D_1}\tilde{Y}) + g(\nabla_{D_1}\tilde{Z}, \nabla_{D_1}\tilde{Z})\}\circ \Psi(0,0,t)\, dt$$

$$- \int_0^1 H'(X,Y)\circ \Psi(0,0,t)\, dt,$$

$$\Delta e^*(0,0) = \int_0^1 \{g(\nabla_{D_1}Y, \nabla_{D_1}Y) + g(\nabla_{D_1}Z, \nabla_{D_1}Z)\}\circ \Psi(0,0,t)\, dt$$

$$- \int_0^1 H'(X,Y)\circ \Psi(0,0,t)\, dt,$$

where $\tilde{Y} = Y - g(X,Y)X$.

As for notations see [4](p. 122).

P r o o f. We prove only the assertions for d^*. The proofs for e^* are almost the same and may be omitted. By using the first Synge's formula, we get the assertions in (i). By using the second Synge's formula for $\partial^2 d^*/\partial \xi^2(0,0)$ and $\partial^2 d^*(0,0)/\partial \eta^2(0,0)$ and adding them, we get $\Delta d^*(0,0)$ (see [4], p. 122). In order to prove the assertions it is sufficient to prove the following terms vanish at p and r, respectively:

$$\{g(\nabla_{D_2}Y,X) + g(\nabla_{D_3}Z,X)\}\circ \Psi(0,0,t).$$

We only prove at p. We choose normal coordinates $z^1, z^2, \ldots, z^n$. We write $z^i = x^i + \sqrt{-1}\, x^{n+i}$ and $X_i = \partial/\partial x^i$. From the definition we see that

$$\nabla_{D_2}Y = \sum \frac{\partial^2 \Psi^i}{\partial \xi^2}(0,0,0)\, X_i, \quad \text{where} \quad \Psi^i = x^i \circ \Psi.$$

In the same manner we obtain

$$\nabla_{D_3}Z = \frac{\partial^2 \Psi^i}{\partial \eta^2}(0,0,0)\, X_i .$$

Since Ψ is holomorphic in λ for fixed t, we get the assertions

P r o o f of (2.5). We give proofs only for d^*. From (i) in (2.6) and from (I) in (1.2) we get the conclusion (i). We prove (ii). We write

$$\nabla_{D_1}\tilde{Y} = \sum U^i\, X_i \circ \Psi + \sum V^i\, Y_i \circ \Psi .$$

By using (II), (III) in (1,2)(see (2.12), (2.13) in [7]), we get

$$U^1 = \frac{1}{l}\,\mathcal{O}(t^3) + \mathcal{O}(t^2),$$

$$\begin{aligned} U^i = {} & \frac{1}{l}\,\mathrm{Re}\, a_i + \sum_{k=2}^{n-q} \frac{t^2}{l}\,\mathrm{Re}\, a_k\; 2\mathrm{Re}\, K_{\bar{1}1\bar{1}k}(0) \\ & + \sum_{k=1}^{q} t\,\mathrm{Re}\, a_{n-q+k}\; 2\mathrm{Re}\, K_{\bar{1}1\bar{1}n-q+k}(0) \\ & - \sum_{k=2}^{n-q} \frac{t^2}{l}\,\mathrm{Im}\, a_k\; 2\mathrm{Im}\, K_{\bar{1}1\bar{1}k}(0) \\ & - \sum_{k=1}^{q} t\,\mathrm{Im}\, a_{n-q+k}(0)\; 2\mathrm{Im}\, K_{\bar{1}1\bar{1}n-q+k}(0) \\ & + \mathcal{O}(t^2) + \mathcal{O}(t^3) \qquad (i = 2,3,\ldots, n-q), \end{aligned}$$

$$U^{n-q+k} = \frac{1}{l}\,\mathcal{O}(t^2) + \mathcal{O}(t) \qquad (k = 1,2,\ldots,q),$$

$$\begin{aligned} V^i = {} & \frac{1}{l}\,\mathrm{Im}\, a_i + \frac{t^2}{l}\sum_{k=2}^{n-q} \mathrm{Re}\, a_k\; 2\mathrm{Im}\, K_{\bar{1}1\bar{1}k}(0) \\ & + t\sum_{k=1}^{q} \mathrm{Re}\, a_{n-q+k}\; 2\mathrm{Im}\, K_{\bar{1}1\bar{1}n-q+k}(0) \\ & + \frac{t^2}{l}\sum_{k=2}^{n-q} \mathrm{Im}\, a_k\; 2\mathrm{Re}\, K_{\bar{1}1\bar{1}k}(0) \\ & + t\sum_{k=1}^{q} \mathrm{Im}\, a_{n-q+k}\; 2\mathrm{Re}\, K_{\bar{1}1\bar{1}n-q+k}(0) \\ & + \frac{1}{l}\,\mathcal{O}(t^3) + \frac{1}{l^2}\,\mathcal{O}(t^3) + \mathcal{O}(t^2) \qquad (i = 1,2,\ldots,n-q), \end{aligned}$$

$$V^{n-q+k} = \frac{1}{l}\,\mathcal{O}(t^2) + \mathcal{O}(t) \qquad (k = 1,2,\ldots,q).$$

From U^i and V^i we get $g(\nabla_{D_1}\tilde{Y}, \nabla_{D_1}\tilde{Y})\circ\Psi(0,0,t)$. From (2.5) we obtain $g(\nabla_{D_1}\tilde{Z}, \nabla_{D_1}\tilde{Z})\circ\Psi(0,0,t)$ by replacing $\mathrm{Re}\, a_i$ (resp. $\mathrm{Im}\, a_i$) by $-\mathrm{Im}\, a_i$ (resp. $\mathrm{Re}\, a_i$) in $g(\nabla_{D_1}\tilde{Y}, \nabla_{D_1}\tilde{Y})\circ\Psi(0,0,t)$. By using $\mathrm{Re}\, K_{i\bar{k}1\bar{1}}(0) = \mathrm{Re}\, K_{\bar{1}1\bar{1}k}(0)$, $\mathrm{Im}\, K_{\bar{1}1\bar{1}k}(0) = -\mathrm{Im}\, K_{\bar{k}1\bar{1}i}(0)$ and $\mathrm{Im}\, K_{\bar{1}k\bar{1}1}(0) = \mathrm{Im}\, K_{\bar{1}1\bar{1}k}(0)$ (see IV in (1.2)), we obtain

$$(2.8)\quad \int_0^l \{g(\nabla_{D_1}\tilde{Y}, \nabla_{D_1}\tilde{Y}) + g(\nabla_{D_1}\tilde{Z}, \nabla_{D_1}\tilde{Z})\}\circ\Psi(0,0,t)\, dt$$

$$= \frac{A}{l} + \frac{A'}{l} - \frac{1}{2}\int_0^l H'(X,U^0)\,dt - 2\int_0^l H'(X,U^1)\,dt$$

$$- C_o + \mathcal{O}(l^2),$$

where

$$C_o = 2l \sum_{i=2}^{n-q}\sum_{k=1}^{q} \text{Re } a_i \text{ Re } a_{n-q+k}\; 2\text{Re } K_{\bar{i}1n-q+k\bar{1}}(0)$$

$$- 2l \sum_{i=2}^{n-q}\sum_{k=1}^{q} \text{Re } a_i \text{ Im } a_{n-q+k}\; 2\text{Im } K_{\bar{i}1n-q+k\bar{1}}(0)$$

$$+ 2l \sum_{i=2}^{n-q}\sum_{k=1}^{q} \text{Im } a_i \text{ Re } a_{n-q+k}\; 2\text{Im } K_{\bar{i}1n-q+k\bar{1}}(0)$$

$$+ 2l \sum_{i=2}^{n-q}\sum_{k=1}^{q} \text{Im } a_i \text{ Im } a_{n-q+k}\; 2\text{Re } K_{\bar{i}1n-q+k\bar{1}}(0)$$

$$+ \frac{1}{3}\{(\text{Re } a_1)^2 + (\text{Im } a_1)^2\}\; 2\text{Re } K_{\bar{1}11\bar{1}}(0)$$

$$+ \mathcal{O}(l^2).$$

Next we calculate $H'(X,Y)$. By using $\text{Re } K_{\bar{k}1i\bar{1}}(0) = \text{Re } K_{\bar{i}1k\bar{1}}(0)$ and $\text{Im } K_{\bar{k}1i\bar{1}}(0) = -\text{Im } K_{\bar{i}1k\bar{1}}(0)$, we obtain

$$(2.9)\quad \int_0^l H'(X,Y)\circ\Psi(0,0,t)\,dt = \sum_{i=0}^{2}\int_0^l H'(X,U^i)\circ\Psi(0,0,t)\,dt + C_o + \mathcal{O}(l^2).$$

Combining (2.8) with (2.9) we get the conclusions of (ii).

R e m a r k. The formulas in (2.7) contain only the first derivatives with respect to t, ξ and η ; we see the following: In order to calculate the hessian of the length functions or energy functions, it is sufficient to do so for every complex line.

3. Some q-convexity properties on kähler manifolds

Let M be a kähler manifold and let A be a compact analytic set in M. We consider the following function:

$$\Phi(p) = -\log d(p,A), \quad \text{where} \quad d(p,A) = \inf_{r\in A} d(p,r).$$

We prove the following theorem:

THEOREM I. Let M be a kähler manifold with positive holomorphic bisectional curvature and let A be a compact analytic set whose irreducible components $A_j (j = 1,2,\ldots,r)$ are non-singular. Then there exists a neighbourhood U of A such that $\Phi(p)$ is a q-convex function on $U - A$, where $q = \max_{j=1,\ldots,r} \operatorname{codim} A_j$.

Remark. Theorem I includes the result of W. Barth ([1]) in the case where M is the comlex projective space.

We give the proof of Theorem I in the following two cases:

(i) The case where the kähler metric is real analytic. Let U be a small neighbourhood of A such that for any point $p \in U$ there exists (may be not unique) a geodesic σ from p to $r \in A$ whose length is identical with $d(p,A)$. We assume that $r \in A_j$ for some j. Then we see that $d(p,A) = d(p,A_j)$. Now for σ and r we choose local coordinates $z^1,z^2,\ldots,z^n$ at r as in Lemma (1.1). Making U smaller, if necessary, we may assume that p is contained in the above coordinates. We choose a complex line L through p which is denoted by (2.2). In this section, we assume the normalized condition $\sum|a_i|^2=1$. Restricting the function $d(p,A)$ on L, we obtain a function $d(\xi,\eta)$. As in (2.3) we construct Ψ between L and A_j. We consider the following functions:

$$\Phi(\xi,\eta) = -\log d(\xi,\eta) \quad \text{and} \quad \Phi^*(\xi,\eta) = -\log d^*(\xi,\eta).$$

Then we see that

$$\Phi(0,0) = \Phi^*(0,0) \quad \text{and} \quad \Phi(\xi,\eta) \geqq \Phi^*(\xi,\eta).$$

Hence we see that

$$\Delta_L \Phi(0,0) \geqq \Delta_L \Phi^*(0,0).$$

Therefore we obtain

$$H\Phi(p) \geqq H\Phi^*(p).$$

Hence to prove the assertion, it is sufficient to show that Φ^* is an n-q-convex function on U. For this we prove that Φ^* is $(n-q_j)$-convex at p, where $q_j = \dim A_j$. Now we use (i),(ii) in (2.5). Restricting L in

$$L_j = \{(z^1,z^2,\ldots, z^n): z^2 = z^3,\ldots = z^{n-q_j} = 0\},$$

we see that

$$\Delta_L \Phi^*(0,0) = \frac{2}{l}\int_0^l H'(X,Y)\circ\Psi(0,0,t)\,dt$$
$$-\frac{1}{l}\int_0^l H'(X,U^2)\circ\Psi(0,0,t)\,dt$$
$$-\frac{1}{2l}\int_0^l H'(X,U^0)\circ\Psi(0,0,t)\,dt + O(l).$$

Since $L \subset L_j$ we see that

$$H'(X,Y) = H'(X,U^0) + H'(X,U^2) + \frac{1}{l^2}O(t^3) + \frac{1}{l}O(t) + O(t).$$

Hence we see that

$$\Delta_L \Phi^*(0,0) \geqq \frac{1}{l}\int_0^l H'(X,Y)\circ\Psi(0,0,t)\,dt + O(l).$$

Let ρ_m be the infimum of the holomorphic bisectional curvature on U. Then we obtain

$$\Delta_L \Phi^*(0,0) \geqq \frac{\rho_m}{3} + O(l).$$

Referring to $|O(l)| \leqq M'l$, where M' depends only on M_0 and δ, and making U smaller, we get

$$H_{L_j}\Phi^*(p) \geqq \frac{\rho_m}{6} \ (>0),$$

which prove the assertion.

(ii) The case where the kähler metric is of C^∞-class. By using the result in [8] (see (1.3) in Section 2), we can approximate the metric by real analytic kähler metrics on a local coordinate neighborhood as many orders of differentiation of the metric tensor as possible Therefore by using the result in (i), we can also prove the assertion in this case.

Now we proceed to consider q-convex domains on a kähler manifold with positive holomorphic bisectional curvature. Let D be a relatively compact domain on M. We write

$$\Phi(p) = -\log d(p,\partial D), \quad \text{where} \quad d(p,\partial D) = \inf_{r\in\partial D} d(p,r).$$

We write $D_\delta = \{p \in D : d(p,\partial D) < \delta\}$. Then we obtain

THEOREM II. Let M be a kähler manifold with positive holomorphic bisectional curvature and let K be a compact set in M. Let D be a domain in K which is written locally as an intersection

<u>of finite</u> C^∞-q-<u>convex domains</u>. <u>Then there exists a constant</u> δ <u>such that</u> $\Phi(p)$ <u>is a strongly</u> q-<u>convex function on</u> D_δ. <u>Moreover</u>, δ <u>depends only on</u> K.

<u>Remark.</u> In the case where $q = 1$ Theorem II is a partial result of G.Elencwajg [2].

<u>Proof of Theorem II</u>. At first we assume that the metric is real analytic. As in the proof of Theorem I it is sufficient to show for a C^∞-q-convex domain. Since a C^∞-q-convex function on a domain D in $\mathbb{C}^n$ can be approximated by strongly C^∞-q-convex functions on a compact set in D, hence we may prove our theorem for a strongly q-convex smooth domain (see [9], Lemma 4 and Theorem 1). As for a smooth domain, see (2) in (1.7). We choose δ such that there exists a geodesic σ from a point $p \in D_\delta$ to a point $r \in \partial D$ whose length is identical with $d(p,\partial D)$ for every point $p \in D_\delta$. We fix a point $p \in D_\delta$ and write $d(p,\partial D) = 1$. As in [9](see Lemma 3) we can find a neighborhood $U(r)$ and a non-singular q-codimensional analytic set such that

$$D \cap S = \emptyset \quad \text{and} \quad \partial D \cap S = \{r\} \quad \text{in} \quad U(r).$$

We choose local coordinates $z^1, z^2, \ldots, z^n$ for r and σ as in (1.1). We may assume that S can be written as in the beginning of Sec. 2 (see (2.1)). We take a complex line L at p. Since $d(x,\partial D) \leqq d(x,S)$ for $x \in L$ and $d(p,\partial D) = d(p,S)$, we see that

$$\Delta_L \Phi(p) \geqq \Delta_L \Phi_S(p) \quad \text{where} \quad \Phi_S(p) = -\log d(p,S).$$

Hence from Theorem I we get the conclusion of our Theorem. In the case where the metric is of C^∞ we get the conclusion by using approximation technique as in the proof of Theorem I.

4. A differential geometric meaning of holomorphic bisectional curvature

In this section we give a differential geometric meaning of holomorphic bisectional curvature. For simplicity sake, we assume that the metric is real analytic. Let $\underline{X}$ and $\underline{Y}$ be two unit vectors in $T_p(M)$. We set $\underline{Y}^* = \underline{Y} + \underline{Y}'$ where $\underline{Y}' = g(\underline{Y},\underline{X})\underline{X} + g(\underline{Y},J\underline{X})J\underline{X}$. We take a complex imbedding τ with $\tau_*(\partial/\partial\lambda)_{\lambda=0} = \underline{Y}^*$. Let σ be a normal geodesic whose tangent vector at p is $\underline{X}$, whose parameter is denoted by t. We take a point q on σ with $d(p,q) = l$. Let L_τ denote the complex line associated to τ (see (1.4)). We make a family of geodesics joining a point λ in τ and q, which is parametrized by t. We consider the energy function $e(\lambda)$ for C_λ (see (2.4)). We do the same process for $\underline{X}$ and $\underline{Y}$ on $T_p(M)$ which has the canonical Euclidian metric and get the energy function $e_o(\lambda)$. Then we can get the following

<u>THEOREM III</u>. <u>The holomorphic bisectional curvature with respect to $\underline{X}$ and $\underline{Y}$ is given by</u>

$$(4.1) \qquad H(\underline{X},\underline{Y})_p = \lim_{l\to o} \frac{3}{4} \frac{\Delta_\tau e(0) - \Delta_\tau e_o(0)}{l}.$$

In order to show Theorem we prepare a lemma. For σ and p we choose local coordinates $z^1, z^2, \ldots, z^n$ as in (1.1). We consider the following family of lines:

$$\Psi: \begin{cases} z^1 = \frac{t}{l}(1 + a_1\lambda) \\ z^i = \frac{t}{l}(a_i\lambda) \qquad (i \neq 1), \end{cases}$$

where $L_\tau : z^1 = 1 + a_1\lambda$ and $z^i = a_i\lambda$ $(i \neq 1)$. We denote the energy function for Ψ by $e^*(\lambda)$. Then we can show

<u>LEMMA (4.2)</u>. <u>If l is very small, then</u>

$$\Delta_{L_\tau} e(0) = \Delta_{L_\tau} e^*(0) + O(l^2).$$

<u>Proof of Theorem III</u>. By (4.2) and the remark at the end of Section 2 it is sufficient to show

$$H(\underline{X},\underline{Y})_p = \lim_{l\to 0} \frac{3}{4} \frac{\Delta_{L_\tau} e^*(0) - \Delta_{L_\tau} e_o(0)}{l}$$

From (ii) in (2.6) we see that

$$\Delta_{L_\tau} e^*(0) = 2\frac{A}{l} - 2\int_0^l \{H'(X,Y) + H'(X,U^1)\}\circ\Psi(0,0,t)\,dt + O(l^2).$$

It is easily seen that

$$\lim_{l\to o} \frac{1}{l}\int_0^l \{H'(X,Y) + H'(X,U^1)\}\circ\Psi(0,0,t)\,dt = \frac{1}{3}H(\underline{X},\underline{Y}).$$

Hence we get the conclusion of Theorem III

Now we give the proof of Lemma (4.2). We write

$$\theta^1 = 1 + a_1\lambda \quad \text{and} \quad \theta^i = a_i\lambda \qquad (i = 2,3,\ldots,n),$$

$$\theta^{n+i} = \bar{\theta}^i \qquad (i = 1,2,\ldots,n).$$

Then we see that

$$(4.3) \quad e^*(\xi, \eta) = \frac{1}{I}\int_0^1 \sum g_{\alpha\bar{\beta}}\, \theta^\alpha \bar{\theta}^\beta\, dt.$$

Next we calculate $d(\xi, \eta)$. The geodesic is given by

$$(4.4) \quad \frac{d^2 z^\alpha}{dt^2} + \sum \Gamma^{*\alpha}_{\gamma\delta} \frac{dz^\gamma}{dt}\frac{dz^\delta}{dt} = 0 \qquad (\alpha = 1,2,\ldots, n).$$

We solve the equation under the following boundary conditions:

$$(4.5) \quad \begin{cases} z^\alpha(0) = 0 & (\alpha = 1,2,\ldots,n), \\ z^\alpha(1) = \theta^\alpha & (\alpha = 1,2,\ldots,n). \end{cases}$$

Since the metric is real analytic, we can expand the solution in the following form for small t:

$$z^\alpha = a^{(\alpha)}_1 t + a^{(\alpha)}_2 t^2 + \ldots \qquad (\alpha = 1,2,\ldots,n).$$

From (4.4) we obtain

$$a^{(\alpha)}_k = \sum c^{(\alpha)}_{i_1 i_2 \ldots i_k}\, \beta^{i_1}\beta^{i_2}\ldots\beta^{i_k} \qquad (\alpha = 1,2,\ldots,n),$$

where

$$\beta^i = a^{(i)}_1 \quad \text{and} \quad \beta^{n+i} = \overline{a^{(i)}_1} \qquad (i = 1,2,\ldots,n).$$

Since $\Gamma^{*\alpha}_{\beta\gamma}(0) = 0$ we see that

$$a^{(\alpha)}_2 = 0 \qquad (\alpha = 1,2,\ldots,n).$$

Hence in order to solve the boundary value problem (4.5) it is sufficient to solve the following equation:

$$(4.6) \quad \theta^i = \beta^i 1 + \sum_{k=3}^{\infty} c^{(i)}_{i_1 i_2 \ldots i_k}\beta^{i_1}\beta^{i_2}\ldots\beta^{i_k} 1^k \qquad (i = 1,2,\ldots,n).$$

Letting $\gamma^i = \beta^i 1$ we can write (4.6) as follows:

$$\theta^i = \gamma^i + \sum_{k=3}^{\infty} c^{(i)}_{i_1 i_2 \ldots i_k}\gamma^{i_1}\gamma^{i_2}\ldots\gamma^{i_k} \qquad (i = 1,2,\ldots,n).$$

By using the implicit function theorem we obtain

$$(4.7)\quad \gamma^i = \theta^i + \sum_{k=3}^{\infty} C^{(i)*}_{i_1 i_2 \dots i_k}\, \theta^{i_1}\theta^{i_2}\dots\theta^{i_k} \qquad (i = 1,2,\dots,n).$$

We easily see that

$$(4.8)\quad C^{(i)*}_{i_1 i_2 i_3} = -\,C^{(i)}_{i_1 i_2 i_3} \qquad (i = 1,2,\dots,n).$$

By using (4.7) we obtain the solution of (4.5) as follows:

$$(4.9)\quad z^{\alpha}(\xi,\eta,u) = \gamma^{\alpha} u + \sum_{k=3}^{\infty} C^{(\alpha)}_{i_1 i_2 \dots i_k}\, \gamma^{i_1}\dots\gamma^{i_k} u^k \qquad (\alpha = 1,2,\dots,n),$$

where we set $u = t/l$. We denote one of θ^i $(i\neq 1,n)$ by θ. Then (4.7) can be written as

$$(4.10)\quad \gamma^i = \theta^i + \sum_{i_1,i_2,i_3} C^{(i)*}_{i_1 i_2 i_3}\, \theta^{i_1}\theta^{i_2}\theta^{i_3} + A\, l^2\theta^2 + B\, l^3 + C\theta^3,$$

where A,B,C are polynomials of l and θ. In the same manner we see that

$$(4.11)\quad \sum g_{\alpha\bar\beta}\bar\gamma^{\beta}\gamma^{i_1}\gamma^{i_2}\gamma^{i_3} = \sum g_{\alpha\bar\beta}\bar\theta^{\beta}(\theta^{i_1}\theta^{i_2}\theta^{i_3}) + B' l^3 + C'\theta^3,$$

where B', C' are polynomials of l and θ. By using (4.7) – (4.11) we calculate

$$J = \sum g_{\alpha\bar\beta}\frac{dz^{\alpha}}{du}\frac{d\bar z^{\beta}}{du}.$$

From (4.11) we see that

$$J = J_1 + (\mathrm{Re}\, J_2)3u^2 + O(l^3) + O(\theta^3),$$

where

$$(4.12)\quad \begin{aligned} J_1 &= \sum g_{\alpha\bar\beta}\gamma^{\alpha}\bar\gamma^{\beta},\\ J_2 &= 2\sum g_{\alpha\bar\beta}\bar\theta^{\beta}\Big(\sum C^{(\alpha)}_{i_1 i_2 i_3}\theta^{i_1}\theta^{i_2}\theta^{i_3}\Big). \end{aligned}$$

From (4.10) we see that

$$J_1 = J_o + 2\mathrm{Re}(\Sigma g_{\alpha\bar\beta}\bar\theta^\beta(C^{(\alpha)*}_{i_1 i_2 i_3}\theta^{i_1}\theta^{i_2}\theta^{i_3}))$$

$$+ \mathcal{O}(l^3) + \mathcal{O}(\theta^3),$$

where

$$J_o = \Sigma g_{\alpha\bar\beta}\theta^\alpha\bar\theta^\beta.$$

From (4.8) we see that

$$J_1 = J_o - \mathrm{Re}\, J_2 + \mathcal{O}(l^3) + \mathcal{O}(l^3) + \mathcal{O}(\theta^3).$$

Finally we obtain

(4.13) $J = J_o - \mathrm{Re}\, J_2 + (\mathrm{Re}\, J_2)3u^2 + \mathcal{O}(l^3) + \mathcal{O}(\theta^3).$

Now we calculate $\Delta_L e(0,0)$. Since $J = l^2$ on $\mathcal{G}$, we have

$$\frac{\partial^2 e}{\partial\xi^2}(0,0) = \frac{1}{2l}\int_0^l \left(\frac{\partial^2 J}{\partial\xi^2}\right)_{\lambda=0} du.$$

From (4.13) we see that

$$\frac{\partial^2 e}{\partial\xi^2}(0,0) = \frac{\partial^2 e^*}{\partial\xi^2}(0,0) + \mathcal{O}(l^2).$$

In the same manner we see that

$$\frac{\partial^2 e}{\partial\eta^2}(0,0) = \frac{\partial^2 e^*}{\partial\eta^2}(0,0) + \mathcal{O}(l^2).$$

From these qualities we get the conclusion of Lemma (4.2).

References

[1] W. BARTH: Der Abstand von einer algebraischen Manigfaltigkeit im komplexe-projektiven Raum, Math. Ann. 187 (1970), 150-162.

[2] G. ELENCWAJG: Pseudo-convexité local dans les variétés kählerien-nes, Ann. Inst. Fourier 25 (1975), 295-314.

[3] S.I. GOLDBERG and S. KOBAYASHI: On holomorphic bisectional curvature, J. Differential Geometry 1 (1967), 225-233.

[4] D. GROMOLL, W. KLINGENBERG, and W. MEYER: Riemannsche Geometrie im Großen, Lecture notes in Mathematics 55, 1968, Springer, Berlin-Heidelberg-New York.

[5] S. KOBAYASHI and K. NOMIZU: Foundations of differential geometry, I and II, Interscience Publishers, 1963 and 1969, John Wiley and Sons, New York-London.

[6] J.W. MILNOR: Lectures on Morse Theory, Ann. Math. Studies No. 51, 1963, Princeton Univ. Press, Princeton, New Jersey.

[7] O. SUZUKI: Pseudoconvex domains on a kähler manifold with positiv holomorphic bisectional curvature, Publ. RIMS, Kyoto Univ. 12 (1976), 191-214.

[8] ——: Supplement to "Pseudoconvex domains on a kähler manifold with positive holomorphic bisectional curvature", ibid. 12 (1976), 439-445.

[9] A. TAKEUCHI: Domains pseudoconvexes sur les varietés kähleriennes, J. Math. Kyoto Univ. 6 (1967), 323-357.

Department of Mathematics
College of Humanities and
Sciences, Nihon University
Setagaya, Sakurajosui 3-25-40
Tokyo, 156, Japan

ON THE GRONWALL'S PROBLEM FOR SOME CLASSES OF UNIVALENT FUNCTIONS

Anna Szynal, Jan Szynal (Lublin), and Stanisław Wajler (Kielce)

1. Let S denote the class of holomorphic and univalent functions f in the unit disk $K = \{z : |z| < 1\}$ of the form

$$(1) \qquad f(z) = z + a_2 z^2 + \dots, \qquad z \in K.$$

The Gronwall's problem for the class S solved by Jenkins (cf. [2]) consists in giving sharp bound for $|f(z)|$ (when $z \in K$ is fixed) depending on $|a_2|$.

In this note we solve the Gronwall's problem for the class S^o of convex univalent functions in K (by different method then in [5]) as well as for the class G^M of quasi-starlike functions [1].

From the last result in the limit case $M \longrightarrow \infty$, $a \longrightarrow 1$ as a corollary we get the corresponding results from [1], [8].

Our method is based on the Löwner type equation for the above classes and shows that the special form of the Löwner equation can also be useful.

2. We start with some notations. By $\mathcal{P}_a$ we denote the class of functions p holomorphic in K which have the form:

$$(2) \qquad p(z) = 1 + 2az + p_2 z^2 + \dots,$$

and satisfying the condition $\operatorname{Re} p(z) > 0$, $z \in K$.

We assume that the coefficient $p_1 = 2a$ is fixed, and with no loss of generality we may assume that $a \in [0,1]$.

Further on by S_a^o (S_a^*) we denote the class of convex (starlike) univalent functions of the form (1) with fixed second coefficient $a_2 = a$ $(a_2 = 2a)$, $a \in [0,1]$, respectively.

We have the following

<u>THEOREM 1. A function</u> f <u>belongs to</u> S_a^o <u>iff</u>

(3) $$f(z) = \lim_{x\to\infty} (1+x)\, w(z,x),$$

where $w = w(z,x)$ *is the solution of the differential equation*

(4) $$\frac{dw}{dx} = -w\,\frac{1}{1+xp(w)}, \quad p \in \mathcal{P}_a,$$

with the initial condition $w(z,0) = z$.

The proof of Theorem 1 can be obtained in similar way as in [3] or [4]. It is only necessary to take into account that if we fix $a_2 = 2a$ for convex functions, then the correspdnding coefficient for the function p in (4) is $p_1 = 2a$.

Now we can prove

THEOREM 2. *If* $w = w(z,x)$ *is a solution of* (4), *then the following inequality holds:*

(4') $$\frac{1+2a|w|+|w|^2}{|w|\sqrt{1-a^2}}\left\{\operatorname{arc\,tg}\frac{r+a}{\sqrt{1-a^2}} - \operatorname{arc\,tg}\frac{|w|+a}{\sqrt{1-a^2}}\right\} \leq x$$
$$\leq \frac{(1-|w|)^{1+a}(1+|w|)^{1-a}}{|w|}\{\Phi(r) - \Phi(w)\},$$

where $\Phi(\tau) = \int_0^\tau \frac{dt}{(1-t)^{1+a}(1+t)^{1-a}}, \quad a \in [0,1].$

P r o o f. From equation (4) we obtain the equality

(5) $$\frac{dx}{d|w|} = -\frac{1}{|w|\operatorname{Re} Q(w)}, \qquad Q(w) = \frac{1}{1+xp(w)}.$$

In order to estimate $\operatorname{Re} Q(w)$ we use the region of valued of $\{p(w);\ p \in \mathcal{P}_a\}$ when w is a fixed point of K.

Namely in [6] (or [7]) it was proved that the region of values of $\{p(w);\ p \in \mathcal{P}_a\}$ is the closed disk with centre

$$w_0 = \frac{(1-|w|^2)(1-|w|^2 + 2ia\operatorname{Im} w) + 2|w|^2(1-a^2)}{(1-|w|^2)(1+|w|^2 - 2a\operatorname{Re} w)}$$

and radius

$$R_0 = \frac{2|w|^2(1-a^2)}{(1-|w|^2)(1+|w|^2 - 2a\operatorname{Re} w)}.$$

After straightforward calculations we obtain sharp bounds

(6) $$\left[1 + x\,\frac{1+2a|w|+|w|^2}{1-|w|^2}\right]^{-1} \leq \operatorname{Re} Q(w) \leq \left[1 + x\,\frac{1-|w|^2}{1+2a|w|+|w|^2}\right]^{-1}.$$

The signs of equality hold for the functions

$$p_1(w) = \frac{1+2aw+w^2}{1-w^2}, \quad p_2(w) = \frac{1-w^2}{1+2aw+w^2}.$$

Inequality (6) together with (5) imply

$$(7) \qquad -\left[1+x\,\frac{1+2a|w|+|w|^2}{1-|w|^2}\right]\cdot\frac{1}{|w|} \le \frac{dx}{d|w|} \le -\left[1+x\,\frac{1-|w|^2}{1+2a|w|+|w|^2}\right]\cdot\frac{1}{|w|}.$$

After integrating (7) and using the initial condition we get (4'), what ends the proof.

Now if we multiply the inequality (4') by $(1+x)$ and afterwards use the relation (3), we get

<u>THEOREM 3</u>. <u>If</u> $f \in S_a^o$, <u>then for</u> $z \in K$, $|z| = r < 1$, <u>the following sharp estimates hold</u>:

$$(8) \qquad |f(z)| \ge \frac{1}{\sqrt{1-a^2}}\left\{\operatorname{arc\,tg}\frac{r+a}{\sqrt{1-a^2}} - \operatorname{arc\,tg}\frac{a}{\sqrt{1-a^2}}\right\},$$

$$(9) \qquad |f(z)| \le \int_0^r \frac{dt}{(1-t)^{1+a}(1+t)^{1-a}}.$$

<u>The extremal functions are</u>:

$$f(z) = \int_0^z \frac{dt}{1-2at+t^2} \quad \underline{\text{in}} \quad (8),$$

$$f(z) = \int_0^z [(1-t)^{1+a}(1+t)^{1-a}]^{-1}dt \quad \underline{\text{in}}\ (9).$$

From (8) and (9) in the limit case $a \longrightarrow 1$ we get the corresponding estimates in the whole class of convex univalent functions in K.

The estimates were obtained in [5] by a different method.

<u>3</u>. Now we recall the definition of quasi-starlike functions [1].

We say that g which is holomorphic in K is <u>quasi-starlike</u> in K if it satisfies the equation

$$(10) \qquad F[g(z)] = \frac{1}{M}\cdot F(z),$$

where $M \ge 1$ is a fixed number and F is arbitrary starlike function of the form (1).

The functions of the form $f(z) = M\,g(z)$ are called <u>normalized quasi-starlike functions</u>.

If in the definition (10) we take a starlike function F with second coefficient $a_2 = 2a$, $a \in [0,1]$, then we get the class G_a^M of quasi-starlike functions g with fixed second coefficient:

$$A_2 = 2a(M^{-1} - M^{-2}).$$

In an analogous way as in [1] one can prove

THEOREM 4. *A function g belongs to G_a^M iff it has the form*

$$g(z) = \lim_{x \to \log M} w(z,x), \tag{11}$$

where $w = w(z,x)$ is the solution of the equation

$$\frac{dw}{dx} = -w\,p(w), \quad p \in \mathcal{P}_{-a}, \tag{12}$$

with the initial condition $w(z,0) = z$.

Now we have

THEOREM 5. *If $g \in G_a^M$, then for $z \in K$, $|z| = r < 1$, the following inequalities hold:*

$$\frac{|g(z)|}{(1-|g(z)|)^{1+a}(1+|g(z)|)^{1-a}} \le \frac{1}{M}\,\frac{r}{(1-r)^{1+a}(1+r)^{1-a}}, \tag{13}$$

$$|g(z)| \ge \frac{1}{2r} \cdot [1 - 2ar - \sqrt{4r^2(a^2-1) + \kappa(\kappa-4a)}\,], \tag{14}$$

where $\kappa = M(1 + 2ar + r^2)$. Inequalities (13) and (14) are sharp and the extremal functions are given by the equations

$$g(z)[(1-g(z))^{1+a}(1+g(z))^{1-a}]^{-1} = \frac{1}{M} z[(1-z)^{1+a}(1+a)^{1-a}]^{-1},$$

$$g(z)[1-2a\,g(z) + g^2(z)]^{-1} = \frac{1}{M} z[(1 - 2az + z^2)]^{-1},$$

respectively.

If in (13) and (14) we take normalized quasi-starlike function $f(z) = M\,g(z)$ and put $M \longrightarrow \infty$, then we obtain the corresponding results from [8].

COROLLARY 1. *If $f \in S_a^*$ then for $z = re^{i\varphi}$, $0 < r < 1$, the following sharp estimates hold:*

$$\frac{r}{1+2ar+r^2} \le |f(z)| \le \frac{r}{(1-r)^{1+a}(1+r)^{1-a}}. \tag{15}$$

Putting $a = 0$ in Theorem 5 we obtain

COROLLARY 2. *If g is an odd quasi-starlike function then for $z = re^{i\varphi}$, $0<r<1$, we have sharp estimates*

$$\frac{2r}{M(1+r^2)+\sqrt{M^2(1+r^2)^2-4r^2}} \le |g(z)| \le \frac{2r}{M(1-r^2)+\sqrt{M^2(1-r^2)^2+4r^2}}.$$

The extremal functions satisfy the equations

$$g(z)[1\pm g^2(z)]^{-1} = \frac{1}{M} z(1\pm z^2)^{-1}.$$

Also from Theorem 5 one may obtain the exact bounds for $|g(z)|$ in whole class of quasi-starlike functions [1].

4 In quite similar way (only with some technical difficulties) the Gronwall's problem for quasi-α-starlike functions may be solved as well as for meromorphic quasi-starlike functions.

We state the following

THEOREM 6. *If f is a quasi-1/2-starlike function [1] with fixed coefficient $A_2 = a(M^{-1} - M^{-2})$, then for $z = re^{i\varphi}$, $0<r<1$, the following sharp estimates hold:*

$$\frac{|f(z)|}{(1-|f(z)|)^{\frac{1+a}{2}}(1+|f(z)|)^{\frac{1-a}{2}}} \le \frac{1}{M}\cdot\frac{r}{(1-r)^{\frac{1+a}{2}}(1+r)^{\frac{1-a}{2}}},$$

(16)

$$\frac{|f(z)|}{\sqrt{1+2a|f(z)|+|f(z)|^2}} \ge \frac{1}{M}\cdot\frac{r}{\sqrt{1+2ar+r^2}}.$$

Putting in (15), (16), and (8) $r \longrightarrow 1$ we get

COROLLARY 3. *The Koebe constans for the classes of starlike, 1/2-starlike and convex functions with fixed second coefficients* ($a_2 = 2a$, $a_2 = a$, $a_2 = a$, $a \in [0,1]$) *are equal to*

$$\frac{1}{2(1+a)}, \quad \frac{1}{\sqrt{2(1+a)}}, \quad \frac{1}{\sqrt{1-a^2}} \operatorname{arc\,tg} \sqrt{\frac{1-a}{1+a}},$$

respectively.

References

[1] BAZILEVICH, I.E. and I.DZIUBIŃSKI: Löwner's general equation for quasi-α-starlike functions, Bull. Acad. Polon. Sci. Sér. Sci. Math. Astronom. Phys. 21 (1973), 823-831.

[2] ГОЛУЗИН, Г.М.: Геометрическая теория функций комплексного переменного, Изд. "Наука", Москва 1966.

[3] LEWANDOWSKI, Z. and S.WAJLER: L'equation de Löwner généralisée pour certaines sous-classes de fonctions univalents, Bull. Acad. Polon. Sci. Sér. Sci. Math. Astronom. Phys. 24 (1973), 223-229.

[4] LIBERA, R. and E.ZŁOTKIEWICZ: Loewner-type approximation for convex functions, Colloq, Math. 26 (1976), 143-151.

[5] PADMANABHAN, K.S.: Estimates of growth for certain convex and close-to-convex functions, J. Indian Math. Soc. 33 (1969), 37-48.

[6] SAKAGUCHI, K.: On extremal problems in the classes of functions with positive real part and typically real ones I, Bull. Nara U. Educ. 18 (1969), 1-6.

[7] SZYNAL, A. and J.SZYNAL: On some problems concerning subordination and majorization of functions, Demonstration Math. 11 (1978), 331-350.

[8] TEPPER, D.E.: On the radius of convexity and boundary distortion of schlicht functions, Trans. Amer. Math. Soc. 150 (1970), 519-528.

Institute of Mathematics
Maria-Curie Skłodowska University
Nowotki 10, PL-20-031 Lublin, Poland

THE ALGEBRA OF BOUNDED HYPER-ANALYTIC FUNCTIONS ON THE BIG DISC HAS NO CORONA

Toma V. Tonev (Sofia)

Let H^∞ be the algebra of bounded analytic functions on the open unit disc Δ, provided with the norm $\|f\|_\infty = \sup_\Delta |f|$, and $H(\Delta)$ is the algebra of all analytic functions on Δ. If R_d is the group of real rational numbers, equipped with discrete topology, R_+ denotes nonnegative of them, $G = \hat{R}_d$ denotes the compact group of characters on R_d and Δ_G denotes the "big disc", i.e. the cone over G : $\Delta_G = G \times [0,1)/G \times \{0\}$ with the peak $* = G \times \{0\}/G \times \{0\}$. For any $p \in R_+$ the character $\chi_p(g) = g(p)$ of G is extendable on $\bar{\Delta}_G$ as follows: $\tilde{\chi}_p(\lambda,g) = \lambda^p \chi_p(g)$ for $p \neq 0$ and $\lambda \neq 0$; $\tilde{\chi}_p(*) = 0$ for any $p \neq 0$ and $\tilde{\chi}_0 \equiv 1$ on $\bar{\Delta}_G$. In 1956 R.Arens and I.Singer [1] introduced the algebra A_G of so called generalized analytic functions on Δ_G. Namely, a function $f \in C(\bar{\Delta}_G)$ is called <u>generalized analytic function</u> (relative to the group R_d), iff f is uniformly approximable on $\bar{\Delta}_G$ by hyper-polynomials (i.e. by finite linear combinations $\tilde{\chi}_p$), [4].

<u>Definition 1</u>. We call <u>hyper-analytic</u> any function $f \in C(\Delta_G)$ which is uniformly approximable on Δ_G by functions $h \circ \tilde{\chi}_{1/n}$, where $n \in Z_+$ and h is an analytic function on Δ.

Let H_G^∞ be the algebra of bounded hyper-analytic functions on the big disc Δ_G. Provided with the norm

$$\|f\| = \sup_{\substack{|\lambda|<1 \\ g \in G}} |f(\lambda,g)|$$

and with the pointwise operations, H_G^∞ is a commutative Banach algebra with unit. What can be said about maximal ideal space sp H_G^∞ of algebra H_G^∞? It is clear, that every point $(\lambda,g) \in \Delta_G$ induces a linear and multiplicative functional on H_G^∞, namely $\varphi_{(\lambda,g)} f = f(\lambda,g)$, where $(0,g) = \{*\}$ for any $g \in G$. The Gelfand transform $f \longrightarrow \hat{f}$ is one-to-one and isometric: if $\hat{f} \equiv 0$, then $\hat{f}(\varphi_{(\lambda,g)}) = f(\lambda,g) = 0$ for any

$\lambda < 1$ and $g \in G$, i.e. $f \equiv 0$; $\max_{spH_G^\infty} |\hat{f}| \geq \sup_{\Delta_G} |\hat{f}(\varphi_{(\lambda,g)})| = \|f\| \geq \max_{spH_G^\infty} |\hat{f}|$. Hence H_G^∞ is isometrically isomorphic to algebra $\hat{H}_G^\infty$ like in the classical case [2].

THEOREM 1. *There exists a continuous mapping* τ *from* $\mathrm{sp}\, H_G^\infty$ *onto* $\bar{\Delta}_G$, *which is one-to-one and homeomorphic on* $\tau^{-1}(\Delta_G)$.

P r o o f. Let $\varphi \in \mathrm{sp}\, H_G^\infty$. Evidently $\varphi = \varphi_*$ if $\varphi(\tilde{\chi}_p) = 0$ for every $p \in R_+ \setminus \{0\}$ and then we define $\tau(\varphi_*) = \{*\}$. If $\varphi(\tilde{\chi}_a) \neq 0$ for some $a \in R_+ \setminus \{0\}$, then $\varphi(\tilde{\chi}_p) \neq 0$ for any $p \in R_+$ (see e.g. [5]). The function $\gamma_\varphi : R \longrightarrow \mathbb{C} : \gamma_\varphi(p) = \varphi(\tilde{\chi}_p)/|\varphi(\tilde{\chi}_p)|$ for $p \geq 0$, and $\gamma_\varphi(p) = \overline{\gamma_\varphi(-p)}$ for $p \leq 0$ is a character on R_d, i.e. γ_φ is an element of G. Now we define $\tau(\varphi) = (\lambda_\varphi, \gamma_\varphi)$, where $\lambda_\varphi = |\varphi(\tilde{\chi}_1)|$. It is clear, that $\tau(\varphi) \in \bar{\Delta}_G$ for any $\varphi \in \mathrm{sp}\, H_G^\infty$. Let $(\mu,h) \in \Delta_G$. If $\mu = 0$, i.e. if $(\mu,h) = \{*\}$, then $\tau(\varphi) = \{*\} = (\mu,h)$ by definition. If $\mu \neq 0$, we have also: $\tau(\varphi_{(\mu,h)}) = (\lambda,\gamma) \in \bar{\Delta}_G$. By definition $\lambda = |\varphi_{(\mu,h)}(\tilde{\chi}_1)| = \mu^1 |\tilde{\chi}_1(h)| = \mu$; $\gamma_\varphi(p) = \varphi_{(\mu,h)}(\tilde{\chi}_p)/|\varphi_{(\mu,h)}(\tilde{\chi}_p)|$ $= \tilde{\chi}_p(\mu,h)/|\tilde{\chi}_p(\mu,h)| = \mu^p \chi_p(h)/\mu^p = \chi_p(h) = h(p)$. Consequently $\tau(\varphi_{(\mu,h)}) = (\mu,h)$ for any $(\mu,h) \in \Delta_G$, i.e. $\Delta_G \subset \tau(\mathrm{sp}\, H_G^\infty)$. In order to prove that τ is continuous, we take $\varphi_\alpha \longrightarrow \varphi_0 \in \mathrm{sp}\, H_G^\infty$, so that $\varphi_\alpha(\tilde{\chi}_p) \longrightarrow \varphi_0(\tilde{\chi}_p)$ for every $p \in R_+$. If $\varphi_0(\tilde{\chi}_p) = 0$, we see, applying again the argument from [5], that $\varphi_0(\tilde{\chi}_p) = 0$ for every $p \in R_+ \setminus \{0\}$, i.e. that $\varphi_0 = \varphi_*$. Consequently $\varphi_\alpha(\tilde{\chi}_p) \longrightarrow 0$ for every $p \in R_+ \setminus \{0\}$, and $\varphi_\alpha(\tilde{\chi}_1) \longrightarrow 0$ in particular, i.e. $(\mu_\alpha, \gamma_\alpha) = \tau(\varphi_\alpha)$, so that $\mu_\alpha = |\varphi_\alpha(\tilde{\chi}_1)| \longrightarrow 0$ and $\tau(\varphi_\alpha) = (\mu_\alpha, \gamma_\alpha) \longrightarrow \{*\} = \tau(\varphi_0)$. If $\varphi_0(\tilde{\chi}_p) \neq 0$ for every $p \in R_+ \setminus \{0\}$, then $\mu_\alpha = |\varphi_\alpha(\tilde{\chi}_1)| \longrightarrow |\varphi_0(\tilde{\chi}_1)|$ $= \mu_0$, and $\gamma_\alpha(p) = \varphi_\alpha(\tilde{\chi}_p)/|\varphi_\alpha(\tilde{\chi}_p)| \longrightarrow \varphi_0(\tilde{\chi}_p)/|\varphi_0(\tilde{\chi}_p)| = \gamma_0(p)$, hence $\tau(\varphi_\alpha) = (\mu_\alpha, \gamma_\alpha) \longrightarrow (\mu_0, \gamma_0) = \tau(\varphi_0)$. Because of the compactness of $\mathrm{sp}\, H_G^\infty$, $\tau(\mathrm{sp}\, H_G^\infty)$ is a compact subset of $\bar{\Delta}_G$, containing of course Δ_G, which yields $\tau(\mathrm{sp}\, H_G^\infty) = \bar{\Delta}_G$. Let now $0 \leq \mu < 1$, $h \in G$ and $(\mu,h) = \tau(\varphi)$ for some $\varphi \in \mathrm{sp}\, H_G^\infty$. If $\mu = |\varphi(\tilde{\chi}_1)| = 0$, then $\tau(\varphi) = \{*\}$, and φ can be only φ_* (by definition of τ). If $\mu \neq 0$, $h(p) = \varphi(\tilde{\chi}_p)/|\varphi(\tilde{\chi}_p)|$, form where $\varphi(\tilde{\chi}_p) = |\varphi(\tilde{\chi}_p)| . h(p) = |\varphi(\tilde{\chi}_1)|^p \chi_p(h) = \mu^p \chi_p(h) = \tilde{\chi}_p(\mu,h)$ for every $p \in R_+$. Because the equality $\varphi(f) = f(\mu,h)$ holds for every hyper--polynomial f, it holds also for every hyper-analytic function f by continuity. Consequently $\varphi = \varphi_{(\mu,h)}$, i.e. τ is one-to-one mapping on $\tau^{-1}(\Delta_G)$. Finally, $\tau|_{\tau^{-1}(\Delta_G)}$ is an homeomorphism, because τ is homeomorphic on any compact neighbourhood in $\tau^{-1}(\Delta_G)$. The theorem is proved.

Theorem 1 allows us to extend some results from $H^\infty(\Delta)$-theory to the case of algebra H_G^∞ of bounded hyper-analytic functions on the big disc Δ_G. Like in the classical situation, there is no reason to expect τ to be a homeomorphism as a map of the whole set $\mathrm{sp}\, H^\infty_G$ and there arise the natural question: has the set $\mathrm{sp}\, H^\infty_G \setminus \Delta_G$ nonempty interior ("corona") or not? The following theorem gives the answer:

THEOREM 2. (for H^∞_G-corona property). The big disc Δ_G is dense in $\mathrm{sp}\, H^\infty_G$.

For the proof we need the following

LEMMA. Δ_G is dense in $\mathrm{sp}\, H^\infty_G$ iff there exist bounded hyper-analytic functions $g_1,\dots,g_n$ on Δ_G for any given collection $f_1,\dots,f_n$ of functions $f_j \in H^\infty_G$, $j = 1,\dots,n$, with $|f_1(\lambda,g)| + \dots + |f_n(\lambda,g)| \geq \delta > 0$, $|\lambda| < 1$, $g \in G$, such that $f_1 \cdot g_1 + \dots + f_n \cdot g_n \equiv 1$.

The proof is similar to the proof of classical version of the same lemma (see [2]), and will be omitted.

Proof of Theorem 2. The corona theorem of Carleson [3] asserts that if $f_1,\dots,f_k \in H^\infty$, $\|f_j\|_\infty \leq 1$ and $|f_1| + \dots + |f_k| \geq \sigma > 0$ on Δ, where $0 < \sigma \leq 1/2$, then there exist $g_1,\dots,g_k \in H^\infty$ and constant $C(k,\sigma) > 0$, such that $\|g_j\|_\infty \leq C(k,\sigma)$ and $f_1 \cdot g_1 + \dots + f_k \cdot g_k \equiv 1$. Without loss of generality we may assume that there are given $f_1,\dots,f_n \in H^\infty_G$ with $\|f_j\| \leq 1/2$, $|f_1(\lambda,g)| + \dots + |f_n(\lambda,g)| \geq \delta > 0$ on Δ_G, where $0 < \delta \leq 1/2$. Let $c = C(n,\delta/2) \geq 1$ be the corresponding constant of Carleson. For every $j = 1,\dots,n$ there exist a function $f'_j \in H^\infty$ and a nonnegative integer m_j, such that for $F_j = f'_j \circ \tilde{\chi}_{1/m_j}$ holds: $\|f_j - F_j\| = \sup_{\Delta_G} |f_j - F_j| < \delta/2nc$. Without loss of generality we may assume that m_j does not depend on j (forinstance by replacing them by their product $m = m_1 \cdot m_2 \cdot \dots \cdot m_n$). Now $|F_1| + \dots + |F_n| \geq \Sigma_{j=1}^n |f_j| - \Sigma_{j=1}^n |f_j - F_j| \geq \delta - \delta/2c \geq \delta/2 > 0$ on Δ_G and for bounded analytic functions $f'_1,\dots,f'_n$ on Δ with $\|f'_j\|_\infty = \|F_j\| \leq \|f_j\| + \|f_j - F_j\| = \|f_j\| + \delta/2nc \leq 1$, we have that $|f'_1| + \dots + |f'_n| \geq \delta/2 > 0$ on Δ. According to Carleson's theorem there exist functions $h_1,\dots,h_n \in H^\infty$, $\|h_j\|_\infty \leq C(n,\delta/2) = c$, such that $f'_1 \cdot h_1 + \dots + f'_n \cdot h_n \equiv 1$ on Δ. If $H_j = h_j(\tilde{\chi}_{1/m})$, we see that $\|H_j\| \leq c$ and $F_1 \cdot H_1 + \dots + F_n \cdot H_n = f'_1(\tilde{\chi}_{1/m}) \cdot h_1(\tilde{\chi}_{1/m}) + \dots + f'_n(\tilde{\chi}_{1/m}) \cdot h_n(\tilde{\chi}_{1/m}) \equiv 1$ on Δ_G. Let F denotes the bounded hyper-analytic function $F = f_1 \cdot H_1 + \dots + f_n \cdot H_n$. Now $\|1 - F\| = \|\Sigma F_j H_j - \Sigma f_j H_j\| \leq \|\Sigma F_j - f_j\| \cdot \|H_j\| \leq (\delta/2c) \cdot c = \delta/2 \leq 1/4$. Consequently F is invertible in H^∞_G and the equality $f_1 \cdot g_1 + \dots + f_n \cdot g_n \equiv 1$ holds for the functions $g_j = H_j \cdot F^{-1} \in H^\infty_G$. The theorem is proved.

References

[1] R.ARENS and I.SINGER: Generalized analytic functions, Trans. Amer. Math. Soc. 81 (1956), 379-393.

[2] K.HOFFMAN: Banach spaces of analytic functions, Prentice-Hall inc., Englewood Cliffs, N.J., 1962.

[3] L.CARLESON: Interpolation by bounded analytic functions and the corona problem, Ann. of Math. 76 (1962), 542-559.

[4] T.GAMELIN: Uniform algebras, Prentice-Hall inc., Englewood Cliffs, N.J., 1969.

[5] T.TONEV: Algebras of generalized analytic functions, Banach Center Publ., Warsaw 1977 (to appear).

Institute of Mathematics of the
Bulgarian Academy of Sciences
and Faculty of Mathematics and Mechanics
of the Sofia University
BG-1090 Sofia, P.O.Box 373, Bulgaria

INTEGRAL REPRESENTATIONS OF HOLOMORPHIC FUNCTIONS BY HOLOMORPHIC DENSITIES AND THEIR APPLICATIONS

Nazaret Ervandovič Tovmasjan (Erevan)

Integral representations of holomorphic functions with real or complex densities were obtained by I. N. Vekua, N. I. Mushelišvili, N. P. Vekua, G. S. Litivinčuk and others. By means of these integral representations boundary value problems for holomorphic functions are reduced to singular integral equations. In this paper the author obtained some new integral representations of holomorphic functions by holomorphic densities. They allow to reduce the general boundary value problems of retarded conjugations for holomorphic functions to well-known boundary value problems of nonretarded conjugation. It gives possibilities to solve effectively the boundary value problems of retarded conjugation.

1°. Let us denote by $H(G)$ the class of functions which fulfil on G the Hölder condition. Let D^+ and D_1^+ be simply connected domains with smooth boundaries Γ and Γ_1 on the complex plane, and D^- and D_1^- are the complements to the whole plane of $D^+ \cup \Gamma$ and $D_1^+ \cup \Gamma$, respectively. We consider as positive directions on Γ and Γ_1 those by which the domains D^+ and D_1^+ remain on the left side.

Let $\alpha(t)$ be a sense-preserving homeomorphism of Γ onto Γ_1, $\beta(t)$ – be a sense-reversing homeomorphism of Γ onto Γ_1, and $\gamma(t)$ – a given function on Γ, where $\alpha'(t)$, $\beta'(t)$ and $\gamma(t)$ are different from zero on Γ and belong to the class $H(\Gamma)$. Let m denote the index of $\gamma(t)$ on Γ.

The following integral representations hold:

THEOREM 1. *If $\varphi_1(z)$ is a holomorphic function in the domain D_1^+, $\varphi_1(z) \subset H(D_1^+ \cup \Gamma_1)$, then the functions can be represented in the form*

$$(1) \qquad \varphi_1(z) = \frac{1}{2\pi i} \int_{\Gamma} \frac{\gamma(t)\, \varphi(t)\, dt}{\alpha(t) - z} + P_{m-1}(z), \quad z \in D_1^+,$$

Here $P_{m-1}(z)$ *is a polynomial of degree not greater than* $m-1$, *and*

$\varphi(z)$ is a holomorphic function in the domain D^+, of the class $H(D^+ \cup \Gamma)$. For $m<0$ the derivatives of $\varphi(z)$ should fulfil additional conditions

(2) $\varphi^{(j)}(z_0) = b_j \quad (j=0,\dots,-m-1),$

where z_0 is a fixed point in the domain D^+, b_j are arbitrarily chosen numbers, and $P_{m-1}(z)$ is identically equal to zero.

THEOREM 2. If $\varphi_1(z)$ is a holomorphic function in the domain D_1^+, $\varphi_1(z) \in H(D_1^+ \cup \Gamma_1)$, then the function can be represented as

(3) $$\varphi_1(z) = \frac{1}{2\pi i} \int_\Gamma \frac{\gamma(t)\,\varphi(t)\,dt}{\beta(t) - z}, \quad z \in D_1^+,$$

where $\varphi(z)$ is a holomorphic function in D which in the neighbourhood of infinity fulfils the estimate

(4) $$|\varphi(z)| \leq \text{const}\cdot|z|^{-m-2}.$$

THEOREM 3. If $\varphi_1(z)$ is a holomorphic function in the domain D_1^+ containing the origin of the coordinates, $\varphi_1^{(n)}(z) \in H(D_1^+ \cup \Gamma_1)$, then the function can be represented in the form

(5) $$\varphi_1(z) = \frac{1}{2\pi i} \int_\Gamma \gamma(t)\,\varphi(t)\left[1 - \frac{z}{\alpha(t)}\right]^{n-1} \ln\left[1 - \frac{z}{\alpha(t)}\right] dt + P_m(z),$$

Here $\varphi(z)$ is a holomorphic function in the domain D^+, belonging to the class $H(D \cup \Gamma)$, which for $m<0$ fulfils the additional conditions (2) and $P_m(z)$ is a polynomial of degree not greater than m (with $P_m(z) = \text{const}$ for $m \leq 0$).

THEOREM 4. If $\varphi_1(z)$ is a holomorphic function in the domain D_1^+ containing the origin of coordinates, $\varphi_1^{(n)}(z) \in H(D_1^+ \cup \Gamma_1)$, then the function can be represented in the form

(6) $$\varphi_1(z) = \frac{1}{2\pi i} \int_\Gamma \gamma(t)\,\varphi(t)\left(1 - \frac{z}{\beta(t)}\right)^{n-1} \ln\left(1 - \frac{z}{\beta(t)}\right) dt + c$$

where $\varphi(z)$ is a holomorphic function in the domain D^- and in the neighbourhood of infinity it fulfils the estimate (4); $c = \text{const}$.

In the representations (1), (3), (5), and (6) the functions $\varphi(z)$, $P_{m-1}(z)$, $P_m(z)$, and c are uniquely determined by $\varphi_1(z)$.

The representation (1) holds also in multiply connected domains.

Proof of Theorem 1. Let $m \geq 0$. By the change of the variable

$t=\alpha_{-1}(\zeta)$, where $\alpha_{-1}(\zeta)$ is the inverse mapping to $\zeta=\alpha(t)$, and the Cauchy's integral formula for the holomorphic function $\varphi_1(z)$ and polynomial $P_{m-1}(z)$, the equality (1) can be rewritten in the form

$$(7)\qquad \frac{1}{2\pi i}\int_{\Gamma_1}\frac{\beta_1(\zeta)\,\varphi(\alpha_{-1}(\zeta))-P_{m-1}(\zeta)-\varphi_1(\zeta)}{\zeta-z}\,d\zeta=0,\quad z\in D_1^+,$$

where $\beta_1(\zeta)=\gamma(\alpha_{-1}(\zeta))\,\alpha'_{-1}(\zeta)$.

The equality (7) is equvalent (cf. [1]) to the equality

$$(8)\qquad \beta_1(\zeta)\,\varphi(\alpha_{-1}(\zeta))-P_{m-1}(\zeta)-\varphi_1(\zeta)=\Phi(\zeta),\quad \zeta\in\Gamma_1,$$

where $\Phi(\zeta)$ is the boundary value of a holomorphic function $\Phi(z)$ in D_1^-, vanishing at infinity.

Multiplying the both sides of (8) by $(\zeta-\zeta_0)^{-m}$, where ζ_0 is a fixed point in D_1^+, we obtain

$$(9)\qquad \beta_1(\zeta)(\zeta-\zeta_0)^{-m}\varphi(\alpha_{-1}(\zeta))=\Phi_0(\zeta)+(\zeta-\zeta_0)^{-m}\varphi_1(\zeta),\quad \zeta\in\Gamma_1,$$

where

$$(10)\qquad \Phi_0(z)=(z-z_0)^{-m}(\Phi(z)+P_{m-1}(z)),\quad z\in D_1^-.$$

Obviously $\operatorname{ind}(\beta_1(\zeta)(\zeta-\zeta_0)^{-m})=0$ on Γ_1 and $\Phi_0(z)$ is a holomorphic function D_1^- where $\Phi_0(\infty)=0$.

The boundary value problem (9) is a problem of retarded conjugation with slide with respect to the holomorphic functions $\varphi(z)$ and $\Phi_0(z)$, which possess a unique solution (cf. [1], p. 470). Hence, by the equality (10), we obtain Theorem 1 for $m\geq 0$. Let now $m<0$. From (2) we have

$$(11)\qquad \varphi(z)=\sum_{j=0}^{-m-1}\frac{b_j}{j!}(z-z_0)^j+(z-z_0)^{-m}\,\psi(z),$$

where $\psi(z)$ is a holomorphic function in the domain D^+. By (11) the case $m<0$ reduces to the case $m=0$.

<u>P r o o f of Theorem 3</u>. Let $m\geq 0$. First we prove the uniqueness of the representation (5), i.e., if $\varphi_1(z)\equiv 0$, then $\varphi(z)\equiv 0$ and $P_m(z)\equiv 0$.

Expanding the right-hand side of (5) in powers of z in a neighbourhood of the point $z=0$ and equating the coefficients to zero, we obtain

$$(12)\qquad \int_{\Gamma}\frac{\gamma(t)\,\varphi(t)\,dt}{\alpha^k(t)}=0,\quad k=m+1,\ m+2,\ldots$$

Hence

$$\int_{\Gamma} \frac{\gamma(t)}{\alpha^{m}(t)} \frac{\varphi(t)dt}{\alpha(t)-z} = 0, \quad z \in D_1^+; \tag{13}$$

to this end it is sufficient to expand the left-hand side of (13) in powers of z in a neighbourhood of the point $z=0$ and take (12) into account. Since $\operatorname{ind}(\gamma(t)\alpha^{-m}(t))=0$ on Γ, then from the uniqueness of (1) it follows that $\varphi(z)\equiv 0$, which proves our assertion on the uniqueness of (5) for $m \geq 0$.

Now we shall prove the possibility of representation (5) for $m \geq 0$. From Theorem 1 it follows that the function $\varphi_1^{(n)}(z)$ can be represented in the form

$$\varphi_1^{(n)}(z) = \frac{(-1)^n(n-1)!}{2\pi i} \int_{\Gamma} \frac{\gamma(t)\alpha^{1-n}(t)\,\psi(t)dt}{\alpha(t)-z} + P(z), \tag{14}$$

where $\psi(z)$ is a holomorphic function in D^+ and $P(z)$ is the polynomial of degree not greater than $m-n$ ($P(z)\equiv 0$ for $m<n$).

Let now $P_0(z)$ be the polynomial of degree not greater than m, whose n-th derivative equals $P(z)$. Let us denote by $\varphi_0(z)$ the right-hand side of (5) for $\varphi(z)=\psi(z)$ and $P_m(z)=P_0(z)$; then the equality (14) takes the form $\varphi_1^{(n)}(z)=\varphi_0^{(n)}(z)$. Hence we see that

$$\varphi(z) = \varphi_0(z) + Q_{n-1}(z), \tag{15}$$

where $Q_{n-1}(z)$ is a polynomial of degree not greater than $n-1$.

From the equality (15) follows the validity of (5) for $m \geq n-1$. For the completeness of the proof for $0 \leq m \leq n-2$ it is sufficient to show that the polynomial $Q_{n-1}(z)$ can be represented in the form (5).

Using Theorem 1 for $\varphi_1(z)\equiv 0$ we conclude that there exist $n-m-1$ linearly independent holomorphic functions $\psi_j(z)$ $(j=1,\dots,n-m-1)$ in the domain D^+, fulfilling the equality

$$\frac{1}{2\pi i} \int_{\Gamma} \frac{\gamma(t)\alpha^{1-n}(t)\,\psi_j(t)dt}{\alpha(t)-z} = 0, \quad z \in D_1^+, \quad (0 \leq m \leq n-2). \tag{16}$$

Let us denote by $\Phi_1(z),\dots,\Phi_n(z)$ the right-hand sides of (5) where $\psi_j(z)$ and 0 $(j=0,1,\dots,n-m-1)$; 0 and z^j $(j=0,1,\dots,m)$ are substituted for $\varphi(z)$ and $P_m(z)$. From (16) it follows that $\Phi_j^{(n)}(z) = 0$ $(j=1,\dots,n)$ i.e. $\Phi_j(z)$ are polynomials of degree not greater than $n-1$. From the uniqueness of the representation (5) it follows that the polynomials are linearly independent. Hence any polynomial of degree not greater than $n-1$ can be represented in the form of a linear combination of the polynomials. Thus Theorem 3 for $m \geq 0$ is proved. By the equality (11) Theorem 3 for $m<0$ reduces to the one for $m=0$.

The representations (3) and (6) can be obtained from the representations (1) and (5) by substitution $\tau = 1/(t-z_0)$ and $\Phi(z) = \varphi(1/(z-z_0))(z-z_0)^{-m-2}$.

Let now $\varphi_1(z)$ be a holomorphic function in the domain D_1^- including the infinity. Writing for the holomorphic function $\psi_1(\ ) = \varphi_1(1/(\zeta-z_0))$ the representations (1), (3), (5), and (6) and substituting $z=1/(\zeta-z_0)$ for z we obtain integral representations of the holomorphic function $\varphi_1(z)$ in the infinite domain D_1^-.

It is well known that an arbitrary holomorphic function in a multiply connected domain is the sum of holomorphic functions in simply connected domains. Using for each of these functions the above given representations we shall obtain the analogous integral representations in multiply connected domains.

2°. Generalized Hilbert's problem with slide. Let D, D_k, and G_k be simply connected domains with smooth boundaries Γ, Γ_k, and l_k, respectively. Let $\alpha_k(t)$ be a sense-preserving homeomorphism and $\beta_k(t)$ a sense-reversing homeomorphism of Γ onto Γ_k and l_k, respectively, where $\alpha_k'(t), \beta_k'(t) \in H(\Gamma)$, $\alpha_k'(t) \neq 0$, $\beta_k'(t) \neq 0$ $(k=1,\dots,n)$.

Let us consider the following problem: find in D_k and G_k holomorphic functions $\varphi_k(z)$ and $\psi_k(z)$ $(k=1,\dots,n)$ fulfilling the boundary value condition

$$(17)\qquad \sum_{k=1}^{n}\Big[\sum_{j=0}^{m_k} a_{kj}(t)\varphi_k^{(j)}(\alpha_k(t)) + \sum_{j=1}^{r_k} b_{kj}(t)\psi_k^{(j)}(\beta_k(t))\Big] = f(t),\quad t\in\Gamma,$$

where $a_{kj}(t)$, $b_{kj}(t)$ and $f(t)$ are given n-dimensional vector functions of the class $H(\Gamma)$, and

$$\varphi_k^{(m_k)}(z) \in H(D_k \cup \Gamma_k),\quad \psi_k^{(r_k)}(z) \in H(G_k \cup l_k).$$

Let $A_0(t)$ and $B_0(t)$ be matrices with columns

$$a_{1m_1}(t),\dots,a_{nm_n}(t) \quad\text{and}\quad b_{1r_1}(t),\dots,b_{nr_n}(t),$$

respectively. It is assumed that

$$(18)\qquad \det A_0(t) \neq 0 \quad\text{and}\quad \det B_0(t) \neq 0 \quad\text{on}\quad \Gamma.$$

Without any loss of generality we may assume that D, D_k and G_k $(k=1,\dots,n)$ contain the origin of the coordinates. Accoridng to the representations (1), (3), (5), and (6), the holomorphic functions $\varphi_k(z)$ and $\psi_k(z)$ appearing in the boundary condition (17) can be uniquely represented in the form

$$(19)\qquad \varphi_k(z) = \frac{1}{2\pi i}\int_\Gamma \frac{\Phi_k^+(t)\, d\alpha_k(t)}{\alpha_k(t) - z} \qquad \text{for } m_k = 0,$$

$$(20)\qquad \psi_k(z) = \frac{1}{2\pi i}\int_\Gamma \frac{t\,\Phi_k^-(t)\, d\beta_k(t)}{\beta_k(t) - z} \qquad \text{for } r_k = 0,$$

$$(21)\qquad \varphi_k(z) = \frac{1}{2\pi i}\int_\Gamma \Phi_k^+(t)\left(1 - \frac{z}{\alpha_k(t)}\right)^{m_k-1} \ln\left(1 - \frac{z}{\alpha_k(t)}\right) d\alpha_k(t) + c_k$$

$$\text{for } m_k \geq 1,$$

$$(22)\qquad \psi_k(z) = \frac{1}{2\pi i}\int_\Gamma t\,\Phi_k^-(t)\left(1 - \frac{z}{\beta_k(t)}\right)^{r_k-1} \ln\left(1 - \frac{z}{\beta_k(t)}\right) d\beta_k(t) + d_k$$

$$\text{for } r_k \geq 1,$$

Here $\Phi_k(z)$ is a piecewise holomorphic function on the whole complex plane with the line of discontinuity Γ, vanishing at the infinity, c_k and d_k are constants, $\Phi_k^+(t)$ (respectively $\Phi_k^-(t)$) is the limit value of $\Phi_k(z)$ at the point $t \in \Gamma$ from the interior (respectively from the exterior) of the domain D, $\Phi_k^+(t)$ and $\Phi_k^-(t) \in H(\Gamma)$.

Applying the representations $\varphi_k(z)$ and $\psi_k(z)$ of (19) - (22) in the boundary condition (17), we obtain

$$(23)\qquad \sum_{k=1}^{n} [A_k(t)\,\Phi_k^+(t) - B_k(t)\,\Phi_k^-(t) + \int_\Gamma (M_k(t,\tau)\,\Phi_k^+(\tau)$$

$$+ N_k(t,\tau)\,\Phi_k^-(\tau))d\tau + a_{ko}(t)c_k + b_{ko}(t)d_k] = f(t), \quad t \in \Gamma,$$

where

$$A_k(t) = (-1)^n(n-1)!\, a_{km_k}(t)\alpha_k^{1-n}(t),$$

$$B_k(t) = (-1)^n(n-1)!\, b_{kr_k}(t)\beta_k^{1-n}(t),$$

with $c_k = 0$ for $m_k = 0$ and $d_k = 0$ for $r_k = 0$. Here $M_k(t,\tau)$ and $N_k(t,\tau)$ are completely defined n-dimensional vector-functions of the variables t and τ on Γ, having at the point $t = \tau$ a singularity not greater than of the logarithmic type.

Consequently, the problem (17) is reduced to the well known problem of the retarded conjugation with slides. Using the results of the monograph [1] concerning the boundary value problem (23) we conclude that the boundary value problem (17) is of the Noether type and its index $\varkappa$ is defined by the formula

$$\varkappa = \operatorname{ind}(\det B_0(t)) - \operatorname{ind}(\det A_0(t)) + \sum_{k=1}^{n} (m_k + r_k) + n.$$

In the case where all the domains D, D_k, G_k are identical and in the boundary condition (17) there do not appear the derivatives of the searched functions, the problem (17) was investigated in the monograph [2], where the searched solutions were represented by means of the integrals with complex densities and the problem was reduced to singular integral equations.

References

[1] MUSHELIŠVILI, N.I.: Singular integral equations [in Russian], Izdat. "Nauka", Moscow 1968.

[2] VEKUA, N.P.: Systems of singular integral equations [in Russian], Izd. "Nauka", Moscow 1970.

Chair of Applied Mathematics
Polytechnical Institute
105, Teryan; Erevan 09, USSR

REDUCTION OF THE PROBLEM OF LINEAR CONJUGATION FOR FIRST ORDER NONLINEAR ELLIPTIC SYSTEMS IN THE PLANE TO AN ANALOGOUS PROBLEM FOR HOLOMORPHIC FUNCTIONS

Wolfgang Tutschke (Halle an der Saale)

Contents

Introduction

In this paper we shall construct solutions of general first order elliptic systems of type

(1) $H_j(x,y,u,v,\frac{\partial u}{\partial x},\frac{\partial u}{\partial y},\frac{\partial v}{\partial x},\frac{\partial v}{\partial y}) = 0, \quad j = 1,2.$

Denote $x+iy$ by z, $x-iy$ by z^*, and $u+iv$ by w. Then the system (1) can be written in the complex form

(2) $\frac{\partial w}{\partial z^*} = F(z,w,\frac{\partial w}{\partial z})$

(see [7]), what allows to apply methods of complex analysis in order to solve (1).

In the first part of the paper it will be proved that one can reduce a side-condition $\mathcal{B}w = g$ for (2) to a fixed-point problem for a corresponding operator. We shall show that this method is applicable for very general side-conditions. We need the following property only: For arbitrary g_1, g_2 the equation

(3) $\mathcal{B}(\Phi + g_1) = g_2$

possesses a uniquely determined holomorphic solution Φ (in order to prove the existence of a fixed point some estimates about the holomorphic solution of (3) are needed additionally).

In the second part of the paper the following side-condition will be considered: We look for the solution w fulfilling both a Rieman-Hilbert boundary condition and a condition of linear conjugation on a closed inner curve[1]). For holomorphic functions this mixed problem was solved by E. Obolašvili [5]. For general nonlinear systems (2) in the paper [6] the following special case of the above side-condition was studied: On the boundary a Dirichlet boundary condition is prescribed. A general condition of linear conjugation was replaced by prescibing a jump of the solution on a closed inner curve.

In order to solve the corresponding auxiliary problem for holomorphic functions we shall apply the theory of Cauchy type integrals[2]). In order to conclude that the side-condition has uniquely determined solution we should prescribe the value of the solution in certain points too.

1. A general method for solving first order nonlinear systems in the plane with side-conditions

For a given bounded domain G on $\mathbb{C}$ let T_G and Π_G be the operators defined by

$$T_G\, g[z] = -\frac{1}{\pi}\iint\limits_G \frac{g(\zeta)}{\zeta - z}\, d\xi\, d\eta, \quad \Pi_G\, g[z] = -\frac{1}{\pi}\iint\limits_G \frac{g(\zeta)}{(\zeta-z)^2}\, d\xi\, d\eta .$$

Then the following relations in the sense of Sobolev are fulfilled:

$$(4) \quad \frac{\partial}{\partial z^*} T_G\, g = g, \quad \frac{\partial}{\partial z} T_G\, g = \Pi_G\, g$$

(see, for instance, [1]).

For a given solution w of (2) we define

$$(5) \quad \Phi = w - T_G F(\cdot, w, \frac{\partial w}{\partial z}).$$

By virtue of (4) one obtains immediately

$$\frac{\partial \Phi}{\partial z^*} = \frac{\partial w}{\partial z^*} - F(\cdot, w, \frac{\partial w}{\partial z}) = 0$$

(in the sense of Sobolev), so, by Weyl's lemma, the function Φ is

1) The problem of solving the general side-condition for nonlinear systems of type (1) was proposed by L. Obolašvili during a seminar in the I.N. Vekua Institute for Applied Mathematics of Tbilisi University. During this seminar, in September 1978, the author gave a lecture on a more special side-condition [6].

2) Another method of solving the Riemann-Hilbert boundary value problem is its reduction to the Dirichlet boundary value problem (see [8,9,10]).

proved to be a classical holomorphic function.

Denote $(\partial/\partial z)w$ by h. Differentiating (5) with respect to z we get the following result:

For each given solution w of the equation (2) there exists a holomorphic function Φ such that (w,h) is a solution of the system of singular integral equations

$$(6)\quad \begin{aligned} w &= \Phi + T_G F(\cdot,w,h),\\ h &= \Phi' + \Pi_G F(\cdot,w,h). \end{aligned}$$

Now we shall develop a general concept for solving boundary and other problems for (2) under the assumption that the corresponding problem for holomorphic functions has a unique solution. To this end we have to assume, in addition, that the solution of the corresponding holomorphic problem admits certain a-priori-estimates.

Suppose that the side condition can be written in the form

$$(7)\quad \mathfrak{B} w = g,$$

where $\mathfrak{B}$ is any operator. For example $\mathfrak{B}$ is an operator acting on the boundary ∂G of a given domain G.

In order to construct a solution of (2) satisfying the given side condition (7) we define an operator $\tilde{T}$, with help of the right-hand side of (6), mapping a certain function space $\mathfrak{R}$ into itself:

For a given element (w,h) belonging to a set $\mathfrak{D}$ of $\mathfrak{R}$ a map $\tilde{T}$ will be defined by

$$(8)\quad \begin{aligned} W &= \Phi_{(w,h)} + T_G F(\cdot,w,h),\\ H &= \Phi'_{(w,h)} + \Pi_G F(\cdot,w,h), \end{aligned}$$

where $\Phi_{(w,h)}$ is a holomorphic function depending on (w,h), satisfying the side condition[3])

$$(9)\quad \mathfrak{B}(\Phi_{(w,h)} + T_G F(\cdot,w,h)) = g.$$

Since it was assumed that the equation (9) possesses a uniquely determined holomorphic solution, the operator $\tilde{T}$ is uniquely determined too.

<u>THEOREM</u>. <u>Let $\mathfrak{B}$ be an operator for which there exists a unique-</u>

---

[3]) Further, if we assume that $\mathfrak{B}$ is linear, then the side condition (9) for $\Phi_{(w,h)}$ is equvalent to

$$\mathfrak{B}\Phi_{(w,h)} = g - \mathfrak{B} T_G F(\cdot,w,h).$$

<u>ly determined holomorphic function Φ fulfilling the side condition</u>

$$\mathfrak{B}(\Phi + g_1) = g_2,$$

<u>where g_1, g_2 are arbitrary functions. Then, with given g, the operator $\tilde{T}$, defined with the help of (8) and (9), has the following property:</u>

<u>If (w,h) is a fixed point, then w is a solution of the given differential equation (2). Moreover, w fulfils the side condition</u>

$$\mathfrak{B}w = g.$$

From the definition of $\Phi_{(w,h)}$ it follows immediately that w indeed fulfils the side condition. From the computation similar to the one at the beginning of this section we see that w fulfils the differential equation (2).

The existence of a fixed point follows from the fixed-point theorems. They are applicable if the holomorphic solution of the side-condition (9) admits some estimates. Special nature of such estimates depends on the kind of the considered side condition and, moreover, on the chosen space $\mathfrak{R}$. In Section 3 the corresponding estimates will be regarded in detail for both the Riemann-Hilbert boundary condition and the condition of linear conjugation.

2. Statement of the result

Let G be a bounded simply connected domain on $\mathbb{C}$. Moreover, let a finite number of simple smooth curves $\gamma_1, \ldots, \gamma_p$ in G without common points be given. Their union is denoted by γ, the boundary curve is denoted by γ_0. We assume that all γ_j, $j = 0,1,\ldots,p$, have Hölder-continuous representations with Hölder exponent α, $0 < \alpha < 1$. We assume, furthermore, that Hölder continuously differentiable complex-valued functions $\tilde{g} = \tilde{g}(t)$ and $g = g(t)$ on γ are given (with the same Hölder-exponent α). Analogously it is assumed that on $\gamma_0 = \partial G$ there are given Hölder-continuously differentiable real-valued functions $a = a(t)$, $b = b(t)$, and $c = c(t)$, where $a^2 + b^2 \neq 0$ everywhere on γ_0.

We look for a piecewise continuous solution $w = u + iv$ of (1) fulfilling on γ_0 the Riemann-Hilbert boundary condition

$$\text{(10)} \quad \mathrm{Re}[(a + ib)w] = au - bv = c$$

and on γ the condition of linear conjugation

(11) $w^+ = \tilde{g}\, w^- + g.$

Here w^+ and w^- denote the limits of w on γ from the left and from the right side, respectively (with respect to a given orientation of the curves γ_j, $j = 1,\dots,p$).

The domain G is decomposed into a finite number of subdomains G_k by the given curves γ_j in G. Denote by $\mathcal{C}_\alpha(\bar{G};\gamma_1,\dots,\gamma_p)$ the space containing all piecewise continuous complex-valued functions on $\bar{\Omega}$, in general with jumps on γ_j, where it is required for w to satisfy the Hölder condition with exponent α on each $\bar{G}_k$. The norm is defined by

(12) $$\| w \|_\alpha = \max_k \Big(\sup_{G_R} |w|,\ \sup_{z_1 \neq z_2 \in G_k} \frac{|w(z_2) - w(z_1)|}{|z_2 - z_1|^\alpha}\Big).$$

Denote by $\mathcal{C}^1_\alpha(G;\gamma_1,\dots,\gamma_p)$ the space of all elements of $\mathcal{C}_\alpha(\bar{G};\gamma_1,\dots,\gamma_p)$, where both $(\partial/\partial z)w$ and $(\partial/\partial z^*)w$ belong to $\mathcal{C}_\alpha(G;\gamma_1,\dots,\gamma_p)$. The norm in this space is defined by

$$\| w \|_{1+\alpha} = \max\Big(\| w \|_\alpha,\ \Big\|\frac{\partial w}{\partial z}\Big\|_\alpha,\ \Big\|\frac{\partial w}{\partial z^*}\Big\|_\alpha\Big).$$

Using the derivatives with respect to the length of arcs it is possible to define a norm in $\mathcal{C}^1_\alpha(\gamma_j)$.

The notation $\mathcal{C}_\alpha(\bar{G};\gamma_1,\dots,\gamma_p)$ will be used also for the set of all pairs (w,h) for which both w and h belong to $\mathcal{C}_\alpha(\bar{G};\gamma_1,\dots,\gamma_p)$. In this case the norm is defined by

$$\| (w,h) \|_\alpha = \max(\| w \|_\alpha,\ \| h \|_\alpha).$$

Now we consider the bicylinder

$$\mathfrak{D} = \{(w,h) \in \mathcal{C}_\alpha(\bar{G};\gamma_1,\dots,\gamma_p) : \| w \|_\alpha \le R_1,\ \| h \|_\alpha \le R_2\}.$$

On the right-hand-side $F(z,w,h)$ of the differential equation (2) it is assumed firstly that

(13) $$|F(z,w,h) - F(\tilde{z},\tilde{w},\tilde{h})| \le l(|z - \tilde{z}|^\alpha + |w - \tilde{w}| + |h - \tilde{h}|)$$

for $z,\tilde{z} \in G$ and for all w, $\tilde{w}$, h, $\tilde{h}$ such that $|w| \le R_1$, $|\tilde{w}| \le R_1$, $|h| \le R_2$, and $|\tilde{h}| \le R_2$. From this condition it follows that $F(\cdot,w,h)$ belongs to $\mathcal{C}_\alpha(\bar{G};\gamma_1,\dots,\gamma_p)$ if $(w,h) \in \mathcal{C}^\alpha(\bar{G};\gamma_1,\dots,\gamma_p)$. Secondly we shall assume that the right-hand-side of (2) fulfils in $\mathfrak{D}$ the Lipschitz condition

(14) $$\| F(\cdot,w,h) - F(\cdot,\tilde{w},\tilde{h}) \|_\alpha \le L \| (w,h) - (\tilde{w},\tilde{h}) \|_\alpha$$

(with respect to the metric of $\mathcal{C}_\alpha(\bar{G}; \gamma_1, \ldots, \gamma_p)$).

In analogy to the formulae (12) we define the norm of functions which are Hölder-continuous on γ_j or on $\gamma_0 = \partial G$. The norms are denoted by $\|\cdot\|_{\alpha,\gamma_j}$ and $\|\cdot\|_{\alpha,\partial G}$, respectively. The corresponding spaces are denoted by $\mathcal{C}_\alpha(\gamma_j)$ and $\mathcal{C}_\alpha(\partial G)$.

The index $\varkappa$ of the Riemann-Hilbert boundary condition is defined as an increment of $(1/\pi i)\log(a - ib)$ on γ_0. Consequently, it is an even number (see e.g. [3]). We assume that the index $\varkappa$ is non-negative.

In order to formulate our theorem we must consider $1 + \frac{1}{2}\varkappa$ points in G not belonging to any γ_j. The general solution of the Riemann -Hilbert boundary value problem with index $\varkappa$, $\varkappa \geq 0$, contains an arbitrary polynomial of the form

$$\sum_{k=0}^{\varkappa} d_k z^k,$$

where $a_k = a^*_{\varkappa-k}$ for $k = 0,1,\ldots,\varkappa$ (see [3]). The last condition means that $a_0,\ldots,a_{\frac{1}{2}\varkappa-1}$ and $\mathrm{Re}\, a_{\frac{1}{2}\varkappa}$ can be chosen arbitrarily. In order to determine these coefficients with help of the values of the polynomial at given points z_l, $l = 1,\ldots,\frac{1}{2}\varkappa$, and with help of its real part at z_0, we must take into consideration the following condition: the determinant with elements

$$\mathrm{Re}\, z_l^k + \mathrm{Re}\, z_l^{\varkappa-k},\ k = 0,1,\ldots,\tfrac{1}{2}\varkappa,\ l = 0,1,\ldots,\tfrac{1}{2}\varkappa,$$

$$-\mathrm{Im}\, z_l^k + \mathrm{Im}\, z_l^{\varkappa-k},\ k = 0,1,\ldots,\tfrac{1}{2}\varkappa - 1,\ l = 0,1,\ldots,\tfrac{1}{2}\varkappa,$$

$$\mathrm{Im}\, z_l^k + \mathrm{Im}\, z_l^{\varkappa-k},\ k = 0,1,\ldots,\tfrac{1}{2}\varkappa,\ l = 1,\ldots,\tfrac{1}{2}\varkappa,$$

$$\mathrm{Re}\, z_l^k - \mathrm{Re}\, z_l^{\varkappa-k},\ k = 0,1,\ldots,\tfrac{1}{2}\varkappa - 1,\ l = 1,\ldots,\tfrac{1}{2}\varkappa$$

is different from zero. Any points fulfilling this condition are called admissible.

<u>THEOREM</u>. <u>Assume that</u> $\| F(z,0,0) \|_\alpha$ <u>and the Lipschitz constant</u> L <u>are not greater than a certain bound. Then there exists a uniquely determined solution</u> w <u>with</u> $(w,(\partial/\partial z)w) \in D$ <u>of the differential equation</u> (2) <u>satisfying the following side-conditions</u>:

a) <u>on</u> G <u>the Riemann-Hilbert boundary condition</u> (10) <u>is fulfilled</u> (<u>where its index</u> $\varkappa$ <u>is non-negative</u>),

b) <u>on</u> γ <u>the condition</u> (11) <u>of linear conjugation is fulfilled</u>,

c) $\mathrm{Re}\, w(z_0)$ <u>and</u> $w(z_l)$, $l = 1,\ldots,\frac{1}{2}\varkappa$, <u>has arbitrarily prescribed values</u> (<u>here it is assumed that the points</u> $z_0, z_1, \ldots z_{\frac{1}{2}}$ <u>are admis-</u>

sible).

The bounds for $\| F(z,0,0) \|_\alpha$ and L are uniform with respect to all the cases for which

$$\| c \|_{1+\alpha,\partial G}, \quad \| \tilde{g} \|_{1+\alpha,\gamma}, \quad \| g \|_{1+\alpha,\gamma},$$

$$|\mathrm{Re}\, w(z_0)| \quad \text{and} \quad |w(z_l)|, \quad l = 1,\dots,\tfrac{1}{2}\varkappa,$$

are not larger than certain constants depending on the choice of R_1, R_2.

3. Proof of the theorem

Firstly we remark that T_G and Π_G are bounded operators in $\mathcal{C}_\alpha(\bar{G};\, \gamma_1,\dots,\gamma_p)$ (see [1, 6]). Moreover T_G is a bounded operator mapping $\mathcal{C}_\alpha(\bar{G};\, \gamma_1,\dots,\gamma_p)$ into $\mathcal{C}^1_\alpha(\bar{G};\, \gamma_1,\dots,\gamma_p)$. Denote by $I_{\gamma_j} f$ the function defined by

$$I_{\gamma_j} f[z] = \frac{1}{2\pi i} \int_{\gamma_j} \frac{f(\zeta)}{\zeta - z}\, d\zeta .$$

Then I_{γ_j} is a bounded operator mapping $\mathcal{C}^1_\alpha(\gamma_j)$ into itself. On the other hand I_{γ_j} defines a bounded operator mapping $\mathcal{C}^1_\alpha(\gamma_j)$ into $\mathcal{C}^1_\alpha(\bar{G};\, \gamma_j)$ (this follows from the properties of Cauchy-type integrals).

Without any loss of generality we can assume that G is the unit disk. Let (w,h) be an arbitrary element belonging to the bicylinder D, where the composed function with the values $F(z,w(z),h(z))$ is denoted by $\mathfrak{G}$.

According to Section 1 we have to choose a function $\Phi_{(w,h)}$ in such a manner that all images W fulfil the given side condition. Since

$$\Phi_{(w,h)} = W - T_G \quad ,$$

one gets for $\Phi_{(w,h)}$ the following two conditions:

$$\text{(15)} \quad \mathrm{Re}[(a + ib)\,\Phi_{(w,h)}] = c - \mathrm{Re}[(a + bi)T_G\mathfrak{G}],$$

$$\text{(16)} \quad \Phi^+_{(w,h)} = \tilde{g}\,\Phi^-_{(w,h)} + (g + (\tilde{g} - 1)T_G\mathfrak{G}).$$

It is easy to check that the new coefficients in the conditions (15) and (16) have the same properties as the initial coefficients in the conditions (10) and (11). Let $X = X(z)$ be the well-known canonical solution fulfilling the homogenous condition of linear conjugation, which is given explicitly by a Cauchy-type integral over $\gamma = \gamma_1 \cup \dots \cup \gamma_p$. This solution is different from zero everywhere in $\overline{G}$. Multiplying this function $X = X(z)$ by a constant we can assume that $X(z_o) = 1$. On the other hand the general solution of the inhomogenous problem (16) of linear conjugation has the representation

$$(17) \quad \Phi_{(w,h)} = X(\Psi_o + \Psi),$$

where $\Psi_o = I_\gamma[g + (\tilde{g} - 1)\, T_G \sigma]$ and Ψ is an arbitrary holomorphic function in G. Substituting the expression (17) in (15) one gets an analogous Riemann-Hilbert boundary condition for Ψ. Since $X = X(z)$ can be represented as $X(z) = \exp \Gamma(z)$ with uniquely determined $\Gamma(z)$, the considered substitution does not change the index of the boundary condition.

Finally the values $W(z_o)$ and $W(z_l)$, $l = 1,\dots,\frac{1}{2}\varkappa$, are prescribed. From formula (17) we get, consequently, the conditions

$$\begin{aligned}(18) \quad \operatorname{Re} \Psi(z_o) &= \operatorname{Re} \Phi_{(w,h)}(z_o) - \operatorname{Re} \Psi_o(z_o) \\ &= \operatorname{Re} W(z_o) - \operatorname{Re} T_G \sigma[z_o] - \operatorname{Re} \Psi_o(z_o)\end{aligned}$$

and

$$(19) \quad \Psi(z_l) = (X(z_l))^{-1}\, (W(z_l) - T_G \sigma[z_l]) - \Psi_o(z_l).$$

Therefore the images W fulfil all prescribed conditions if the holomorphic function $\Phi_{(w,h)}$ realizes the analogous conditions (15), (16), (18), (19). Since the auxiliary problem for holomorphic functions has a uniquely determined solution, the operator $\tilde{T}$ is uniquely defined.

Using the assumption (14) one gets

$$\begin{aligned}\| T_G \sigma \|_{1+\alpha} &\le \text{const}\, \| \sigma \|_\alpha = \text{const}\, \| F(\cdot,w,h)\|_\alpha \\ &\le \text{const}\, (\| F(z,0,0)\|_\alpha + \| F(z,w,h) - F(z,0,0)\|_\alpha) \\ &\le \text{const}\, (\| F(z,0,0)\|_\alpha + L(R_1 + R_2))\end{aligned}$$

for all $(w,h) \in \mathfrak{D}$.

From the boundedness of I_{γ_j} it follows that

$$\| \Psi_o \|_{1+\alpha} \le \text{const}\, \| g + (\tilde{g} - 1)\, T_G \sigma \|_{1+\alpha,\gamma}.$$

This means that the operator $\tilde{T}$ maps $\mathfrak{D}$ into itself if $\| g \|_{1+\alpha,\gamma}$, $\| \tilde{g} \|_{1+\alpha,\gamma}$, $|\operatorname{Re} w(z_o)|$, $|w(z_l)|$ for $l = 1,\dots,\frac{1}{2}\varkappa$, $\| F(z,0,0)\|_\alpha$, and L are not greater than certain constants.

Let $(\tilde{w},\tilde{h})$ be another element belonging to $\mathfrak{D}$. Setting $F(\cdot,\tilde{w},\tilde{h}) = \tilde{\sigma}$ we get

$$\|\sigma - \tilde{\sigma}\|_\alpha \leq L(\| w - \tilde{w} \|_\alpha + \| h - \tilde{h} \|_\alpha)$$
$$\leq 2L \|(w,h) - (\tilde{w},\tilde{h})\|_\alpha .$$

Since $\| \tilde{T}(w,h) - \tilde{T}(\tilde{w},\tilde{h})\|_\alpha$ can be estimated by $\|\sigma - \tilde{\sigma}\|_\alpha$, the operator $\tilde{T}$ is contractive if L is small enough. Then the existence of a fixed point of $\tilde{T}$ follows from the Banach fixed-point theorem. On the contrary, the function $\Phi_{(w,h)} = w - T_G \sigma$ fulfils the conditions (15), (16) if w is any solution with $(w, (\partial/\partial z)w) \in \mathfrak{D}$. In this way it appears that for a given solution w the pair $(w, (\partial/\partial z)w)$ is a fixed point of $\tilde{T}$. The theorem is proved.

4. Concluding remarks

Instead of the assumption (14) one can assume that the right-hand side of (2) satisfies a weaker condition:

$$\| F(\cdot,w,h) - F(\cdot,\tilde{w},\tilde{h})\|_\alpha \leq L_1 \| w - \tilde{w}\|_\alpha + L_2 \| h - \tilde{h}\|_\alpha .$$

Then one can replace the bound for L_1 by a bound for the diameter of the regarded domains (see [7] where the case of domains with variable diameter was studied in the Dirichlet boundary value problem).

Applying the Schauder fixed point theorem instead of the Banach theorem one gets still weaker limitations for the given boundary values.

The proved theorem contains as its special case $(g = 1,\ g = 0)$ the solvability of the Riemann-Hilbert boundary value problem (with an index $\varkappa \geq 0$).

All considerations are valid in the case of vector-differential equations, too. In the case of Section 1 this is immediately clear. Using the concept of canonical matrices the considerations of the other sections are applicable in the case of vector-differential equations.

Bibliography

[1] И. Н. ВЕКУА: Обобщенные аналитические функции, Москва 1959.

[2] Ф. Д. ГАХОВ: Краевые задачи, Изд. второе, Москва 1963.

[3] Н. И. МУСХЕЛИШВИЛИ: Сингулярные интегральные уравнения, Изд. второе, Москва 1962.

[4] W. POGORZELSKI: Integral equations and their applications, Warszawa 1966.

[5] Е. Л. ОБОЛАШВИЛИ: Безмоментные оболочки положительной кривизны, на которых действуют разрывные внешние силы, Труды конференции по теории пластин и оболочек, Казань 1960.

[6] W. TUTSCHKE: Die neuen Methoden der komplexen Analysis und ihre Anwendung auf nichtlineare Differentialgleichungssysteme, Sitzungsberichte d. Akad. d. Wiss. d. DDR, Heft 17N, 1976.

[7] ——: Vorlesungen über partielle Differentialgleichungen, Leipzig 1978.

[8] ——: Задача Римана-Гильберта для нелинейных систем дифференциальных уравнений в плоскости, in "Комплексный анализ", Сборник посвящ. И. Н. Векуа, Москва 1978.

[9] ——: О задаче Римана-Гильберта для нелинейных систем уравнений первого порядка эллиптического типа, Труды института прикладной математики им. И. Н. Векуа, Тбилиси (in print).

[10] WENDLAND, W.: Elliptic systems in the plane, Oxford 1979.

Sektion-Mathematik der Martin-
Luther-Universität Halle-Wittenberg
Universitätsplatz 8/9
DDR-402 Halle an der Saale, DDR

G-INDEX OF AN INVARIANT DIFFERENTIAL OPERATOR AND ITS APPLICATIONS

Antoni Wawrzyńczyk (Warszawa)

Contents

Introduction

The present paper is a contribution to the global theory of differential equations on non-compact homogenous manifolds. Given an invariant differential operator we define its G-index in terms of group representation theory.

In the case of a compact group G the G-index coincides with homogenous index considered by R. Bott [3] and our main theorem (Thm 2.2) reduces to the Bott index theorem.

The example of symmetric space of noncompact type G/K provide evidence of efficiency of our result. When applied to Cauchy-Riemann operator on a hermitian symmetric space or to the Dirac operator, it shows that the space of L^2-solutions of the equation $Df = 0$ is non-zero. In particular our method can be used for the investigation of the discrete series representations of G.

1. Invariant differential operators

In this section we deal with an arbitrary unimodular Lie group G countable at infinity. Let K be a compact subgroup of G. Given a finite-dimensional unitary representation (τ, E) of K we denote

* This project was supported by funds from the Maria Skłodowska-Curie Fund established by the contributions of the United States and Polish Governments.

by $\mathbb{E}_\tau$ the homogenous vector bundle associated to the principal bundle $G \longrightarrow G/K$. The bundle $\mathbb{E}$ admits a hermitian structure invariant with the group action. We denote by $C^\infty(\mathbb{E}_\tau)$ the G-invariant space of infinitely differentiable sections of the bundle and by $L^2(\mathbb{E}_\tau)$ the space of sections, which are square-integrable with respect to the G-invariant measure on G/K.

The representation of G on $L^2(\mathbb{E}_\tau)$ is unitary and equivalent to the representation induced by (τ, E) in the sense of Mackey. This representation will be denoted by U^τ. It can be constructed equivalently in the following way. Define a unitary action of K in the space $L^2(G)\otimes E = L^2(G,E)$ as follows:

$$kf(g) := \tau(k)(f(gk)).$$

This action commutes with the translations of f by the left. Denote by dk the Haar measure of the group K. The operator

$$(1.1)\quad P_\tau f := \int_K kf\, dk$$

projects the space $L^2(G,E)$ onto Mackey's space H^τ and interwines the regular representation of G on $L^2(G,E)$ with the representation U^τ.

Now, let $D : C^\infty(\mathbb{E}_\tau) \longrightarrow C^\infty(\mathbb{E}_\sigma)$ be a differential operator commuting with the action of G on corresponding bundles. Recall that after choosing hermitian structure on $\mathbb{E}_\tau$, $\mathbb{E}_\sigma$ we have uniquely determined formally adjoint operator $D^+ : C^\infty(\mathbb{E}_\sigma) \to C^\infty(\mathbb{E}_\tau)$ satisfying the condition:

$$(1.2)\quad (Ds\,|\,\varphi)_{L^2(\mathbb{E}_\sigma)} = (s\,|\,D^+\varphi)_{L^2(\mathbb{E}_\tau)}$$

for all $s \in C^\infty(\mathbb{E}_\tau)$ and compactly supported $\varphi \in C^\infty(\mathbb{E}_\sigma)$.

The aim of this section is to define in $L^2(\mathbb{E}_\tau)$ a dense G-invariant domain of the operator D. It can be done in various ways. The construction chosen is especially convenient for our purposes explained in Section 2.

Define on $L^2(G,E)$ the representation of the group $G\times G$ by the formula:

$$(1.3)\quad U_{(g,h)}f(x) := L_g R_h f(x) = f(g^{-1}xh).$$

The representation U is equivalent to the one induced by the trivial representation on E of the subgroup $G := \{(g,g^{-1}) : g \in G\}$. $L^2(G,E)_\infty$ will denote the space of smooth vectors of this representation. Recall, that according to Goodman's description of the space of smooth vectors of an induced representation we have [12]:

$$L^2(G,E)_\infty = \{f \in C^\infty(G,E) : XfY \in L^2(G,E),\ X,Y \in \mathcal{G}\}.$$

We have denoted by $\mathcal{G}$ the universal enveloping algebra of the Lie algebra $\mathfrak{g}$ of G. The element X (resp.Y) acts on f as a right (left) invariant operator on G.

Now, define a G-invariant subspace of $H^\tau \in L^2(\mathbb{E}_\tau)$ consisting of differentiable functions (sections):

$$\Phi_\tau := P_\tau L^2(G,E)_\infty .$$

Since the space of smooth vectors of a representation is dense in the representation space, we observe that Φ_τ is dense in H^τ.

PROPOSITION 1.1. $D(\Phi_\tau) \subset \Phi_\sigma \subset L^2(\mathbb{E}_\sigma)$.

Proof. We appeal to the description of a differential operator by means of universal derivative (see Palais [9]). Denote by ∇^0 the covariant derivative in the bundle $\mathbb{E}_\tau$ and by ∇^k the covariant derivative in $(\overset{k}{\otimes} T^*) \otimes E$ determined by the G-invariant hermitian structures. To each $s \in C^\infty(\mathbb{E}_\tau)$ we associate a section $J^k s$ of the bundle $\bigoplus_{i=0}^{k} (\overset{i}{\otimes} T^* \otimes \mathbb{E}_\tau)$ whose i-th component is equal to $\nabla^{i-1} \dots \nabla^0 s$, 0-component $= s$. Then for every k-th order differential operator $D: C^\infty(\mathbb{E}_\tau) \longrightarrow C^\infty(\mathbb{E}_\sigma)$ there exists a bundle homomorphism $A: \bigoplus_{i=0}^{k} (\overset{i}{\otimes} T^*) \otimes \mathbb{E}_\tau \longrightarrow \mathbb{E}_\sigma$ such that

(1.4) $\quad Ds = A \circ J^k s.$

The G-invariant covariant derivative in an arbitrary induced bundle can be described as an operator on E-valued functions on G. Let $\mathfrak{g} = \mathfrak{k} + \mathfrak{p}$ be the K-invariant decomposition of the Lie algebra $\mathfrak{g}$ of G into the direct sum of the Lie algebra $\mathfrak{k}$ of K and the subspace $\mathfrak{p}$, which can be identified with the tangent space T of the manifold G/K at the point eK. There exists a K-invariant inner product on $\mathfrak{p}$. Chose an orthonormal base $\{X_i\}$ in $\mathfrak{p}$ and the dual base $\{X_i^*\}$ in $\mathfrak{p}^* = T^*$. Then the covariant derivative acts on a differentiable E-valued function by the formula:

(1.5) $\quad \nabla f = \sum_{i=1}^{\dim \mathfrak{p}} fX_i \otimes X_i^* .$

Comparing the formulas (1.4) and (1.5) we observe that

$$Df \in L^2(G,F)_\infty \cap H^\sigma = \Phi_\sigma .$$

In virtue of Proposition 1.1 we can treat the operator D as a densely defined G-invariant operator from $L^2(\mathbb{E}_\tau)$ into $L^2(\mathbb{F}_\sigma)$.

2. Discrete part of the induced representation

G-index theorem

An irreducible representation of the group G is called square-integrable if it is a direct summand of the left regular representation of G in $L^2(G)$. The set of all classes of square-integrable irreducible representations of G is denoted by $\hat{G}_d$.

According to the Frobenius reciprocity the discrete part of the representation U ($= L \times R$) of $G \times G$ is equal to

$$(2.1)\quad L^2(G)_d = \bigoplus_{[\pi]\in\hat{G}_d} H_\pi \otimes \bar{H}_\pi ,$$

$$U^d = \bigoplus_{[\pi]\in\hat{G}_d} \pi \otimes \bar{\pi} .$$

(π, H_π) denotes a square-integrable irreducible representation of the class $[\pi] \in \hat{G}_d$.

The discrete part of the representation U^τ in $L^2(\mathbb{E})$ is then of the form

$$P_\tau(L^2(G)_d \otimes E) = \bigoplus_{[\pi]\in\hat{G}_d} H_\pi \otimes P_\tau(\bar{H}_\pi \otimes E).$$

The action of the operator P_τ on $\bar{H}_\pi \otimes E$ is given by the formula:

$$P_\tau(f \otimes e) = \int_K \pi(k) f \otimes \tau(k) e \, dk.$$

The tensor product $\bar{H}_\pi \otimes E$ can be identified with the space $L(H_\pi, E)$ of all linear homomorphisms and under this identifications the subspace $P_\tau(\bar{H}_\pi \otimes E)$ corresponds to the space $L_K(H_\pi, E)$ of all operators interwining for $\pi|K$ and τ. Thus we can write:

$$(2.2)\quad L^2(\mathbb{E}_\tau)_d = \bigoplus_{[\pi]\in\hat{G}_d} H_\pi \otimes L_K(H_\pi, E) =: \bigoplus_{[\pi]\in\hat{G}_d} H_\pi \otimes V_\pi .$$

Throughout what follows we shall assume that the subgroup K is massive in G, what means that for any $\pi \in \hat{G}$ and $\tau \in \hat{K}$ the dimension of the space $L_K(\pi|K, \tau)$ (= multiplicity of τ in $\pi|K$) is finite. Given a set S we denote by $\mathbb{Z}(S)$ the abelian group of all finite formal combinations of elements of S with coefficients in $\mathbb{Z}$ and by $\mathbb{Z}(S)^*$ the group of all such combinations, that is the space of all functions $S \ni s \longrightarrow n_s \in \mathbb{Z}$.

There exists unique $\mathbb{Z}$-bilinear multiplicity function $\langle : \rangle$ on $\mathbb{Z}(\hat{K})^* \times \mathbb{Z}(\hat{K})$ which satisfies:

$$\langle 1\cdot\tau : 1\cdot\tau_1\rangle = \begin{cases} 0, & \tau \neq \tau_1 \\ 1, & \tau = \tau_1 \end{cases}, \quad \text{for } \tau,\ \tau_1 \in \hat{K}.$$

The assumption K to be massive in G enables us to define a restriction map: $i^* : \mathbb{Z}(\hat{G}_d) \longrightarrow \mathbb{Z}(\hat{K})^*$ defined uniquely by the formula

$$(2.3)\quad i^*(\pi) = \sum_{\tau\in\hat{K}} \langle \pi | K : \tau\rangle \tau \qquad \text{for } \pi \in \hat{G}_d.$$

On the other hand we can introduce a mapping

$$i_* : \mathbb{Z}(K) \longrightarrow \mathbb{Z}(G_d)^*$$

putting:

$$(2.4)\quad i_*(\tau) = \sum_{\pi\in\hat{G}_d} \langle \pi | K : \tau\rangle \pi .$$

The group homomorphism i_* will be referred to as the formal induction mapping. Given a completely reducible representation of G, say $V = \bigoplus_{[\pi]\in\hat{G}_d} n_\pi \pi$, we denote by $[V]$ the formal combination $\sum_{[\pi]\in\hat{G}_d} n_\pi[\pi]$.

Then according to the formula (2.2)

$$(2.5)\quad [U_d^\tau] = \sum_{[\pi]\in\hat{G}_d} \langle \pi | K : \tau\rangle\, [\pi] = i_*(\tau).$$

Return to considering the invariant differential operator D. The domain Φ_τ defined in the previous section has a nontrivial intersection with all of the discrete parts of the representation.

LEMMA 2.1. <u>The space</u> $\Phi_\tau \cap L^2(\mathbb{E}_\tau)_d$ <u>is dense in</u> $L^2(\mathbb{E}_\tau)_d$. <u>Moreover</u> $\Phi_\tau \cap (H_\pi \otimes V_\pi)$ <u>is dense in</u> $H_\pi \otimes V_\pi$ <u>for any</u> $[\pi] \in \hat{G}_d$.

<u>Proof</u>. The set $\Psi := (L^2(G)\otimes E)_\infty \cap (H_\pi \otimes \bar{H}_\pi \otimes E)$ consists of smooth vectors of the representation U of $G\times G$ in $H_\pi \otimes H_\pi \otimes E$ and then is dense in this space. Being so, the space $(H_\pi \otimes V_\pi) \cap \Phi_\tau = P_\tau(\Psi)$ is dense in $P_\tau(H_\pi \otimes \bar{H}_\pi \otimes E) = H_\pi \otimes V_\pi$ and the proof follows.

With the aid of the lemma we can restrict the differential operator D to the space $H_\pi \otimes V_\pi$. The restriction $D_\pi := D | H_\pi \otimes V_\pi$ is a densely defined invariant operator valued on $L^2(\mathbb{E}_\sigma)$.

Using standard arguments we prove D_π to be closeable. In fact, if x_n belongs to the domain of D_π and $x_n \longrightarrow 0$ in $L^2(\mathbb{E}_\tau)$ and $Dx_n \longrightarrow f$ in $L^2(\mathbb{E}_\sigma)$, then for any $\varphi \in C_0(\mathbb{E}_\sigma)$ we have: $(f|\varphi) = \lim_n (Dx_n|\varphi) = \lim_n (x_n|D^+\varphi) = 0$. Thus $f = 0$.

The closure of the operator D_π will be denoted by $\tilde{D}_\pi$. According to von Neumann theorem the operator D admits the polar decomposition $\tilde{D}_\pi = U\,H$, where $H : H_\pi \otimes V_\pi \longrightarrow H_\pi \otimes V_\pi$ is a self-adjoint operator commuting with the action of G and U is partially isometric. The commutand of the factor representation of G in $H_\pi \otimes V_\pi$ is equal to $\mathrm{id} \otimes L(V_\pi)$, hence the operator H is bounded. It means that the operator D is bounded. Let us write

$$(2.6)\quad L^2(\mathbb{F}_\sigma)_d = \bigoplus_{[\pi] \in \hat{G}_d} H_\pi \otimes W_\pi .$$

The operator $\tilde{D}_\pi$ maps continuously $H_\pi \otimes V_\pi$ into $H_\pi \otimes W_\pi$ and intertwines the corresponding actions of G. Hence this operator is of the form

$$(2.7)\quad \tilde{D}_\pi = \mathrm{id} \otimes A_\pi , \qquad A \in L(V_\pi, W_\pi).$$

The adjoint operator $\tilde{D}^*_\pi : H_\pi \otimes W_\pi \longrightarrow H_\pi \otimes V_\pi$ is equal to

$$(2.8)\quad \tilde{D}^*_\pi = \mathrm{id} \otimes A^*_\pi .$$

<u>Definition</u>. By G-index of the operator D we mean the element of $\mathbb{Z}(G_d)^*$ defined by the formula

$$(2.9)\quad \mathrm{index}_G D := \sum_{[\pi] \in \hat{G}_d} ([\ker \tilde{D}_\pi] - [\ker \tilde{D}^*_\pi]).$$

<u>THEOREM 2.2</u>. <u>Let</u> K <u>be a massive subgroup of a unimodular Lie group</u> G. <u>Let</u> $D : C^\infty(\mathbb{E}_\tau) \longrightarrow C^\infty(\mathbb{F}_\sigma)$ <u>be an invariant differential operator</u>. <u>Then</u>

$$\mathrm{index}_G D = i_*([\tau] - [\sigma]).$$

<u>Proof</u>. According to the formulae (2.7), (2.8) we have

$$\ker \tilde{D}_\pi = H_\pi \otimes \ker A_\pi \ ; \quad \ker \tilde{D}^*_\pi = H_\pi \otimes \ker A^*_\pi .$$

Then $[\ker \tilde{D}_\pi] - [\ker \tilde{D}^*_\pi] = (\dim \ker A_\pi - \dim \ker A^*_\pi)[\pi]$ and

$$(\dim V_\pi - \dim W_\pi)[\pi] = (\langle \pi | K : \tau \rangle - \langle \pi | K : \sigma \rangle)[\pi].$$

Finally $\mathrm{index}_G D = \sum_{[\pi] \in \hat{G}_d} (\langle \pi | K : \tau \rangle - \langle \pi | K : \sigma \rangle)[\pi] = i_*([\pi] - [\sigma]).$

3. Applications

Theorem 2.2 will be used in this section to deduce the existence of non-zero L^2-solutions of some invariant differential equations. In order to prove that there exists a smooth function $f \in L^2(\mathbb{E}_\tau)$ which satisfies the equation $Df = 0$, it suffices to find out that the element $i_*([\tau] - [\sigma]) \in \mathbb{Z}(\hat{G}_d)$ contains a component $n_\pi \pi$ with $n_\pi > 0$. The positive result depends on the good knowledge of K-components of the representation $\pi|K$, $\pi \in \hat{G}_d$.

In the case of semisimple Lie groups with finite centre this problem was intensively studied and established in papers by W. Schmid [10], Enright and Varadarajan [11], W.Schmid and Hecht [6].

Let G be a semisimple Lie group with the maximal compact subgroup K. Then it is known that K is massive in G (Harish-Chandra [4]). The discrete series representations of G exist iff rank K = rank G, that is K contains a maximal torus T which is maximal in G also. Let $\mathfrak{g}$, $\mathfrak{k}$, $\mathfrak{t}$ denote Lie algebras of corresponding groups and $\mathfrak{g}^c$, $\mathfrak{k}^c$, $\mathfrak{t}^c$ their complexifications. The Killing form of $\mathfrak{g}$ defines on $\mathfrak{t}$ by restriction a scalar product $\langle\cdot|\cdot\rangle$ and by duality a product on the space $i\mathfrak{t}^*$. The set $\hat{T}$ of characters of T can be identified via exponential map with a lattice $\Lambda \in i\mathfrak{t}^*$. The lattice contains in particular the set of roots of the algebra $\mathfrak{g}$, denoted by Δ. Let $\mathfrak{g}^\alpha$ be the root space of the root $\alpha \in \Delta$.

If $\mathfrak{g} = \mathfrak{k} + \mathfrak{p}$ is the Cartan decomposition of $\mathfrak{g}$, then we have : $\mathfrak{g}^c = \mathfrak{t}^c + \Sigma_{\alpha\in\Delta}\mathfrak{g}^\alpha$, $\mathfrak{k}^c = \mathfrak{t}^c + \Sigma_{\alpha\in\Delta_c}\mathfrak{g}^\alpha$, $\mathfrak{p}^c = \Sigma_{\alpha\in\Delta_n}\mathfrak{g}^\alpha$; the subsets Δ_c, Δ_n are called systems of compact and noncompact roots respectively. Connected components of the set $\{\lambda \in it^* : \langle\alpha|\lambda\rangle \neq 0\}$ are called Weyl chambers. Elements of Weyl chambers are called nonsingular functionals.

Choose some ordering in $i\mathfrak{t}^*$ and denote by ϱ one half of roots which are positive with respect to this ordering. The set $\Lambda' = \Lambda + \varrho$ does not depend on the particular ordering.

Harish-Chandra has defined a mapping which assigns to each nonsingular $\lambda \in \Lambda'$ certain invariant distribution Θ_λ, which is the character of a discrete series representation denoted by π_λ in what follows. This construction exhausts all elements of the space $\hat{G}_d$ and $\pi_\lambda \cong \pi_\mu$ iff $\lambda = s\mu$ for some $s \in W$ - the Weyl group of the pair (K,T). It was W. Schmid who characterized discrete series representation of an algebraic semisimple group of its K-components [10].

Let us fix a nonsingular $\lambda \in \Lambda'$ and introduce in $i\mathfrak{t}^*$ the ordering calling μ positive iff $\langle\mu|\lambda\rangle > 0$.

Denote by P the set of positive roots and $P_n := P \cap \Delta_n$, $P_c := P \cap \Delta_c$. Set $\varrho_n = \frac{1}{2}\sum_{\alpha\in P_n}\alpha$; $\varrho_c = \frac{1}{2}\sum_{\alpha\in P_c}\alpha$; $\varrho = \varrho_c + \varrho_n$.

THEOREM 3.1. The representation π_λ is the one and only irreducible representation of G, whose restriction to K

1) contains the irreducible K-module of the highest weight with respect to P,

2) does not contain any irreducible K-module with highest weight of the form $\lambda - \rho_c + \rho_n - A$ where A stands for a non-empty sum of positive roots.

Example 1

The discrete class π_λ is called "holomorphic" when the positive root systems satisfy the condition $\alpha + \beta \notin P_c$ for $\alpha, \beta \in P_n$. Then the subspaces $\mathfrak{p}_\pm = \sum_{\alpha \in P_n} \mathfrak{g}^{\pm\alpha}$ are K-invariant and determine the unique holomorphic structure on G/K such that $\mathfrak{p}_-$ plays the role of antiholomorphic tangent space. Denote by τ_μ the K-module with the highest weight μ. The bundle $\mathbb{E}_{\tau_\mu}$ is a holomorphic vector bundle. Then we can define the Cauchy-Riemann operator

$$\bar{\partial} : C^\infty(\mathbb{E}_{\tau_{\lambda-\rho_c+\rho_n}}) \longrightarrow C^\infty(\mathbb{E}_{\tau_{\lambda-\rho_c+\rho_n}} \otimes \mathfrak{p}_-) \cong C^\infty(\mathbb{E}_{\tau_{\lambda-\rho_c+\rho_n} \otimes \mathfrak{p}_-}).$$

All the highest weights of the K-module $\tau_{\lambda-\rho_c+\rho_m} \otimes \mathfrak{p}_-$ are of the form $\lambda - \rho_c + \rho_n - \alpha$, $\alpha \in P_n$. According to Thm 3.1 the coefficient n_{π_λ} of the G-index of the operator $\bar{\partial}$ is equal to 1. Consequently, the L^2-kernel of the operator $\bar{\partial}$ is non-zero. In fact it is equal to H_{π_λ}.

Example 2

Now assume that the symmetric space G/K admits G-invariant Spin structure. Denote by $L = L^+ \oplus L^-$ the spin representation. When restricted to K the representation L^+ contains in particular the irreducible component with the highest weight ρ_n. Other weights of $L^\pm$ are of the form $\rho_n - A$, where A is a sum of r positive roots and $(-1)^r = \pm 1$, respectively (cf. [8]).

Consider spin vector bundles $E_{L^+ \otimes \tau_{\lambda-\rho_c}}$ and $E_{L^- \otimes \tau_{\lambda-\rho_c}}$ and the Dirac operator $D^+ : C^\infty(E_{L^+ \otimes \tau_{\lambda-\rho_c}}) \longrightarrow C^\infty(E_{L^- \otimes \tau_{\lambda-\rho_c}})$ (cf. [8]). The representation $L^+ \otimes \tau_{\lambda-\rho_c}$ of K contains the component with the highest weight $\lambda - \rho_c + \rho_n$ and provides the contribution $1 \cdot \pi_\lambda$ to the G-index of the operator D^+. All weights of the tensor product $L^- \otimes \tau_{\lambda-\rho_c}$ are of the form $\lambda + \rho_n - \rho_c - A$ and according to Thm 3.1 $\langle \pi_\lambda | K : L^- \otimes \tau_{\lambda-\rho_c} \rangle = 0$. This shows that the L^2-kernel of the operator D^+ is nontrivial.

References

[1] M.F. ATIYAH: Elliptic operators, discrete groups and von Neumann algebras, Asterique 32/33 (1976), 43-72.

[2] M.F. ATIYAH, W. SCHMID: A geometric construction of the discrete series of semisimple Lie groups, Inv. Math. 42 (1977), 1-62.

[3] R. BOTT: The index theorem for homogenous differential operators, In : Differential and Combinatorial Topology (A symposium in honour of Marston Morse), 167-186, Princeton University Press, Princeton, 1965.

[4] HARISH-CHANDRA: Representations of semisimple Lie groups, IV, V, VI, Amer. J. Math. 77 (1955), 743-777; 78 (1956), 1-41 and 564-628.

[5] HARISH-CHANDRA: Discrete series for semisimple Lie groups, II, Acta Math. 116 (1966), 1-111.

[6] H. HECHT, W. SCHMID: A proof of Blattner conjecture, Inv. Math. 31 (1975), 129-154.

[7] R. HOTTA, R. PARTHASARATHY:Multiplicity formulae for discrete series, Inv. Math. 26 (1974), 133-178.

[8] R. PARTHASARATHY: Dirac operator and the discrete series, Ann. of Math. 96 (1972), 1-30.

[9] R.S. PALAIS et al.: Seminar on the Atiyah-Singer index theorem, Ann. of Math. Studies 57, Princeton University Press, 1965.

[10] W. SCHMID: Some properties of square-integrable representations of semisimple Lie groups, Ann. of Math. 102 (1975), 535-564.

[11] T. J. ENRIGHT, V.S. VARADARAJAN: On an infinitesimal characterization of the discrete series, Ann. of Math. 102 (1975), 1-16.

[12] R. GOODMAN: One parameter groups generated by operators in an enveloping algebra, J. Funct. Anal. 6 (1970), 218.

Department of Mathematical Methods
in Physics, University of Warsaw
Hoża 74, PL-00-682 Warszawa, Poland

ON CONNECTION BETWEEN PROPERTIES OF A COMPACT SET IN $\mathbb{C}^n$ AND ITS CONJUGATE SET

Jurii Borisovič Zelinskii (Kiev)

Consider the n-dimensional complex space $\mathbb{C}^n$. Let $z = (z_1, z_2, \dots, z_n)$, $w = (w_1, w_2, \dots, w_n)$, $\bar{z} = (\bar{z}_1, \bar{z}_2, \dots, \bar{z}_n)$, $zw = z_1w_1 + z_2w_2 + \dots + z_nw_n$.

For an arbitrary set E in $\mathbb{C}^n$ the <u>conjugate</u> set to E [1] is defined as follows:

$$\tilde{E} = \{w \mid wz \neq 1 \quad \text{for every} \quad z \in E\}.$$

A set E is called <u>linearly convex</u> if for every point $z^0 \in \mathbb{C}^n \setminus E$ there exists an $(n-1)$-dimensional analytic hyperplane passing through z^0 and not intersecting E.

The notion of linear convexity is applicable in the theory of holomorphic functions of several complex variables. In particular it is exploited in problems of summation of n-iterated power series [2] and in separation of singularities of meromorphic functions [3]. In terms of linear convexity necessary and sufficient conditions are given [1] for the existence of a development of holomorphic functions into partial fractions. It appears that linearly convex sets which contain the origin $\theta = (0,0,\dots,0)$ are characterized by the property $\tilde{\tilde{E}} = E$ (cf. [4]). In complex analysis the case of linearly convex domains and compact sets with connected conjugate set is of great interest.

In this paper certain relationships between the homological properties of a linearly convex compact set and its conjugate set are given. The main results of this paper without proofs may be found in [5].

<u>PROPOSITION 1</u>. <u>The conjugate sets $\tilde{E}$ and $\tilde{E}'$ are homeomorphic for every elongation $E \to E' = \alpha E$ (α being a positive real number) and every translation $E \to E' = z_0 + E$ such that $\theta \in E \cap E'$.</u>

<u>Proof</u>. We consider the set of all complex analytic hyperplanes

in $\mathbb{C}^n$. It is known that this set forms the Grassman manifold $\mathbb{C}G'(n, n-1)$ [6]. If we identify in this manifold the points which correspond to complex analytic hyperplanes passing through the origin, then the obtained quotient space is homeomorphic with $\mathbb{C}^n$, more precisely with $(\mathbb{C}^n \cup \infty) \setminus \{\theta\}$ according to the following rulr: the point w_0 corresponds to the hyperplane $\{z \mid w_0 z = 1\}$ and the point ∞ corresponds to all the hyperplanes passing through the origin. Obviously, under this correspondence the set $\tilde{E}$ corresponds homeomorphically to the set M_E of all hyperplanes in $\mathbb{C}G'(n, n-1)$, not intersecting E.

The homeomorphic transformation $M_E \to M_{\alpha E}$ is a counterpart of the transformation $E \to \alpha E$ and, consequently, $\tilde{E}$ is homeomorphically transformed into $\alpha\tilde{E}$.

The translation $E \to z_0 + E$ maps M_E into the set $M_{(z_0+E)}$ which does not intersect the set $z_0 + E$. The above factorization moves neither M_E nor $M_{(z_0+E)}$ because of $\theta \in E \cap E'$. Therefore we have the following homeomorphisms: $\tilde{E} \approx M_E \approx M_{(z_0+E)} \approx \tilde{E}'$. The proposition is proved.

Let K be an arbitrary compact set in $\mathbb{C}^n$. Let us consider the following continuous many-valued mapping [7]:

$$\Phi: \mathbb{C}^n \to \hat{\mathbb{C}}^n = \mathbb{C}^n \cup \{\infty\}, \quad \Phi(z) = \{w \mid (wz = 1) \cup (w = \infty)\}.$$

Obviously, $\tilde{K} = \mathbb{C}^n \setminus \Phi(K)$, and therefore the examination of the conjugate set $\tilde{K}$ may be replaced by the examination of its complement $\Phi(K)$.

The homeomorphism

$$f(z) = \begin{cases} (1/|z|^2)(\bar{z}_1, \bar{z}_2, \ldots, \bar{z}_n) & \text{for } z \neq \theta, \\ \infty & \text{for } z = \theta \end{cases}$$

is a continuous section of the many-valued mapping Φ, i.e. $f(z) \subset \Phi(z)$ for every $z \in \mathbb{C}^n$.

LEMMA 1. *The graph of* Φ *is of the form*

$$\Gamma_\Phi(K) = \begin{cases} K \rtimes S^{2n-2}/S^{2n-2} & \text{if } \partial K \text{ contains } \theta \text{ and } K \text{ lies in the half-space } \operatorname{Re} z_n \geq 0, \\ K \rtimes S^{2n-2} & \text{if } \theta \text{ does not belong to } K \end{cases}$$

for every compact set K.

Proof. At first we are going to prove that in a sufficiently small neighbourhood of the origin the graph of Φ forms a trivial bundle [8]. Let $z \neq \theta$ be any point in $\mathbb{C}^n$. Then $\Phi(z)$ is the one-point-compactification of a (n-1)-(complex) dimensional analytic hyperplane. By continuity of Φ there exists a positive number ε_o such that the Hausdorff distance [7] of sets $\Phi(z)\cap T_R$ and $\Phi(z')\cap T_R$ is smaller than ϱ for $|z'-z|<\varepsilon_0$, where T_R is the ball with centre θ and radius $R>|f(z)|+\varrho$. Let us choose $\varepsilon<\min(\varepsilon_0, |z|, |f(z)|)$. For a point z_t such that $|z_t-z|<\varepsilon$, let $g(t):\Phi(z_t)\to\Phi(z)$ be the orthogonal projection of $\Phi(z_t)$ onto $\Phi(z)$. We will show that this projection is a homeomorphism. We notice that there exists a unique (2n-1)-dimensional hyperplane passing through a given (n-1)-(complex) dimensional analytic hyperplane and a given point α which lies beyond the latter. In this (2n-1)-dimensional hyperplane there is exactly one straight line perpendicular to the hyperplane $\Phi(z)$, passing through α. If the projection $\Phi(z_t)\to\Phi(z)$ is not one-to-one, then the hyperplane $\Phi(z_t)$ contains a real line γ which projects onto a single point $\beta\in\Phi(z)$.

Let us consider a complex analytic hyperplane l containing γ. Since the intersection of l with $\Phi(z_t)$ contains the real line γ, then $l\subset\Phi(z_t)$. Besides, $\Phi(z)\cap l=\beta$. Let us choose a point a on γ at the distance ϱ from β. Then the distance between a and $\Phi(z)$ equals ϱ. Let us take $R>\max(|f(z_t)|, |a|)$ over all z_t satisfying $|z_t-z|<\frac{1}{2}|z|$ and choose ε_0 so that $\mathrm{dist}(\Phi(z)\cap T_R, \Phi(z_t)\cap T_R)<\varrho$ (dist denotes the Hausdorff distance) for every z_t satisfying $|z_t-z|<\varepsilon_0$ with $a\in\Phi(z_t)\cap T_R$ and $\mathrm{dist}(a, \Phi(z)\cap T_R)=\varrho$. The contradition obtained shows that $g(t)$ is a homeomorphism.

If we give now a correspondence between the product $B(z)\times\Phi(z)$ and the set $\Gamma_\Phi(B(z))$ as follows: $(z(t), y)\to(f(z(t)), g^{-1}(t)[y])$, then we observe that the graph of Φ is locally a trivial bundle over $\mathbb{C}^n\setminus\{\theta\}$. By linear convexity the compact set K lies in $\mathbb{C}^n\setminus\lambda$, where λ is a ray starting at θ (under the assumption that $\theta\notin K$). The space $\mathbb{C}^n\setminus\lambda$ is contractible. Thus from [8] it follows that the graph $\Gamma_\Phi(\mathbb{C}^n\setminus\lambda)$ is homeomorphic with $(\mathbb{C}^n\setminus\lambda)\times S^{2n-2}$, where the spheres S^{2n-2} may be chosen with a given radius. Since $K\subset\mathbb{C}^n\setminus\lambda$, then $\Gamma_\Phi(K)=K\times S^{2n-2}$ as a restriction of the graph to a subset.

If, however, K is contained in the half-space $\mathrm{Re}\, z_n\geq 0$, then for the graph of Φ over this half-space without the origin, similarly as above, the homeomorphic correspondence $\Gamma_\Phi[(\mathrm{Re}\, z_n\geq 0)\setminus\{\theta\}]\approx[(\mathrm{Re}\, z_n\geq 0)\setminus\{\theta\}]\times S^{2n-2}$ holds, where the spheres in question have a fixed radius r. Let us consider the canonical homeomorphic correspondence

$$[(\mathrm{Re}\, z_n\geq 0)\setminus\{\theta\}]\times S^{2n-2}\approx\bigcup_{z\in[(\mathrm{Re}\, z_n\geq 0)\setminus\{\theta\}]} S_z^{2n-2},$$

where the radius of the sphere S_z^{2n-2} equals $\tau|z|$. This correspondence can be easily extended to a homeomorphism of the spaces $\Gamma_\Phi(\mathrm{Re}\, z_n \geq 0)$ and $\bigcup_{z\in(\mathrm{Re}\, z_n \geq 0)} S_z^{2n-2}$, where $S_\theta^{2n-2} = *$ is a point.

Finally we obtain the conclusion of the lemma restricting Φ to the considered compact set K.

Let us notice that we shall use Lemma 1 to examine the conjugate set $\tilde{K}$ which, by Proposition 1, will not change if we translate the compact set K with $\theta \in K$ in such a way that it will lie in the half-space $\mathrm{Re}\, z_n \geq 0$ and that $\theta \in \partial K$.

Let us denote the i-th (reduced) cohomology group of Aleksandrov-Čech by $H^i(A)$ (resp. $\tilde{H}^i(A)$) and the i-th reduced singular homology group with coefficients in the group $\mathbb{Z}$ of integers by $\tilde{H}_i(A)$ for a set A, and let p and q be the projections of the graph $\Gamma_\Phi(K)$ onto K and $\Phi(K)$, respectively.

<u>LEMMA 2</u>. <u>If a compact set</u> K <u>lies in the half-space</u> $\mathrm{Re}\, z_n \geq 0$ <u>and is linearly convex, then</u>

$$H^{i+2n-2}(\Gamma_\Phi(K)) \approx H^i(K) \quad \underline{\text{for}} \quad i > 0.$$

<u>Proof</u>. Obvious by Lemma 1 and Künneth's formula [9].

Hereafter, without any loss of generality, we assume that the compact set K lies in the half-space $\mathrm{Re}\, z_n \geq 0$ (cf. Proposition 1).

If we examine now the projection $\Gamma_\Phi(K) \xrightarrow{q} \Phi(K)$, then we obtain some results which describe relations between $\Phi(K)$ and K.

Let $f: X \to Y$ be a continuous proper mapping, where X and Y are locally compact spaces. By $d_k(f)$ we denote the relative dimension [10] of the set $B_k = \{y \in Y \mid \tilde{H}^k(f^{-1}y) \neq 0\}$; if $B_k = \emptyset$, we put $d_k = -\infty$. It is well known [11] that if for $k \geq 0$

$$d_k(f) \leq \tau - 1 - k,$$

then the mapping f induces the isomorphism

$$f^*: H^i(fX) \to H^i(X) \quad \text{for} \quad i > \tau$$

and an epimorphism for $i = \tau$.

Under the projection q the inverse image of every point $y \in \Phi(K)$ is homeomorphic with $\Phi^{-1}(y)$. Consequently, we may in an analogous way introduce $d_k(\Phi)$ setting $d_k(\Phi) = d_k(q)$. Let us notice that the inverse image of a point w $\Phi(K)$, $w \neq \infty$, under the mapping Φ is the intersection of K with the plane $\{z \mid wz = 1\}$, and the inverse image of the point ∞ is the whole compact set K.

<u>THE MAIN THEOREM</u>. <u>If</u> K <u>is a linearly convex compact set in</u> $\mathbb{C}^n$ <u>and</u> $d_k(\Phi) \leq 2n-2-k$ <u>for every</u> $k \geq 0$, <u>then</u> $q^*(p^{-1})^*: H^{2n-1}(\Phi(K)) \to H^1(K)$ <u>is an epimorphism and</u> $H^i(K) = 0$ <u>for</u> $i > 1$. <u>If, however,</u> $d_k(\Phi) \leq 2n-3-k$ <u>for every</u> $k \geq 0$, <u>then</u> $H^{2n-1}(\Phi(K)) \approx H^1(K)$.

Proof. According to the theorem of Vietoris-Begle-Skljarenko the inequality $d_k(\Phi) \leq 2n-2-k$ ensures that the projection q induces an isomorphism of $H^i(\Phi(K))$ and $H^i(\Gamma_\Phi(K))$ for $i > 2n-1$ and an epimorphism $H^{2n-1}(\Phi(K)) \to H^{2n-1}(\Gamma_\Phi(K))$. By Lemma 1 we obtain the sequence

$$H^{i+2n-2}(\Phi(K)) \xrightarrow{q^*} H^{i+2n-2}(\Gamma_\Phi(K)) \xrightarrow[\approx]{(p^{-1})^*} H^i(K) \quad \text{for } i > 0,$$

where q^* is an isomorphism for $i > 1$ and an epimorphism for $i = 1$. But $H^{i+2n-2}(\Phi(K)) = 0$ for $i > 1$ because $\Phi(K)$ is a compact set in $\mathbb{C}^n$. It closes up the proof of the first part of the theorem. The second part of the theorem may be proved analogously.

<u>COROLLARY 1</u>. <u>If</u> K <u>is a linearly convex compact set in</u> $\mathbb{C}^n$, $d_k(\Phi) \leq 2n-3-k$ <u>for every</u> $k \geq 0$, <u>and the group</u> $H^1(K)$ <u>is finitely generated, then the number of components of the set</u> $\tilde{K}$ <u>is greater than the rank of</u> $H^1(K)$ <u>by one</u>.

According to the duality theorem of Alexander [9] and our main theorem the following isomorphisms hold: $\tilde{H}_0(\tilde{K}) \approx H^{2n-1}(\Phi(K)) \approx H^1(K)$, where the rank of $\tilde{H}_0(\tilde{K})$ is smaller than the number of components of $\tilde{K}$ by one.

<u>COROLLARY 2</u>. <u>If</u> K <u>is a linearly convex compact set in</u> $\mathbb{C}^n$, $d_k(\Phi) \leq 2n-2-k$ <u>for every</u> $k \geq 0$, <u>and</u> $H^1(K) \neq 0$, <u>then</u> $H^{2n-1}(\Phi(K)) \neq 0$, <u>i.e.</u> $\tilde{K}$ <u>is disconnected</u>.

This follows from the sequence $\tilde{H}_0(\tilde{K}) \approx H^{2n-1}(\Phi(K)) \xrightarrow{q^*(p^{-1})^*} H^1(K)$ since, according to the theorem, $q^*(p^{-1})^*$ is an epimorphism.

<u>COROLLARY 3</u>. <u>If</u> K <u>is a linearly convex compact set in</u> $\mathbb{C}^n$ <u>such that its intersections with analytic complex hyperplanes are acyclic, except for intersections with a finite number of them, then</u> $H^1(K) \approx H^{2n-1}(\Phi(K))$ <u>and</u> $H^i(K) = 0$ <u>for</u> $i > 1$.

Proof. It is obvious that, by the assumptions, we have $d_k(\Phi) = 0$ or $-\infty$ for $0 \leq k \leq 2n-3$ and $d_k(\Phi) = -\infty$ for $k > 2n-3$, since the inverse image of a point $w \in \Phi(K)$, $w \neq \infty$ is the intersection of the hyperplane $\{z \mid wz = 1\}$ with K (a compact subset of a (2n-2)-(real) dimensional hyperplane). If, however, $w = \infty$, then $\Phi^{-1}(\infty) = K$ and, as it was proved in [12], $H^{2n-1}(K) = 0$. If $H^{2n-2}(K) \neq 0$ then, according to the theorem of Vietoris-Begle-Skljarenko, infinite number of hyperplanes among the collection of complex analytic parallel hyperplanes intersect K forming a cyclic set, and this contradicts the assumption. Thus we have $d_k(\Phi) \le 2n-3-k$ for every $k \ge 0$ and the result follows from the main theorem.

Let a compact set K lie in a complex analytic line. Then, applying the main theorem, we arrive at the following

PROPOSITION 2. *If K is a compact set in $\mathbb{C}^n$ which lies in a complex analytic line, then $H^1(K) \approx H^{2n-1}(\Phi(K))$.*

Proof. The set of complex analytic hyperplanes containing the complex line in question is homeomorphic to the projective space $\mathbb{C}P^{n-2}$. The remaining hyperplanes either do not intersect it or intersect it at one point. Hence we obtain $d_0(\Phi), d_1(\Phi) \le \dim \mathbb{C}P^{n-2} = 2n-4$ and $d_k(\Phi) = -\infty$ for $k > 1$. In consequence the inequality $d_k(\Phi) \le 2n-3-k$ holds for every k and, by the main theorem, $H^1(K) \approx H^{2n-1}(\Phi(K))$.

For compact sets which lie in a two-(real) dimensional plane, a weaker result takes place:

PROPOSITION 3. *If K is a compact set in $\mathbb{C}^n$ which lies in a two-(real) dimensional plane T, then the mapping $q^*(p^{-1})^*: H^1(K) \longrightarrow H^{2n-1}(\Phi(K))$ is an epimorphism.*

Proof. Let us suppose that the plane is not identical with any complex analytic plane since that case has been studied before. At the intersection of K with a complex analytic hyperplane γ, we have $H^1(\gamma \cap K) \neq 0$ only in the case when $\gamma \supset T$; the inequality $\tilde{H}^0(\gamma \cap K) \neq 0$ is possible when either $\gamma \supset T$ or $\gamma \cap T = l$, where l is a real line. There exists exactly one two-(complex) dimensional analytic hyperplane λ containing T. The set of complex analytic hyperplanes which contain the plane λ is homeomorphic to the projective space $\mathbb{C}P^{n-3}$. Consequently $d_1(\Phi) \le \dim \mathbb{C}P^{n-3} = 2n-6$.

The set of complex analytic hyperplanes which contain a fixed real line is homeomorphic to the projective space $\mathbb{C}P^{n-2}$. In turn the set of real lines in the plane T forms the Grassman manifold $RG'(2,1)$ with $\dim RG'(2,1) = 2$. In consequence

$$\begin{aligned} d_0(\Phi) &\le \max(\dim \mathbb{C}P^{n-3}, \dim \mathbb{C}P^{n-2} + \dim RG'(2,1)) \\ &= \dim \mathbb{C}P^{n-2} + \dim RG'(2,1) = 2n-2. \end{aligned}$$

Besides, $d_k(\Phi) = -\infty$ for $k > 1$. Therefore $d_k(\Phi) \le 2n-2-k$ for every $k \ge 0$. Thus, by the main theorem, $q^*(p^{-1})^*$ is an epimorphism.

LEMMA 3. Let K be a polyhedral compact set and $\dim K = 1$. Then K is linearly convex.

Proof. Let $z_0 \in \mathbb{C}^n \setminus K$. We shall show that there is a complex analytic hyperplane γ passing through z_0 and such that $K \cap \gamma = \emptyset$. The set K is a polyhedron [6], i.e. it consists of a finite union of line segments and and isolated points.

Let $[a,b]$ be one of the segments forming K and let $z \in [a,b]$. The set of complex analytic hyperplanes which contain z_0 and z forms the projective space $\mathbb{C}P^{n-2}(z)$ with $\dim \mathbb{C}P^{n-2}(z) = 2n-4$. Consequently, the set

$$E = \bigcup_{z \in [a,b]} \mathbb{C}P^{n-2}(z)$$

of such hyperplanes passing through z_0 and intersecting the segment $[a,b]$ is closed and its dimension does not exceed $2n-3$. The set of all complex analytic hyperplanes passing through z_0 and intersecting K is obtained as a finite union of the sets E with respect to all segments contained in K. This union is, according to [10], of dimension not greater than $2n-3$ (isolated points in K are treated as line segments with $b=a$). But since the set of all complex analytic hyperplanes passing through z_0 forms the projective space $\mathbb{C}P^{n-1}$ of dimension $2n-2$, the set of those which do not intersect K is dense in $\mathbb{C}P^{n-1}$.

Remark 1. Suppose that any two points of $\tilde{K}$ different from the origin can be joint by an arc which does not pass through the origin. With any point $w(t)$ we associate the plane $\{z \mid w(t)z = 1\}$. This means that two arbitrary complex analytic hyperplanes corresponding to two different points $w_1, w_2 \in \tilde{K}$ can be joined by an isotopy $F(t,w)$ which for every t represents the hyperplane $\{z \mid w(t)z = 1\}$ and is such that $F(t,z) \cap K = \emptyset$. This condition is always satisfied when $\tilde{K}$ is an open connected set.

Remark 2. Disconnectivity of the conjugate set means that the planes of the form $\{z \mid wz = 1\}$ corresponding to two points w_1, w_2, which belong to different components of $\tilde{K}$, cannot be isotopically transformed one into the other so that for every t the isotopy $F(z,t)$ represent the plane $\{z \mid w(t)z = 1\}$ and

$$F(t, w(t)) \cap K = \emptyset \quad \text{for every } t.$$

If K is a compact set and one of the points, say w_2, belongs to the component of $\tilde{K}$ containing the point θ, then we can choose the second point in this component (according to the previous remark with all these points isotopic hyperplanes are associated) so that the planes $\{z \mid w_1 z = 1\}$ and $\{z \mid w_2 z = 1\}$ intersect, i.e. we can consider them as coming out of the same point. Therefore it is enough to consider the points w of a sphere S^{2n-1} with centre at θ and radius such that $S^{2n-1} \subset \tilde{K}$.

PROPOSITION 4. Let K be a connected polyhedral compact set in $\mathbb{C}^n$ and $\dim K = 1$. If $\tilde{K}$ is connected, then K is acyclic.

Proof. We apply the induction with respect to n. If $n = 1$ then the statement is obvious because $\tilde{K}$ is homeomorphic to the complement of K. Let us suppose that the proposition is proved for $n = m - 1$; we shall prove it for $n = m$.

Suppose that there exists an intersection of K by a complex analytic hyperplane γ, such that $H^1(K \cap \gamma) \neq 0$. We choose a connected component $K' \subset K \cap \gamma$ such that $H^1(K') \neq 0$. Notice that K' is a one-dimensional polyhedron. Hence, by the induction assumption, the set $\tilde{K}'$ is disconnected; otherwise we can obtain our assertion because $\gamma \approx \mathbb{C}^{m-1}$ (the conjugation is taken with respect to the $(m-1)$-(complex) dimensional space γ). Take a point $z_0 \in \gamma \cap (\mathbb{C}^m \setminus K)$. Notice that $z_0 \in \mathbb{C}^m \setminus K$, whence, by Lemma 3, there exists in $\mathbb{C}^n$ a complex analytic hyperplane λ passing through z_0 and not intersecting K. Then $\lambda \cap \gamma$ is a complex analytic hyperplane in γ not intersecting K'. Consequently, $\lambda \cap \gamma$ is determined by a point $w_0 \in \tilde{K}'$. Let the points w_0 and w_1 belong to different components of $\tilde{K}'$ and let l be an analytic complex line in λ such that $\lambda \cap \gamma = z_0$. Let us project K on γ parallelly to the straight line l. Let K_1 be the image of K in γ under this projection. Then K_1 is a one-dimensional connected polyhedron. By Lemma 3 it is linearly convex. Besides, $K_1 \supset K'$, $z_0 \notin K_1$, and $(\lambda \cap \gamma) \cap K_1 = \emptyset$. Obviously $H^1(K_1) \neq 0$ and $\tilde{K}_1$ is disconnected. According to [1] $\tilde{K}_1 \subset \tilde{K}'$, and from Lemma 3 it follows that $\tilde{K}_1$ is an open set, dense everywhere in $\tilde{K}'$.

Let w_1', $w_1' \in \tilde{K}_1$, belong to the component of the set $\tilde{K}$ to which the point w_1 belongs. Then, by Remark 2, the plane $\lambda \cap \gamma$ cannot be isotopically transformed inside γ into the plane λ' corresponding to the point w_1', without intersecting the compact set K_1. Consider now in $\mathbb{C}^m$ the complex analytic hyperplanes λ and $\lambda' \times l$. Obviously they do not intersect K and, in consequence, two points w^1 and w^2 in $\tilde{K}$ correspond to them, respectively. Because of the connectivity of $\tilde{K}$ the points w^1 and w^2 may be joined by an arc $w(t)$ in $\tilde{K}$. Moreover, notice that in $\mathbb{C}^m$ the set P of the points which correspond to hyperplanes parallel with γ is homeomorphic to a complex plane, i e. it is of dimension 2. Consequently the arc $w(t)$, $w(0)=w^1$, $w(1)=w^2$ may be chosen so that it does not intersect P. Then among the planes $\{z \mid w(t)z=1\}$ there do not exist the planes parallel with γ, and therefore for every such plane its intersection with γ is an analytic plane of complex dimension $m-2$. These hyperplanes define an isotopy of the plane $\lambda\cap\gamma$ into the plane λ' not intersecting K'. This contradicts the fact that w_0 and w_1' belong to different components of $\tilde{K}'$.

The contradiction obtained shows that for the intersection of K by an arbitrary complex analytic hyperplane γ we have $H^1(K\cap\gamma)=0$, i.e. these intersections are simply connected. Let us recall, however, that $\Phi^{-1}(\infty)=K$. Therefore it is possible that $H^1(\Phi^{-1}(\infty))\neq 0$. Consequently $d_1(\Phi)\leq 0$ and $d_k(\Phi)=-\infty$ for $k>1$.

Notice now that, since K is a polyhedron, the set of real lines intersecting the compact set K in at least two points is of dimension not greater than two in the set $RG'(2n,1)$ of all real lines of the space $\mathbb{C}^n$. The set M of all complex analytic hyperplanes containing a fixed straight line is homeomorphic to $\mathbb{C}P^{n-2}$, so $\dim M\leq 2n-4$. Hence it follows that the dimension of the set of those hyperplanes which intersect K in at least two points within the set $\mathbb{C}G'(n,n-1)$ is not greater than $2n-2$. Consequently, $d_0(\Phi)\leq 2n-2$. Therefore, by our main theorem,

$$H^{2n-1}(\Phi(K))\longrightarrow H^{2n-1}(\Gamma_K(\Phi))\approx H^1(K)$$

is an epimorphism, where $H^{2n-1}(\Phi(K))\approx\tilde{H}_0(\tilde{K})=0$, and thus $H^1(K)=0$. The proof of Proposition 4 is completed.

Remark 3. The above proof yields that Lemma 3 and Proposition 4 are also valid when the compact set K consists of a countable union of one-dimensional polyhedrons.

We are going to give an example which shows that the assumption $\dim K=1$ in Proposition 4 is essential.

Example 1. We give an example of a connected, linearly convex polyhedral compact set $K\subset\mathbb{C}^2$, for which the set $\tilde{K}$ is connected, but $H^1(K)\neq 0$. The compact K has to be situated in the three-dimensional hyperplane

$$\alpha=\{z\in\mathbb{C}^2 \mid \operatorname{Im}(z_1-z_2)=0\}.$$

Let us choose in α the coordinate system $x_1, x_2, y_1=y_2$. The straight lines in α which are parallel with x_1 or x_2 are traces in α of the analytic planes $z_2=\text{const}$, $z_1=\text{const}$.

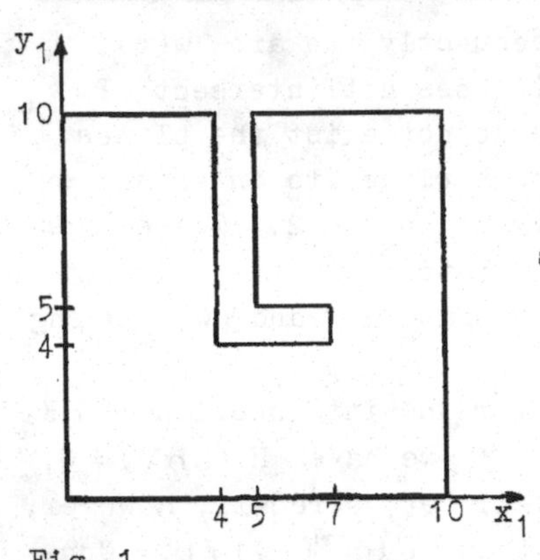

Fig. 1

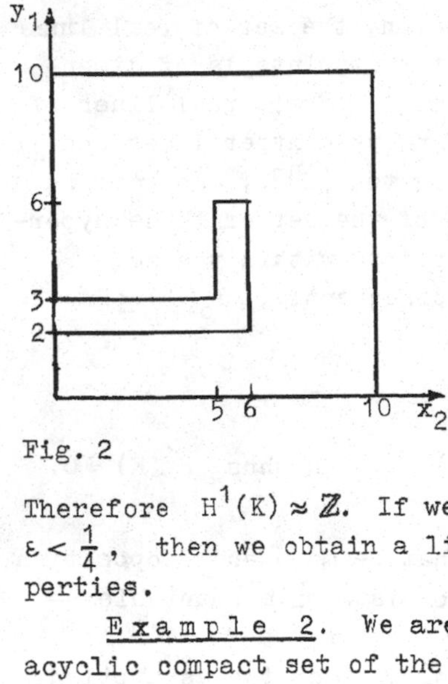

Fig. 2

Let us consider in α the cube

$$A = \{(x_1, x_2, y_1) \mid 0 \le x_1 \le 10,\ 0 \le x_2 \le 10,\ 0 \le y_1 \le 10\}$$

and remove from it the following cuboids:

$$B_1 = \{(x_1, x_2, y_1) \mid 4 \le x_1 \le 5,\ 4 \le y_1 \le 10,\ 0 \le x_2 \le 10\},$$

$$B_2 = \{(x_1, x_2, y_1) \mid 5 \le x_1 \le 7,\ 4 \le y_1 \le 5,\ 0 \le x_2 \le 10\},$$

$$B_3 = \{(x_1, x_2, y_1) \mid 0 \le x_1 \le 10,\ 0 \le x_2 \le 6,\ 1 \le y_1 \le 2\},$$

$$B_4 = \{(x_1, x_2, y_1) \mid 0 \le x_1 \le 10,\ 5 \le x_2 \le 6,\ 2 \le y_1 \le 6\}.$$

The compact set K in question has the property that an arbitrary plane $\{z \mid wz = 1\}$ in $\mathbb{C}^2 \setminus K$ may be transformed isotopically into a plane $z_1 = \text{const}$ or $z_2 = \text{const}$ such that the isotopy does not affect the compact set K. Consequently, the set $\tilde{K}$ is connected. But it is easy to see that K is of homotopy type [6] of a circle. Therefore $H^1(K) \approx \mathbb{Z}$. If we take now an ε-neighbourhood of K, where $\varepsilon < \frac{1}{4}$, then we obtain a linearly convex domain with analogous properties.

E x a m p l e 2. We are going to show that for a one-dimensional acyclic compact set of the form $K = K_1 \times K_2$, where K_1 and K_2 are plane compact sets satisfying the property $H^1(K_1) \approx H^1(K_2) \approx 0$, the set $\tilde{K}$ may be disconnected.

We define the compact sets K_1 and K_2 as follows: $K_1 = A_1 \cup A_2$, where A_1 is a union of segments in the plane $z_2 = 0$:

$$A_1 = \{(x_1, y_1) \mid (0 \le x_1 \le 36,\ y_1 = 6) \cup (2 \le x_1 \le 6,\ y_1 = 4) \cup (2 \le x_1 \le 5,\ y_1 = 1) \cup (0 \le x_1 \le 6,\ y_1 = 0) \cup (x_1 = 0,\ 0 \le y_1 \le 6) \cup (x_1 = 2,\ 1 \le y_1 \le 4) \cup (x_1 = 6,\ 0 \le y_1 \le 4)\},$$

and A_2 is centrally symmetrical with respect to A_1. The compact set K_2 consists of four points in the plane $z_1 = 0$ $(x_2 = 0, 3, 5, 8,$

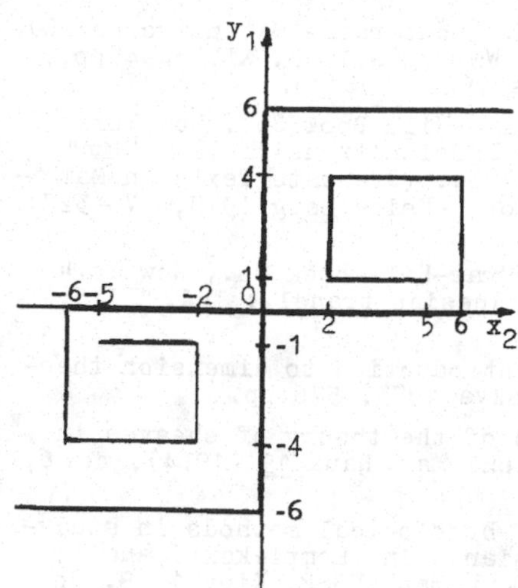

Fig. 3

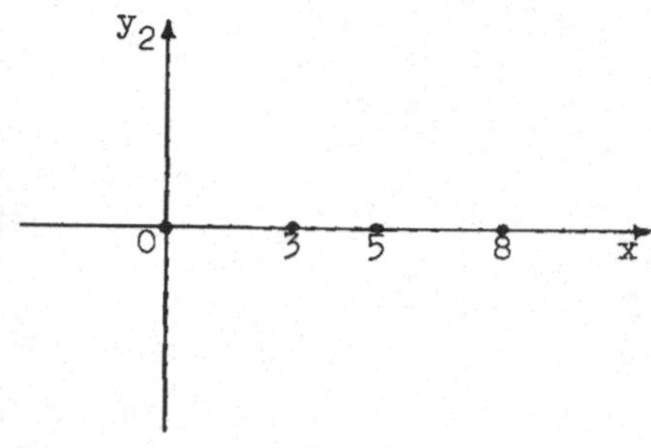

Fig. 4

$y_2 = 0$). In this way we obtain a compact set which is a one-dimensional polyhedron and, consequently, by Lemma 3, it is linearly convex. It is easily shown that the complex analytic line

$$l: \quad z_1 = (z_2 - 4)(3 + 2i)$$

cannot be isotopically transformed into a line of the form $z_2 = \text{const}$ in such a way that every step of the isotopy is a complex analytic line and the isotopy does not affect the compact set K. For this purpose it is enough to see that in the real three-dimensional hyperplane $y_2 = 0$ the straight line

$$2x_1 = 3y_1, \quad x_1 = 3(x_2 - 4)$$

(the trace of the complex line l) cannot be isotopically transformed in any line of the form $x_1 = \text{const}$, $y_1 = \text{const}$ without intersecting $\breve{K}$ and so that every step of the isotopy be a real line. Consequently, the set K is disconnected.

References

[1] L. A. AĬZENBERG: The expansion of holomorphic functions of several complex variables in partial fractions [in Russian], Sibirsk. Mat. Ž. 8 (1967), no. 5, 1124-1142.

[2] L. A. AĬZENBERG, V. M. TRUTNEV: A certain Borel summability method for multiple power series [in Russian], Sibirsk. Mat. Ž. 12 (1971), no. 6, 1398-1404.

[3] L. A. AĬZENBERG: Linear convexity in $\mathbb{C}^n$ and the distribution of the singularities of holomorphic functions [in Russian], Bull. Acad. Polon. Sci. Sér. Sci. Math. Astronom. Phys. 15 (1967), no. 7, 487-495.

[4] L. A. AĬZENBERG, A. P. JUŽAKOV, L. Ja. MAKAROVA: Linear convexity in $\mathbb{C}^n$ [in Russian], Sibirsk. Mat. Ž. 9 (1968), no. 4, 733-746.

[5] Ju. B. ZELINSKIĬ: On application of many-valued mappings in complex analysis [in Russian], Dokl. Akad. Nauk SSSR (1979).

[6] V. A. ROHLIN, D. B. FUKS: Primary course in topology. Geometrical part [in Russian], Izd. "Nauka", Moskva 1977, 488 pp.

[7] C. KURATOWSKI: Topologie I, 4th ed. (Monografie Matematyczne 20), PWN - Polish Scientific Publishers, Warszawa 1958, XII + 494 pp.; Russian transl.: Izd. "Mir", Moskva 1966, 596 pp.

[8] D. HOUSEMOLLER, Fibre bundles, McGraw-Hill Book Co., New York - London - Sydney 1966, XIV + 300 pp.; Russian transl.: Izd. "Mir", Moskva 1970, 444 pp.; 2nd (English) ed. (Graduate Texts in Mathematics 20), Springer-Verlag, New York - Heidelberg 1975, XV + 327 pp.

[9] E. SPANIER: Algebraic topology, McGraw-Hill Book Co., New York - London - Sydney 1966, pp. XIV + 528; Russian transl.: Izd. "Mir", Moskva 1971, 680 pp.

[10] P. S. ALEKSANDROV, B. A. PASYNKOV: Introduction to dimension theory [in Russian], Izd. "Nauka", Moskva 1973, 576 pp.

[11] E. G. SKLJARENKO: Some applications of the theory of sheaves in general topology [in Russian], Uspehi Mat. Nauk 19 (1964), no. 6, 47-70.

[12] Ju. B. ZELINSKIĬ: On application of homological methods in studying conjugate sets in $\mathbb{C}^n$ [in Russian], in: Kompleksnyĭ analiz i mnogoobrazija, Izd. Inst. Mat. Akad. Nauk USSR, Kiev 1978, pp. 149-152.

Инститyт математики
Академии наук УССР
ул. Репина 3
SU-252 601 Киев, СССР